A-Level Year 2

Physics

The Complete Course for AQA

Let's face it, Physics is a tough subject. You'll need to get to grips with a lot of difficult concepts, and have plenty of practical skills up your lab-coat sleeve.

But don't worry — this brilliant CGP book covers everything you'll need for the new AQA course. It's packed with clear explanations, exam practice, advice on maths skills and practical investigations... and much more!

It even includes a free Online Edition to read on your PC, Mac or tablet.

CGP

How to get your free Online Edition

Go to **cgpbooks.co.uk/extras** and enter this code...

4408 5638 5079 1002

Contents

How to use this book

Learning Objectives

- These tell you exactly what you need to learn, or be able to do, for the exams.
- There's a specification reference at the bottom that links to the AQA specification.

Learning Objectives:
- Estimate the radius of a nucleus by calculating the distance of closest approach of an alpha particle using Coulomb's law.
- Be able to determine the radius of a nucleus using electron diffraction.
- Be familiar with the graph of intensity against angle for electron diffraction by a nucleus.
  Specification Reference 3.8.1.5

2. Measuring Nuclear Radius

Rutherford's scattering experiment gives a good estimate of nuclear radii, but to get a really good measurement, you'll need to diffract electrons.

Closest approach of a scattered particle

You can estimate the radius of an atomic nucleus by using Rutherford's scattering experiment (see p.157–158). An alpha particle that 'bounces back' and is deflected through 180° will have stopped a short distance from the nucleus (see Figure 1).

Figure 1: The closest approach of a scattered alpha particle.

The alpha particle does this at the point where its electric potential energy (see page 92) equals its initial kinetic energy. This combined with Coulomb's law (see page 87) gives the equation:

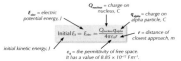

$$\text{Initial } E_k = E_{elec} = \frac{Q_{nucleus}q_{alpha}}{4\pi\varepsilon_0 r}$$

E_{elec} = electric potential energy, J

$Q_{nucleus}$ = charge on nucleus, C

q_{alpha} = charge on alpha particle, C

r = distance of closest approach, m

initial kinetic energy, J

ε_0 = the permittivity of free space. It has a value of 8.85×10^{-12} Fm⁻¹.

This is just conservation of energy — and if you know the initial kinetic energy of the alpha particle you've fired, you can use it to find how close the particle can get to the nucleus.

To find the charge of a nucleus you need to know the atom's **proton number**, Z — that tells you how many protons are in the nucleus (surprisingly). A proton has a charge of $+e$ (where e is the magnitude of the charge on an electron), so the charge of a nucleus must be $+Ze$.

The distance of closest approach is an estimate of nuclear radius — it gives a maximum value for it. However electron diffraction (which you'll meet on the next page) gives much more accurate values for nuclear radii.

Example — Maths Skills

An alpha particle with an initial kinetic energy of 6.0 MeV is fired at a gold nucleus. Estimate the radius of the nucleus by finding the closest approach of the alpha particle to the nucleus. ($Z_{gold} = 79$)

Initial particle energy = 6.0 MeV = 6.0×10^6 eV

Convert energy into joules: $6.0 \times 10^6 \times 1.60 \times 10^{-19} = 9.6 \times 10^{-13}$ J

Exam Tip
The value of ε_0 will be in the data and formulae booklet you'll be given in the exam, so don't worry about memorising it.

Tip: The magnitude of the charge on an electron, e, is 1.60×10^{-19} C, but an electron has a negative charge of $-e$.

Tip: Alpha particles contain 2 protons, so they have a charge of $+2e$.

Tip: To convert from electronvolts (eV) to joules (J), just multiply by the charge on an electron, e.

Exam Tips

There are tips throughout the book to help with all sorts of things to do with answering exam questions.

Tips

These are here to help you understand the theory.

Example

The energy diagram below represents one possible decay of cobalt-60 into nickel-60 via a beta-minus and a gamma decay. The order of the decays on the diagram shows that the beta-minus decay occurred before the gamma decay.

The cobalt is transformed into nickel by the beta-minus decay — but the nucleus is excited and unstable at this energy level. It becomes de-excited once it has emitted a gamma ray. The nuclide notation for the nickel isotope only needs to be written on one of the levels.

Tip: From this diagram, you can work out that the total decay energy to get from cobalt-60 to nickel-60 is: 1.49 + 1.33 = 2.82 MeV.

Practice Questions — Application

Q1 Complete the following nuclear equation for the beta-minus decay of caesium-137: $^{137}_{55}Cs \longrightarrow$?Ba + ? + ?

Q2 Write out the nuclear equation for the alpha decay of At-211 ($Z = 85$) to an isotope of bismuth. (The chemical symbol of bismuth is Bi.)

Q3 Rubidium-83 decays via electron capture to krypton. Write out the nuclear equation for this process. (Rubidium has 37 protons and the chemical symbol Rb, krypton has 36 protons and the chemical symbol Kr.)

Q4 A simple energy level diagram for the decay of an iodine nucleus to a xenon nucleus is shown below.
   a) Describe the decay process that is represented by this diagram.
   b) What does ΔE represent in these types of diagram?

Tip: Remember that the number in the name of an isotope is the nucleon number.

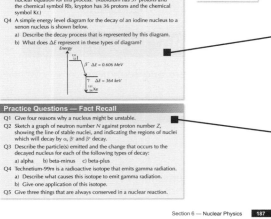

Practice Questions — Fact Recall

Q1 Give four reasons why a nucleus might be unstable.

Q2 Sketch a graph of neutron number N against proton number Z, showing the line of stable nuclei, and indicating the regions of nuclei which will decay by α, β^- and β^+ decay.

Q3 Describe the particle(s) emitted and the change that occurs to the decayed nucleus for each of the following types of decay:
   a) alpha   b) beta-minus   c) beta-plus

Q4 Technetium-99m is a radioactive isotope that emits gamma radiation.
   a) Describe what causes this isotope to emit gamma radiation.
   b) Give one application of this isotope.

Q5 Give three things that are always conserved in a nuclear reaction.

Examples

These are here to help you understand the theory.

Practice Questions — Application

- Annoyingly, the examiners expect you to be able to apply your knowledge to new situations — these questions are here to give you plenty of practice at doing this.
- All the answers are in the back of the book (including any calculation workings).

Practice Questions — Fact Recall

- There are a lot of facts you need to learn — these questions are here to test that you know them.
- All the answers are in the back of the book.

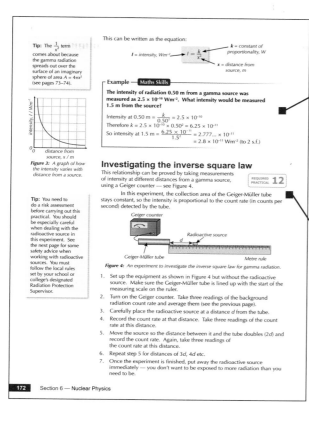

Tip: The $\frac{1}{x^2}$ term comes about because the gamma radiation spreads out over the surface of an imaginary sphere of area $A = 4\pi x^2$ (see pages 73-74).

This can be written as the equation:

$$I = \frac{k}{x^2}$$

I = intensity, Wm^{-2}
k = constant of proportionality, W
x = distance from source, m

Example — Maths Skills

The intensity of radiation 0.50 m from a gamma source was measured as 2.5×10^{-10} Wm^{-2}. What intensity would be measured 1.5 m from the source?

Intensity at 0.50 m = $\frac{k}{0.50^2}$ = 2.5×10^{-10}
Therefore $k = 2.5 \times 10^{-10} \times 0.50^2 = 6.25 \times 10^{-11}$
So intensity at 1.5 m = $\frac{6.25 \times 10^{-11}}{1.5^2}$ = $2.777... \times 10^{-11}$
$= 2.8 \times 10^{-11}$ Wm^{-2} (to 2 s.f.)

Figure 3: A graph of how the intensity varies with distance from a source.

Investigating the inverse square law

This relationship can be proved by taking measurements of intensity at different distances from a gamma source, using a Geiger counter — see Figure 4.

REQUIRED PRACTICAL **12**

In this experiment, the collection area of the Geiger-Müller tube stays constant, so the intensity is proportional to the count rate (in counts per second) detected by the tube.

Tip: You need to do a risk assessment before carrying out this practical. You should be especially careful when dealing with the radioactive source in this experiment. See the next page for some safety advice when working with radioactive sources. You must follow the local rules set by your school or college's designated Radiation Protection Supervisor.

Figure 4: An experiment to investigate the inverse square law for gamma radiation.

1. Set up the equipment as shown in Figure 4 but without the radioactive source. Make sure the Geiger-Müller tube is lined up with the start of the measuring scale on the ruler.
2. Turn on the Geiger counter. Take three readings of the background radiation count rate and average them (see the previous page).
3. Carefully place the radioactive source at a distance d from the tube.
4. Record the count rate at that distance. Take three readings of the count rate at this distance.
5. Move the source so the distance between it and the tube doubles ($2d$) and record the count rate. Again, take three readings of the count rate at this distance.
6. Repeat step 5 for distances of $3d$, $4d$ etc.
7. Once the experiment is finished, put away the radioactive source immediately — you don't want to be exposed to more radiation than you need to be.

Maths Skills Examples

There's a range of maths skills you could be expected to apply in your exams. Examples that show these maths skills in action are marked up like this.

Maths Skills

There's a whole Maths Skills section on pages 400-412 that's packed with plenty of maths you'll need to need to be familiar with.

Required Practicals

There are some key practicals that you'll be expected to do throughout your course. You'll need to know all about them for the exams. They're all marked up throughout the book with stamps.

Practical Skills

There are some key practical skills you'll not only need to use in your required practicals, but you could be tested on in the exams too. There's a Practical Skills section to cover these skills at the front of the book.

Exam-style Questions

- Practising exam-style questions is really important — you'll find some at the end of each section.

- They're the same style as the ones you'll get in the real exams — some will test your knowledge and understanding and some will test that you can apply your knowledge.

- All the answers are in the back of the book, along with a mark scheme to show you how you get the marks.

Exam Help

There's a section at the back of the book stuffed full of things to help with your exams.

Glossary

There's a glossary at the back of the book full of all the definitions you need to know for the exam, plus loads of other useful words.

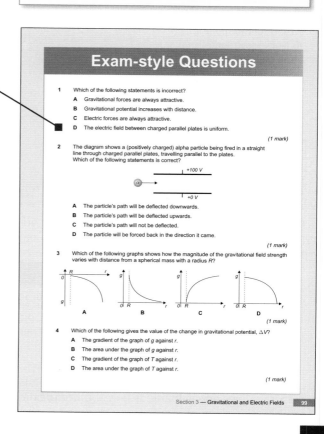

Exam-style Questions

1 Which of the following statements is incorrect?

A Gravitational forces are always attractive.
B Gravitational potential increases with distance.
C Electric forces are always attractive.
D The electric field between charged parallel plates is uniform.

(1 mark)

2 The diagram shows a (positively charged) alpha particle being fired in a straight line through charged parallel plates, travelling parallel to the plates. Which of the following statements is correct?

+100 V
+0 V

A The particle's path will be deflected downwards.
B The particle's path will be deflected upwards.
C The particle's path will not be deflected.
D The particle will be forced back in the direction it came.

(1 mark)

3 Which of the following graphs shows how the magnitude of the gravitational field strength varies with distance from a spherical mass with a radius R?

A   B   C   D

(1 mark)

4 Which of the following gives the value of the change in gravitational potential, ΔV?

A The gradient of the graph of g against r.
B The area under the graph of g against r.
C The gradient of the graph of T against r.
D The area under the graph of T against r.

(1 mark)

Published by CGP

Editors:
Emily Garrett, David Maliphant, Rachael Marshall, Sam Pilgrim, Frances Rooney, Charlotte Whiteley,
Sarah Williams, Jonathan Wray.

Contributors:
Mark A. Edwards, Duncan Kamya, Barbara Mascetti, John Myers, Moira Steven, Tony Winzor.

ISBN: 978 1 78294 328 0

With thanks to Mark Edwards and Glenn Rogers for the proofreading.
With thanks to Jan Greenway for the copyright research.

Printed by Elanders Ltd, Newcastle upon Tyne.
Clipart from Corel®

1. Experiment Design

Before you can even think about picking up some apparatus, you need to design every aspect of your experiment. The information on these pages should help you answer questions on experimental methods in the exam.

Planning experiments

Scientists solve problems by asking questions, suggesting answers and then testing them to see if they're correct. Planning an experiment is an important part of this process to help get accurate and precise results (see p.15). There's plenty of info on the next few pages, but here's a summary of how you go about it...

- Make a **prediction** — a specific testable statement about what will happen in the experiment, based on observation, experience or a **hypothesis** (a suggested explanation for a fact or observation).

- Think about the aims of the experiment and identify the independent, dependent and control variables (see below).

- Select appropriate equipment (see page 3) that will give you accurate and precise results.

- Do a risk assessment and plan any safety precautions (see p.3).

- Decide what data to collect and how you'll do it (p.4-5).

- Write out a clear and detailed method — it should be clear enough that anyone could follow it and repeat your experiment exactly.

- Carry out tests — to provide evidence that will support the prediction or refute it.

Tip: You might be asked to design a physics experiment to investigate something or answer a question.

It could be a lab experiment that you've seen before, or something applied, like deciding which building material is best for a particular job. Either way, you'll be able to use the physics you know and the skills in this topic to figure out the best way to investigate the problem.

Variables

You probably know this all off by heart but it's easy to get mixed up sometimes. So here's a quick recap.

A **variable** is a quantity that has the potential to change, e.g. mass. There are two types of variable commonly referred to in experiments:

Independent variable — the thing that you change in an experiment.

Dependent variable — the thing that you measure in an experiment.

It's important to control the variables in an experiment. Keeping all variables constant apart from the independent and dependent variables means that the experiment is a **fair test**.

This means you can be more confident that any effects you see are caused by changing the independent variable. The variables that are kept constant (or at least monitored) in an experiment are called **control variables**.

Tip: When drawing graphs, the dependent variable should go on the (vertical) *y*-axis, and the independent variable on the (horizontal) *x*-axis. The main exceptions to this are load-extension (or force-extension) and stress-strain graphs.

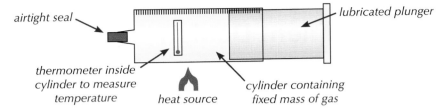

Example

You could investigate the effect of varying the temperature on the volume of some gas in a gas syringe at a fixed pressure.

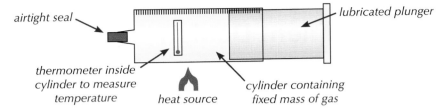

airtight seal

lubricated plunger

thermometer inside cylinder to measure temperature

heat source

cylinder containing fixed mass of gas

Figure 1: *Experimental setup for investigating the effect of temperature on the volume of a gas.*

- The independent variable will be the temperature inside the cylinder.
- The dependent variable will be the volume of the gas in the cylinder.
- All the other variables must be kept the same. These include the pressure inside the cylinder (done by making sure the plunger is well lubricated), the type of gas used and the mass of the gas.

Apparatus and techniques

You need to think about what units your measurements of the independent and dependent variables are likely to be in before you begin (e.g. millimetres or metres, milliseconds or hours).

Think about the range you plan on taking measurements over too — e.g. if you're measuring the effect of increasing the mass on a spring undergoing simple harmonic motion, you need to know whether you should increase the mass in steps of 1 gram, 10 grams or 100 grams. Sometimes, you'll be able to estimate what effect changing your independent variable will have, or sometimes a pilot experiment might help.

Example

A student is investigating the time period of the oscillations of a mass on a spring using the apparatus in Figure 2. She will increase the mass on the end of the spring and measure the change in the period of the oscillations by measuring the time taken to complete ten oscillations, using a stopwatch and a reference point.

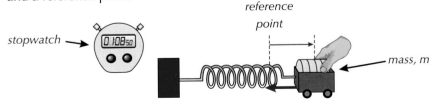

reference point

stopwatch

mass, m

Figure 2: *Experimental setup for investigating the time period of oscillations for a mass on a spring.*

She decides to do a pilot experiment to decide on the increments in which she should increase the mass. She is aiming to find out how much the mass needs to be increased by in order for the time taken for ten oscillations to increase by a sensible amount. The amount it increases by needs to be measurable with the stopwatch.

Considering your measurements before you start will also help you choose the most appropriate apparatus and techniques for the experiment. You want to pick the apparatus that will give you the best results. For example, using equipment with a high **resolution** can reduce uncertainty in your measurements (see page 10). Using the most appropriate equipment will also help to ensure you conduct the experiment safely.

Examples

- If you're measuring the length of a spring that you're applying a force to, you might need a ruler. If you're measuring the diameter of a wire, you'd be better off with a micrometer.

- If you're measuring a time interval, you could use a stopwatch. If the time is really short (for example if you're investigating the acceleration due to gravity of an object as it falls to the floor), you might need something more sensitive, like light gates.

Exam Tip
There's a whole range of apparatus and techniques that could come up in your exam. Make sure you know how to use all the ones you've come across in class.

Whatever apparatus and techniques you use, make sure you use them correctly. E.g. if you're measuring a length, make sure your eye is level with the ruler when you take the measurement.

Risks, hazards and ethical considerations

You'll be expected to show that you can identify any risks and hazards in an experiment. You'll need to take appropriate safety measures depending on the experiment.

For example, anything involving lasers will usually need special laser goggles, and to work with radioactive substances you'll probably need to wear gloves.

You need to make sure you're working ethically too — you've got to look after the welfare of any people or animals in an experiment to make sure they don't become ill, stressed or harmed in any way.

You also need to make sure you're treating the environment ethically, e.g. making sure not to destroy habitats when doing outdoor experiments.

Figure 3: *Goggles should be used in most experiments to protect the eyes from any moving objects, snapping wires, chemicals and other dangers.*

Evaluating experiment designs

If you need to evaluate an experiment design, whether it's your own or someone else's, you need to think about the following things:

- Does the experiment actually test what it sets out to test?
- Is the method clear enough for someone else to follow?
- Apart from the independent and dependent variables, is everything else going to be properly controlled?
- Are the apparatus and techniques appropriate for what's being measured? Will they be used correctly?
- Will the method give precise results? E.g. are repeat measurements going to be taken in order to calculate a mean value (see p.5)?
- Is the experiment going to be conducted safely and ethically?

2. Data

When you're planning an experiment, you need to think about the best way to record and present your data. It all depends on what sort of data you've got...

Types of data

Experiments always involve some sort of measurement to provide data. There are different types of data — and you need to know what they are.

1. Discrete data

You get discrete data by counting — e.g. the number of turns of a coil in a transformer (see Figure 1). You couldn't really have 1.25 turns — it wouldn't make sense. Shoe size is another good example of a discrete variable — only certain values are allowed.

There are lots of ways to present discrete data, depending on what other data sets you've recorded. Scatter graphs (p.6) and bar charts are often used.

2. Continuous data

A continuous variable can have any value on a scale. For example, the potential difference across a wire or the gravitational field strength caused by a mass. You can never measure the exact value of a continuous variable.

The best way to display two sets of continuous data is a line graph or a scatter graph. To draw a line graph, plot all the points and join them up with straight lines.

3. Categoric data

A categoric variable has values that can be sorted into categories. For example, types of material might be brass, wood, glass or steel.

If one of your data sets is categoric, a pie chart or a bar chart is often used to present the data — see Figure 2.

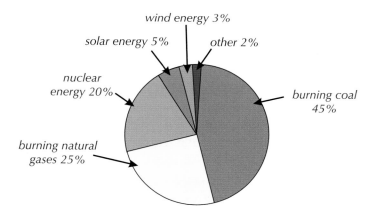

wind energy 3%
solar energy 5%
other 2%
nuclear energy 20%
burning coal 45%
burning natural gases 25%

Figure 2: *Categoric data of different types of energy production in a particular country presented on a pie chart.*

4. Ordered (ordinal) data

Ordered data is similar to categoric data, but the categories can be put in order. For example, if you classified frequencies of light as 'low', 'fairly high' and 'very high' you'd have ordered data.

A bar chart is often used if one of your data sets is ordered — see Figure 3.

Figure 1: *A transformer is used to increase or decrease an alternating voltage. The number of turns in each coil are discrete data, but the input and output voltage are continuous data.*

Tip: For line-graphs, e.g. a displacement-time graph for a journey, you join the points up rather than drawing a line of best fit. This is because there isn't a general trend, but a journey with different stages.

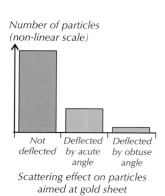

Number of particles (non-linear scale)

| Not deflected | Deflected by acute angle | Deflected by obtuse angle |

Scattering effect on particles aimed at gold sheet

Figure 3: *Ordered data of particle scattering effects presented on a bar chart.*

Tables of data

When you do an experiment you'll need to make a table to write your results in. You should include:

- Space for your independent variable and your dependent variable. You should specify the units in the headers, not within the table itself.
- Space to repeat each test at least three times to see how precise and repeatable your results are (see page 15).
- Space for any data processing you need to do, e.g. calculating an average from repeats, or calculating speed from measurements of distance and time.

Figure 4 (below) is the sort of table you might end up with when you investigate the current flowing through a circuit connected to a discharging capacitor over time.

Time / s	Current / A Run 1	Current / A Run 2	Current / A Run 3	Mean current / A (to 3 d.p.)
0.0	0.370	0.368	0.371	**0.370**
1.0	0.218	0.219	0.172	**0.219**
2.0	0.118	0.122	0.116	**0.119**
3.0	0.067	0.071	0.069	**0.069**
4.0	0.022	0.015	0.017	**0.018**
5.0	0.010	0.007	0.007	**0.008**

Figure 4: *Table of results showing the current flowing through a circuit connected to a discharging capacitor over time.*

Most of the time, you'll be recording numerical values (quantitative data). Occasionally, you may have to deal with data that can be observed but not measured with a numerical value. This is known as qualitative data, e.g. categoric or ordered data. It's still best to record this kind of data in a table to keep your results organised, but the layout may be a little different.

Calculating a mean

For many experiments, you'll need to calculate the arithmetic mean (average) of some repeated measurements:

$$\text{arithmetic mean (average) of a measurement} = \frac{\text{sum of your repeated measurements}}{\text{number of measurements taken}}$$

Watch out for **anomalous results**. These are ones that don't fit in with the other values and are likely to be wrong. They're usually due to experimental errors, such as making a mistake when measuring. You should ignore anomalous results when you calculate averages.

Example — Maths Skills

Look at the table in Figure 4 again — the current at 1.0 s in Run 3 looks like it might be an anomalous result. It's much lower than the values in the other two runs. It could have been caused by the wires in the circuit being hotter for Run 3 than they were for the first two runs.

The anomalous result should be ignored to calculate the average:

With anomalous result: $(0.218 + 0.219 + 0.172) \div 3 = 0.203$ (to 3 d.p.)

Without anomalous result: $(0.218 + 0.219) \div 2 = 0.219$ (to 3 d.p.)

So the average current at 1.0 s should be 0.219 A (rather than 0.203 A).

Tip: You should give all of your data to the number of decimal places that you measured to — e.g. 0.000 A, not just 0 A. When processing data, you should give your result to the least number of significant figures (p.11) found in your original data.

Tip: See below for more on finding the mean of a set of experimental results.

Tip: See pages 111-112 for more on discharging capacitors.

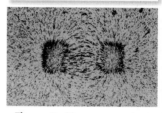

Figure 5: *You can use iron filings to show the magnetic field lines around the ends of a horseshoe magnet. Measuring how these magnetic field lines vary for different magnets could involve recording qualitative data. The field lines can be observed, but have no numerical value.*

Tip: The arithmetic mean is usually just called the mean (or the average when it's clear which average is meant).

Tip: Just because you ignore anomalous results in your calculations, you shouldn't ignore them in your write-up. Try to find an explanation for what went wrong so that it can be avoided in future experiments.

3. Graphs

You'll usually be expected to make a graph of your results.
Graphs make your data easier to understand if done the right way.

Scatter graphs

Scatter graphs, like Figure 1, are great for showing how two sets of data are related (or correlated — see below for more on correlation). Don't try to join all the points — draw a **line of best fit** to show the trend.

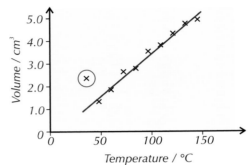

Figure 1: Scatter graph showing the relationship between the volume and temperature of an ideal gas at constant pressure.

Scatter graphs and correlation

Correlation describes the relationship between two variables — usually the independent one and the dependent one. Data can show positive correlation, negative correlation or no correlation (see Figure 2). There's more about this on the next page.

Positive correlation
As one variable increases, the other also increases.

Negative correlation
As one variable increases, the other decreases.

No correlation
There is no relationship between the variables.

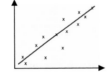  

Figure 2: Scatter graphs showing positive, negative and no correlation.

Curved graphs and lines of best fit

Sometimes your variables will have a non-linear relationship (see page 8) — the plotted points will form a curve. Your line of best fit will need to be a smooth curve that should have about the same number of points on either side of it (see Figure 3).

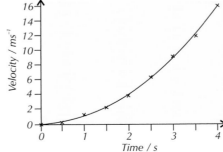

Figure 3: Line of best fit on a non-linear graph.

Correlation and cause

Ideally, only two quantities would ever change in any experiment — everything else would remain constant. But in experiments or studies outside the lab, you can't usually control all the variables. So even if two variables are correlated, the change in one may not be causing the change in the other. Both changes might be caused by a third variable.

Tip: If an experiment really does confirm that changing one variable causes another to change, we say there's a <u>causal link</u> between them.

┌─ **Example** ─────────────────────────────

Some studies have found a correlation between exposure to the electromagnetic fields created by power lines and certain ill health effects. So some people argue that this means we shouldn't live close to power lines, or build power lines close to homes. But it's hard to control all the variables between people who live near power lines and people who don't. Ill health in people living near power lines could be affected by many lifestyle factors or even genetics. Also, people living close to power lines may be more likely to believe that any ill health they suffer is due to the electromagnetic fields from the power lines if they are aware of the studies.

Tip: Watch out for bias too — for instance, a neighbourhood campaigning against unsightly power lines being built nearby may want to show that they are a health danger.

Straight-line graphs

If you plot two variables that have a linear relationship, you'll get a straight-line graph. The equation of a straight line is $y = mx + c$, where m = gradient (slope of the line) and c = y-intercept. This means you can use your graph to work out certain values, and the relationship between your variables.

Tip: x and y have a linear relationship if $y = mx + c$, where m and c are constants.

Proportionality

If you plot two variables against each other and get a straight line that goes through the origin, the two variables are **directly proportional**. The y-intercept, c, is 0, so the equation of the straight line is $y = mx$ where m is a constant. The constant of proportionality, m, is the gradient of the graph.

Tip: Two variables are directly proportional if one variable = constant × other variable.

┌─ **Example** ─── **Maths Skills** ──────────

The potential difference across a charging capacitor is directly proportional to the charge stored on it. You can see this using the circuit in Figure 4. Use the variable resistor to keep the charging current constant for as long as possible. Take readings of the potential difference across the capacitor until it's close to the battery p.d., and find the value of Q at each of these points in time using $Q = It$, where I is current and t is the time since charging started. Once you have all the data, plot a graph of Q against V (Figure 5). The graph you'll get is a straight line going through the origin — so charge and potential difference are directly proportional.

Tip: For more on this experiment, see p.103.

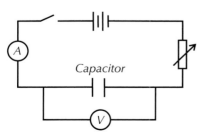

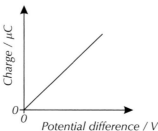

Figure 4: Circuit for showing the proportional relationship between Q and V for a capacitor.

Figure 5: A Q-V graph for a capacitor.

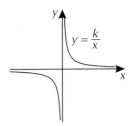

Figure 6: An inverse proportionality relationship between x and y.

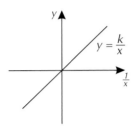

Figure 7: If $y = \frac{k}{x}$, plotting y against $\frac{1}{x}$ gives a straight line through the origin.

Tip: You met displacement-time graphs in year 1 of A-level.

Tip: For more on work functions, see page 380.

Tip: If you were to calculate the gradient, you should get h, the Planck constant.

Tip: Here, $y = E_{K(max)}$, $m = h$, $x = $ frequency and $c = -\phi$.

Non-linear relationships

Straight line graphs are really easy to work with, but some variables won't produce a straight-line graph if you plot them against each other. You can sometimes change what you plot on the axes so that you get one though.

For example, say two variables are inversely proportional, then $y = \frac{k}{x}$, where k is a constant. If you plot y against x, you'll get a curved graph which shoots off to infinity (Figure 6). It's not very easy to work out the value of k from this graph, but if you plot y against $\frac{1}{x}$ you'll get a lovely straight-line graph (Figure 7) with a constant gradient of k that goes through the origin. This is because the graph plotted is $y = k\left(\frac{1}{x}\right)$, which is just the equation of a straight line in form $y = mx + c$, where $m = k$, $c = 0$ and $\frac{1}{x}$ is on the x-axis.

Finding the gradient and y-intercept

If you've plotted a straight-line graph, you can calculate the gradient and read off the y-intercept. This means you can work out certain quantities from your graph.

┌─ **Example** ── **Maths Skills** ─────────────────────

If an object is travelling at a constant velocity, you can find its velocity by calculating the gradient of a displacement-time graph showing its motion.

Velocity is the rate of change of displacement, which can be written as $v = \frac{\Delta s}{\Delta t}$, so $\Delta s = v\Delta t$. This is in the form $y = mx + c$, where $y = \Delta s$, $m = v$, $x = \Delta t$ and $c = 0$. So the gradient $(m) = v$. You can find the gradient of the straight-line graph by dividing the change in y (Δy) by the change in x (Δx).

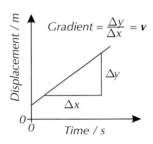

Figure 8: Using a Q-V graph to work out the capacitance of a capacitor.

┌─ **Example** ── **Maths Skills** ─────────────────────

The solid line in the graph in Figure 9 shows how the maximum kinetic energy of the electrons on a metal surface varies with the frequency of the light shining on it. You can use a graph like this to find the value of the work function of the metal (ϕ) by extending the graph back to the y-axis.

Rearranging the equation $hf = \phi + E_{K(max)}$ gives $E_{K(max)} = hf - \phi$. Since h and ϕ are constants, $E_{K(max)} = hf - \phi$ is just the equation of a straight line (in the form: $y = mx + c$). You can just read ϕ from the graph — it's the intercept on the vertical axis. You'll just need to continue the line back to the y-axis to find the intercept, then the value of the y-intercept will be $-\phi$.

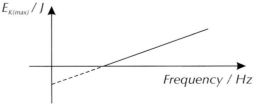

Figure 9: You can extend the line on a graph to find the y-intercept.

4. Error Analysis

Scientists always have to include the uncertainty of a result, so you can see the range the actual value probably lies within. Knowing how to calculate and minimise error and uncertainty in your results is an important skill

Types of error

Every measurement you take has an experimental uncertainty. Say you measure the temperature of a beaker of water with an analogue thermometer. You might think you've measured its temperature as 21 °C, but at best you've probably measured it to be 21 ± 0.5 °C. And that's without taking into account any other errors that might be in your measurement.

The ± bit gives you the range in which the true temperature (the one you'd really like to know) probably lies — 21 ± 0.5 °C tells you the true temperature is very likely to lie in the range of 20.5 to 21.5 °C. The smaller the uncertainty, the nearer your value must be to the true value, so the more accurate your result (see page 15 for more on accuracy).

If you measure a length of something with a ruler, you actually take two measurements, one at each end of the object you're measuring. There is an uncertainty in each of these measurements. E.g. a length of 17.0 cm measured using a mm ruler will have an uncertainty of 0.05 + 0.05 = 0.1 cm.

There are two types of error:

Random errors

Random errors cause readings to be spread about the true value due to the results varying in an unpredictable way. They affect precision (see p.15).

They can just be down to noise, or because you're measuring a random process such as nuclear radiation emission. You get random errors in any measurement and no matter how hard you try, you can't correct them.

For example, if you measured the length of a wire 20 times, the chances are you'd get a slightly different value each time, e.g. due to your head being in a slightly different position when reading the scale. It could be that you just can't keep controlled variables exactly the same throughout the experiment. Or it could just be the wind was blowing in the wrong direction at the time.

> **Example**
>
> You could investigate the local value of *g* (see page 75) using a pendulum, as shown in Figure 2.
>
> The pendulum's motion might be affected by vibrations in the room from people walking by, heavy vehicles driving past, etc.
> This introduces random errors.
>
>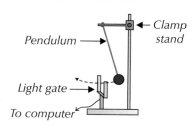
> Pendulum → Clamp stand
> Light gate →
> To computer
>
> *Figure 2: An experiment to find the value of g.*

Systematic errors

Systematic errors usually cause each reading to be different to the true value by the same amount — i.e. they shift all of your measurements. They affect the accuracy of your results (see p.15).

You get systematic errors not because you've made a mistake in a measurement, but because of the environment, the apparatus you're using, or your experimental method — e.g. using an inaccurate clock.

Figure 1: This thermometer measures to the nearest °C. Any measurement you take using it will have an uncertainty of ± 0.5 °C.

Tip: To get the lowest possible value, subtract the value after the ± sign from your measurement, and to get the highest possible value, add it to your measurement.

Tip: It's not just rulers that have two errors in each measurement, it's anything where you have to take a start and end measurement, for example calipers, protractors and stopwatches.

Tip: A common source of error is changes in temperature — all sorts of things are affected by temperature, from the properties of a material to the current flowing in a circuit.

Tip: A newton meter that always measures values 1 N greater than they should be will shift all your results up by 1 N — this would introduce a systematic error due to the apparatus used.

The problem is often that you don't know systematic errors are there. You've got to spot them first to have any chance of correcting for them. They're annoying, but there are things you can do to reduce them if you manage to spot them (see below).

If you suspect a systematic error, you should repeat the experiment with a different technique or apparatus and compare the results.

┌─ **Example — continued** ────────────────────────
Look back at the investigation of the value of *g* on the previous page (see Figure 2). If there is any friction in the system, or if the length of the pendulum is measured inaccurately (you should measure it to the centre of mass of the bob), all the readings will be shifted in the same direction. This would introduce a systematic error due to the experimental set-up.

You can **calibrate** your apparatus by measuring a known value. If there's a difference between the measured and known value, you can use this to correct the inaccuracy of the apparatus, and so reduce your systematic error.

Calibration can also reduce **zero errors** (caused by the apparatus failing to read zero when it should do, e.g. when no current is flowing through an ammeter) which can cause systematic errors.

Reducing uncertainty

There are a few different ways you can reduce the uncertainty in your results:

Repeating and averaging

One of the easiest things you can do is repeat each measurement several times. The more repeats you do, and the more similar the results of each repeat are, the more precise the data.

By taking an average (mean) of your repeated measurements (see page 5), you will reduce the random error in your result. The more measurements you average over, the less random error you're likely to have.

Repeating also allows you to check your data for any anomalous results, for example, a measurement that is ten times smaller than all of your other data values. You should not include anomalous results when you take averages.

Using appropriate equipment

You can also cut down the uncertainty in your measurements by using the most appropriate equipment. The smallest possible uncertainty in a measurement is usually taken to be ± half the smallest interval that the measuring instrument can measure. A micrometer scale has smaller intervals than a millimetre ruler. So by measuring a wire's diameter with a micrometer instead of a ruler, you instantly cut down the random error in your experiment.

Computers and data loggers can often be used to measure smaller intervals than you can measure by hand and reduce random errors, e.g. timing an object's fall using a light gate rather than a stop watch. You also get rid of any human error that might creep in while taking the measurements.

There's a limit to how much you can reduce the random uncertainties in your measurements, as all measuring equipment has a **resolution** — the smallest change in what's being measured that can be detected by the equipment.

Tip: Another possible source of a systematic error for an experiment like this could be if the light gate or the computer wasn't calibrated properly.

Tip: To calibrate a set of scales you could weigh a 10.0 g mass and check that it reads 10.0 g. If these scales are precise to the nearest 0.1 g, then you can only calibrate to within 0.05 g. Any measurements taken will have an uncertainty of ± 0.05 g.

Tip: See the next page for finding the error / uncertainty on a mean value.

Figure 3: *A micrometer is very precise — it gives readings to within 0.01 mm.*

Averaging and uncertainty

In the exam, you might be given a graph or table of information showing the results for many repetitions of the same experiment, and asked to estimate the true value and give an uncertainty in that value. Here's how to go about it:

1. Estimate the true value by finding the mean of the results you've been given — just like in the example on page 5.
2. To get the uncertainty on the mean, you just need to find the range of the repeated measurements and halve it.

Example — **Maths Skills**

A class measure the resistance of a component to 2 s.f. and record their results on the bar chart shown below. Estimate the resistance of the component, giving a suitable range of uncertainty in your answer.

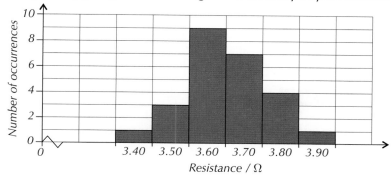

There were 25 measurements, so taking the mean:

$$\frac{(3.40 + (3.50 \times 3) + (3.60 \times 9) + (3.70 \times 7) + (3.80 \times 4) + 3.90)}{25} =$$

$$= 3.652$$
$$= 3.66 \ \Omega \ \text{(to 3 s.f.)}$$

Tip: Just add up the heights of all the bars to find the total number of measurements.

The maximum value found was 3.90 Ω, the minimum value was 3.40 Ω. The range is 3.90 – 3.40 = 0.50, so the uncertainty is 0.5 ÷ 2 = 0.25 Ω.

So the answer is 3.66 ± 0.25 Ω (to 3 s.f.).

Significant figures and uncertainty

You always have to assume the largest amount of uncertainty in data.

Whether you're looking at experimental results or just doing a calculation question in an exam, you must round your results to the same number of significant figures as the given data value with the fewest significant figures. Otherwise you'd be saying there is less uncertainty in your result than in the data used to calculate it.

If no uncertainty is given for a value, the number of significant figures a value has gives you an estimate of the uncertainty.

Example — **Maths Skills**

2 N only has 1 significant figure, so without any other information you know this value must be 2 ± 0.5 N — if the value was less than 1.5 N it would have been rounded to 1 N (to 1 s.f.), if it was 2.5 N or greater it would have been rounded to 3 N (to 1 s.f.).

---

5. Uncertainty Calculations

You'll often be given the uncertainty of a measurement, or asked to work it out (see pages 9-11). But if you need to process your data to calculate a result, finding the uncertainty of that result is a bit trickier.

Uncertainties

Uncertainties come in absolute amounts, fractions and percentages.

- **Absolute uncertainty** is the uncertainty of a measurement given as a certain fixed quantity.
- **Fractional uncertainty** is the uncertainty given as a fraction of the measurement taken.
- **Percentage uncertainty** is the uncertainty given as a percentage of the measurement.

An uncertainty could also include a level of confidence, to indicate how likely the true value is to lie in the interval. E.g. '5.0 ± 0.4 Ω at a level of confidence of 80%' means you're 80% sure that the true value is within 0.4 Ω of 5.0 Ω.

Tip: You won't need to calculate the level of confidence.

┌─ **Example** ── **Maths Skills** ──────────

The resistance of a resistor is given as 5.0 ± 0.4 Ω. Give the absolute, fractional and percentage uncertainties for this measurement.

- The absolute uncertainty is given in the question — it's 0.4 Ω.
- To calculate fractional uncertainty, divide the uncertainty by the measurement and simplify:

  The fractional uncertainty is $\frac{0.4}{5.0} = \frac{4}{50} = \frac{2}{25}$

- To calculate percentage uncertainty, divide the uncertainty by the measurement and multiply by 100:

  The percentage uncertainty is $\frac{2}{25} \times 100 = 8\%$

└──────────────────────────────

***Figure 1:** You'd assume an absolute uncertainty of ±2 V in this voltmeter reading of 24 V (as there is an assumed ±1 V uncertainty at each end). If you were measuring a potential difference of e.g. 2 V, the percentage uncertainty would be much higher, and you might consider using apparatus with a higher resolution.*

You can decrease the percentage uncertainty in your data by taking measurements of large quantities or by using measuring instruments that can measure in smaller increments (see page 10).

Say you take measurements with a thermometer which measures to the nearest ± 0.5 °C. The percentage error in measuring a temperature of 1 °C will be ± 50%, but using the same thermometer to measure a temperature of 100 °C will give a percentage error of only ± 0.5%.

┌─ **Example** ── **Maths Skills** ──────────

Two students measure the diameter of a wire to be 1 mm. Student A uses a ruler that can measure to the nearest mm and student B uses a micrometer that can measure to the nearest 0.01 mm. Calculate the percentage uncertainty of each measurement.

Student A's ruler has an uncertainty at both ends (see p.9):

Absolute uncertainty = 2 × 0.5 mm, so % uncertainty = $\frac{1.0}{1} \times 100 = 100\%$

Student B's micrometer also has two uncertainties:

Absolute uncertainty = 2 × 0.005 mm, so % uncertainty = $\frac{0.01}{1} \times 100 = 1\%$

└──────────────────────────────

Combining uncertainties

When you do calculations involving values that have an uncertainty, you have to combine the uncertainties to get the overall uncertainty for your result.

Adding or subtracting

When you're adding or subtracting data, you add the absolute uncertainties.

> **Example — Maths Skills**
>
> **The mass of a bob on the end of a pendulum is increased from 12 ± 1 g to 20 ± 1 g. Calculate change in mass of the bob.**
>
> First subtract the masses without the uncertainty values:
> $$20 - 12 = 8 \text{ g}$$
> Then find the total uncertainty by adding the individual absolute uncertainties:
> $$1 + 1 = 2 \text{ g}$$
> So, the mass has been increased by 8 ± 2 g.

Tip: Be very careful. Even if you subtract the data, you add the uncertainties (combining uncertainties should always make the uncertainty increase).

Multiplying or dividing

When you're multiplying or dividing data, you add the percentage uncertainties.

> **Example — Maths Skills**
>
> **A passenger car on a Ferris wheel has a linear velocity of 2.4 ± 3% ms⁻¹. The Ferris wheel has radius 32 ± 2 m. Calculate the angular velocity of the passenger car and state the percentage uncertainty in this value.**
>
> First calculate the angular velocity without uncertainty:
> $$\omega = v \div r = 2.4 \div 32 = 0.075 \text{ rad s}^{-1}$$
> Next, calculate the percentage uncertainty in the radius:
> $$\% \text{ uncertainty in } r = \frac{2}{32} \times 100 = 6.25\%$$
> Add the percentage uncertainties in the velocity and radius values to find the total uncertainty in the angular velocity:
> $$\text{Total uncertainty} = 3 + 6.25 = 9.25\%$$
> So, the angular velocity = 0.075 ± 9.25% rad s⁻¹

Tip: See pages 19-25 for more on circular motion.

Tip: Don't forget to convert all the uncertainties to percentages before you combine by multiplying or dividing — see previous page for how.

Raising to a power

When you're raising data to a power, you multiply the percentage uncertainty by the power.

> **Example — Maths Skills**
>
> **The radius of a circle is $r = 40 \pm 2.5\%$ cm. What will the percentage uncertainty be in the area of this circle, i.e. πr^2?**
>
> The radius will be raised to the power of 2 to calculate the area.
> So, the percentage uncertainty will be 2.5 × 2 = 5%

Error bars

Most of the time, you work out the uncertainty in your final result using the uncertainty in each measurement you make. When you're plotting a graph, you show the uncertainty in each measurement by using **error bars** to show the range the point is likely to lie in. In exams, you might have to analyse data from graphs with and without error bars — so make sure you really understand what error bars are showing.

The error in measuring the volume of a gas container can be found using the error bars in the graph below.

Tip: Your line of best fit (p.6) should always go through all of the error bars (excluding any anomalous data, see page 5).

Tip: Be careful — sometimes error bars are calculated using a set percentage uncertainty for each measurement, and so will change depending on the measurement.

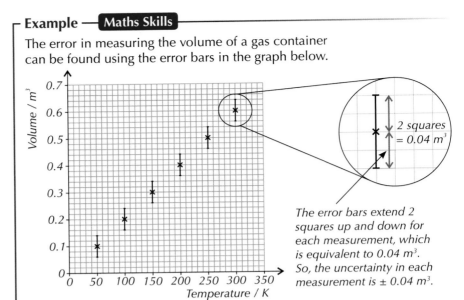

The error bars extend 2 squares up and down for each measurement, which is equivalent to 0.04 m³. So, the uncertainty in each measurement is ± 0.04 m³.

Figure 2: A graph of volume against temperature for a fixed mass of gas.

Measuring the uncertainty of final results

Normally when you draw a graph you'll want to find the gradient or intercept. For example, you can calculate v, the velocity of a moving object, from the gradient of the graph in Figure 3 — here it's about 2 ms^{-1}. You can find the uncertainty in that value by using **worst lines**.

Tip: If you have a graph showing the relation between two variables that are directly proportional, you must treat the origin as a normal point whilst drawing a line of best fit. The origin will be a measurement with no uncertainty and so all lines of best fit must pass through it.

Draw lines of best fit which have the maximum and minimum possible slopes for the data and which should go through all of the error bars (see the pink and blue lines in Figure 3). These are the worst lines for your data.

Calculate the worst gradient — the gradient of the slope that is furthest from the gradient of the line of best fit. In Figure 3, the gradient of the blue line is about 2.1 ms^{-1} and the gradient of the pink line is about 1.9 ms^{-1}, so you can use either.

The uncertainty in the gradient is given by the difference between the best gradient (of the line of best fit) and the worst gradient — here it's 0.1 ms^{-1}. So this is the uncertainty in the value of the velocity. For this object, the velocity is 2 ± 0.1 ms^{-1} (or 2 ms^{-1} ± 5%).

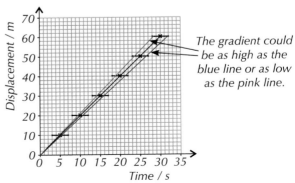

The gradient could be as high as the blue line or as low as the pink line.

Figure 3: The maximum and minimum slopes possible through the error bars.

Similarly, the uncertainty in the y-intercept is just the difference between the best and worst intercepts (although there's no uncertainty in Figure 3 since the best and worst lines both go through the origin).

6. Evaluating and Concluding

Once you've got results, you can use them to form a conclusion. But be careful... your conclusion must be supported by your results, and you should keep in mind how much you can trust your results by evaluating them.

Evaluations

Now that you can measure uncertainty, you'll need to evaluate your results to see how convincing they are. You need to be careful about what words you use — valid, accurate, precise, repeatable and reproducible may all sound similar, but they all say different things about your results.

1. Precise results

The smaller the amount of spread of your data from the mean, the more precise it is. Precision only depends on the amount of random error in your readings.

2. Repeatable results

Results are repeatable if you can repeat an experiment multiple times and get the same results.

3. Reproducible results

Results are reproducible if someone else can recreate your experiment using different equipment or methods, and get the same results you do.

4. Valid results

A valid result arises from a suitable procedure to answer the original question. If you don't keep all variables apart from the ones you're testing constant, you haven't only tested the variable you're investigating and so the results aren't valid.

5. Accurate results

An accurate result is really close to the true answer.
You can only comment on how accurate a result is if you know the true value of the result.

There's normally a lot of different things to say when you're looking at data. Have a think about:

- What patterns or trends, if any, the results show.
- Whether the experiment managed to answer the question it set out to answer. If it did, was it a valid experiment, and if not, why not? How precise was the data?
- How close the results are to the true value.
- Whether the measuring instruments had enough resolution.
- Any anomalies in the results and the possible causes of them.
- How large the uncertainties are. If the percentage uncertainty is large, this suggests the data is not precise and a strong conclusion cannot be made.

If you're asked to analyse data in the exam, look at how many marks the question is worth — the more marks allocated to the question in the exam, the more detail you have to go into.

Tip: Precision is sometimes called reliability — but you shouldn't use this term.

Figure 1: *Newton's famous experiment to show that white light is made up of a spectra of colours has been reproduced by scientists and students all over the world. The results are now accepted.*

Tip: It's possible for results to be precise but not accurate, e.g. a balance that weighs to 1/1000th of a gram is likely to give precise results, but if it's not calibrated properly the results won't be accurate.

Evaluating methods

When you're evaluating experiment design, make sure that everything that could have been done to reduce uncertainties has been done.

You should make sure you think about:

- Whether all the variables were controlled. If not, how they could have been controlled, and could they have produced some random error in the results?
- Whether any anomalous results have crept into the data and could have been prevented.
- Whether the uncertainty could have been reduced (see page 10).

Drawing conclusions

The data should always support the conclusion. This may sound obvious but it's easy to jump to conclusions. Conclusions have to be specific and supported by the data — not make sweeping generalisations.

Your conclusion is only valid if it is supported by valid data, known as **evidence**.

Example

A capacitor was discharged through a fixed resistor, and the charge on the plates was measured at times 0.0, 0.2, 0.4, 0.6, 0.8 and 1.0 s. Each time reading had an estimated error of 0.01 s and each charge reading had an estimated error of 10 μC. All other variables were kept constant. The results are shown in Figure 2.

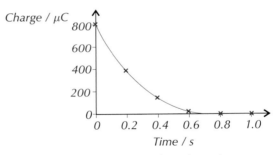

Figure 2: Graph to show charge on the plates of a capacitor over time.

A study concluded from this data that this type of capacitor discharges to 5% of its maximum charge in 0.5 s. This could be true — but the data doesn't support this. Because time increments of 0.2 s were used and the charge at in-between times wasn't measured, you can't tell when the charge becomes 5% of its maximum charge from this data. All you know is that the charge reaches this level at some point between 0.4 and 0.6 seconds.

Also, the graph only gives information about this particular experiment. You can't conclude that the discharge time would be the same for all capacitors of this type and all circuits — only this one. In truth, the discharge time of a capacitor depends on the circuit it's connected to.

You must also consider the error in the charge readings. The error in each reading is 10 μC, which gives a percentage uncertainty of around 50% for the lowest charge reading. This means the results might not be accurate.

Tip: Remember, you should evaluate your method before you begin the experiment (p.3), but sometimes it's impossible to predict what will go wrong — you should talk about things you didn't plan for (and how to correct them) in your evaluation.

Tip: Taking more readings can allow you to make stronger conclusions. If a measurement had been taken at a time of 0.5 s here, you'd be able to say more about when the charge reached 5% of its maximum value. It's a good idea to do a practice experiment to get an idea of the time span over which you need to record data.

Tip: Whoever funded the research (e.g. a circuit components company) may have some influence on what conclusions are drawn from the results, but scientists have a responsibility to make sure that the conclusions they draw are supported by the data.

7. The Practical Endorsement

Alongside your A-level exams, you have to do a separate 'Practical Endorsement'. This assesses practical skills that can't be tested in a written exam.

What is the Practical Endorsement?

The Practical Endorsement is assessed slightly differently to the rest of your course. Unlike the exams, you don't get a mark for the Practical Endorsement — you just have to get a pass grade.

In order to pass the Practical Endorsement, you have to carry out at least twelve practical experiments and demonstrate that you can:

- use a range of specified apparatus, (e.g. be able to use a micrometer),
- carry out a range of specified practical techniques, (e.g. you must be able to correctly construct a circuit from a circuit diagram (see page 126), including circuit components that rely on polarity),
- increase the accuracy of measurements in various experiments (e.g. recording several oscillations over time to find the time taken for one oscillation),
- use ICT to collect and process data (e.g. using a data logger or spreadsheet modelling).

The twelve practicals that you do are the twelve Required Practicals that form part of the AQA A-level Physics course — these cover all the techniques you need to be able to demonstrate for your Practical Endorsement. You may carry out other practicals as well, which may also count towards your Practical Endorsement. You'll do the practicals in class, and your teacher will assess you as you're doing them.

You'll need to keep a record of all your assessed practical activities. Required Practicals 1 to 6 are all part of the Year 1 material (Sections 1 to 4) so you may have already done some experiments that count towards the Practical Endorsement in Year 1 of the course. This book contains information about Required Practicals 7 to 12 (which are all part of the Year 2 material).

Assessment of the Practical Endorsement

When assessing your practical work, your teacher will be checking that you're able to do five things:

1. Follow written methods and instructions

Make sure you read any instructions you are given fully before you start work and that you understand what you're about to do. This will help you to avoid missing out any important steps or using the wrong piece of apparatus.

2. Use apparatus and investigative methods correctly

You need to be able to demonstrate that you can use apparatus and materials — this includes the use of computers to record data. You also need to show that you can carry out practical techniques correctly. This means being able to do a procedure in the right order, as well as being able to fix problems when you come across them.

You'll also need to show that you can identify which variables need to be controlled in an experiment and how to control them (see p.1). If you can't easily control all the variables, you'll need to show that you have recognised this and, if possible, found a way around it. Finally, you'll need to demonstrate that you can decide which apparatus to use to get the most accurate results and what measurements you'll take with it (see p.3 for more).

Exam Tip
You could also get asked questions about the Required Practicals in your written exams.

Tip: Throughout this book, examples of methods you could use for the Required Practicals are marked with a big stamp, like this one:

REQUIRED PRACTICAL **9**

Tip: You won't necessarily need to demonstrate each of these things every time you carry out a practical. You'll also be given the opportunity to build up to some of the more difficult skills.

Tip: The CLEAPSS®
website has a database
with details of the
potential harm that
hazardous substances
you're likely to come
across could cause.
It also has student
safety sheets, and
your school or college
may have CLEAPSS®
hazcards you can use.
These are all good
sources of information
if you're writing a risk
assessment.

3. Use apparatus and materials safely

This means being able to carry out a risk assessment (see page 3) to identify the dangers in your experiment and what you can do to reduce the risks associated with those dangers. You'll also have to show that you can work safely in the lab, using appropriate safety equipment to reduce the risks you've identified, and that you adapt your method as you go along if necessary.

4. Make observations and record results

You need to show that you can collect data that's valid, accurate and precise. When you record the data, e.g. in a table, you need to make sure you do so to an appropriate level of accuracy and you include things like the units. There's more on recording data on page 5.

5. Carry out supporting research and write reports using appropriate references

You need to be able to write up an investigation properly. As well as reporting your results and drawing a conclusion about your findings, you'll need to describe the method you used and any safety precautions you took.

In your report, you'll need to make sure you display your results in the most sensible way — for example, if you had a continuous dependent variable and a continuous independent variable, a scatter graph would probably be best (see page 6). You'll also need to show that you can use computer software to process your data — e.g. drawing a graph of your results using the computer.

You should also write up any research you've done (e.g. to help you with planning a method or to draw your conclusions) and properly cite the sources that you've used.

Research, references and citations

You can use books or the Internet to carry out research, but there a few things you'll need to bear in mind:

Tip: If you're unsure
whether the information
on a website is true
or not, try and find
the same piece of
information in a different
place. The more sources
you can find for the
information, the more
likely it is to be correct.

- Not all the information you find on the Internet will be true. It's hard to know where information comes from on forums, blogs and websites that can be edited by the general public, so you should avoid using these. Websites of organisations such as the Institute of Physics (IOP) and the National Health Service (NHS) provide lots of information that comes from reliable scientific sources. Scientific papers and textbooks are also good sources of reliable information.

- It may sound obvious, but when you're using the information that you've found during your research, you can't just copy it down word for word. Any data you're looking up should be copied accurately, but you should rewrite everything else in your own words.

Tip: There are lots of
slightly different ways of
referencing sources, but
the important thing is
that it's clear where you
found the information.

- When you've used information from a source, you need to cite the reference properly. **Citations** allow someone else to go back and find the source of your information. This means they can check your information and see you're not making things up out of thin air. Citations also mean you've properly credited other people's data that you've used in your work. A citation for a particular piece of information may include the title of the book, paper or website where you found the information, the author and/or the publisher of the document and the date the document was published.

1. Circular Motion

You might have met radians before, but only as an alternative option to using degrees. In circular motion, you'll see that all angle measurements use radians, so make sure you're comfortable with them before tackling this topic.

Radians

Objects in circular motion travel through angles — these angles are usually measured in radians.

The angle in **radians**, θ, is equal to the arc-length divided by the radius of the circle (see Figure 1).

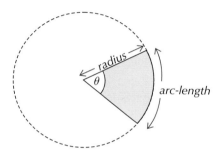

Figure 1: *The radius of a circle, and the angle and arc-length of a sector.*

For a complete circle (360°), the arc-length is just the circumference of the circle ($2\pi r$). Dividing this by the radius (r) gives 2π. So there are 2π radians in a complete circle. 1 radian is equal to about 57°.

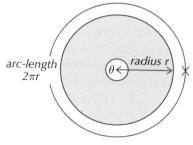

Figure 2: *For a complete circle, the angle $\theta = 2\pi$.*

To convert from degrees to radians, you multiply the angle by $\frac{\pi}{180°}$:

$$\text{angle in radians} = \text{angle in degrees} \times \frac{\pi}{180°}$$

Figure 3 shows some common angles, given in both degrees and radians.

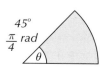
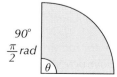
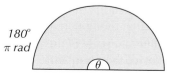

$45°$ $\frac{\pi}{4}$ rad θ $90°$ $\frac{\pi}{2}$ rad θ $180°$ π rad θ

Figure 3: *Angles of sectors in degrees and radians.*

Angular speed

Tip: The symbol for angular speed is the little Greek 'omega', not a w.

Tip: Angular speed is the same at any point on a solid rotating object.

Tip: The linear speed v is how fast an object would be travelling in a straight line if it broke off from circular motion.

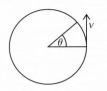

Tip: You don't need to know the derivation of $\omega = \frac{v}{r}$ for your exam.

The **angular speed** is the angle an object rotates through per second. Just as linear speed, v, is defined as distance ÷ time, the angular speed, ω, is defined as angle ÷ time. The unit is rad s^{-1} (radians per second).

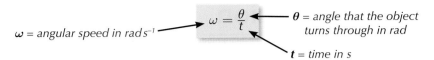

ω = angular speed in rad s^{-1} —— $\omega = \frac{\theta}{t}$ —— θ = angle that the object turns through in rad

t = time in s

The angular speed can be written in terms of the linear speed (sometimes called the tangential speed), v.

Consider an object moving in a circular path of radius r that moves an angle of θ radians in t seconds. The linear speed is equal to distance ÷ time, where the distance travelled is the arc length that the object moves through in its circular motion. So the equation for speed can be written as:

$$v = \frac{\text{arc length}}{t}$$

The arc length of a sector of angle θ was given on the previous page as:

$$\text{arc length} = r\theta$$

Substituting this into the equation for v gives:

$$v = \frac{r\theta}{t}$$

Rearrange to get:

$$\frac{v}{r} = \frac{\theta}{t}$$

You have just seen that $\omega = \frac{\theta}{t}$, so the equation can be written as:

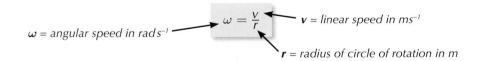

ω = angular speed in rad s^{-1} —— $\omega = \frac{v}{r}$ —— v = linear speed in ms^{-1}

r = radius of circle of rotation in m

Example

A cyclotron is a type of particle accelerator. Particles start off in the centre of the accelerator, and electric and magnetic fields cause them to move in circles of increasing size.

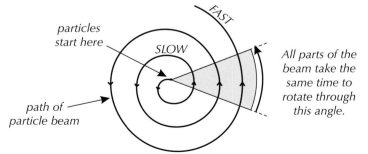

particles start here

FAST

SLOW

All parts of the beam take the same time to rotate through this angle.

path of particle beam

Figure 4: *The path of a particle in a cyclotron.*

Different parts of the particle beam are rotating at different linear speeds, v. But all the parts rotate through the same angle in the same time — so they have the same angular speed, ω.

Figure 5: *One of the first cyclotrons, built by Ernest Lawrence at the University of California.*

A cyclist is travelling at a speed of 28.8 km h⁻¹ along a road. The diameter of his front wheel is 68 cm. What is the angular speed of the wheel? How long does it take for the wheel to turn one complete revolution?

Linear speed in ms⁻¹ = 28.8 × 1000 ÷ 3600 = 8.00 ms⁻¹

Angular speed = $\omega = \dfrac{v}{r}$,

where $r = \dfrac{1}{2} \times$ diameter

$= \dfrac{1}{2} \times 0.68$

$= 0.34$ m

So angular speed:

$\omega = \dfrac{8.00}{0.34}$

$= 23.5294...$

$= 24$ rad s⁻¹ (to 2 s.f.)

Time to complete one revolution can be found by rearranging $\omega = \dfrac{\theta}{t}$:

$t = \dfrac{\theta}{\omega}$

$= \dfrac{2\pi}{23.5294...}$

$= 0.2670...$

$= 0.27$ s (to 2 s.f.)

Tip: The cyclist's speed is the same as the linear speed of the tyre, as long as there's enough friction between the tyre and ground so that the wheel doesn't slip.

Frequency and period

Circular motion has a frequency and period. The **frequency**, f, is the number of complete revolutions per second (rev s⁻¹ or hertz, Hz). The **period**, T, is the time taken for a complete revolution (in seconds). Frequency and period are linked by the equation:

f = frequency in rev s⁻¹ ⟶ $f = \dfrac{1}{T}$ ⟵ T = period in s

For a complete circle, an object turns through 2π radians in a time T. So the equation for angular speed becomes:

ω = angular speed in rad s⁻¹ ⟶ $\omega = \dfrac{2\pi}{T}$ ⟵ T = period in s

By replacing $\dfrac{1}{T}$ with frequency, f, you get an equation that relates ω and f:

ω = angular speed in rad s⁻¹ ⟶ $\omega = 2\pi f$ ⟵ f = frequency in rev s⁻¹

Exam Tip
Working with angular speed can be confusing when you're used to linear speed, so make sure you learn these equations off by heart to help you. The equation of $\omega = 2\pi f$ will be given in your data and formulae booklet though, just make sure you know how to use it.

Tip: The radii of the Earth and Moon are much smaller than the distance between them, so you can just treat them as two points.

Tip: 'rpm' is 'revolutions per minute'.

Tip: The equation for kinetic energy is $\frac{1}{2}mv^2$.

Example ── Maths Skills

A wheel is turning at a frequency of 20 rev s⁻¹. Calculate the period and angular speed of its rotation. Leave your answer in terms of π.

$f = \frac{1}{T}$, so $T = \frac{1}{f} = \frac{1}{20} = 0.05$ s

Angular speed $= 2\pi f = 2 \times \pi \times 20 = 40\pi$ rad s⁻¹

Practice Questions ── Application

Q1 The Moon orbits the Earth at a distance of roughly 384 000 km. If it takes the Moon 28 days to complete a full orbit, what are its angular and linear speeds?

Q2 An observation wheel of diameter 125 m turns exactly four times every hour. What angle does it rotate through each minute? What is the linear speed of one of the passenger cars?

Q3 A CD is spinning at a frequency of 460 rpm. What are the angular and linear speeds at points 2.0 and 4.0 cm from the centre of the CD?

Q4 A ball with mass m is spun at a constant speed in a circle on the end of a string of length l with time period T. Find the kinetic energy of the ball in terms of m, l and T.

Practice Questions ── Fact Recall

Q1 How would you convert an angle from degrees to radians?

Q2 What is the definition of angular speed?

Q3 Write down the formula that links angular speed and linear speed.

Q4 What is meant by the period and frequency of rotation?

Q5 How would you calculate the angular speed of an object in circular motion using its frequency?

2. Centripetal Force and Acceleration

Learning Objectives:

- Understand that an object travelling in a circle is accelerating towards the centre of the circle.
- Understand that centripetal acceleration is due to a centripetal force.
- Know how to use the equations for centripetal acceleration and force.

Specification Reference 3.6.1.1

Objects moving in a circle are accelerating even if their speed isn't changing. This might sound strange, but it's because velocity is a vector quantity and the direction of the object's velocity is constantly changing.

Centripetal acceleration

Objects travelling in circles are accelerating since their velocity is changing. Even if the car shown in Figure 1 below is going at a constant speed, its velocity is changing since its direction is changing. Since acceleration is defined as the rate of change of velocity, the car is accelerating even though it isn't going any faster. This acceleration is called the **centripetal acceleration** and is always directed towards the centre of the circle.

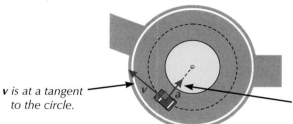

v is at a tangent to the circle.

The acceleration of the car is directed towards the centre of the circle.

Figure 1: *A car moving in a circle.*

Derivation of formula for centripetal acceleration

A ball is moving at a constant speed in a circle (see Figure 2). Because the ball is always changing direction, the **linear velocity** is always changing. However, the magnitude of the linear velocity, v, is always the same (sometimes this is called linear speed). During time Δt, the ball moves from a point on the circle, A, to another point, B. The ball turns through the angle θ (the angle between the lines OA and OB).

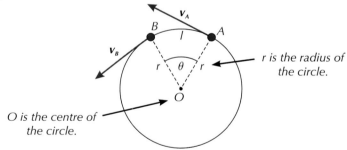

r is the radius of the circle.

O is the centre of the circle.

Figure 2: *A ball travelling in a circular path travels through angle θ when it moves from point A to point B.*

Tip: You don't need to know this derivation for the exam, it's just here so you can see where the equations for centripetal acceleration come from. These equations are on the next page.

The distance, l, that the ball travels along the circle from A to B is equal to the ball's linear speed multiplied by the time it takes to move that distance:

$$l = v \, \Delta t$$

At point A the ball has linear velocity v_A and at point B it has linear velocity v_B. The change in linear velocity, Δv, is:

$$\Delta v = v_B - v_A$$

You can draw a triangle made up of the velocity vectors v_A, v_B and Δv (see Figure 3). The linear velocity is always at a tangent to the radius, so the angle between v_A and v_B is also θ. This triangle is the same shape as the triangle ABO, since both are isosceles triangles with the same angle θ between the two sides of identical length.

Tip: Linear velocity has the same magnitude as linear speed, but is a vector that is a tangent to the circle. v_A and v_B have the same magnitude but different directions.

Tip: Isosceles triangles have two sides the same length.

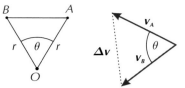

Figure 3: The velocity vectors v_A, v_B and Δv form an isosceles triangle.

Tip: These two triangles are similar — they're the same shape, just different sizes and orientations.

Tip: If θ is small, $l \approx AB$.

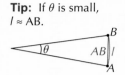

Tip: The length of each of the vectors v_A and v_B is just v.

Tip: Remember — linear speed and the magnitude of the linear velocity are the same thing.

Exam Tip
You'll be given both of these formulas in the exam data and formulae booklet.

Tip: $\omega = \frac{v}{r}$, so $v^2 = \omega^2 r^2$. This gives $\frac{v^2}{r} = \frac{\omega^2 r^2}{r} = \omega^2 r$.

Tip: You met Newton's laws of motion in year 1 of A-level.

Tip: Don't confuse the centripetal force with the centrifugal force. The centrifugal force is the outwards reaction force you experience when you're spinning.

If θ is small, then the length of the straight line AB is approximately equal to the length of the arc l. Because the two triangles in Figure 3 have the same shape, the ratio of l and r is the same as the ratio of Δv and v_A.

$$\frac{l}{r} = \frac{\Delta v}{v_A} = \frac{\Delta v}{v}$$

The previous equation, $l = v\,\Delta t$, can be substituted in:

$$\frac{v\Delta t}{r} = \frac{\Delta v}{v} \Rightarrow \frac{\Delta v}{\Delta t} = \frac{v \times v}{r} = \frac{v^2}{r}$$

Since acceleration, a, is equal to the change in velocity over time, this gives:

$$a = \frac{\Delta v}{\Delta t} = \frac{v^2}{r}$$

The formula for centripetal acceleration can then be written in terms of either the magnitude of linear velocity (which is linear speed) or angular speed:

a = centripetal acceleration in ms^{-2} $\longrightarrow$ $a = \dfrac{v^2}{r}$ $\longleftarrow$ v = magnitude of linear velocity in ms^{-1}

r = radius in m

a = centripetal acceleration in ms^{-2} $\longrightarrow$ $a = \omega^2 r$ $\longleftarrow$ r = radius in m

ω = angular speed in rad s^{-1}

Centripetal force

Newton's first law of motion says that an object's velocity will stay the same unless there's a force acting on it. Since an object travelling in a circle has a centripetal acceleration, there must be a force causing this acceleration. This force is called the **centripetal force** and acts towards the centre of the circle.

Newton's second law says $F = ma$, so substituting this into the equations above for the centripetal acceleration gives you equations for the centripetal force:

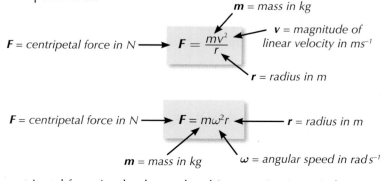

m = mass in kg

F = centripetal force in N $\longrightarrow$ $F = \dfrac{mv^2}{r}$ $\longleftarrow$ v = magnitude of linear velocity in ms^{-1}

r = radius in m

F = centripetal force in N $\longrightarrow$ $F = m\omega^2 r$ $\longleftarrow$ r = radius in m

m = mass in kg $\qquad$ ω = angular speed in rad s^{-1}

The centripetal force is what keeps the object moving in a circle — remove the force and the object would fly off at a tangent with velocity v.

Example — Maths Skills

A car of mass 890 kg is moving at 15 ms⁻¹. It travels around a circular bend of radius 18 m. Calculate the centripetal acceleration and centripetal force experienced by the car.

Acceleration = $a = \frac{v^2}{r} = \frac{15^2}{18} = 12.5 = 13$ ms⁻² (to 2 s.f.)

Force = $F = \frac{mv^2}{r} = ma = 890 \times 12.5 = 11\,125 = 11\,000$ N (to 2 s.f.)

Example — Maths Skills

A satellite orbits the Earth twice every 24 hours. The acceleration due to the Earth's gravity is 0.57 ms⁻² at the satellite's orbiting altitude. How far above the Earth's surface is it orbiting? The radius of the Earth is 6400 km.

1 orbit takes 12 hours, so period in seconds = $T = 12 \times 3600 = 43\,200$ s

Angular speed = $\omega = \dfrac{2\pi}{T}$

$\qquad\qquad = \dfrac{2\pi}{43\,200} = 1.4544... \times 10^{-4}$ rad s⁻¹

Acceleration $a = \omega^2 r$ so radius $r = \dfrac{a}{\omega^2}$

$\qquad\qquad = \dfrac{0.57}{(1.4544... \times 10^{-4})^2} = 26\,945.2...$ km

Height above Earth = radius of orbit − radius of Earth
$\qquad\qquad = 26\,945.2... - 6400 = 20\,545.2... = 21\,000$ km (to 2 s.f.)

Figure 4: *When a hammer thrower is spinning, the centripetal force acting along the chain keeps the hammer moving in a circle. When the thrower lets go of the chain, this force vanishes and the hammer flies off with an initial velocity equal to its linear velocity.*

Tip: The acceleration due to the Earth's gravity decreases as you move further away from the centre of the planet — only at the surface is it 9.81 ms⁻². See page 76 for more.

Practice Questions — Application

Q1 On a ride at a theme park, riders are strapped into seats attached to the edge of a horizontal wheel of diameter 8.5 m. The wheel rotates 15 times a minute. Calculate the force felt by a rider of mass 60.0 kg.

Q2 A planet follows a circular orbit of radius r around a star. The planet experiences a constant centripetal acceleration a. How long does it take for the planet to orbit the star 3 times? Choose the correct option.

A $3\pi\sqrt{\dfrac{a}{r}}$ B $6\pi\sqrt{\dfrac{a}{r}}$ C $6\pi\sqrt{\dfrac{r}{a}}$ D $\dfrac{2}{3}\pi\sqrt{ar}$

Q3 a) A car is being driven in a circle of radius 56.8 m with a linear speed of 31.1 ms⁻¹. Calculate the centripetal acceleration of the car.

     b) The linear speed of the car decreases to half its original value. If the original centripetal force experienced by the car was F, what is the new centripetal force experienced by the car in terms of F?

Q4 A biker rides in a vertical circle around the inside of a cylinder of radius 5.0 m (so he's upside down at the top of the cylinder). For the biker to not fall off at the top of the loop, he must be going fast enough that his centripetal acceleration does not drop below the acceleration due to gravity, 9.81 ms⁻¹. What's the minimum speed he must be travelling at the top of the loop?

Tip: For Q2, start by finding expressions for ω in terms of a and r and in terms of T. You can then rearrange and combine them to find the answer.

Practice Questions — Fact Recall

Q1 What is meant by centripetal acceleration and centripetal force?

Q2 Give the equation for centripetal acceleration in terms of angular speed. Define all the symbols you use.

Q3 Give the equation for centripetal force in terms of linear speed. Define all the symbols you use.

- Know the characteristics of simple harmonic motion (SHM), oscillation with acceleration proportional to the displacement from the equilibrium, $a \propto -x$.

- Be able to sketch the graphs of displacement, velocity and acceleration for an object moving with SHM as a function of time, and understand the phase difference between them.

- Know that velocity is given by the gradient of a displacement-time graph, and that acceleration is the gradient of a velocity-time graph.

- Be able to describe how kinetic, potential and mechanical energy change with displacement for an object moving with SHM and also how kinetic and potential energy change with time.

- Know that the maximum speed of an object moving with SHM is given by: $v_{max} = \omega A$.

Specification References 3.6.1.2 and 3.6.1.3

Tip: As acceleration is proportional to displacement, the force is also proportional to displacement.

3. Simple Harmonic Motion

A swinging pendulum moves with simple harmonic motion — this topic is all about what simple harmonic motion is and where you might see it occurring.

What is simple harmonic motion?

An object moving with **simple harmonic motion** (SHM) oscillates to and fro, either side of an equilibrium position (see Figure 1). This equilibrium position is the midpoint of the object's motion. The distance of the object from the equilibrium is called its displacement.

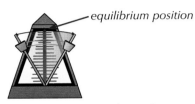

Figure 1: *A metronome moves with simple harmonic motion about an equilibrium position.*

There is always a restoring force pulling or pushing the object back towards the equilibrium position. The size of the restoring force depends on the displacement (see Figure 2). The restoring force makes the object accelerate towards the equilibrium.

Figure 2: *The size of the restoring force for an object moving with simple harmonic motion depends on its displacement from its equilibrium position.*

SHM can be defined as:

> An oscillation in which the acceleration of an object is directly proportional to its displacement from its equilibrium position, and is directed towards the equilibrium.

This definition can be written as the relation:

$$a \propto -x$$

Where *a* is the acceleration and *x* is the displacement. The minus sign shows that the acceleration is always opposing the displacement (that's the "directed towards the equilibrium" bit in the definition).

Graphs of simple harmonic motion

You can draw graphs to show how the displacement, velocity and acceleration of an object oscillating with SHM change with time (see Figure 3).

Displacement

Displacement, x, varies as a cosine or sine wave with a maximum value, A (the amplitude).

Velocity

Velocity, v, is the gradient of the displacement-time graph. It has a maximum value of ωA (where ω is the angular frequency of the oscillation). Remember that ω is given by $\omega = 2\pi f$ (see page 21).

Acceleration

Acceleration, a, is the gradient of the velocity-time graph. It has a maximum value of $\omega^2 A$.

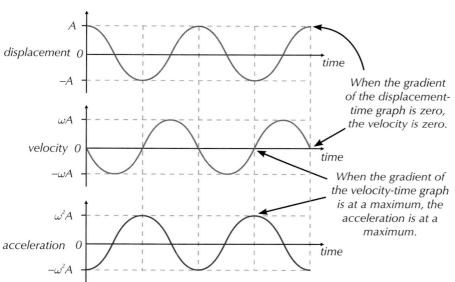

When the gradient of the displacement-time graph is zero, the velocity is zero.

When the gradient of the velocity-time graph is at a maximum, the acceleration is at a maximum.

Figure 3: *Graphs showing how displacement, velocity and acceleration of an object experiencing SHM change with time.*

It might seem odd that the acceleration should be a maximum at zero velocity, but it might be helpful to remember that acceleration is the rate of change of velocity — acceleration is found from the gradient of the velocity-time graph. The steepest part of the velocity-time graph (and therefore the largest acceleration) is when the curve is passing through zero.

Also, remember from the previous page that $a \propto -x$. So the maximum acceleration occurs at the maximum negative displacement (the turning points of the displacement-time graph), which coincides with the points at which velocity is zero.

Phase difference

Phase difference is a measure of how much one wave lags behind another wave, and can be measured in radians, degrees, or fractions of a cycle. Two waves that are in phase with each other have a phase difference of zero (or 2π radians) — i.e. the maxima and minima in each wave occur at the same time. If two waves are exactly out of phase ('in antiphase'), they have a phase difference of π radians or $180°$ — one wave's maximum occurs at the same time as the other's minimum.

Tip: Sine and cosine waves are graphs plotted of the functions sin and cos of some changing value (like time).

Tip: ω is called angular frequency in SHM, and angular speed in circular motion (see page 20).

Tip: The velocity-time graph is derived from the gradient of the displacement-time graph because $v = \frac{\Delta x}{\Delta t}$. Similarly, $a = \frac{\Delta v}{\Delta t}$.

Tip: See the next topic for where these maximum values of velocity and acceleration come from.

Tip: Kinetic energy (E_K) will vary in a similar way to v, since $E_K = \frac{1}{2}mv^2$ — except the graph for E_K will also be positive (as v^2 is always positive).

Tip: These graphs can be plotted from experiments by using position sensors and data loggers to measure the motion of an object undergoing SHM.

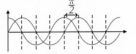
The velocity is a quarter of a cycle in front of the displacement in the velocity-time and displacement-time graphs for an object in SHM (see Figure 3) — it's $\frac{\pi}{2}$ radians out of phase. The acceleration is another $\frac{\pi}{2}$ radians ahead of the velocity, and so is in antiphase with the displacement.

Frequency and period

From maximum positive displacement (e.g. maximum displacement to the right) to maximum negative displacement (e.g. maximum displacement to the left) and back again is called a cycle of oscillation. The frequency, f, of the SHM is the number of cycles per second (measured in hertz, Hz). The period, T, is the time taken for a complete cycle (in seconds).

The amplitude of an oscillation is the maximum magnitude of the displacement. In SHM, the frequency and period are independent of the amplitude (i.e. they're constant for a given oscillation). So a pendulum clock will keep ticking in regular time intervals even if its swing becomes very small (see page 38).

Potential and kinetic energy

An object in SHM exchanges **potential energy** and **kinetic energy** as it oscillates. The type of potential energy (E_p) depends on what it is that's providing the restoring force. This will be gravitational E_p for pendulums and elastic E_p (elastic strain energy) and possibly gravitational E_p for masses on springs.

- As the object moves towards the equilibrium position, the restoring force does work on the object and so transfers some E_p to E_K.
- When the object is moving away from the equilibrium, all that E_K is transferred back to E_p again.
- At the equilibrium, the object's E_p is said to be zero and its E_K is maximum — therefore its velocity is maximum.
- At the maximum displacement (the amplitude) on both sides of the equilibrium, the object's E_p is maximum, and its E_K is zero — so its velocity is zero.

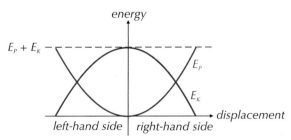

Figure 5: Graph to show how the kinetic and potential energy of an object in SHM change with displacement.

Figure 4: An oscillating pendulum moves fastest when it passes through its equilibrium position — when all its energy is kinetic.

The sum of the potential and kinetic energy is called the **mechanical energy** and stays constant (as long as the motion isn't damped — see page 41). The energy transfer for one complete cycle of oscillation (see Figure 6) is: E_p to E_K to E_p to E_K to E_p ... and then the process repeats...

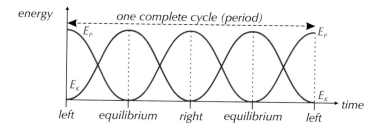

Figure 6: *Graph to show how the kinetic and potential energy of an object in SHM change with time.*

Exam Tip
Make sure you know the graphs in Figures 5 and 6 well — you could be asked to recognise, sketch and apply them to other things.

Practice Questions — Application

Q1 A 50 kg (to 2 s.f.) mass on the end of a spring is undergoing simple harmonic motion. The potential energy of the mass is 28 J at maximum displacement. Calculate the maximum velocity that the mass reaches whilst oscillating. (Assume there are no energy losses to the surroundings).

Q2 A girl with a mass of 35 kg is sitting on a swing, which is undergoing simple harmonic motion. To start the motion, she raised the swing 0.40 m above its lowest position, then lifted her feet off the ground.

a) Find the maximum velocity the girl reaches on the swing.

b) Sketch a graph of the girl's kinetic energy against time for one complete oscillation of the swing.

Q3 A pendulum is undergoing simple harmonic motion. The maximum displacement of the pendulum is 0.60 m and the maximum velocity of the pendulum is 0.90 ms⁻¹.

a) Calculate the frequency of the oscillation.

b) Calculate the time taken for the pendulum to complete one oscillation.

c) Calculate the maximum acceleration of the pendulum.

Tip: Gravitational potential energy is equal to mgh, where h is the height from rest position, and kinetic energy is equal to $\frac{1}{2}mv^2$. For SHM, the sum of these two is a constant, i.e. $mgh + \frac{1}{2}mv^2$ = constant.

Tip: Remember that velocity has a maximum value of ωA or $2\pi fA$.

Practice Questions — Fact Recall

Q1 What is the definition of simple harmonic motion?

Q2 What is meant by the frequency and period of an oscillation?

Q3 What is the phase difference between the displacement and the velocity of an object moving with SHM?

Q4 For an object undergoing simple harmonic motion, state whether the following occur at equilibrium or maximum displacement:

a) maximum velocity

b) minimum velocity

c) maximum acceleration

d) minimum acceleration

Q5 Sketch the graphs of displacement, velocity and acceleration against time for an object moving with SHM. Mark on your graphs the maximum and minimum values that each can take.

Q6 Describe how the kinetic and potential energy of an object moving with SHM change with time.

- Know that the displacement of an object moving with SHM is given by: $x = A\cos(\omega t)$.
- Know that the defining equation of SHM is: $a = -\omega^2 x$.
- Know that the magnitude of the maximum acceleration of an object moving with SHM is: $a_{max} = \omega^2 A$.
- Know that the velocity of an object moving with SHM is given by: $v = \pm\omega\sqrt{A^2 - x^2}$.
- Know that the maximum speed of an object moving with SHM is given by: $v_{max} = \omega A$.

Specification Reference 3.6.1.2

Tip: Don't forget, A is the maximum displacement — it's not acceleration.

Tip: See pages 19-22 for more about circular motion and where these equations come from.

Tip: Angular frequency (ω) and frequency (f) are similar, but are not the same. ω is measured in rads^{-1} and f is measured in s^{-1}. The relation between the two is $\omega = 2\pi f$.

Exam Tip
Remember to set your calculator to radians when using this equation.

4. Calculations with SHM

Now you've met SHM, this topic's all about the maths you can use to describe what's going on.

Equations for simple harmonic motion

To understand the maths behind SHM, it's useful to think about it as the 'projection' of circular motion onto a horizontal plane (see Figure 1). Imagine a ball spinning in a circle along the same plane. From above it will look like it's following a circular path. From the side (i.e. in the plane of rotation) it will look like it's oscillating from side to side, and moving with SHM. Its speed, v, will appear fastest when $x = 0$ and appear slowest when $x = \pm A$.

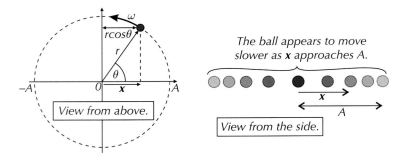

The ball appears to move slower as x approaches A.

Figure 1: *A ball following a circular path, pictured from above (left) and from the side (right).*

Displacement

- Figure 1 shows that the displacement, x, of the ball is just equal to the horizontal component of the ball's position, i.e. $x = r\cos\theta$.
- The radius of the circle, r, is the same as the amplitude of the ball's simple harmonic motion (as viewed from the side), A. This is the maximum displacement, x, of the ball.
- The angular speed, ω, is defined as $\omega = \frac{\theta}{t}$, so $\theta = \omega t$. The equivalent 'component' to ω in SHM is called **angular frequency**. It still has the same value, the same units and the same symbol, it's just got a different name.
- Combining these equations gives you an expression for the displacement, x, of any object undergoing SHM:

$$x = r\cos\theta = A\cos(\omega t)$$

x = displacement in m $\longrightarrow$ $x = A\cos(\omega t)$ $\longleftarrow$ ω = angular frequency in rads^{-1}

A = amplitude in m $\qquad$ t = time in s

To use this equation you need to start timing when the pendulum is at its maximum displacement — i.e. when $t = 0$, $x = \pm A$.

Acceleration

The acceleration of an object moving with SHM can also be related to the acceleration of an object in circular motion.

- For circular motion, $a = \omega^2 r$.
- The acceleration of an object in SHM is then the horizontal component of this (in the same plane as the displacement x, as shown above):

$$a = -\omega^2 r \cos\theta$$

Tip: You don't need to know the derivations of all these equations — they're just here to help you understand SHM a little more.

There's a minus sign because the acceleration is always acting towards the centre of the circle. $r\cos\theta$ is equal to the horizontal component of the ball's position, x, so the acceleration of the object is:

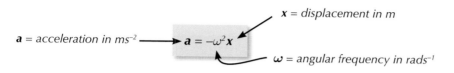

x = displacement in m

a = acceleration in ms^{-2} → $a = -\omega^2 x$

ω = angular frequency in rads^{-1}

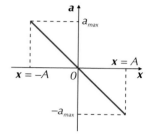

Figure 2: Graph showing how acceleration varies with displacement for an object moving with SHM.

This is the defining equation of SHM — remember $a \propto -x$ from page 26.

The object's maximum acceleration occurs when it's at its maximum magnitude of displacement — i.e. when $x = \pm A$:

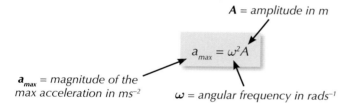

A = amplitude in m

$a_{max} = \omega^2 A$

a_{max} = magnitude of the max acceleration in ms^{-2}

ω = angular frequency in rads^{-1}

Tip: You'll be given all these equations in the data and formulae booklet, but you need to understand what they mean and how to use them.

Velocity

You also need to know the equation for the velocity of an object moving with SHM. Don't worry about the derivation of v, it's pretty complicated. You just need to know that it is given by:

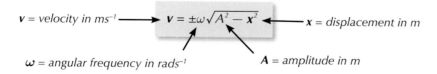

v = velocity in ms^{-1} → $v = \pm\omega\sqrt{A^2 - x^2}$ ← x = displacement in m

ω = angular frequency in rads^{-1}

A = amplitude in m

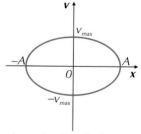

Figure 3: Graph showing how velocity varies with displacement for an object moving with SHM.

The velocity is positive if the object is moving in the positive direction (e.g. to the right), and negative if it's moving in the negative direction (e.g. to the left) — that's why there is a ± sign in there. The maximum speed (magnitude of velocity) is when the object is passing through the equilibrium, where $x = 0$:

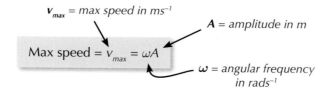

v_{max} = max speed in ms^{-1}

A = amplitude in m

Max speed = $v_{max} = \omega A$

ω = angular frequency in rads^{-1}

Tip: Remember — ω is called angular <u>speed</u> in circular motion and angular <u>frequency</u> in SHM. It's just the name that's different, so the equation for ω is always: $\omega = 2\pi f$.

Example — Maths Skills

A mass is attached to a horizontal spring. It is pulled 7.5 cm from its equilibrium position and released. It begins oscillating with SHM, and takes 1.2 s to complete a full cycle.
a) What is the frequency of oscillation of the mass?
b) What is the magnitude of its maximum acceleration?
c) What is its speed when it is 2.0 cm from its equilibrium position?

a) Frequency $= \dfrac{1}{T} = \dfrac{1}{1.2} = 0.8333... = 0.83$ s^{-1} (to 2 s.f.)

b) Amplitude in m $= 7.5 \div 100 = 0.075$ m
$\omega = 2\pi f = 2 \times \pi \times 0.8333... = 5.2359...$ rads^{-1}
$a_{max} = \omega^2 A = (5.2359...)^2 \times 0.075 = 2.056... = 2.1$ ms^{-2} (to 2 s.f.)

c) Speed at $x = 0.02$ m
$v = \omega\sqrt{A^2 - x^2} = 5.2359... \times \sqrt{0.075^2 - 0.02^2} = 0.3784...$
$= 0.38$ ms^{-1} (to 2 s.f.)

Practice Questions — Application

Q1 A pendulum oscillating with SHM has an angular frequency of 1.5 rads^{-1} and an amplitude of 1.6 m.

    a) Calculate its acceleration when it has a displacement of 1.6 m.

    b) How long does it take to complete 15 oscillations?

Q2 A mass attached to a spring oscillates with SHM. It has a period of 0.75 s, and moves with speed 0.85 ms^{-1} when passing through its equilibrium position.

    a) What is the amplitude of its oscillation?

    b) What will its velocity be when it is 8.0 cm to the right of its equilibrium position?

Q3 A pendulum is pulled a distance 0.45 m from its equilibrium position and is released at time $t = 0$. If it takes 15.5 s to complete exactly 5 oscillations, how far will it be from its equilibrium position after 10.0 s?

Q4 A mechanical metronome produces a ticking sound every time a pendulum arm moving with SHM passes through its equilibrium position. Its maximum displacement from its equilibrium position is 6.2 cm. If it is set to produce 120 ticks per minute, what is the magnitude of the arm's maximum acceleration?

Practice Questions — Fact Recall

Q1 What is the equation for the displacement of an object moving with SHM as a function of time?

Q2 Give the equations for the acceleration and velocity of an object moving with SHM as a function of displacement.

Q3 What is the magnitude of the maximum acceleration of an object moving with SHM? Choose the correct option.
    A ω^2 B $\omega^2 A$ C ωA D $\omega^2 A^2$

Q4 What is the maximum speed of an object moving with SHM?

Q5 Sketch the graph of velocity as a function of displacement for an object moving with SHM.

Q6 Sketch the graph of acceleration as a function of displacement for an object moving with SHM.

5. The Mass-Spring System as a Simple Harmonic Oscillator

Learning Objectives:

- Know that a mass on a spring is a simple harmonic oscillator.
- Know the formula for the period of a mass-spring system.
- Know how to verify how the period of oscillation for a mass on a spring depends on mass, spring constant and amplitude using an experiment (Required Practical 7).

Specification Reference 3.6.1.3

Simple harmonic oscillators are systems that oscillate with SHM. There are two common types you need to know about — masses on springs, and pendulums. This topic gives all the details you need to know for a mass-spring system.

A mass on a spring

A mass on a spring is a **simple harmonic oscillator** (SHO). When the mass is pushed or pulled either side of the equilibrium position, there's a restoring force exerted on it (see Figure 1).

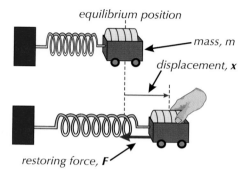

equilibrium position

mass, m

displacement, x

restoring force, F

Figure 1: A mass on a spring is pulled from its equilibrium position and displaced by amount **x**.

The size and direction of this restoring force is given by Hooke's law:

F = restoring force in N ⟶ $F = k\Delta L$ ⟵ **k** = spring constant in Nm⁻¹

ΔL = displacement in m

Exam Tip
You met Hooke's law in year 1 of A-level, and you'll find the formula for it in materials sections of the exam data and formulae booklet.

For a displacement **x**, the restoring force becomes **F** = −k**x**. The negative sign is because the force acts in the opposite direction to the displacement — back towards the equilibrium position.

Newton's second law states that the resultant force on an object equals the mass of the object times its acceleration, **F** = m**a**. Inserting this into the equation **F** = −k**x**, and replacing **a** with the acceleration for an object oscillating with SHM, **a** = −ω^2**x** = −(2πf)²**x**, gives:

$$-(2\pi f)^2 m\mathbf{x} = -k\mathbf{x}$$

Exam Tip
You don't need to know this derivation — just make sure you know how to use the equation at the end for the period, **T**.

The **x**'s cancel, and rearranging this equation gives the frequency of a mass oscillating on a spring:

$$f = \frac{1}{2\pi}\sqrt{\frac{k}{m}}$$

Using $f = \frac{1}{T}$ (see page 21) gives the period of a mass oscillating on a spring:

T = period of oscillation in s ⟶ $T = 2\pi\sqrt{\frac{m}{k}}$ ⟵ **m** = mass in kg

k = spring constant in Nm⁻¹

Tip: Remember, the period is the time it takes for one complete oscillation (maximum left position, to maximum right position, back to maximum left position).

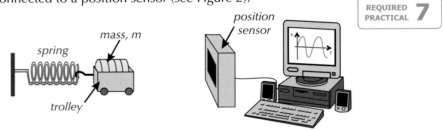

Example — **Maths Skills**

A trolley of mass 2.5 kg is attached to a horizontal spring. The trolley is pulled past its equilibrium position and released and begins oscillating with SHM. The period of the oscillation is measured as 1.2 s.
What is the spring constant?

Period $= T = 2\pi\sqrt{\dfrac{m}{k}}$ so rearranging for k gives:

$$k = \frac{4\pi^2 m}{T^2} = \frac{4\pi^2 \times 2.5}{1.2^2} = 68.53... = 69 \text{ Nm}^{-1} \text{ (to 2 s.f.)}$$

Tip: Make sure you do a risk assessment before carrying out any experiments.

Tip: Instead of a position sensor and data logger, you could also use a reference point and a stopwatch to measure the time it takes for the mass to complete one oscillation. Timing several oscillations and dividing by the number that you timed will increase the accuracy of your measurements.

Tip: You could also do this with a vertical mass-spring system above a position sensor.

Tip: The symbol $\propto$ in an equation means "is proportional to."

Investigating the mass-spring system

This experiment backs up the relationships shown by the formulas from the previous page. It's not too tricky — you just have to change one variable at a time and see what happens. Attach a trolley to a spring, pull it to one side by a certain amount and then let go. The trolley will oscillate back and forth as the spring pulls and pushes it in each direction. You can measure the period, T, by getting a computer to plot a displacement-time graph from a data logger connected to a position sensor (see Figure 2).

REQUIRED PRACTICAL **7**

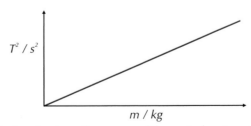

Figure 2: An experiment to investigate how the period of an oscillating mass on a spring varies with mass, amplitude and spring constant.

Variable: mass

Change the mass, m, by loading the trolley with masses — don't forget to include the mass of the trolley in your calculations. Since $T \propto \sqrt{m}$, the square of the period, T^2, should be proportional to the mass (see Figure 3).

Figure 3: A graph showing how the square of the period varies with increasing mass.

Variable: spring constant

Change the spring constant, k, e.g. by using different combinations of springs (see Figure 4).

Tip: When you place springs in parallel (side by side), spring constants add normally: $(k_{total} = k_1 + k_2 + ...)$.
When they're in series (end to end), the inverses of the spring constants are added:
$\left(\dfrac{1}{k_{total}} = \dfrac{1}{k_1} + \dfrac{1}{k_2} + ...\right)$

Figure 4: Using different combinations of springs to change the spring constant.

Since $T \propto \sqrt{\frac{1}{k}}$, the square of the period, T^2, should be proportional to the inverse of the spring constant (see Figure 5).

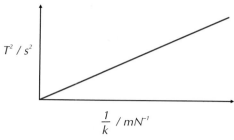

Figure 5: *A graph showing how the square of the period varies with the inverse of the spring constant.*

Variable: amplitude

Change the amplitude, A, by pulling the trolley across by different amounts. Since T doesn't depend on amplitude, A, there should be no change in the period (see Figure 6).

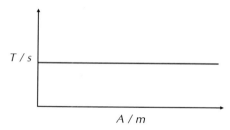

Figure 6: *A graph showing how the period is independent of amplitude.*

> **Tip:** You don't need to remember these graphs off by heart, but make sure you can sketch them, using the equation for T to work out their shapes.

Practice Questions — Application

Assume the motion in these questions is SHM.

Q1 A trolley is connected between two horizontal springs, as shown below, one with spring constant 25 Nm⁻¹ and the other with spring constant 45 Nm⁻¹. What will the size of the resultant force on the trolley be if it is pulled 2.5 cm from its equilibrium position?

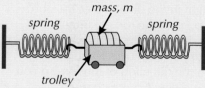

Q2 A 1.0 kg mass hung on a spring is pulled 45 mm from its equilibrium position. Doing this requires a force of 18 N. If the mass is then released, what will the period of its oscillation be?

Practice Question — Fact Recall

Q1 a) What's the formula for the period of oscillation of a mass on a spring?

  b) Describe an experiment that you could do with a mass on a spring to investigate the relationship of the period of oscillation with the mass and spring constant.

Learning Objectives:

- Know that a simple pendulum is a simple harmonic oscillator.
- Know the formula for the period of a simple pendulum.
- Know how to verify how the period of oscillation for a pendulum depends on pendulum length, mass and amplitude using an experiment (Required Practical 7).
- Be able to apply the SHM equations to other systems such as a liquid in a U-tube.

Specification Reference 3.6.1.3

6. The Simple Pendulum and Other Types of SHO

The simple pendulum is another type of SHO that you need to know about. You also need to know how to apply the SHM equations on pages 30-31 to different types of SHOs, for example how water oscillates in a U-tube.

The simple pendulum

The simple pendulum is the classic example of an SHO. The formula for the period of an oscillating pendulum is similar to the one for a spring.

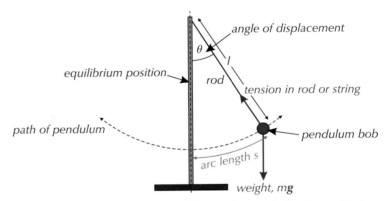

T = period, s $\longrightarrow$ $$T = 2\pi\sqrt{\frac{l}{g}}$$ $\longleftarrow$ *l* = length of pendulum, m

g = gravitational field strength, 9.81 Nkg^{-1}

To derive this formula, you need to think about the forces acting on the bob (the weight at the end of the pendulum (see Figure 1)).

Tip: You don't need to remember this derivation — it's just here so you understand where the equation for the period of a pendulum comes from.

Figure 1: The forces acting on a pendulum bob at angle θ to its equilibrium position.

Tip: Have a look back at mechanics in year 1 of A-level if you're struggling to follow the bits about vector components.

Tip: In case you were wondering, the other component of the weight (parallel to the rod), balances out the tension force.

Tension acts up along the rod or string, while the bob's weight, *mg*, acts vertically downwards. The component of the weight in the direction of the bob's motion (and perpendicular to the rod or string) is the restoring force, $F = mg\sin\theta$ (see Figure 2), which acts towards the equilibrium position. This is the only component that acts in the direction of motion, so you can use Newton's second law, $F = ma$, to write:

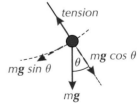

Figure 2: The bob's weight can be resolved into components.

$$F = ma = -mg\sin\theta$$

The masses cancel on either side of the equation, so the acceleration is $a = -g\sin\theta$. When the angle of displacement, θ, is only small, $\sin\theta$ is approximately θ (in radians). The formulas derived here depend on this approximation.

The ratio of the angle θ to the total angle in a circle, 2π, is equal to the ratio of the length of the arc, s, to the circumference of the circle, $2\pi r$. This lets you derive a formula for the acceleration in terms of s and l (here $l = r$):

$$s = r\theta \Rightarrow \theta = \frac{s}{r} \quad \text{and} \quad a = -g\sin\theta = -g\theta \Rightarrow a = -g\frac{s}{l}$$

The formula for acceleration in SHM is $a = -\omega^2 x = -(2\pi f)^2 x$, so $\frac{a}{x} = -(2\pi f)^2$. For small angles the arc length, s, is the same as the displacement, x. So:

$$-\frac{a}{x} = -\frac{a}{s} = (2\pi f)^2 = \frac{g}{l}$$

Rearranging this for frequency, and then taking the inverse gives the period of an oscillating pendulum — the formula given on the previous page:

$$T = 2\pi\sqrt{\frac{l}{g}}$$

Tip: Because we took the approximation that $\sin\theta \approx \theta$, this formula only works for small angles, about 10° or less.

Example — **Maths Skills**

A clock maker wants to build a grandfather clock with a pendulum that swings with frequency 1.0 Hz. How long must the pendulum be?

Period = 1 ÷ frequency, so the period = 1.0 s.

$T = 2\pi\sqrt{\frac{l}{g}}$, so rearranging for length:

$l = \frac{gT^2}{4\pi^2} = \frac{9.81 \times 1.0^2}{4 \times \pi^2} = 0.2484... = 0.25$ m (to 2 s.f.)

Investigating the formula experimentally

You can use a simple pendulum attached to an angle sensor and computer (see Figure 3) to test this formula.

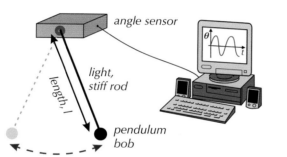

angle sensor

light, stiff rod

length, l

pendulum bob

REQUIRED PRACTICAL **7**

Tip: You should be made aware of any safety risks before carrying out this experiment.

Figure 3: Apparatus to investigate the formula for the period of an oscillating pendulum.

Use the computer to plot a displacement-time graph and read off the period, T, from it. Make sure you calculate the average period over several oscillations to reduce the percentage error in your measurement.

As for a spring (see pages 34-35), you can change one variable at a time and measure what happens. Since the period, T, is proportional to the square root of the length of the pendulum, $\sqrt{l}$, varying l should show that $T^2 \propto l$ (see Figure 4).

Tip: You could also measure the period of a pendulum using a stopwatch. It's sensible to measure the time taken for e.g. five oscillations, then divide by the numer of oscillations to get an average, as it'll reduce the random error in your result. You can use a fiducial marker to make sure you start and stop timing at the same point in the pendulum's swing.

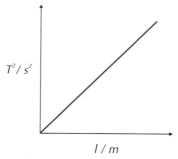

T^2 / s^2

l / m

Figure 4: Graph showing how period squared varies with pendulum length.

T is independent of the mass of the bob, *m*, and the amplitude of the oscillation, *A*, so varying these should not change *T* (see Figures 5 and 6).

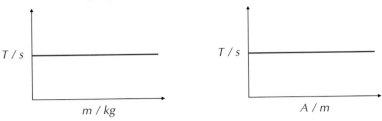

Figure 5: *Graph showing how period is independent of mass.*

Figure 6: *Graph showing how period is independent of amplitude.*

You can also do this experiment by hanging the pendulum from a clamp and timing the oscillations using a stop watch. Mark a reference point to tell when the pendulum has reached the mid-point of its oscillation.

Another example of SHO

Simple harmonic motion can be investigated using other setups, such as a U-tube containing some water. When at equilibrium, the levels of the water on either side of the tube are equal. When the water is pushed down on one side of the tube, the water level on the opposite side rises. When the pressure is then released, the water undergoes SHO, as the water levels on each side rise and fall.

Figure 7 shows the setup for water in a U-tube, where the water has a density ρ, volume *V*, cross-sectional area *A* and length *L*. To start the SHO, the water is displaced by x_{max}. The period of oscillation of the water can be found using conservation of energy.

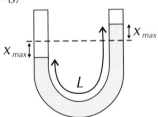

Figure 7: *Diagram showing SHO for water in a U-tube. The red dashed line indicates the equilibrium position.*

As the water oscillates, it will exchange kinetic energy and potential energy (see page 28). At the equilibrium position (when the water levels are aligned), the kinetic energy will be at a maximum and you can say that the potential energy of the water is zero.

At the magnitude of maximum displacement (when one side is x_{max} above the equilibrium level — see Figure 7), potential energy will be at its maximum and kinetic energy will be zero.

Using conservation of mechanical energy and assuming that friction is negligible, the change in kinetic energy is equal to the change in the potential energy, so you can write:

$$E_{K\ (max)} = E_{P\ (max)}$$

The maximum kinetic energy of the water during SHM is:

$$E_{K\ (max)} = \frac{1}{2}mv_{max}^2$$

where *m* is the mass of the whole body of water and v_{max} is the maximum speed of the water.

Tip: Measuring the period manually might affect the results by introducing more random errors (see page 9).

Tip: You don't need to know this example of SHO, but it is an example of how the equation of SHM can be applied to situations other than a pendulum and a mass on a spring.

You could get asked any example in the exam, but you'll be given all the info you need to be able to answer the question.

Tip: The cross-sectional area of the column of water is *A*, and the length of the column of water is *L*. So the volume of water is $V = AL$.

Tip: This experiment assumes that friction is negligible.

Using $\rho = \frac{m}{V}$, the mass of water is $m = V\rho = AL\rho$. Substituting this back into the equation for maximum kinetic energy, you get:

$$E_{K\,(max)} = \tfrac{1}{2}AL\rho v_{max}^{2}$$

Next, the maximum potential energy can be calculated. At maximum displacement, the gravitational potential energy of the water is the energy needed to lift a section of water of height x_{max} above the equilibrium position, which is mgx_{max}. This section of water has length x_{max}, so its mass can be calculated as:

$$m = V\rho = Ax_{max}\rho$$

which when substituted back into the equation for potential energy gives:

$$E_{P\,(max)} = Ax_{max}\rho gx_{max} = A\rho gx_{max}^{2}$$

Tip: Remember, the equation for gravitational potential energy is $E_p = mgh$.

You now have the equations for maximum kinetic energy and maximum potential energy. So by equating the two, you can write:

$$A\rho gx_{max}^{2} = \tfrac{1}{2}AL\rho v_{max}^{2}$$

Cancelling A and ρ and rearranging gives:

$$\frac{x_{max}}{v_{max}} = \sqrt{\frac{L}{2g}}$$

Remember from SHM that $v_{max} = \omega A$ (A = amplitude, not area). So for this case, $v_{max} = \omega x_{max}$, and so $\frac{x_{max}}{v_{max}} = \frac{1}{\omega}$. Therefore:

$$\frac{1}{\omega} = \sqrt{\frac{L}{2g}}$$

Remember from page 21 that $\omega = \frac{2\pi}{T}$, so $\frac{1}{\omega} = \frac{T}{2\pi}$. Substituting this into the above equation and rearranging gives:

$$T = 2\pi\sqrt{\frac{L}{2g}}$$

Exam Tip
Don't worry if you get a bit of a funny example like this. You'll be given all the information you need. Just make sure you're comfortable with using the SHM equations and conservation of energy.

Practice Questions — Application

Assume the motion in these questions is SHM.

Q1 Find the time period of a pendulum of length 2.50 m ($g = 9.81$ Nkg^{-1}).

Q2 a) Find the maximum kinetic energy of water undergoing SHM in a U-tube if the volume of water is 0.0120 m³ and the maximum velocity is 1.20 ms^{-1}. The density of water is assumed to be 1000 kgm^{-3} (to 3 s.f.).

b) The angular frequency of water in a U-tube is $\omega = \sqrt{\frac{2g}{L}}$.

State the effect on the angular frequency if the following increase:

(i) the maximum displacement
(ii) the volume of water
(iii) the density of water (a larger mass but the same volume)

Tip: You can derive this formula for ω in a similar way to the formula for T above.

Practice Questions — Fact Recall

Q1 a) What's the formula for the period of oscillation for an oscillating pendulum?

b) Describe an experiment you could do to verify the relationship of the period of oscillation of a pendulum with the pendulum length, mass and amplitude.

Q2 Sketch graphs to show how you would expect T^2 (the period squared) for a pendulum to vary with pendulum length, mass and amplitude.

An object can be forced to oscillate by providing a driving frequency. If this is near the object's natural frequency, the object will start to resonate — which can be good or bad news, depending on what the object is needed for.

Free vibrations

Free vibrations involve no transfer of energy to or from the surroundings. If you stretch and release a mass on a spring, it oscillates at its **resonant frequency** (or **natural frequency**). The same happens if you strike a metal object — the sound you hear is caused by vibrations at the object's natural frequency. If no energy's transferred to or from the surroundings, it will keep oscillating with the same amplitude forever. In practice this never happens, but a spring vibrating in air is called a free vibration anyway.

Forced vibrations

Forced vibrations happen when there's an external driving force. A system can be forced to vibrate by a periodic external force. The frequency of this force is called the **driving frequency**. If the driving frequency is much less than the natural frequency then the two are in phase (see page 27). But if the driving frequency is much greater than the natural frequency, the oscillator won't be able to keep up — you end up with the driver completely out of phase with the oscillator (in antiphase).

Resonance

When the driving frequency approaches the natural frequency, the system gains more and more energy from the driving force and so vibrates with a rapidly increasing amplitude. When this happens the system is **resonating**. At resonance the phase difference between the driver and oscillator is 90°. Figures 1 and 2 show how the relationship between amplitude and driving frequency can be investigated by experiment.

Tip: You should carry out a risk assessment before starting this experiment.

Tip: At resonance, the driver displacement is 90° out of phase to the displacement of the oscillator — i.e. when the displacement of the driver is at its maximum, the oscillator is passing through its equilibrium point.

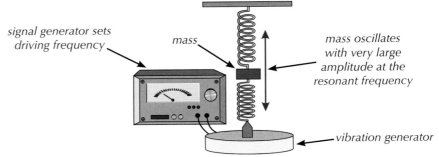

Figure 1: *Using a vibration generator to oscillate a mass-spring system. The system resonates when the driving frequency equals the natural frequency.*

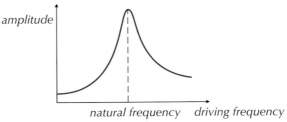

Figure 2: *Graph showing how the amplitude of oscillation of a system changes with driving frequency.*

Here are some examples of resonance:

A radio is tuned so the electric circuit resonates at the same frequency as the radio station you want to listen to.

A glass resonates when driven by a sound wave at the right frequency.

The column of air resonates in an organ pipe, driven by the motion of air at the base. This creates a stationary wave in the pipe.

A swing resonates if it's driven by someone pushing it at its natural frequency.

Tip: Make sure you know some examples of resonance.

Tip: Remember, a stationary wave is created by the superposition of two waves (with the same frequency and amplitude) moving in opposite directions.

Damping

In practice, any oscillating system loses energy to its surroundings. This is usually down to frictional forces like air resistance. These are called **damping** forces. Systems are often deliberately damped to stop them oscillating or to minimise the effect of resonance.

Tip: Another name for damping forces is 'dissipative forces', since they dissipate the energy of the oscillator to its surroundings.

┌ **Example** ──────────────────────────────────

Shock absorbers in a car's suspension provide a damping force by squashing oil through a hole when compressed.

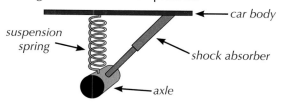

Figure 3: Damping in a car suspension system.

The degree of damping can vary from light damping (where the damping force is small) to overdamping. Damping reduces the amplitude of the oscillation over time. Generally, the heavier the damping, the quicker the amplitude is reduced to zero (although overdamping is an exception). Figure 4 shows how different degrees of damping reduce the amplitude to zero at different speeds.

Light and heavy damping

Lightly damped systems take a long time to stop oscillating, and their amplitude only reduces a small amount each period. Heavily damped systems take less time to stop oscillating, and their amplitude gets much smaller each period.

Exam Tip
Make sure you're able to describe the four different types of damping (the next two are on the next page), and sketch these graphs showing how the amplitude of oscillation changes with time.

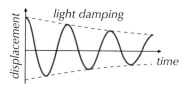
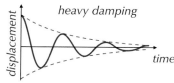

Figure 4: Graphs showing the effect of light and heavy damping.

Example

A pendulum formed of a small bob on a rod is an example of a lightly damped system — air resistance will cause the pendulum to slow down only very slightly each period. If the pendulum bob was removed and replaced with this book, the larger surface area of the book would increase air resistance, and the damping forces would be larger, slowing the oscillation more quickly.

Critical damping

Critical damping reduces the amplitude (i.e. stops the system oscillating) in the shortest possible time (see Figure 5).

Example

Car suspension systems are critically damped so that they don't oscillate but return to equilibrium as quickly as possible.

Overdamping

Systems with even heavier damping are **overdamped**. They take longer to return to equilibrium than a critically damped system (see Figure 5).

Example

Some heavy doors are overdamped, so that they don't slam shut too quickly, but instead close slowly, giving people time to walk through them.

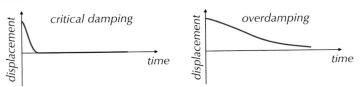

Figure 5: Graphs showing the effects of different amounts of damping.

Plastic deformation of ductile materials reduces the amplitude of oscillations in the same way as damping. As the material changes shape, it absorbs energy, so the oscillation will be smaller.

Damping and resonance peaks

Lightly damped systems have a very sharp resonance peak. Their amplitude only increases dramatically when the driving frequency is very close to the natural frequency. Heavily damped systems have a flatter response. Their amplitude doesn't increase very much near the natural frequency and they aren't as sensitive to the driving frequency. Figure 6 shows the effect of increasing levels of damping on oscillations near the natural frequency.

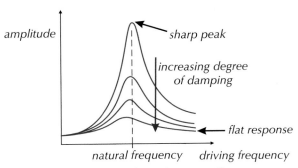

Figure 6: Graph showing how damping affects resonance.

> **Example**
>
> Some structures are damped to avoid being damaged by resonance. Some buildings in regions prone to earthquakes have a large mass called a tuned mass damper. When an earthquake causes the building to shake, the mass moves in the opposite direction to the building, damping its oscillation (see Figure 7). This is an example of critical damping.

> **Example**
>
> Damping can also be used to improve performance. For example, loudspeakers in a room create sound waves in the air. These reflect off of the walls of the room, and at certain frequencies stationary sound waves are created between the walls of the room. This causes resonance and can affect the quality of the sound — some frequencies are louder than they should be. Places like recording studios use soundproofing on their walls which absorb the sound energy and convert it into heat energy.

Figure 7: The Taipei 101 building in Taiwan, standing at 508 m tall. A 660-tonne pendulum suspended down the centre of the building acts as a tuned mass damper, reducing the amplitude of the building's oscillations in earthquakes.

Figure 8 shows apparatus that can be used to demonstrate the effect of damping on the resonance of a spring-mass system. A flat disc is attached to the set-up you saw in Figure 1. As the mass oscillates, air resistance on the disc acts as a damping force, reducing the amplitude of the oscillation. The larger the disc, the larger the damping force and the smaller the amplitude of oscillation of the system at resonance.

Tip: Before carrying out this experiment, you should be aware of any safety risks that are associated with the equipment.

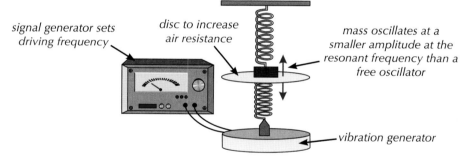

Figure 8: Experiment to show how damping affects resonance.

Practice Questions — Fact Recall

Q1 What is the difference between a free vibration and a forced vibration?

Q2 Describe the phase difference between a driving oscillation and an oscillating object in each of the following cases:

a) The driving frequency is much lower than the natural frequency.

b) The driving frequency is equal to the natural frequency.

c) The driving frequency is much higher than the natural frequency.

Q3 What is resonance, and when does it occur?

Q4 Give three examples of situations where resonance can occur.

Q5 What is meant by a damping force?

Q6 Name and briefly describe the four types of damping.

Q7 Sketch a graph of amplitude against driving frequency for a system to show how the level of damping affects the sharpness of the resonance peak at the natural frequency.

Section Summary

Make sure you know...

- How to use radians as a measure of angles.
- How to convert between radians and degrees.
- That angular speed is the angle an object moving with circular motion rotates through per second.
- How to calculate the angular speed of an object moving with circular motion given its linear speed.
- That the frequency of rotation is the number of complete revolutions per second.
- That the period of rotation is the time taken for one complete revolution in seconds.
- The relationship between frequency and period.
- How to calculate angular speed from the frequency or period.
- That objects moving with circular motion experience a centripetal acceleration towards the centre of the circle.
- That centripetal acceleration is caused by a centripetal force.
- How to calculate centripetal acceleration and force from angular and linear speeds.
- That simple harmonic motion is the oscillation of an object with an acceleration that is proportional to its displacement from its equilibrium position, and that is always directed towards the equilibrium.
- How to sketch the graphs of displacement, velocity and acceleration against time for SHM.
- That the velocity of an object undergoing SHM is given by the gradient of a displacement-time graph and acceleration is given by the gradient of a velocity-time graph.
- The phase differences between displacement, velocity and acceleration for SHM.
- That a cycle of oscillation in SHM is from maximum positive displacement to maximum negative displacement and back again, and how this relates to the frequency and period of SHM.
- That frequency and period are independent of amplitude for an object moving with SHM.
- How the kinetic and potential energy of an object moving with SHM vary with time and displacement.
- That mechanical energy is the sum of the kinetic and potential energy of an object and that it stays constant for undamped oscillations.
- How to use the formula for the displacement of an object in SHM as a function of time, and the formulas for the acceleration and velocity of an object in SHM as functions of displacement.
- How to use the formulas for the magnitude of maximum acceleration and maximum speed.
- How to sketch the graphs of acceleration and velocity of an object in SHM as functions of displacement.
- That a mass on a spring and a simple pendulum are examples of simple harmonic oscillators.
- How to use the formulae for the period of oscillation of a mass on a spring, and of a pendulum, and how to experimentally verify the relationships shown by these formulas.
- How to apply SHM equations to different types of simple harmonic oscillators, e.g. liquid in a U-tube.
- That free vibrations involve no transfer of energy between an object and its surroundings, and that an object oscillating freely does so at its natural frequency.
- That a forced vibration is driven by a periodic, external driving force at a driving frequency.
- That resonance is a rapid increase in the amplitude of oscillation of an object, and that it usually occurs at the object's natural frequency.
- How the phase difference between the driver and driven displacements changes with driving frequency.
- That a damping force causes an oscillator to lose energy to its surroundings, reduce the amplitude of its oscillations and reduce the sharpness of resonance.
- Examples of where damping is used, including systems with stationary waves.

Exam-style Questions

1   A mass on a string is displaying simple harmonic motion. If the frequency of the
       oscillation is doubled, what will happen to the maximum acceleration of the mass?

       A   It will halve.
       B   It will stay the same.
       C   It will double.
       D   It will quadruple.

                                                                              (1 mark)

2   A mass is attached to a spring and set oscillating with simple harmonic motion.
       The system is damped. Which of the following is false?

       A   If there was no damping at all, the mass would oscillate forever.
       B   The maximum displacement of the mass decreases with each oscillation.
       C   The mass would take longer to come to rest if it was critically damped
                compared to if it was overdamped.
       D   When the amplitude decreases, the frequency remains the same.

                                                                              (1 mark)

3   A pendulum is oscillating with simple harmonic motion. At time t = 0, the pendulum
       passes through its equilibrium position, and passes through it again 1.5 s later.
       Which of the following graphs shows how the pendulum's potential energy, E_P,
       varies with time?

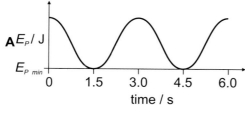

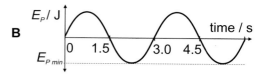

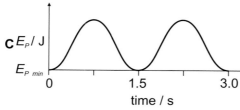

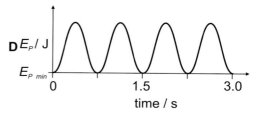

                                                                              (1 mark)

4   A block of wood on a smooth surface is attached to one end of a horizontal spring. The other end of the spring is attached to a wall. The mass oscillates freely back and forth with simple harmonic motion. The mass of the block is 0.60 kg and the spring constant of the spring is 18 Nm⁻¹. The maximum displacement of the block from its equilibrium position is 20 cm.

4.1   Calculate the time taken for the block to complete 5 oscillations.

(2 marks)

4.2   Find the velocity of the block when it is 15 cm away from the equilibrium position.

(3 marks)

4.3   Calculate the maximum kinetic energy of the block.

(3 marks)

4.4   Another block of wood of mass 1.0 kg is attached to the original. Assuming the two masses act as one mass, calculate what the spring constant would need to change to in order for the time period of the oscillations to remain the same.

(2 marks)

5   A horizontal spring is attached to a wall. A mass of 1.8 kg is attached to the other end of the horizontal spring and oscillates with simple harmonic motion. The period of oscillation is 3.2 s, and the oscillation has amplitude 0.22 m.

5.1   Calculate the spring constant of the spring.

(2 marks)

5.2   Calculate the magnitude of the maximum force experienced by the mass.

(2 marks)

5.3   Calculate the kinetic energy of the mass when it is 0.12 m from its equilibrium position.

(4 marks)

5.4   Sketch a graph of the mass's acceleration against its displacement on the axes below. Mark on your graph the maximum values of the acceleration, and the values of displacement these occur at.

(3 marks)

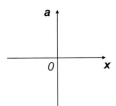

6   A simple pendulum is made to oscillate in a vacuum by a periodic driving force, at a frequency below its natural frequency.

6.1   Describe what happens as the driving frequency is increased to the pendulum's natural frequency. Give the name of this phenomenon.

(2 marks)

6.2   The same pendulum is placed in a tank of water, and made to oscillate again. Describe and explain what happens this time as the driving frequency is increased up to the pendulum's natural frequency.

(3 marks)

1. Thermal Energy Transfer

The energy and temperature of gas particles are closely linked — in fact, one can be used to work out the other. The specific heat capacity of any substance tells you how much energy is needed to raise its temperature by a given amount.

The absolute temperature scale

In thermal physics, temperature is measured using the Kelvin or absolute temperature scale. The unit of this scale is the kelvin (K).

There is a lowest possible temperature that any object can theoretically have, called **absolute zero** — around −273 °C. Absolute zero is given a value of zero kelvins, written 0 K, on the absolute temperature scale. At 0 K all molecules have zero kinetic energy — everything theoretically stops. At higher temperatures, molecules have more energy. In fact, with the Kelvin scale, a molecule's energy is proportional to its temperature (see p.65).

The Kelvin scale is named after Lord Kelvin, who first suggested it. A change of 1 K equals a change of 1 °C. To change from degrees Celsius into kelvins you add 273 (or 273.15 if you need to be really precise).

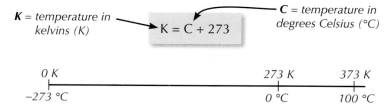

$$K = C + 273$$

K = temperature in kelvins (K) *C = temperature in degrees Celsius (°C)*

0 K		273 K	373 K
−273 °C		0 °C	100 °C

Figure 1: *Equivalent temperatures in the Kelvin and degrees Celsius scales.*

Internal energy of a gas

The particles in a gas don't all travel at the same speed. The speed distribution of gas particles depends on the temperature. Some particles will be moving fast but others much more slowly. Most will travel around the average speed. The shape of the speed distribution (see Figure 3) depends on the temperature of the gas. As the temperature of the gas increases:

- The average particle speed increases.

- The average kinetic energy of the particles increases.

- The distribution curve becomes more spread out.

Particles in gases also have potential energy (unless it is an ideal gas — see page 65). The amount of potential energy each of the particles has is randomly distributed and depends on their relative positions. Combining potential and kinetic energies gives the internal energy:

> The **internal energy** of a body is the sum of the randomly distributed kinetic and potential energies of all its particles.

Learning Objectives:

- Know that there is an 'absolute zero' of temperature, given as 0 K or −273 °C.

- Know that the internal energy of a body is the sum of the randomly distributed kinetic and potential energies of its particles.

- Know how the internal energy of a system is increased when energy is transferred to it (and vice versa).

- Be able to do calculations involving transfer of energy.

- Be able to calculate the change in temperature using $Q = mc\Delta\theta$.

- Be able to do calculations involving continuous flow.

- Understand that during a change of state the potential energies of the particles are changing but not the kinetic energies.

- Know that $Q = ml$ for a change of state.

Specification References 3.6.2.1 and 3.6.2.2

Tip: You don't use the degrees sign (°) when writing temperatures in kelvins.

Figure 2: The energy transfer from a hot drink to its surroundings can be monitored by measuring the temperature inside the cup.

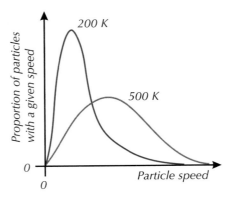

Figure 3: The distribution of particle speeds for a gas at 200 K and at 500 K.

Tip: Changing internal energy through heating or doing work is described by the first law of thermodynamics (p.328).

Tip: The same is true for hot substances in a cooler room. A hot drink at 100 °C will transfer heat to the surrounding air molecules much quicker than a hot drink at 50 °C.

Energy changes between particles

A system is just a group of bodies considered as a whole. A **closed system** is one which doesn't allow any transfer of matter in or out. For a closed system, the total internal energy is constant, as long as it isn't heated or cooled, and no energy is transferred to or from the system.

Energy is constantly transferred between particles within a system, through collisions between the particles. But the total combined energy of all the particles doesn't change during these collisions.

The energy of an individual particle changes at each collision, but the total internal energy of the system doesn't change. So, the average speed of the particles will stay the same provided the temperature of the closed system stays the same and no work is done on the system.

However, the internal energy of the system can be increased by heating it, or by doing work to transfer energy to the system (e.g. changing its shape). In this case, the average speed of the particles will increase.

The opposite is also true — the internal energy can be reduced by cooling the system, or by doing work to remove energy from the system. In such a change, the average kinetic and/or potential energy of the particles will decrease as a result of energy being transferred from the system.

Heat transfer

Heat is always transferred from hotter substances to cooler substances. In particle terms, the particle with more energy transfers some energy to the particle with less energy. The higher the difference in temperature between two substances, the faster the heat transfer between substances will happen.

This is why ice will melt if you leave it out of the freezer — gas particles collide with ice particles and transfer some energy to them. The hotter the room, the faster the ice will melt. The gas in the room will also get colder, but you don't notice the difference unless there's a huge amount of ice.

Heat is also transferred by radiation, and hotter substances radiate heat quicker than cooler substances.

Specific heat capacity

When you heat something, its particles get more kinetic energy and its temperature rises.

The **specific heat capacity** (c) of a substance is the amount of energy needed to raise the temperature of 1 kg of the substance by 1 K (or 1°C).

Or put another way:

energy change = mass × specific heat capacity × change in temperature

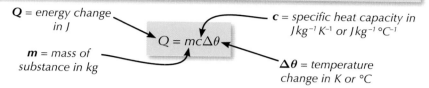

Q = energy change in J

m = mass of substance in kg

$Q = mc\Delta\theta$

c = specific heat capacity in $J\,kg^{-1}\,K^{-1}$ or $J\,kg^{-1}\,°C^{-1}$

$\Delta\theta$ = temperature change in K or °C

Tip: You might see $\Delta\theta$ written as ΔT.

Example — Maths Skills

The specific heat capacity of water is 4180 $J\,kg^{-1}\,K^{-1}$. If 172 kJ of energy is supplied to 5.00 kg of water at 300 K (correct to 3 significant figures), what will its final temperature be?

First find the value of $\Delta\theta$:

$$Q = mc\Delta\theta \Rightarrow \Delta\theta = \frac{Q}{mc} = \frac{172 \times 10^3}{5.00 \times 4180}$$
$$= 8.229...\,K$$

Then add this to the initial temperature:
final temperature $= 300 + 8.229... = 308.229...$
$= 308\,K$ (to 3 s.f.)

Tip: Energy is supplied to the water, so the temperature goes up.

Investigating factors that affect the change in temperature

You can use this experiment to investigate how changing the mass of an object, changing the material the object is made from (and so changing the specific heat capacity) or changing the rate of energy transfer affects the change in temperature of that object.

The method is the same for solids and liquids, but the set-up is a little bit different:

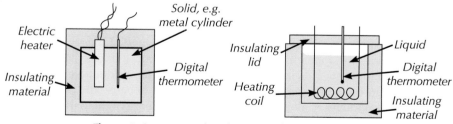

Electric heater

Solid, e.g. metal cylinder

Insulating material

Digital thermometer

Insulating lid

Heating coil

Liquid

Digital thermometer

Insulating material

Figure 4: *Set-ups used to determine factors that affect change in temperature of a solid or liquid.*

Tip: As always, do a full risk assessment before beginning this experiment, and take extra care if you are placing an electrical heater in water.

Decide which factor to investigate and keep all others unchanged. Then heat the water for a set amount of time and record the change in temperature. Repeat the experiment for different values of the factor you are changing, and compare the changes in temperature.

For example, to investigate how the rate of energy transfer affects the change in temperature, you could vary the current supplied to the heater and compare the change in temperature after 3 minutes of heating at each current.

Tip: The value you end up with for c will probably be too high by quite a long way. That's because some of the energy from the heater gets transferred to the air and the container. You can reduce the error by starting below and finishing above room temperature to cancel out gains and losses.

Finding specific heat capacity using a continuous-flow calorimeter

Continuous-flow heating is when a fluid flows continuously over a heating element. As it flows, energy is transferred to the fluid. You can use this to find the specific heat capacity using a **continuous-flow calorimeter**:

- Set up the experiment as shown in Figure 5 and let water flow at a steady rate until the water out is at a constant temperature.
- Record the flow rate of the water and the duration of the experiment, t, (to find the mass of water). You also need to measure the temperature difference, $\Delta\theta$, (of the water from the point that it flows in to the point that it flows out) between the thermometers. Also record the current, I, and potential difference, V.
- The energy supplied to the water is $Q = mc\Delta\theta + H$, where H is the heat lost to the surroundings.

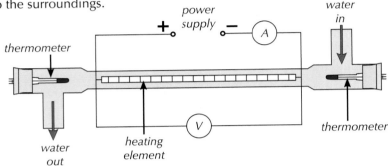

Figure 5: Continuous-flow calorimeter

- Repeat the experiment changing only the p.d. of the power supply and the flow rate (mass) so $\Delta\theta$ remains constant. You should now have an equation for each experiment: $Q_1 = m_1c\Delta\theta + H$ and $Q_2 = m_2c\Delta\theta + H$.
- The values of c, $\Delta\theta$ and H are the same, so $Q_2 - Q_1 = (m_2 - m_1)c\Delta\theta$. Rearranging gives: $c = \dfrac{Q_2 - Q_1}{(m_2 - m_1)\Delta\theta}$.
- Q is just the electrical energy supplied over time t in each case, so you can use $Q = VIt$ to find Q_1 and Q_2, and therefore c, the **specific heat capacity** of water.

Changes of state

A change of state, sometimes also referred to as a change of phase, occurs when a substance changes between a solid, liquid or gas. When a substance changes state its internal energy changes but its kinetic energy and temperature stay the same (see Figure 7). This is because the potential energy of the particles is altered while the kinetic energy of the particles stays constant.

As a liquid turns into a gas (for example, boiling water becoming steam) its potential energy increases even though the water molecules in both states are at 100 °C.

You can demonstrate this using a digital thermometer attached to a data logger as energy is supplied to an insulated body at a constant rate. As it undergoes a change of state, the rate of temperature increase recorded by the thermometer will decrease.

Figure 6: When boiling water, the energy is used to convert the water into steam rather than to increase the temperature of the water.

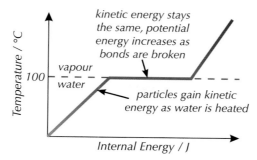

Figure 7: Graph of temperature against internal energy for a sample of water being heated at a constant rate to its boiling point and beyond.

Specific latent heat

To melt a solid or boil or evaporate a liquid, you need energy to break the bonds that hold the particles in place. The energy needed for this is called latent heat. The larger the mass of the substance, the more energy it takes to change its state. That's why the **specific latent heat** is defined per kg:

> The specific latent heat (l) of fusion or vaporisation is the quantity of thermal energy needed to be gained or lost to change the state of 1 kg of a substance.

Which gives:

> energy change = specific latent heat × mass of substance changed

Q = energy change in J ⟶ $Q = ml$ ⟵ m = mass of substance in kg
l = specific latent heat in J kg^{-1}

┌─ **Example** ── **Maths Skills** ──────────────

Find the energy needed to turn 1.00 kg of water at 90.0 °C to 1.00 kg of steam at 110.0 °C. c_{water} = 4180 J kg^{-1}K^{-1}, c_{steam} = 1890 J kg^{-1}K^{-1}, and the latent heat of vaporisation of water is 2.26 × 10^6 J kg^{-1}.

First find the energy needed to heat the water by 10.0 °C:

$$Q = mc\Delta\theta = 1.00 \times 4180 \times 10.0$$
$$= 41\ 800 \text{ J (or 41.8 kJ)}$$

Then find the energy needed to turn the water to steam:

$$Q = ml = 1.00 \times (2.26 \times 10^6)$$
$$= 2.26 \times 10^6 \text{ J (or 2260 kJ)}$$

Then find the energy needed to heat the steam by 10.0 °C:

$$Q = mc\Delta\theta = 1.00 \times 1890 \times 10.0$$
$$= 18\ 900 \text{ J (or 18.9 kJ)}$$

Then just add all these numbers together:

$$\text{Energy needed} = 41.8 + 2260 + 18.9 = 2320.7$$
$$= 2320 \text{ kJ (to 3 s.f.)}$$

Tip: The specific latent heat of a substance changes slightly depending on its temperature.

Tip: The specific latent heat of fusion is used when a substance is melting or freezing. The specific latent heat of vaporisation is used when a substance is boiling or condensing.

Tip: Energy has to be added for a material to melt or boil. When a material freezes or condenses, energy is lost from the material.

Tip: You'll usually see the latent heat of vaporisation written l_v and the latent heat of fusion written as l_f.

Tip: Water turns to steam at 100 °C.

Tip: Head back to page 47 for a reminder about the Kelvin temperature scale.

Q1 How much energy is required to raise the temperature of 0.45 kg of a liquid with specific heat capacity 244 J kg^{-1} K^{-1} by 3.0 kelvins?

Q2 a) A student carries out the continuous-flow experiment to determine the specific heat capacity of a liquid, as shown on page 50. Give one potential source of systematic error and random error, and suggest a way to reduce each.

   b) In the first run of the experiment, 5511 J of energy is transferred to 0.330 kg of vegetable oil to increase the temperature by 10.0 K. In the second run, 7515 J of energy is transferred to 0.450 kg of the vegetable oil to increase the temperature by the same amount. Use these values to estimate the specific heat capacity of vegetable oil. Both runs have the same time period.

Q3 A bowl containing 100.0 g of water at 25.0 °C is placed in a freezer. A few hours later, all of the water has become ice at −5.00 °C.

   a) Find how much energy the water lost to its surroundings. c_{water} = 4180 J kg^{-1} K^{-1}, c_{ice} = 2110 J kg^{-1} K^{-1}, l_f = 334 000 J kg^{-1}.

   b) The freezer temperature is increased from −20 °C to −10 °C and an identical bowl of water at 25.0 °C is placed inside the freezer. After the same amount of time, not all of the water has frozen. Why is this?

Practice Questions — Fact Recall

Q1 a) What is the lowest possible temperature an object can theoretically reach called?

   b) What is the value of this temperature in kelvins and degrees Celsius?

Q2 What is meant by the internal energy of a body?

Q3 a) Sketch a graph to show the speed distribution of the molecules of a gas at 100 K and at 250 K.

   b) State two ways in which increasing the temperature of a gas affects the speeds of the gas molecules.

Q4 How is energy transferred between gas molecules?

Q5 Sketch and label a graph of temperature against internal energy to show how the internal energy of water changes as it is heated at a constant rate to its boiling point and beyond. Your labels should explain the shape of the graph.

Q6 What is the specific latent heat of fusion?

Tip: It's easier to draw the graph in Q5 using degrees Celsius, but you could also use kelvins.

2. The Three Gas Laws

An 'ideal gas' follows the three gas laws, which describe how a fixed mass of gas behaves when you change its temperature, pressure or volume. Be careful though — they work on the Kelvin temperature scale, not traditional °C.

Boyle's law

Each of the three gas laws was worked out independently by careful experiment with fixed masses of gases.

The first of them is **Boyle's law**, which says that:

> At a constant temperature the pressure *p* and volume *V* of a gas are inversely proportional.

Inversely proportional means that as one variable increases, the other decreases by the same proportion, i.e. $p \propto \frac{1}{V}$.

For example, if you reduce the volume of a gas, its particles will be closer together and will collide with each other and the container more often, so the pressure increases.

An **ideal gas** is a (theoretical) gas that obeys Boyle's law at all temperatures. Boyle's law means that at any given temperature the product of *p* and *V* will always be the same:

$$pV = \text{constant}$$

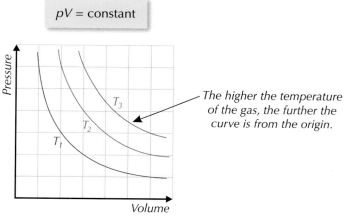

The higher the temperature of the gas, the further the curve is from the origin.

Figure 1: *Graphs of pressure against volume for an ideal gas at different temperatures.*

> REQUIRED PRACTICAL **8**

Experiment to investigate Boyle's law

You can investigate the effect of **pressure on volume** by setting up the experiment shown in Figure 3.

- The oil traps a pocket of air in a sealed tube with fixed dimensions.
- Use a tyre pump to increase the pressure on the oil.
- Use the Bourdon gauge to record the pressure. As the pressure increases, more oil will be pushed into the tube, the oil level will rise, and the air will compress. The volume occupied by air in the tube will reduce.

Learning Objectives:

- Know the three gas laws for a fixed mass of an ideal gas.
- Know that the gas laws are experimental relationships between *p, V, T* and mass.
- Be able to explain the relationships between *p, V* and *T* in terms of a simple molecular model.
- Be able to investigate Boyle's law and Charles' law experimentally (Required Practical 8).

Specification References 3.6.2.2 and 3.6.2.3

Tip: If the pressure doubles, the volume halves, and so on.

Tip: Most gases can usually be assumed to act as ideal gases — see the assumptions on pages 63-64

Tip: Do a full risk assessment before you start. This experiment involves pressurising a glass container — so wear goggles and don't use too high a pressure.

- Measure the volume of air when the system is at atmospheric pressure by multiplying the length of the part of the tube containing air by π × radius of tube squared.
- Gradually increase the pressure by a set interval, keeping the temperature constant.
- Note down both the pressure and the volume of air as it changes. Multiplying these together at any point should give the same value.
- Repeat the experiment twice more and take a mean for each reading.

If you plot a graph of p against $\frac{1}{V}$ you should get a straight line.

Figure 2: *Irish chemist and physicist Robert Boyle.*

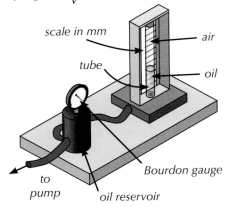

Figure 3: *Experimental set-up for investigating Boyle's Law.*

Charles' law

Charles' law states that:

> At constant pressure, the volume V of a gas is directly proportional to its absolute temperature T.

Tip: If the temperature doubles, the volume doubles, and so on.

Tip: In practice a real (non-ideal) gas would condense before reaching 0 K.

An ideal gas also obeys Charles' law. V and T are directly proportional, i.e. $V \propto T$. At the lowest theoretically possible temperature (0 K, see page 47) the volume is zero. If Charles' law is obeyed, the volume divided by the temperature is a constant:

$$\frac{V}{T} = \text{constant}$$

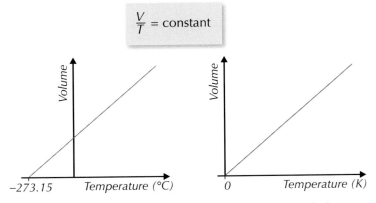

Figure 5: *Graphs of volume against temperature for an ideal gas at constant pressure on the degrees Celsius scale and on the Kelvin scale.*

Figure 4: *French inventor, scientist, mathematician and balloonist Jacques Charles.*

When you heat a gas, the particles gain kinetic energy (page 47) and move more quickly. At a constant pressure, this means they move further apart, and so the volume of the gas increases.

Experiment to investigate Charles' law

REQUIRED PRACTICAL **8**

You can investigate the effect of temperature on volume by setting up the experiment shown in Figure 6.

- For this experiment, you will need a capillary tube containing a drop of concentrated sulfuric acid positioned halfway up the tube. The tube should be sealed at the bottom, so that a small column of air is trapped between the bottom of the tube and the acid drop.

- Place the capillary tube in a beaker of hot water. Position a ruler behind the capillary tube so that you can measure the length of the column of air trapped between the bottom of the tube and the drop of sulfuric acid (see Figure 6).

- As the water cools, regularly record the temperature of the water and the length of the air column. You can assume the air pressure from the air above the droplet is constant.

- Repeat the experiment with fresh near-boiling water twice more, allowing the capillary tube to adjust to the new temperature between each repeat, so that you have three readings for each temperature. Take an average of the length of the air column for each temperature.

- You should see that the length of the trapped air column decreases as the water temperature decreases.

- Plot your results on a graph of length against temperature and draw a line of best fit — you should get a clear straight line. This shows that the length of the air column is proportional to the temperature.

- The volume of the column of air is equal to the volume of a cylinder, $V = \pi r^2 l$, where r is the radius of the column and l is its length. r^2 remains constant in this experiment, which means the length is proportional to the volume of the air column, and so the volume also is proportional to the temperature. This agrees with Charles' law.

Tip: This experiment involves acid and hot water — use suitable safety equipment and do a full risk assessment before you start work.

Tip: It's safest to use a kettle or urn to provide your hot water, but you could also use a Bunsen burner. Either way, take care.

Tip: If the markings on your measuring equipment are quite far apart, you can often interpolate between them. This just means making an estimate of the measurement to get a value that isn't shown by the scale you're reading from. E.g. if the temperature is halfway between the markings for 24 °C and 25 °C you could record it as 24.5 °C. But it's better to use something with a finer scale if you can.

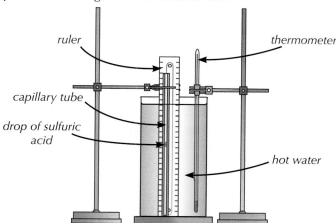

Figure 6: *Experimental set-up for investigating Charles' law.*

The pressure law

The pressure law states that:

> At constant volume, the pressure p of an ideal gas is directly proportional to its absolute temperature T.

Tip: If the temperature doubles, the pressure doubles, and so on.

For example, if you heat a gas, the particles gain kinetic energy. This means they move faster. If the volume doesn't change, the particles will collide with each other and their container more often and at higher speeds, increasing the pressure inside the container.

At absolute zero the pressure is also zero. If the pressure law is obeyed, the pressure divided by the temperature is a constant:

$$\frac{p}{T} = \text{constant}$$

So a graph of pressure against temperature will be a straight line (its gradient will be constant). This is shown in Figure 7.

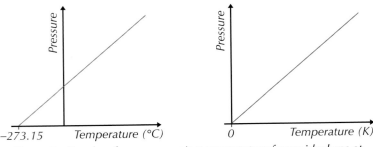

Figure 7: Graphs of pressure against temperature for an ideal gas at constant volume on the degrees Celsius scale and on the Kelvin scale.

Practice Question — Application

Q1 A gas syringe is filled with 30.0 cm³ of an ideal gas at 27 °C and the pressure inside the syringe is measured as 1.4×10^5 Pa.

   a) The syringe is pushed in so that its volume becomes 15 cm³. If the temperature remains constant, what will the pressure inside the syringe be?

   b) The syringe is returned to its starting position and then cooled to −173 °C. Assuming the pressure inside the syringe remains constant, what will its new volume be?

Practice Questions — Fact Recall

Q1 What must be true of a gas's mass for it to obey the three gas laws?

Q2 a) What's Boyle's law?

   b) Sketch a graph of pressure against volume for an ideal gas at fixed temperature T.

   c) Describe an experiment you could use to investigate Boyle's law. You may draw a diagram to help your description.

Q3 a) What's Charles' law?

   b) Sketch a graph of volume against temperature (in °C) for an ideal gas at constant pressure p.

Q4 a) What's the pressure law?

   b) Sketch a graph of pressure against temperature (in °C) for an ideal gas at constant volume V.

3. The Ideal Gas Equation

You can use the gas laws from the last topic to form the ideal gas equation. It's pretty important in thermal physics, so you'll probably see it quite a lot.

Molecular mass

Molecular mass is the sum of the masses of all the atoms that make up a single molecule. Periodic tables usually include the **relative atomic mass** for every element — this is an average mass on a scale where a carbon-12 atom has a mass of 12.

Relative molecular mass is just the sum of the relative atomic masses of all the atoms making up a molecule. For example, a carbon dioxide molecule is made up of a carbon atom (relative atomic mass = 12.0) and two oxygen atoms (relative atomic mass = 16.0). So the relative molecular mass of carbon dioxide is $12.0 + 16.0 + 16.0 = 44.0$.

Relative molecular mass is often just called "molecular mass".

The Avogadro constant

Because gas molecules are tiny, it's easier to consider them in large groups when carrying out calculations. Amedeo Avogadro had an idea in the 19th century that a fixed volume of gas at a fixed temperature and pressure contains the same number of gas molecules, regardless of what type of gas it is. The mass, however, is proportional to its molecular mass, i.e. to the relative mass of the molecules of gas. So 1.0 g of hydrogen (relative atomic mass = 1.0) and 16.0 g of oxygen (relative atomic mass = 16.0) occupy the same volume at the same temperature and pressure.

This gave rise to the **Avogadro constant**, N_A, defined as the number of atoms in exactly 12 g of the carbon isotope $^{12}_{6}C$. This number gives the number of atoms (or molecules) in any volume of substance whose mass, in grams, is the same as its relative atomic (or molecular) mass. The value of Avogadro's constant is 6.02×10^{23} mol^{-1}.

Molar mass

A substance containing N_A atoms or molecules, all of which are identical, is defined as 1 **mole** of that substance. The **molar mass** of a substance is the mass that 1 mole of the substance would have (usually in grams), and is equal to its relative atomic or relative molecular mass. So, for example, the molar mass of helium (relative atomic mass = 4.0) is 4.0 g.

For gases with only one type of element, the mass of one mole is often twice their relative atomic mass because many gases (e.g. oxygen) exist as molecules, which are two (or more) atoms bonded together. This is why the molar mass of oxygen (32.0 g) is twice its relative atomic mass (16.0). Helium exists as single atoms, so its molar mass is the same as its relative atomic mass. The gas laws all apply whether the gas is made up of molecules or atoms.

The number of moles in a substance is usually given by n, and its units are 'mol'. The number of molecules in a mass of gas is given by the number of moles, n, multiplied by the Avogadro constant. So the number of molecules, $N = nN_A$.

Learning Objectives:

- Know what is meant by molecular mass and molar mass.
- Know what the Avogadro constant, N_A, the molar gas constant, R, and the Boltzmann constant, k, are.
- Be able to use the ideal gas equation $pV = nRT$ for n moles.
- Be able to use the ideal gas equation $pV = NkT$ for N molecules.
- Know that work done on a gas of constant pressure as it changes volume is given by $p\Delta V$.

Specification Reference 3.6.2.2

Figure 1: *19th-century Italian scientist Amedeo Avogadro.*

Tip: 1 mole of a gas at room temperature and atmospheric pressure (~101 kPa) takes up around 0.0224 m^3.

The ideal gas equation

The three gas laws from the previous topic can be used to derive the **ideal gas equation**. Start by combining the three laws to get:

$$\frac{pV}{T} = \text{constant}$$

Tip: You don't need to know how to combine the gas laws, you just need to know that combining them gives $\frac{pV}{T}$ = constant.

Putting in values for 1 mole of an ideal gas at room temperature and atmospheric pressure gives the constant a value of $8.31 \, \text{JK}^{-1}\text{mol}^{-1}$. This is the **molar gas constant**, R.

The value of $\frac{pV}{T}$ increases or decreases if there's more or less gas present — the more gas you have, the more space it takes up. The amount of gas is measured in moles, n, so the constant in the equation above becomes nR, where n is the number of moles of gas present. Plugging this into the equation gives:

$$\frac{pV}{T} = nR$$

Which can be rearranged to give the **ideal gas equation for n moles**:

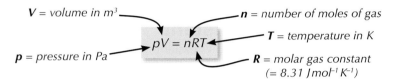

V = volume in m^3
n = number of moles of gas
$$pV = nRT$$
T = temperature in K
p = pressure in Pa
R = molar gas constant ($= 8.31 \, Jmol^{-1}K^{-1}$)

This equation works well (i.e., a real gas approximates to an ideal gas) for gases at low pressure and a fairly high temperature.

Example — **Maths Skills**

What's the volume occupied by exactly one mole of an ideal gas at room temperature (20.0 °C) and atmospheric pressure (1.00 × 10⁵ Pa)?

First rearrange the ideal gas equation to make V the subject:

$$pV = nRT \Rightarrow V = \frac{nRT}{p}$$

Then put the correct numbers in:

$$V = \frac{nRT}{p} = \frac{1 \times 8.31 \times (20.0 + 273)}{1.00 \times 10^5} = 0.0243483$$
$$= 2.43 \times 10^{-2} \, m^3 \, \text{(to 3 s.f.)}$$

Tip: Don't forget to convert the units — remember the Kelvin scale is used for temperature in ideal gas calculations.

The Boltzmann constant

The Boltzmann constant, k, is equivalent to $\frac{R}{N_A}$ — you can think of the Boltzmann constant as the gas constant for one molecule of gas, while R is the gas constant for one mole of gas. The value of the Boltzmann constant is $1.38 \times 10^{-23} \, \text{JK}^{-1}$.

If you combine $N = nN_A$ from the previous page and $k = \frac{R}{N_A}$ you'll see that $Nk = nR$. First rearrange to make N_A the subject of the first equation:

$$N = nN_A \Rightarrow N_A = \frac{N}{n}$$

Tip: Remember N is just the number of molecules in a mass of gas.

Then substitute this into the second equation:

$$k = \frac{R}{N_A} \Rightarrow k = \frac{R}{\left(\frac{N}{n}\right)} = \frac{nR}{N}$$

$$\Rightarrow nR = Nk$$

Substituting this into the ideal gas equation (from the previous page) gives the **ideal gas equation for *N* molecules**:

V = volume in m³

p = pressure in Pa

N = number of molecules of gas

$$pV = NkT$$

T = temperature in K

k = the Boltzmann constant
(= 1.38 × 10⁻²³ JK⁻¹)

Tip: You might also see this equation referred to as the equation of state of an ideal gas.

┌─ **Example** ── **Maths Skills** ──────────────────

An ideal gas at 303 K and 1.00 × 10⁵ Pa occupies 23.2 litres. Find how many molecules of the gas are present.

Start by rearranging the equation to make *N* the subject:

$$pV = NkT \Rightarrow N = \frac{pV}{kT}$$

Then put the correct numbers in:

$$N = \frac{pV}{kT} = \frac{(1.00 \times 10^5) \times (23.2 \times 10^{-3})}{(1.38 \times 10^{-23}) \times 303} = 5.5483... \times 10^{23}$$

$$= 5.55 \times 10^{23} \text{ (to 3 s.f.)}$$

Tip: There are 1000 litres in 1 m³.

Work done

For a gas to expand or contract at constant pressure, work must be done — i.e. there must be a transfer of energy. This normally involves the transfer of heat energy — e.g. if you heat a gas-filled balloon, it will expand. Remove the heat source and it will contract back to its original size as the heat is transferred back to its surroundings.

Tip: This is just Charles' law — see page 54.

The work done in changing the volume of a gas at a constant pressure is given by:

p = pressure in Pa

work done = $p\Delta V$

ΔV = change in volume in m³

The area under a graph of pressure against volume shows the energy transferred to change the volume of the gas.

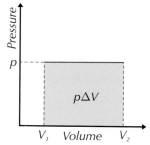

Figure 3: The area under a pressure-volume graph represents the work done.

A gas cylinder is heated so that its volume increases from 0.320 m³ to 0.875 m³. Assuming the pressure remains constant at 1.15×10^5 Pa, calculate the energy transferred to the gas.

energy transferred = work done, so use:

$W = p\Delta V$
$\quad = (1.15 \times 10^5) \times (0.875 - 0.320)$
$\quad = 63\ 825 = 63\ 800$ J (to 3 s.f.)

Tip: A carbon dioxide molecule contains 1 carbon atom (relative atomic mass = 12.0) and 2 oxygen atoms (relative atomic mass = 16.0).

Tip: $R = 8.31$ J mol⁻¹ K⁻¹ and $k = 1.38 \times 10^{-23}$ J K⁻¹.

Tip: Head to page 57 for a reminder on what the molar mass is.

Practice Questions — Application

Q1 What's the molar mass of carbon dioxide (CO_2)?

Q2 Find the volume occupied by 23 moles of an ideal gas at 25 °C and a pressure of 2.4×10^5 Pa.

Q3 A sealed, airtight container contains 8.21×10^{24} molecules of an ideal gas at a fixed volume of 4.05 m³. The gas is heated to 500 K (correct to 3 s.f.). What's the pressure inside the container?

Q4 A gas doubles in volume to 2.0 m³ at a constant pressure of 1.15×10^5 Pa. How much energy is transferred to the gas?

Q5 A sealed, airtight container is filled with 1.44×10^{25} molecules of an ideal gas and kept at a constant pressure of 1.29×10^5 Pa. The container is heated so that its volume expands to 0.539 m³. What temperature is the gas inside the container at this point?

Q6 A sealed, airtight container with a fixed volume of 0.39 m³ is filled with 0.88 kg of an ideal gas whose molar mass is 44 g. The maximum pressure the container can withstand is 2.3×10^5 Pa.

a) How many moles of gas are inside the container?

b) What's the highest temperature the container can be heated to before the pressure on it would be too high?

Practice Questions — Fact Recall

Q1 What is the relative molecular mass of a substance?

Q2 What is the Avogadro constant?

Q3 What is the molar mass of a substance?

Q4 Write down the ideal gas equation for n moles, and state what each term in the equation represents.

Q5 What is the Boltzmann constant?

Q6 Write down the ideal gas equation for N molecules, and state what each term in the equation represents.

Q7 Write down the equation for work done when changing the volume of a gas at a fixed pressure, and state what each term in the equation represents.

4. Kinetic Theory and the Pressure of an Ideal Gas

Learning Objectives:

- Know how the formula $pV = \frac{1}{3}Nm(c_{rms})^2$ is derived for an ideal gas using algebra and conservation of momentum.

- Be able to explain the relationships between p, V and T in terms of a simple molecular model.

- Know the assumptions made about an ideal gas in kinetic theory.

Specification Reference 3.6.2.3

The equation for the pressure exerted by an ideal gas can be derived by considering the gas as a large number of individual particles. This is kinetic theory. The following few pages are rather tricky, so pay close attention.

Deriving the pressure of an ideal gas

You need to be able to derive the pressure of an ideal gas. Start by deriving the pressure on one wall of a box — in the x direction. Imagine a cubic box with sides of length l containing N molecules each of mass m.

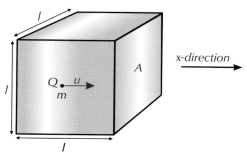

Figure 1: A cubic box with sides of length l, containing N molecules each of mass m.

Say molecule Q moves directly towards wall A with velocity $\boldsymbol{u}$. Its momentum approaching the wall is $m\boldsymbol{u}$. It strikes wall A. Assuming the collisions are perfectly elastic, the wall pushes back on the molecule so it rebounds and heads back in the opposite direction with momentum $-m\boldsymbol{u}$ (because momentum is always conserved). So the change in momentum is:

$$m\boldsymbol{u} - (-m\boldsymbol{u}) = 2m\boldsymbol{u}$$

Assuming Q suffers no collisions with other molecules, the time between collisions of Q and wall A is $\frac{2l}{\boldsymbol{u}}$. The number of collisions per second is therefore $\frac{\boldsymbol{u}}{2l}$. This gives the rate of change of momentum as:

$$2m\boldsymbol{u} \times \frac{\boldsymbol{u}}{2l}$$

Force equals the rate of change of momentum (Newton's second law), so the force exerted on the wall by this one molecule is:

$$\frac{2m\boldsymbol{u}^2}{2l} = \frac{m\boldsymbol{u}^2}{l}$$

Molecule Q is only one of many in the cube. Each molecule will have a different velocity $\boldsymbol{u}_1$, $\boldsymbol{u}_2$ etc. towards A. The total force, $\boldsymbol{F}$, of all these molecules on wall A is:

$$\boldsymbol{F} = \frac{m(\boldsymbol{u}_1^2 + \boldsymbol{u}_2^2 + \text{etc.})}{l}$$

You can define a quantity called the mean square speed, $\overline{\boldsymbol{u}^2}$ as:

$$\overline{\boldsymbol{u}^2} = \frac{\boldsymbol{u}_1^2 + \boldsymbol{u}_2^2 + \text{etc.}}{N}$$

Tip: Kinetic theory is the term given to explaining an object's properties by considering the motion of its particles.

Tip: The wall pushes back because of Newton's third law of motion — every action has an equal and opposite reaction.

Tip: Time = $\frac{\text{distance}}{\text{speed}}$, and $2l$ is the distance a molecule would travel between collisions.

Tip: You should remember this definition of Newton's second law from year 1 of A-level.

If you put that into the equation before it, you get:

$$F = \frac{Nm\overline{u^2}}{l}$$

So, the pressure of the gas on end A is given by:

V = volume of the cube in m^3

$$\text{pressure, } p = \frac{\text{force}}{\text{area}} = \frac{\left(\frac{Nm\overline{u^2}}{l}\right)}{l^2} = \frac{Nm\overline{u^2}}{l^3} = \frac{Nm\overline{u^2}}{V}$$

Tip: The volume of the cube is given by the length of the side cubed, i.e. l^3.

A gas molecule can move in three dimensions, so for the general equation you need to think about all 3 directions — x, y and z. You can calculate its speed, c, from Pythagoras' theorem: $c^2 = u^2 + v^2 + w^2$ where u, v and w are the components of the molecule's velocity in the x, y and z directions.

Tip: You've probably met Pythagoras' theorem before in two dimensions, where $c^2 = a^2 + b^2$. It also works in 3 dimensions, as long as they're at right angles to each other.

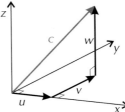

Figure 2: The speed of a molecule, c, can be found by using Pythagoras' theorem on its velocity in the x, y and z directions.

If you treat all N molecules in the same way, this gives an overall **mean square speed**, $\overline{c^2}$, of:

$$\overline{c^2} = \overline{u^2} + \overline{v^2} + \overline{w^2}$$

Since the molecules move randomly, $\overline{u^2} = \overline{v^2} = \overline{w^2}$. So $\overline{c^2} = 3\overline{u^2}$. You can substitute this into the equation for pressure that you derived above:

Tip: The mean square speed is the average of the squares of the speeds of molecules.

$$p = \frac{Nm\overline{u^2}}{V} \quad \Rightarrow \quad p = \frac{1}{3}\frac{Nm\overline{c^2}}{V}$$

$\overline{c^2}$ = mean square speed of gas molecules in $m^2 s^{-2}$

$$pV = \frac{1}{3}Nm\overline{c^2}$$

p = pressure in Pa

V = volume in m^3

N = number of molecules of gas

m = mass of a gas molecule in kg

Root mean square speed

As you've seen, it often helps to think about the motion of a typical molecule in kinetic theory. $\overline{c^2}$ is the mean square speed and has units $m^2 s^{-2}$. $\overline{c^2}$ is the average of the squares of the speeds of molecules — the square root of it gives you the typical speed.

This is called the **root mean square speed** or, usually, the r.m.s. speed. The unit is the same as any speed — ms^{-1}.

Tip: Although this derivation assumed the container was cubic, the result is valid for any shape of container.

$$\text{r.m.s. speed} = \sqrt{\text{mean square speed}} = \sqrt{\overline{c^2}} = c_{rms}$$

So you can write the equation above as:

$$pV = \frac{1}{3}Nm\left(c_{rms}\right)^2$$

Exactly 1 mole of hydrogen gas occupies 2.5×10^{-2} m³ at a pressure of 1.0×10^5 Pa. Find the root mean square speed (c_{rms}) of the hydrogen molecules. The mass of a hydrogen molecule is 3.3×10^{-27} kg.

First find the number of hydrogen molecules present:

$$N = nN_A = 1 \times 6.02 \times 10^{23} = 6.02 \times 10^{23}$$

Then rearrange the formula to make c_{rms} the subject:

$$pV = \frac{1}{3}Nm(c_{rms})^2 \Rightarrow c_{rms} = \sqrt{\frac{3pV}{Nm}}$$

And put in the correct numbers:

$$c_{rms} = \sqrt{\frac{3pV}{Nm}} = \sqrt{\frac{3 \times (1.0 \times 10^5) \times (2.5 \times 10^{-2})}{(6.02 \times 10^{23}) \times (3.3 \times 10^{-27})}} = 1943.0...$$
$$= 1900 \text{ ms}^{-1} \text{ (to 2 s.f.)}$$

Explaining Charles' law and the pressure law

You can use the derivation for the pressure of an ideal gas from the last two pages to explain why the volume or pressure must increase if the temperature of a fixed mass of ideal gas is increased.

> **Tip:** Head to pages 54-55 for a reminder of Charles' law and the pressure law.

Temperature is related to the kinetic energy of the molecules (page 47) — as the temperature increases, the average speed of the molecules increases. This means the rate of change of momentum of the molecules colliding with the walls of the container increases, and so the force on the walls of the container increases.

If the volume of the container is fixed, this will result in an increased pressure inside the container for two reasons:

- There will be more collisions between the molecules and the walls of the container in a given amount of time.

- On average, a collision will result in a larger change in momentum, and so exert a larger force on the walls of the container.

If the pressure inside the container remains constant, the volume of the container will increase to compensate for the temperature change for two reasons:

- If the volume is larger, there will be a longer time between molecule-wall collisions, and so the rate of change of momentum and therefore the force on the walls of the container will be reduced.

- As the volume increases, the surface area of the walls increases. Pressure is defined as the force per unit area, and so increasing the area stops the pressure from increasing.

> **Tip:** There's a longer time between collisions because the walls are further apart, so a molecule takes longer to travel between them.

Assumptions in kinetic theory

In **kinetic theory**, physicists picture gas molecules moving at high speeds in random directions. To get equations like the one derived earlier in this topic though, some simplifying assumptions are needed:

- All molecules of the gas are identical.

- The gas contains a large number of molecules.

- The molecules have a negligible volume compared with the volume of the container (i.e. they act as point masses).

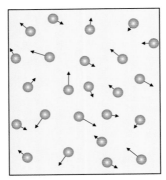

Figure 3: *A visual representation of an ideal gas. All molecules are identical, have negligible volume and move randomly.*

Tip: Remember that the direction matters for velocity. In this case u is in the direction normal (at 90°) to the wall.

- The molecules continually move about randomly.
- Newtonian mechanics apply (i.e. the motion of the molecules follows Newton's laws).
- Collisions between molecules themselves or at the walls of a container are perfectly elastic (i.e. kinetic energy is conserved).
- The molecules move in a straight line between collisions.
- The forces that act during collisions last for much less time than the time between collisions.

A gas obeying these assumptions is called an ideal gas. Real gases behave like ideal gases as long as the pressure isn't too big and the temperature is reasonably high (compared with their boiling point), so they're useful assumptions.

Practice Questions — Application

Q1 a) Show that the change in momentum of an ideal gas molecule colliding with the wall of a cubic container is equal to $2mu$, where m is the mass of the molecule and u is its velocity normal to the wall.

b) Hence show that the force exerted on the wall of the container by the molecule of gas is equal to $\frac{mu^2}{l}$, where l is the length of the edges of the container.

c) Hence show that the total pressure exerted by all the molecules of an ideal gas on one wall of a cubic container is equal to $\frac{Nm\overline{u^2}}{V}$, where N is the number of molecules and $\overline{u^2}$ is the mean square speed of the molecules in the direction normal to the wall.

d) Hence show that $pV = \frac{1}{3}Nm(c_{rms})^2$, where $(c_{rms})^2$ is the mean square speed of the molecules.

Q2 A sealed, rigid container with a volume of 1.44 m³ is filled with exactly 5 moles of an ideal gas. If the molecules of the gas have a mean square speed of 8.11×10^6 m²s⁻² and a mass of 5.31×10^{-26} kg, find the pressure inside the container.

Q3 Explain using kinetic theory why increasing the temperature of a fixed mass of a gas in a container must lead to an increase in its volume or pressure.

Practice Questions — Fact Recall

Q1 What does c_{rms} represent?

Q2 Give two reasons why increasing the temperature of a gas increases the force it exerts on the walls of its container.

Q3 A sealed container contains a fixed mass of an ideal gas. The temperature of the gas inside the container is increased, which causes another property of the gas to increase. Give two properties this could be.

Q4 State four assumptions made about an ideal gas in kinetic theory.

5. Kinetic Energy of Gas Molecules

Kinetic energy is all about the energy in moving particles. It's daft to work out the individual kinetic energies of every single particle in a body of gas, but you can find the average and total kinetic energy quite easily.

Average kinetic energy of gas molecules

For an ideal gas, you can assume that all the internal energy is in the form of kinetic energy. This means you can use the product pV to find the average and total kinetic energy.

There are two equations for the product pV of a gas. You can equate these to get an expression for the average kinetic energy.

The ideal gas equation is:

$$pV = nRT$$

Using kinetic theory (see pages 61-62), you can show that:

$$pV = \frac{1}{3}Nm(c_{rms})^2$$

Equating these two gives:

$$\frac{1}{3}Nm(c_{rms})^2 = nRT$$

Multiplying by $\frac{3}{2}$ gives:

$$\frac{3}{2} \times \frac{1}{3}Nm(c_{rms})^2 = \frac{3nRT}{2}$$

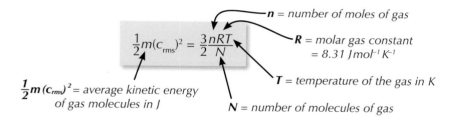

$$\frac{1}{2}m(c_{rms})^2 = \frac{3}{2}\frac{nRT}{N}$$

n = number of moles of gas

R = molar gas constant $= 8.31\ J\,mol^{-1}\,K^{-1}$

T = temperature of the gas in K

N = number of molecules of gas

$\frac{1}{2}m(c_{rms})^2$ = average kinetic energy of gas molecules in J

You can substitute Nk for nR, where k is the Boltzmann constant (see page 58), to show that the average kinetic energy of a molecule is directly proportional to T (the absolute temperature).

You can use $\frac{3}{2}kT$ as an approximation for the average kinetic energy of the molecules in any substance.

$$\frac{1}{2}m(c_{rms})^2 = \frac{3}{2}kT$$

k = the Boltzmann constant $= 1.38 \times 10^{-23}\ JK^{-1}$

Finally, the Boltzmann constant is equivalent to $\frac{R}{N_A}$ (see page 58), so you can substitute this for k in the equation above, to get:

$$\frac{1}{2}m(c_{rms})^2 = \frac{3}{2}\frac{RT}{N_A}$$

N_A = Avogadro's constant $= 6.02 \times 10^{23}\ mol^{-1}$

Example ── Maths Skills ──────────────────────

What's the average kinetic energy of the molecules in an ideal gas at 100 °C (correct to 3 significant figures)?

Just put the numbers into the equation on the previous page, making sure to convert the temperature to an absolute temperature first.

$$\frac{1}{2}m\,(c_{rms})^2 = \frac{3}{2}kT$$
$$= \frac{3}{2} \times (1.38 \times 10^{-23}) \times (100 + 273)$$
$$= 7.7211 \times 10^{-21}$$
$$= 7.72 \times 10^{-21} \text{ J (to 3 s.f.)}$$

Total kinetic energy of gas molecules

Once you've found the average kinetic energy of the molecules in a gas, you can find the total kinetic energy of the molecules as long as you know how many there are. Just multiply the average kinetic energy by the total number of molecules present.

Exam Tip
You can use any of the formulas on the previous page to calculate the average kinetic energy of gas molecules.

Tip: Look back to page 57 for the relationship between N, n and N_A.

Example ── Maths Skills ──────────────────────

What's the total kinetic energy of 3.00 moles of an ideal gas at 341 K?

First find the average kinetic energy of the molecules:

$$\frac{1}{2}m\,(c_{rms})^2 = \frac{3}{2}\frac{RT}{N_A}$$
$$= \frac{3}{2} \times \frac{8.31 \times 341}{6.02 \times 10^{23}} = 7.06... \times 10^{-21} \text{ J}$$

Then multiply this by the number of molecules of gas:

$$\text{total kinetic energy} = n \times N_A \times (7.06... \times 10^{-21})$$
$$= 3.00 \times (6.02 \times 10^{23}) \times (7.06... \times 10^{-21})$$
$$= 12\,800 \text{ J (to 3 s.f.)}$$

Practice Questions — Application

Q1 What's the average kinetic energy of the molecules of an ideal gas at 112 K?

Q2 2.44 moles of an ideal gas are heated from 250 K to 290 K. Find how much energy is supplied to the gas to cause this change in temperature, assuming all of the energy is converted to kinetic energy of the gas molecules.

Practice Questions — Fact Recall

Q1 For an ideal gas, what can you assume about the internal energy?

Q2 Write an equation for calculating the average kinetic energy of ideal gas molecules.

Q3 Write an equation for calculating the total kinetic energy of n moles of an ideal gas.

6. Development of Theories

Our current understanding of the behaviour of gases is the result of many theoretical derivations and empirically-derived laws produced over a long period of time.

Learning Objectives:
- Understand that the gas laws are empirical and the kinetic theory model is theoretical.
- Know how knowledge and understanding of the behaviour of gases has changed over time.
- Know what Brownian motion is and how it provides evidence for the existence of atoms.

Specification Reference 3.6.2.3

Theories and Empirical laws

Empirical laws are based on observations and evidence. This means that they can predict what will happen but they don't explain why. For example, the gas laws (pages 53-56) and the ideal gas equation (page 58) are all based on observations of how a gas responds to changes in its environment. They were discovered by scientists making direct observations of the gases' properties and can be proven with simple experiments.

Kinetic theory (pages 61-64) is based on **theory**. This means it's based on assumptions and derivations from knowledge and theories we already had, and will both predict and explain why a change will occur.

Development of the gas laws

Our knowledge and understanding of gases has developed over thousands of years. The gas laws in this section (see pages 53-56) were developed by lots of different scientists.

Ancient Greek and Roman philosophers including Democritus had ideas about gases 2000 years ago, some of which were quite close to what we now know to be true.

Robert Boyle discovered the relationship between pressure and volume at a constant temperature in 1662 — this is Boyle's law (page 53). This was followed by Charles' law (page 54) in 1787 when Jacques Charles discovered that the volume of a gas is proportional to temperature at a constant pressure. The pressure law (page 55) was discovered by Guillaume Amontons in 1699, who noticed that at a constant volume, temperature is proportional to pressure. It was then re-discovered much later by Joseph Louis Gay-Lussac in 1809.

In the 18th century, a physicist called Daniel Bernoulli explained Boyle's Law by assuming that gases were made up of tiny particles — the beginnings of kinetic theory. But it took another couple of hundred years before kinetic theory became widely accepted.

Then in 1827 Robert Brown discovered Brownian motion (see next page), which helped support kinetic theory (pages 61-64).

Figure 1: *Mathematician and physicist Daniel Bernoulli.*

Acceptance of scientific ideas

You might have thought that when Bernoulli published his work on kinetic theory everyone would immediately agree with it. Not so. The scientific community only accepts new ideas when they can be independently **validated** — that is, other people can reach the same conclusions. Otherwise anyone could make up any old nonsense.

In the case of kinetic theory, most physicists thought it was just a useful hypothetical model and atoms didn't really exist. It wasn't until the 1900s, when Einstein was able to use kinetic theory to make predictions for Brownian motion, that atomic and kinetic theory became widely accepted.

Tip: Validation is the process of repeating an experiment done by someone else, using the theory to make new predictions, and then testing them with new experiments in order to support or refute the theory.

Brownian motion and kinetic theory

In 1827, botanist Robert Brown noticed that pollen grains in water moved with a zigzag, random motion. This type of movement of any particles suspended in a fluid is known as **Brownian motion**.

Albert Einstein later showed that Brownian motion supports the kinetic theory model of the different states of matter. He explained how the random motion of the pollen grains was a result of collisions with fast, randomly-moving particles in the fluid.

You can see this when large, heavy particles (e.g. smoke) are moved with Brownian motion by smaller, lighter particles (e.g. air) travelling at high speeds — it is why smoke particles in air appear to move around randomly when you observe them in the lab. This is evidence that the air is made up of tiny atoms or molecules moving really quickly.

Figure 2: *British botanist Robert Brown, after whom Brownian motion is named.*

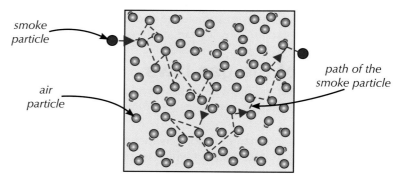

smoke particle

air particle

path of the smoke particle

Figure 3: *Brownian motion of a smoke particle in air.*

So Brownian motion really helped the idea that everything is made from atoms (i.e. **kinetic theory**) gain acceptance from the scientific community.

Practice Questions — Fact Recall

Q1 Give an example of both a theoretical and an empirical concept, and explain the difference between the two.

Q2 Describe how our understanding of the gas laws has developed over time, with reference to the discoveries made.

Q3 Briefly explain how Brownian motion helped kinetic theory gain acceptance in the scientific community.

Section Summary

Make sure you know...

- That temperature can be measured on the Kelvin (or absolute) scale.
- That absolute zero temperature is equal to 0 K or −273 °C.
- That the internal energy of a gas is the sum of the randomly distributed kinetic and potential energies of its particles.
- That the internal energy of a system is increased when energy is transferred to it (and vice versa)
- That (heat) energy can be transferred between substances by the collision of particles.
- What specific heat capacity is, and how to use the formula $Q = mc\Delta\theta$.
- How to do calculations including continuous flow, for example using a continuous flow calorimeter.
- That the potential energies of the particles in a substance change when the substance changes state but the kinetic energies stay the same.
- What is meant by specific latent heat and how to use $Q = ml$.
- Boyle's law, which states that for an ideal gas at constant temperature the pressure p and volume V of a gas are inversely proportional.
- Charles' law, which states that for an ideal gas at constant pressure the volume V is directly proportional to the absolute temperature T.
- The pressure law, which states that for an ideal gas at constant volume, the pressure p is directly proportional to its absolute temperature T.
- How to explain the relationships between p, V and T as demonstrated by the gas laws in terms of a simple molecular model.
- Experiments to investigate Boyle's law and Charles' law.
- What is meant by molecular mass and molar mass.
- That the Avogadro constant N_A is the number of atoms in exactly 12 g of the carbon isotope $^{12}_{6}C$.
- What the molar gas constant R and the Boltzmann constant k are.
- How to use the ideal gas equations for n moles and N molecules.
- How to calculate the work done on a gas.
- How to derive the equation for the pressure exerted by an ideal gas using algebra and conservation of momentum.
- Why increasing the temperature of a fixed mass of an ideal gas causes either its volume or the pressure inside its container (or both) to increase.
- The assumptions made about an ideal gas in kinetic theory.
- That the internal energy of an ideal gas is its kinetic energy.
- How to calculate the average kinetic energy of the molecules in an ideal gas using $\frac{1}{2}m(c_{rms})^2 = \frac{3}{2}kT = \frac{3RT}{2N_A}$.
- That the gas laws are empirical and kinetic theory is theoretical.
- How knowledge and understanding of the behaviour of gases has changed over time.
- What Brownian motion is, and how it provides evidence for the existence of atoms.

Exam-style Questions

1   Which of the following is not an assumption made about the motion of molecules in the kinetic theory model of an ideal gas?

   A   Except for during collisions, molecules always move in straight lines.

   B   The molecules continually move about randomly.

   C   The gas contains a small number of molecules.

   D   Collisions between molecules themselves or at the walls of a container are perfectly elastic.

   (1 mark)

2   State which of the following investigations would produce a straight line graph with a positive gradient when volume is plotted against temperature.

   A   An investigation into Charles' law.

   B   An investigation into the pressure law.

   C   An investigation into Boyle's law.

   D   None of the above, the graph is from a different investigation.

   (1 mark)

3   A sealed container with a fixed volume of 0.51 m³ is filled with an ideal gas at low temperature. The gas is heated so that the pressure inside the container increases. At 0.0 °C the pressure inside the container is 8.1×10^5 Pa.

   3.1   By considering the molecules of the gas, explain why increasing the temperature leads to an increase in pressure inside the container.

   (3 marks)

   3.2   Using the axes below, show how the pressure of the gas varies with temperature between −273 °C and 100 °C.

   (2 marks)

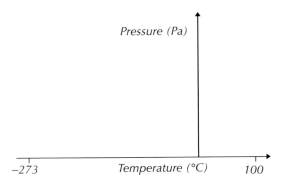

   3.3   Calculate the number of molecules of gas present in the container.

   (2 marks)

   3.4   The specific heat capacity of the gas is 2.2×10^3 J kg⁻¹K⁻¹ and the mass of a molecule of the gas is 2.7×10^{-26} kg. If the gas was heated from −150 °C to 0.0 °C, calculate the energy transferred to the gas.

   (2 marks)

4   A glass beaker contains a cylindrical block of ice at −25 °C. A heating element is placed inside the block of ice and turned on, supplying energy at a rate of 50.0 J s⁻¹. The ice has a mass of 92 g, a specific heat capacity of 2110 J kg⁻¹ K⁻¹ and a specific latent heat of fusion of 3.3×10^5 J kg⁻¹.

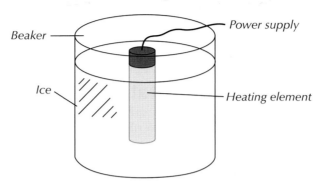

4.1   Calculate the energy needed to heat the ice from −25 °C to its melting point, 0.0 °C.

(1 mark)

4.2   Calculate the energy needed to melt all of the ice once it reaches 0.0 °C.

(1 mark)

4.3   Assuming the ice is heated uniformly and there is no heat transfer between the ice and the surroundings, calculate how long it will take for all of the ice to melt.

(2 marks)

4.4   If the experiment was carried out in a room with air temperature of 25 °C, state whether insulating the beaker would speed up or slow down the melting of the ice. Explain your answer.

(2 marks)

5   Briefly describe what is meant by the following terms.

5.1   Absolute zero temperature

(1 mark)

5.2   The Avogadro constant N_A

(1 mark)

5.3   Molar mass

(1 mark)

5.4   A sealed container with a fixed volume of 4.18 m³ is filled with 54.0 moles of an ideal gas.
Calculate the total number of molecules of gas in the container.

(1 mark)

5.5   Calculate the temperature of the gas if the pressure is 1.00×10^5 Pa.

(2 marks)

5.6   Calculate the total kinetic energy of the gas molecules at this temperature.

(3 marks)

Learning Objectives:

- Be able to define a force field.

- Know that force fields are present in the interaction between masses and static or moving charges.

- Understand that gravity acts as an attractive force between all matter.

- Know that force fields can be represented as vectors.

- Be able to represent gravitational fields with gravitational field lines.

- Know how to calculate the magnitude of the force due to gravity between two point masses using Newton's law of gravitation.

- Know that G is the gravitational constant.

- Know that the gravitational law is an inverse square law.

Specification References 3.7.1, 3.7.2.1 and 3.7.2.2

Tip: The smaller mass, m, has a gravitational field of its own. This doesn't have a noticeable effect on Earth though, because the Earth is so much more massive.

Tip: Any object with mass has a gravitational field (even you have a gravitational field).

1. Gravitational Fields

So far you've probably only considered forces acting at a specific point, with a specific cause (e.g. the pushing of a swing). Fields, on the other hand, are regions in which a charge or mass will experience a force.

What is a gravitational field?

Any object with mass will experience an attractive force if you put it in the gravitational field of another object. A gravitational field is a **force field**.

> A force field is a region in which a body experiences a non-contact force.

Force fields arise from interactions between objects or particles — e.g. between static or moving charges (p. 87), or in the case of gravity, between masses.

Only objects with a large mass, such as stars and planets, have gravitational fields that produce a significant effect. Smaller objects do still have gravitational fields that attract other masses, but the effect is too weak to detect without specialised equipment.

Representation of gravitational fields

Force fields can be represented as vectors, showing the direction of the force they would exert on an object placed in that field. Gravitational field lines, or "lines of force", are arrows showing the direction of the force that masses would feel in a gravitational field.

If you put a small mass, m, anywhere in the Earth's gravitational field, it will always be attracted towards the Earth. The Earth's gravitational field is radial — the lines of force meet at the centre of the Earth (see Figure 1).

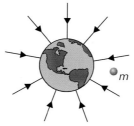

Figure 1: A small mass in Earth's gravitational field.

If you move mass m further away from the Earth — where the lines of force are further apart — the force it experiences decreases. The lines can be used to show the strength of the field at each point, where a higher line density shows a stronger gravitational field. Close to Earth's surface, the field is (almost) uniform — the field lines are (almost) parallel and equally spaced.

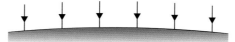

Figure 2: The gravitational field at Earth's surface is roughly uniform.

Newton's law of gravitation

The force experienced by an object in a gravitational field is always attractive. It depends on the masses involved and the distances between them. It's easy to work this out for **point masses** — or objects which behave as if all their mass is concentrated at the centre, e.g. uniform spheres. You just put the numbers into this equation, known as Newton's law of gravitation:

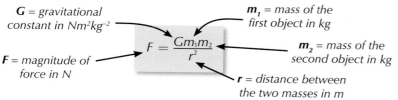

G = gravitational constant in Nm²kg⁻²

m_1 = mass of the first object in kg

$$F = \frac{Gm_1m_2}{r^2}$$

F = magnitude of force in N

m_2 = mass of the second object in kg

r = distance between the two masses in m

F is the <u>magnitude</u> of the force. Remember, force is a vector quantity — in this case, its direction is always towards the centre of the mass which is causing the gravitational force. G is the gravitational constant — 6.67×10^{-11} Nm²kg⁻². Don't get this confused with g, the gravitational field strength (see page 75). The force on m_1 due to m_2 is equal and opposite to the force on m_2 due to m_1.

We sometimes only consider the force acting on the smaller object because that's the one that experiences a greater acceleration — $a = \frac{F}{m}$, so as m becomes bigger, a becomes smaller. Also note that r is the distance between the centres of the objects, not the edges.

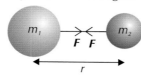

Figure 4: The forces acting on the two masses are equal but opposite.

Example — Maths Skills

Two planets have masses of 7.55×10^{24} kg and 9.04×10^{24} kg respectively. If the force due to gravity between the two planets is 6.69×10^{17} N, how far apart are the planets?

You want to find r, so start by rearranging Newton's law of gravitation:

$$F = \frac{Gm_1m_2}{r^2} \quad \Rightarrow \quad r^2 = \frac{Gm_1m_2}{F} \quad \Rightarrow \quad r = \sqrt{\frac{Gm_1m_2}{F}}$$

Then just put the numbers in:

$$r = \sqrt{\frac{Gm_1m_2}{F}} = \sqrt{\frac{(6.67 \times 10^{-11}) \times (7.55 \times 10^{24}) \times (9.04 \times 10^{24})}{6.69 \times 10^{17}}}$$

$$= 8.2491... \times 10^{10} \text{ m}$$
$$= 82.5 \text{ million km (to 3 s.f.)}$$

Inverse square laws

The law of gravitation is an **inverse square law**: $F \propto \dfrac{1}{r^2}$

This means if the distance r between the masses increases, then the force F will decrease. Because it's r^2 and not just r, if the distance doubles then the force will be one quarter the strength of the original force.

The law of gravitation is an inverse square law because it is radial. The force on any point a distance r from the mass will be the same — if you draw an imaginary sphere with radius r (see Figure 5) then the force will be the same at any point on its surface.

Figure 3: The gravitational fields of the Moon and the Sun are noticeable here on Earth — they're the main causes of our tides.

Tip: m_1 is usually the larger mass, but it doesn't really matter which way you label them.

Tip: This is why we don't notice Earth's acceleration towards us when we're falling to the ground.

Exam Tip
An exam question might ask you to work out the mass of one of the objects first. E.g. you might need to use density = $\dfrac{\text{mass}}{\text{volume}}$.

Tip: The $\propto$ symbol means "is proportional to".

If you double the distance from a mass, the area on the surface of the imaginary sphere of equal force covered by a particular group of radial field lines will be 4 times greater — see Figure 5.

At a distance r, a group of radial field lines passes through a square area with a side length X and an area A = X².

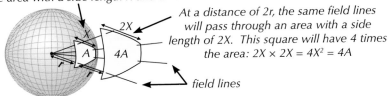

At a distance of 2r, the same field lines will pass through an area with a side length of 2X. This square will have 4 times the area: 2X × 2X = 4X² = 4A

field lines

Tip: The shapes don't look like squares because they've been stretched across the sphere's surface.

Figure 5: *Radial gravitational field lines at a distance of r and 2r from a mass.*

The spread of field lines indicates the strength of a force felt in that field, so here the force will be 4 times weaker at a distance of $2r$ than at r for a given mass. Knowing that $F \propto \dfrac{1}{r^2}$ can come in handy when answering questions.

┌─ **Example** ──[Maths Skills]──────────────────────────

The gravitational force between two objects 10 m apart (to 2 s.f.) is 0.291 N. What will the gravitational force between them be if they move to 25 m apart?

25 m is $\dfrac{25}{10}$ = 2.5 times larger than 10 m. So to find the new gravitational force, divide 0.291 N by 2.5² (because of the inverse square law):

$$\frac{0.291}{2.5^2} = 4.656 \times 10^{-2} \text{ N} = 4.7 \times 10^{-2} \text{ N (to 2 s.f.)}$$

Practice Questions — Application

Q1 Two stars are orbiting each other with a constant separation of 100 million km. If their masses are 2.15×10^{30} kg and 2.91×10^{30} kg, show that the magnitude of the force they are exerting on each other is approximately 4.2×10^{28} N.

Q2 Two asteroids 2.5 km apart exert a gravitational force on each other of 25 N. Calculate the magnitude of the force they will exert on each other when they're 0.5 km apart.

Q3 An aircraft with a mass of 2500 kg is hovering 10 000 m above ground.
  a) If the radius of Earth is 6370 km, how far is the aircraft from the centre of the Earth?
  b) How much upwards force must the aircraft engine be providing to be able to keep it hovering at a constant altitude? The mass of Earth is 5.98×10^{24} kg.

Tip: Don't forget to convert any distance measurements into metres before using Newton's law of gravitation.

Tip: Remember, you have to assume that the Earth is a point mass, so r is the distance of the aircraft from the Earth's centre.

Practice Questions — Fact Recall

Q1 What type of object will experience an attractive force due to gravity?

Q2 What's a gravitational field?

Q3 Draw a diagram showing the gravitational field lines of Earth:
  a) Looking at Earth from a distance.
  b) At Earth's surface.

Q4 a) Give the formula for Newton's law of gravitation.
  b) What does G represent in this formula?
  c) What must be true of the masses in this formula?

Section 3 — Gravitational and Electric Fields

2. Gravitational Field Strength

It's no use being able to draw gravitational field lines if you can't use them to work anything out. The gravitational field strength, g, tells you how strong the force due to gravity is at any point in a gravitational field.

The gravitational field strength, *g*

Gravitational field strength, g, is the force per unit mass. Its value depends on where you are in the field. There's a really simple equation for working it out:

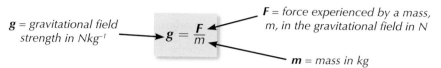

$$g = \frac{F}{m}$$

g = gravitational field strength in Nkg⁻¹

F = force experienced by a mass, m, in the gravitational field in N

m = mass in kg

g is a vector quantity — it has a magnitude and a direction. It's always pointing towards the centre of mass of the object whose field you're describing. Since the gravitational field is almost uniform at the Earth's surface, you can assume **g** is constant near the Earth's surface.

The value of **g** at the Earth's surface is approximately 9.81 Nkg⁻¹

g can also be seen as the acceleration of a mass in a gravitational field. It's often called the acceleration due to gravity. On Earth, this is approximately 9.81 ms⁻².

Tip: From **F** = m**a** you can get **a** = **F**/m, which has the units Nkg⁻¹ — just another way of measuring acceleration.

Example — **Maths Skills**

An 80.0 kg astronaut feels a force of 130.0 N due to gravity on the Moon. What's the value of g on the moon?

Just put the numbers into the formula:

$$g = \frac{F}{m} = \frac{130}{80} = 1.625 = 1.63 \, \text{Nkg}^{-1} \text{ (to 3 s.f.)}$$

Radial fields

Point masses have radial gravitational fields (see page 72). The magnitude of g depends on the distance *r* from the point mass *M*.

G = gravitational constant in Nm²kg⁻²

M = mass of object creating the gravitational field in kg

$$g = \frac{GM}{r^2}$$

g = magnitude of gravitational field strength in Nkg⁻¹

r = distance from the point mass in m

You can derive this formula by looking at Newton's law of gravitation (p. 73):

Start with $F = \frac{Gm_1 m_2}{r^2}$. Replace m_1 with M and m_2 with m:

$$F = \frac{GMm}{r^2}$$

Then substitute this into $g = \frac{F}{m}$, cancelling down where possible:

$$g = \frac{F}{m} = \frac{\left(\frac{GMm}{r^2}\right)}{m} = \frac{GM}{r^2}$$

Tip: If you're not dealing with a point mass, for example if you're dealing with a planet, r is the distance from the centre of the planet with mass M.

Example — **Maths Skills**

The mass of the Earth is 5.98×10^{24} kg and its radius is 6.37×10^6 m. Find the value of g at the Earth's surface.

Just put the numbers into the equation:

$$g = \frac{GM}{r^2} = \frac{(6.67 \times 10^{-11}) \times (5.98 \times 10^{24})}{(6.37 \times 10^6)^2}$$

$$= 9.8298...$$
$$= 9.83 \text{ Nkg}^{-1} \text{ (3 s.f.)}$$

Tip: As you probably know by now, we usually use 9.81 Nkg^{-1} as the value of g — the difference here is just because of the number of significant figures given in the question.

Figure 1: *The value of g on top of a mountain is slightly lower than at sea level.*

This is another case of the inverse square law (page 73) — as r doubles, g decreases to a quarter of its original value.

If you plot a graph of g against r for the Earth, you get a curve like this:

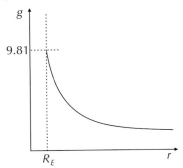

Figure 2: *Graph showing the relationship between g and r for Earth.*

It shows that g is greatest at the surface of the Earth (R_E), but decreases rapidly as r increases and you move further from the centre of the Earth. The area under this curve can be used to find the gravitational potential, V (p. 78).

Combining gravitational field strengths

Gravitational fields are vector fields, which means you can add them up to find the combined effect of more than one object. And remember, vector fields means the direction matters.

Example — **Maths Skills**

What's the value of the gravitational field strength at the point P below?

Tip: Remember, an object's gravitational field points towards the object — gravity's an attractive force.

Take the right direction to be positive and the left direction to be negative, then add up the effect of each sphere:

$$g = \sum \frac{GM}{r^2} = \frac{G(4m)}{(2x)^2} - \frac{Gm}{x^2}$$

$$= \frac{4}{4}\frac{Gm}{x^2} - \frac{Gm}{x^2}$$

$$= 0 \text{ Nkg}^{-1}$$

Practice Questions — Application

Q1 A 105 kg object experiences an attractive force due to gravity of 581 N. What's the gravitational field strength?

Q2 Why would an astronaut find it easier to pick up a rock with a mass of 20 kg on the Moon than a rock with a mass of 20 kg on Earth?

Q3 Mars has a radius of 3390 km and a mass of 6.42×10^{23} kg. Calculate the magnitude of the gravitational field strength at the surface of Mars.

Q4 a) A person standing on the surface of the Moon experiences a gravitational force of magnitude 105 N. If the mass of the Moon is 7.34×10^{22} kg and the mass of the person is 65 kg, what is the radius of the Moon?

   b) What is the magnitude of the gravitational field strength 640 km above the surface of the Moon?

Q5 Find the gravitational field strength at point P in the diagram below.

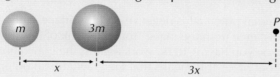

Tip: The mass of an object is always the same wherever it is.

Tip: In some questions you'll need to give the final answer in terms of variables (e.g. m), so don't panic if you can't work out how to get rid of all the letters in your answer.

Practice Questions — Fact Recall

Q1 Other than the acceleration due to gravity, how is g defined and what are its units?

Q2 What does M represent in the formula for gravitational field strength in a radial field?

Q3 Which of the following shows how gravitational field strength changes with distance r?

a)
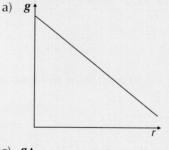

b)
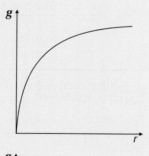

c)
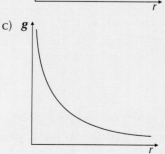

d)
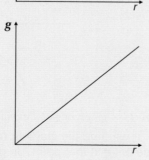

3. Gravitational Potential

Learning Objectives:

- Know what's meant by gravitational potential, including why it is zero at infinity.
- Be able to calculate gravitational potential using the equation $V = -\frac{GM}{r}$ and understand the significance of the negative sign.
- Be able to sketch the graph of V against r in a radial gravitational field.
- Be able to calculate the value of g at a given point using the gradient of a graph of V against r and the equation $g = -\frac{\Delta V}{\Delta r}$.
- Know and use the fact that ΔV can be found from the area under the graph of g against r.
- Know what gravitational potential difference is.
- Know that the work done to move a mass is given by $\Delta W = m\Delta V$.
- Understand what equipotential surfaces are.

Specification Reference 3.7.2.3

All objects in a gravitational field have a gravitational potential that increases the further they are from the centre of the field. It can be hard to get your head round at first, but try not to get it confused with gravitational potential energy.

What is gravitational potential?

The **gravitational potential**, V, at a point is the gravitational potential energy that a unit mass at that point would have. For example, if a 1 kg mass has -10 J of potential energy at a point Z, then the gravitational potential at Z is -10 Jkg^{-1}. In a radial field (like the Earth's), the equation for gravitational potential is:

G = gravitational constant in Nm²kg⁻²

V = gravitational potential in Jkg⁻¹

$$V = -\frac{GM}{r}$$

M = mass of the object causing the gravitational field in kg

r = distance from the centre of the object in m

Gravitational potential is negative on the surface of the mass and increases with distance from the mass. You can think of this negative energy as being caused by you having to do work against the gravitational field to move an object out of it. This means that the gravitational potential at an infinite distance from the mass will be zero. **Gravitational potential energy** is also negative — you might have worked out positive values in the past (from mgh), but this is just the gain in potential energy. Potential energy becomes less negative as the object moves upwards.

Example — Maths Skills

Find the gravitational potential V at the surface of the Earth.
The Earth's mass is 5.98×10^{24} kg and its radius is 6.37×10^{6} m.

Just put the numbers into the equation:

$$V = -\frac{GM}{r} = -\frac{(6.67 \times 10^{-11}) \times (5.98 \times 10^{24})}{6.37 \times 10^{6}}$$
$$= -6.2616... \times 10^{7}$$
$$= -6.26 \times 10^{7} \text{ Jkg}^{-1} \text{ (to 3 s.f.) } (= -62.6 \text{ MJkg}^{-1})$$

The gravitational potential can be plotted against the distance from the centre of the object, as shown in Figure 1.

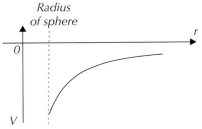

Figure 1: *A graph of gravitational potential against distance for a sphere.*

If you find the gradient of this graph at a particular point, you get the value of $-g$ at that point. In other words:

g = gravitational field strength in Nkg⁻¹

$$g = -\frac{\Delta V}{\Delta r}$$

ΔV = change in gravitational potential in Jkg⁻¹

Δr = change in distance from the centre of the object in m

Tip: This means a 1 kg mass needs 62.6 MJ of energy to be able to fully escape the Earth's gravitational pull.

Tip: Note that V is proportional to $1/r$, not $1/r^2$ like the gravitational field strength.

Example — Maths Skills

The graph below shows the gravitational potential V against the distance r from the centre of a planet. Find the gravitational field strength g at $r = 15 \times 10^6$ m.

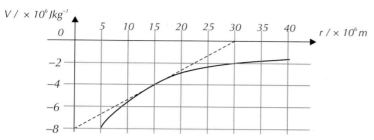

The gravitational field strength is given by the gradient:

$$g = -\frac{\Delta V}{\Delta r} = -\frac{0-(-8 \times 10^6)}{(30 \times 10^6)-0} = -0.2666 = -0.3 \, \text{Nkg}^{-1} \text{ (to 1 s.f.)}$$

Tip: g is negative because it points 'downwards' towards the centre of the planet.

You can also find the area under a g-r graph to give you ΔV, the change in gravitational potential between two radial distances.

Example — Maths Skills

The graph shows the magnitude of the gravitational field strength g of a planet as a function of the distance r away from the centre of the planet. The radius of the planet is approximately 2×10^6 m. Use the graph to estimate the change in gravitational potential between the surface of the planet and a point 3×10^6 m above the planet's surface.

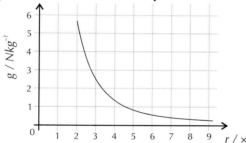

Tip: Rearranging the equation of $g = -\frac{\Delta V}{\Delta r}$ gives $-\Delta V = g\Delta r$. From this, you can see that the area under a g–r graph will give the value of ΔV. (The minus sign can be ignored because you are looking at the magnitude of g and not the vector g.)

The change in gravitational potential is equal to the area under the graph. The question is asking for an estimate, so the best way to do this is to count the number of squares. 3×10^6 m above the surface is equal to 5×10^6 m from the planet's centre.

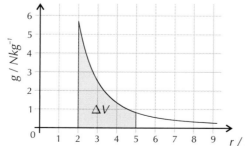

The total number of squares under the graph between 2×10^6 m and 5×10^6 m is approximately 6.5.
So the change in gravitational potential $\approx 6.5 \times 10^6 \, \text{Jkg}^{-1}$.

Tip: The number of squares is multiplied by 10^6 because each square going across is equal to 1×10^6 m.

Gravitational potential difference

Tip: 'Change in gravitational potential' is the difference in gravitational potential between two points in the gravitational field of an object, where zero gravitational potential is at infinity. 'Gravitational potential difference' is the <u>energy needed</u> to move a unit mass between two points in a gravitational field. Both are denoted by ΔV and are pretty much the same thing — they're just defined slightly differently.

Gravitational potential difference is the energy needed to move a unit mass. Two points at different distances from a mass will have different gravitational potentials (because gravitational potential increases with distance) — this means that there is a gravitational potential difference between these two points.

When you move an object you do work against gravity — the amount of energy you need depends on the mass of the object and the gravitational potential difference you move it through:

ΔW = work done in J —→ $$\Delta W = m\Delta V$$ ←— ΔV = gravitational potential difference in Jkg^{-1}

m = mass of the object in kg

You can also use this relationship to describe an object's (change in) gravitational potential energy (E_p) at a given gravitational potential:

E_p = gravitational potential energy in J —→ $$(\Delta)E_p = m(\Delta)V$$ ←— V = gravitational potential in Jkg^{-1}

m = mass of the object in kg

Substituting in the formula for V from page 78, the gravitational potential energy of an object of mass m is:

$$E_p = m\left(-\frac{GM}{r}\right) = -\frac{GMm}{r}$$

Tip: In physics, doing work means using a force to transfer energy from one type to another. For example, if you drop a ball from a height, gravitational potential energy is converted into kinetic energy.

┌─ **Example** ── **Maths Skills** ─────────────

Show that the energy needed to move a mass m through a gravitational potential difference of ΔV can be derived from the equation for gravitational field strength.

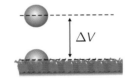

The gravitational field strength is given by:

g in a uniform field —→ $$g = -\frac{\Delta V}{\Delta r} = \frac{F}{m}$$ ←— g defined as force per unit mass (page 75).

Which rearranges to give:

$$m\Delta V = -F\Delta r$$

Using the definition for work done as force × distance moved, you can write:

$$\text{work done} = -F\Delta r = m\Delta V$$

So the energy needed to move a mass m against a gravitational potential difference is the same as the work done, which is given by $m\Delta V$.

Tip: F is negative because you're working against gravity.

┌─ **Example** ── **Maths Skills** ─────────────

A forklift truck does 2120 J of work to increase a pig's gravitational potential by 26.5 Jkg^{-1}. What's the mass of the pig?

You're given the work done and the change in gravitational potential and you need to find the mass, so rearrange the formula at the top of this page to make mass the subject:

$$\Delta W = m\Delta V \Rightarrow m = \frac{\Delta W}{\Delta V} = \frac{2120}{26.5} = 80 \,\text{kg}$$

Figure 2: *Rollercoasters convert gravitational potential energy to kinetic energy and back.*

Equipotentials

Equipotentials are lines (in 2D) and surfaces (in 3D) that join together all of the points with the same gravitational potential, V. This means that as you travel along an equipotential, your potential doesn't change — you don't lose or gain energy. So as you're moving along the equipotential, the gravitational potential difference is zero: $\Delta V = 0$. As $\Delta W = m\Delta V$, this means that the amount of work done is also zero.

For a uniform spherical mass, the equipotentials are spherical surfaces. The equipotentials and the field lines are always perpendicular. Figure 3 shows the equipotentials and field lines around Earth. This is a 2D image of a 3D shape though — in reality, the equipotentials would look like spherical shells that go all the way around the planet.

Tip: A satellite travelling in a circular orbit is an example of an object which travels along an equipotential surface. It stays at the same distance from the Earth's centre, and so no work is done in keeping it in orbit.

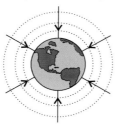

Figure 3: *Dotted lines indicate equipotentials for the gravitational potential of Earth.*

Practice Questions — Application

Q1 A satellite is orbiting Earth. What's the effect (if any) on the following values of halving the satellite's orbital radius? Explain your answers.

    a) G b) V c) g d) m

Q2 A 1.72 kg brick is dropped off the side of a cliff. If its gravitational potential changes by 531 Jkg^{-1}, how much work is done by gravity on the brick?

Q3 The graph below shows the magnitude of the gravitational field strength of a planet against the radial distance from the centre of the planet. Use the graph to estimate the change in gravitational potential between 2×10^6 m and 5×10^6 m.

Practice Questions — Fact Recall

Q1 What is gravitational potential, V? What are its units?

Q2 What is the significance of the negative sign in the equation $V = -\dfrac{GM}{r}$?

Q3 Sketch a graph of the gravitational potential (V) against distance (r) for an object in Earth's gravitational field.

Q4 What does the gradient of a graph of V against r tell you?

Q5 What does ΔW represent in the equation $\Delta W = m\Delta V$?

Q6 What are equipotentials?

Tip: Make sure your graph for Q3 starts at the radius of Earth and not at its centre.

4. Orbits

You saw circular motion in the mechanics section, and here it's covered again for objects in gravitational fields. Gravity provides the centripetal force that keeps objects in orbit around a much larger body.

Satellites

Planets and **satellites** are kept in orbit by gravitational forces. A satellite is just any smaller mass which orbits a much larger mass — the Moon is a satellite of the Earth, planets are satellites of the Sun, etc.

Satellites are kept in orbit by the gravitational 'pull' of the mass they're orbiting. In our Solar System, the planets have nearly circular orbits, so you can use the equations of circular motion (pages 19-25) to investigate their **orbital speed** and **orbital period** (see below).

Orbital period and speed

Any object undergoing circular motion (e.g. a satellite) is kept in its path by a centripetal force. What causes this force depends on the object — in the case of satellites it's the gravitational attraction of the mass they're orbiting. This means that in this case the centripetal force is the gravitational force.

You saw on page 24 that the magnitude of the force acting on an object in circular motion is given by:

$$F = \frac{mv^2}{r}$$

And on page 73, you saw that the magnitude of the force of attraction due to gravity between two objects is given by:

$$F = \frac{GMm}{r^2}$$

Which means you can set the two expressions to be equal to each other and rearrange to find the orbital speed, v, of a satellite in a gravitational field:

$$\frac{mv^2}{r} = \frac{GMm}{r^2} \quad \Rightarrow \quad v^2 = \frac{GM\cancel{m}r}{r^2\cancel{m}} \quad \Rightarrow \quad v = \sqrt{\frac{GM}{r}}$$

So the orbital speed of a satellite is inversely proportional to the square root of its orbital radius, or $v \propto \frac{1}{\sqrt{r}}$.

The time taken for a satellite to make one orbit is called the orbital period, T. Remember, speed $= \frac{\text{distance}}{\text{time}}$, and the distance for a circular orbit is $2\pi r$, so $v = \frac{2\pi r}{T}$:

$$v = \frac{2\pi r}{T} \quad \Rightarrow \quad T = \frac{2\pi r}{v}$$

Then substitute the expression for v found above and rearrange:

$$T = \frac{2\pi r}{v} = \frac{2\pi r}{\left(\sqrt{\frac{GM}{r}}\right)} = \frac{2\pi r\sqrt{r}}{\sqrt{GM}} = \sqrt{\frac{4\pi^2 r^3}{GM}}$$

This shows that the period of an orbit squared is proportional to the radius cubed:

$$T^2 \propto r^3 \qquad \text{or} \qquad T \propto \sqrt{r^3}$$

The greater the radius of a satellite's orbit, the slower it will travel and the longer it will take to complete one orbit.

Tip: Need a reminder on circular motion? Flick back to pages 19-25.

Tip: Here the square root was taken of everything to make the expression neater. That's why anything that wasn't already rooted got squared (e.g. 2 → 4).

Example — Maths Skills

The Moon takes 27.3 days to orbit the Earth. Calculate its distance from the Earth. Take the mass of the Earth to be 5.98×10^{24} kg.

You're trying to find the radius of the orbit, r. Use the formula for period, T, and rearrange for r^3:

$$T = \sqrt{\frac{4\pi^2 r^3}{GM}} \Rightarrow T^2 = \frac{4\pi^2 r^3}{GM} \Rightarrow r^3 = \frac{T^2 GM}{4\pi^2}$$

Then just put the numbers in:

$$r^3 = \frac{(2.35... \times 10^6)^2 \times (6.67 \times 10^{-11}) \times (5.98 \times 10^{24})}{4\pi^2}$$

$$= 5.6210... \times 10^{25}\,\text{m}^3$$
$$r = \sqrt[3]{5.6210... \times 10^{25}}$$
$$= 3.8306... \times 10^8\,\text{m} = 3.83 \times 10^5\,\text{km} \ \ (\text{to 3 s.f.})$$

Tip: This is the distance from the <u>centre</u> of the Earth to the <u>centre</u> of the Moon.

Tip: T is given in days, so you need to convert it to seconds first. 27.3 days is $2.35... \times 10^6$ s

Tip: Just take the cube root of r^3 here — you'll probably have a button for it on your calculator ($\sqrt[3]{\square}$).

Example — Maths Skills

Planets A and B are orbiting the same star. Planet A has an orbital radius of 8.0×10^{10} m and a period of 18 hours. Planet B has an orbital radius of 1.0×10^{12} m. Calculate the orbital period of planet B in hours.

$T^2 \propto r^3$, so $\dfrac{T^2}{r^3} = \text{constant}$

Therefore $\dfrac{T_A^2}{r_A^3} = \dfrac{T_B^2}{r_B^3}$ and so $T_B^2 = \dfrac{T_A^2 r_B^3}{r_A^3}$

$$T_B = \sqrt{\frac{T_A^2 r_B^3}{r_A^3}} = \sqrt{\frac{(18)^2 \times (1.0 \times 10^{12})^3}{(8.0 \times 10^{10})^3}}$$

$$= \sqrt{632\,812.5} = 795.495...$$
$$= 800 \text{ hours (to 2 s.f.)}$$

Tip: You can leave the orbital period of A in hours in your calculation, as you're asked to give your answer in hours too.

Graphs of orbital period

Sketching T against r is not always useful, as the numbers involved are so big and can vary a lot between planets. Astronomers often use logarithmic scales to plot T against r as a way of comparing data for different planets. For example, it would be very difficult to plot the graph in Figure 1 if it wasn't on a logarithmic scale.

Tip: Logarithmic scales don't count up in equal steps like normal graphs do. If you're a bit unsure about them, they are covered in more detail on page 410.

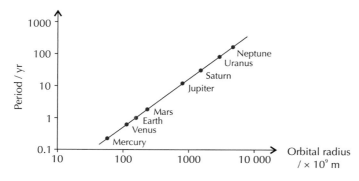

Figure 1: *Graph showing the relation between T and r for the planets in the solar system on a logarithmic scale.*

Tip: Figure 1 assumes that the planets have a circular orbit. In reality their orbits are slightly elliptical, and so the orbital radius is not constant.

Tip: An elliptical orbit is like a squashed circle.

Notice that the object the satellite is orbiting isn't at the centre of the ellipse, it's over to one side. This always happens in an elliptical orbit.

Tip: Remember, gravitational potential energy, $E_p = -\dfrac{GMm}{r}$.

Tip: Saturn is much further from the Sun than the Earth is, so the orbital period of Saturn is a lot longer than 1 Earth year.

Kinetic and potential energy of satellites

An orbiting satellite has kinetic and potential energy — its total energy (i.e. kinetic and potential) is always constant. In a circular orbit, a satellite's orbital speed and distance above the mass it's orbiting are constant. This means that its kinetic energy and potential energy are also both constant.

In an elliptical orbit, a satellite will speed up as its orbital radius decreases (and slow down as its orbital radius increases). This means that its kinetic energy increases as its potential energy decreases (and vice versa), so the total energy remains constant. This matches the description on page 82 — the greater the orbital radius, the slower the orbital speed (and the smaller the orbital radius, the faster the orbital speed).

Examples — Maths Skills

a) Saturn is travelling in an elliptical orbit and has potential energy -5.57×10^{34} J when it is at the point closest to the Sun. The mass of Saturn is 5.68×10^{26} kg and the mass of the Sun is 1.99×10^{30} kg. Calculate the distance between the Sun and Saturn when they're closest to each other.

$E_p = -\dfrac{GMm}{r}$, so rearrange for r:

$r = -\dfrac{GMm}{E_P}$

$= -\dfrac{(6.67 \times 10^{-11}) \times (1.99 \times 10^{30}) \times (5.68 \times 10^{26})}{(-5.57 \times 10^{34})}$

$= 1.3535... \times 10^{12}$

$= 1.35 \times 10^{12}$ m (to 3 s.f.)

b) After 1 Earth year, Saturn has moved to a different point on its orbital path. Will the speed of Saturn be higher or lower than in part a)?

After 1 Earth year, Saturn won't have completed a full orbit, and so won't yet be back at its closest point to the Sun. When Saturn is at its closest point to the Sun, it has its maximum speed. Therefore its speed must be lower when it is at any other point on the orbital path — i.e. its speed after 1 Earth year is lower than in part a).

Escape velocity

The **escape velocity** is the minimum speed an unpowered object needs in order to leave the gravitational field of a planet and not fall back towards the planet due to gravitational attraction.

To derive the equation for escape velocity, you need to think about the energies involved. At an infinite distance (i.e. the point at which an object has escaped the gravitational field of the planet), the gravitational potential energy of the object is zero. At the surface of the planet, the gravitational potential energy is negative (see page 78).

The increase in the gravitational potential energy of the object (from the surface to infinity) comes from the initial kinetic energy of the object. If the object is given just enough kinetic energy to escape the gravitational field of the planet, you know from conservation of energy that all of its initial kinetic energy will be converted into gravitational potential energy.

So kinetic energy lost = gravitational potential energy gained, and so:

$$\tfrac{1}{2}mv^2 = \dfrac{GMm}{r}$$

Cancelling out *m* and rearranging for *v* gives the escape velocity:

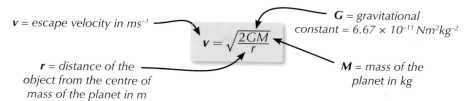

v = escape velocity in ms⁻¹

$$v = \sqrt{\frac{2GM}{r}}$$

G = gravitational constant = 6.67×10^{-11} Nm²kg⁻²

r = distance of the object from the centre of mass of the planet in m

M = mass of the planet in kg

Be careful with the value of *r* in this equation — it's the radial distance from the centre of the planet to the object. So if the object is initially at the planet's surface, then *r* is just the planet's radius. But if the object is initially in orbit, then *r* will be the planet's radius plus the orbital distance.

> **Tip:** The escape velocity equation is not dependent on the mass of the object. So, whether you were throwing a tennis ball or projecting a double decker bus into the air, they'd need the same velocity to escape the Earth's gravitational field (assuming there was no air resistance).

Example — Maths Skills

Calculate the escape velocity of the Earth.

Earth's mass = 5.98×10^{24} kg and Earth's radius = 6.37×10^{6} m.

Substitute these values into the equation for escape velocity:

$$v = \sqrt{\frac{2GM}{r}}$$

$$= \sqrt{\frac{2 \times (6.67 \times 10^{-11}) \times (5.98 \times 10^{24})}{(6.37 \times 10^{6})}}$$

$$= 1.1190... \times 10^{4}$$

$$= 1.12 \times 10^{4} \text{ ms}^{-1} \text{ (to 3 s.f.) (or 11.2 kms}^{-1}\text{)}$$

> **Tip:** The direction of the escape velocity is not important. Although it's called the escape <u>velocity</u>, it's actually a speed (confusing, I know).

Geostationary and low orbiting satellites

A **synchronous orbit** is when an orbiting object has an orbital period equal to the rotational period of the object it is orbiting. **Geostationary satellites** are examples of this — they're always above the same point on Earth. To achieve this their orbit must be in the plane of the equator, so they will always be directly above the equator. Geostationary satellites travel at the same angular speed as the Earth turns below them and in the same direction (west to east). Their orbit takes approximately 24 hours and their orbital radius is about 42 000 km, which is about 36 000 km above the surface of Earth.

These satellites are really useful for sending TV and telephone signals — the satellite is stationary relative to the Earth's surface, so you don't have to alter the angle of your receiver (or transmitter) to keep up.

Low orbiting satellites are defined as any satellites which orbit between 180 and 2000 km above Earth. Satellites designed for low Earth orbits are cheaper to launch and require less powerful transmitters as they're closer, which makes them useful for communications. However, their proximity to Earth and high orbital speed (in comparison to Earth's) means you need multiple satellites working together to maintain constant coverage.

Low orbit satellites are close enough to see the Earth's surface in a high level of detail. Imaging satellites are usually placed in this type of orbit and are used for things like imaging (e.g. mapping and spying) and monitoring the weather. Their orbits usually lie in a plane that includes the north and south pole. As the planet and the satellite rotate at different angular speeds, the satellite doesn't stay over the same part of the Earth, and so the whole of the surface can be scanned.

> **Tip:** A synchronous orbit of Earth would take approximately 24 hours.

Figure 3: *Syncom 2, the first communications satellite to have a synchronous orbit round Earth.*

> **Tip:** The International Space Station is an example of a low orbiting satellite.

Practice Questions — Application

Q1 A satellite with orbital radius r has orbital period T. The satellite is moved so that its orbital radius is halved. What is its new orbital period in terms of T?

Q2 A planet has two satellites. The satellite closest to the planet has orbital period $T = 42$ hours and orbital radius $r = 3.95 \times 10^5$ km. The satellite furthest from the planet has orbital radius $r = 4.84 \times 10^5$ km. Calculate the orbital period of the satellite furthest from the planet. Give your answer in hours.

Q3 Pluto (mass $= 1.31 \times 10^{22}$ kg) has an elliptical orbit around the Sun. It has gravitational potential energy -3.9×10^{29} J and kinetic energy 2.4×10^{29} J when it is travelling closest to the Sun.

   a) Calculate Pluto's velocity when it is closest to the Sun.

   b) When Pluto is at a different distance from the Sun, it has gravitational potential energy -3.2×10^{29} J. What is its kinetic energy at this point?

Q4 If the mass of the Moon is 7.34×10^{22} kg and the radius of the Moon is 1.74×10^6 m, what is the minimum velocity with which an astronaut must project a piece of Moon rock so that it is able to escape the Moon's gravitational field?

Practice Questions — Fact Recall

Q1 Define a satellite.

Q2 How is the orbital speed of a satellite related to the radius of its orbit?

Q3 How is the orbital period of a satellite related to the radius of its orbit?

Q4 When plotting a graph of period, T, against orbital radius, r, what is the best scale to use, and why?

Q5 In terms of energy, why does a satellite in an elliptical orbit move faster when its height, r, is small and slower when its height is large?

Q6 Define escape velocity.

Q7 What is a geostationary satellite and why are they useful for transmitting TV and telephone signals?

Q8 What is a low orbiting satellite and what are they used for?

Tip: Unless a question says otherwise, you can assume that the orbit of a satellite is circular.

5. Electric Fields

Electric fields are a lot like gravitational fields — you might think you're getting déjà vu over the next few pages. But although the concepts are similar, there are still some subtle differences, so make sure you follow carefully.

Electric fields around charged objects

Electric fields can be attractive or repulsive, so they're different from gravitational ones (which are always attractive). It's all to do with charge. Any object with charge has an electric field around it — the region where it can attract or repel other charges.

Electric charge, Q, is measured in coulombs (C) and can be positive or negative. Oppositely charged particles attract each other, and like charges repel each other. If a charged object is placed in an electric field, then it will experience a force.

Coulomb's law

You can calculate the force on a charged object in an electric field using Coulomb's law. It gives the force of attraction or repulsion between two **point charges**, Q_1 and Q_2, in a vacuum:

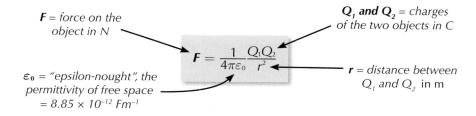

$$F = \frac{1}{4\pi\varepsilon_0}\frac{Q_1 Q_2}{r^2}$$

F = force on the object in N

Q_1 and Q_2 = charges of the two objects in C

r = distance between Q_1 and Q_2 in m

ε_0 = "epsilon-nought", the permittivity of free space = 8.85×10^{-12} Fm^{-1}

The force on Q_1 is always equal and opposite to the force on Q_2 — the direction depends on the charges.

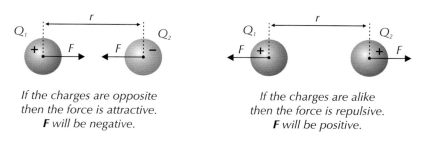

If the charges are opposite then the force is attractive. **F** will be negative.

If the charges are alike then the force is repulsive. **F** will be positive.

Figure 1: *The direction of the forces on two charged objects.*

Coulomb's law is another case of an inverse square law, so $F \propto \frac{1}{r^2}$ (page 73). The further apart the charges, the weaker the force between them. If the point charges aren't in a vacuum, then the size of the force **F** also depends on the permittivity, ε, of the material between them. Air can be treated as a vacuum when using Coulomb's law — so if a question is asking you about force between charges in air, just keep using ε_0 in your equations.

Tip: ε_o is a constant and its units are farads per metre (Fm^{-1}).

Tip: If the electric field is not in air or a vacuum, you'd replace ε_o in Coulomb's law with the ε of the material.

- **Example** ── Maths Skills ──

Find the acceleration experienced by a free electron 2.83 mm from the centre of a 1 mm sphere carrying a charge of +0.510 μC. The charge of an electron is −1.60 × 10⁻¹⁹ C and the mass of an electron is 9.11 × 10⁻³¹ kg.

The acceleration of an object is given by $F = ma \Rightarrow a = \frac{F}{m}$.

You know the mass, so find the force using Coulomb's law:

$$F = \frac{1}{4\pi\varepsilon_0}\frac{Q_1 Q_2}{r^2} = \frac{1}{4\pi\varepsilon_0}\frac{(0.510 \times 10^{-6}) \times (-1.60 \times 10^{-19})}{(2.83 \times 10^{-3})^2}$$
$$= -9.161... \times 10^{-11} \text{ N}$$

Then use this to find the acceleration:

$$a = \frac{F}{m} = \frac{-9.161... \times 10^{-11}}{9.11 \times 10^{-31}} = -1.00564... \times 10^{20}$$
$$= -1.01 \times 10^{20} \text{ ms}^{-2} \text{ (to 3 s.f.)}$$

Tip: The acceleration is negative because it's towards the sphere.

Electric field strength

Electric field strength, E, is defined as the force per unit positive charge. It's the force that a charge of +1 C would experience if it was placed in an electric field.

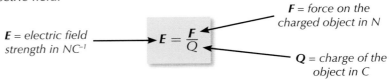

E = electric field strength in NC⁻¹

$E = \frac{F}{Q}$

F = force on the charged object in N

Q = charge of the object in C

Tip: Remember, vector means it has a magnitude <u>and</u> a direction.

E is a vector pointing in the direction that a positive charge would move. The units of E are newtons per coulomb (NC⁻¹). The field strength usually depends on where you are in the field (see next page).

- **Example** ── Maths Skills ──

Find the force acting on an electron in an electric field with a field strength of 5000 NC⁻¹. The charge on an electron is −1.60 × 10⁻¹⁹ C.

Just rearrange the equation for electric field strength and put in the numbers:

$$E = \frac{F}{Q} \Rightarrow F = E \times Q = 5000 \times (-1.60 \times 10^{-19}) = -8 \times 10^{-16} \text{ N}$$

Tip: For parallel plates, there won't necessarily be a positive plate and a negative plate.

Electric field lines

Electric field lines are drawn to show the direction of the force that would act on a positive charge. Point charges have a radial field (see Figure 3). For a positive point charge the field lines point away from the point charge, and for a negative point charge they point towards it. For parallel plates, the field lines point from the plate with the more positive voltage to the plate with the less positive voltage (see Figure 3).

Figure 2: *Electric field lines between two plates shown by the alignment of pepper flakes in oil.*

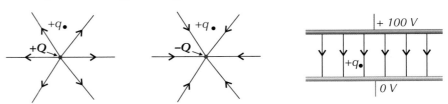

Figure 3: *Electric field lines for a positive point charge and a negative point charge, and between two parallel plates.*

Measuring electric field lines

Conducting paper can be used to map out the field lines of a 2D electric field. The experiment is set up so that a positive charge is on one edge of the paper and a negative charge is on the opposite edge. A voltmeter is then used to measure the potential difference at different points on the paper. Points with the same voltage can be joined up to show equipotential lines. Equipotential lines are always perpendicular to field lines (like gravitational field lines) and so these can also be mapped out.

An **electrolytic tank** can be used in a similar way to plot the field lines of a 3D electric field. The conducting paper is replaced by a tank of water with positive and negative ions dissolved in it. Electrodes are put in the water to create a positive charge on one side of the tank and a negative charge on the other side. A voltmeter is then used to find points within the water where the potential difference is the same. From this, both equipotentials and field lines can be mapped out.

Figure 4: *Conducting paper can be used to map out field lines. The voltage is measured at different points on the paper between two charged edges.*

Electric field strength in radial fields

A point charge — or any body that behaves as if all its charge is concentrated at the centre — has a radial field. In a radial field, the electric field strength, **E**, depends on the distance r from the point charge Q:

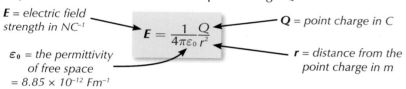

E = electric field strength in NC^{-1}

Q = point charge in C

$$E = \frac{1}{4\pi\varepsilon_0}\frac{Q}{r^2}$$

ε_0 = the permittivity of free space = $8.85 \times 10^{-12}\ Fm^{-1}$

r = distance from the point charge in m

If you have a sphere that has a charge evenly distributed across its surface, you can treat it as if it is a point charge acting at the centre of the sphere.

Example — Maths Skills

The electric field strength 0.15 m away from the centre of a charged sphere is 44 000 NC^{-1}. What's the charge on the sphere? (You may assume the sphere has a radius less than 0.15 m.)

You need to find Q, so make that the subject of the formula and then put the numbers in:

$$E = \frac{1}{4\pi\varepsilon_0}\frac{Q}{r^2} \quad \Rightarrow \quad Q = (4\pi\varepsilon_0 r^2)E$$

$$= 4\pi \times (8.85 \times 10^{-12}) \times 0.15^2 \times 44\,000$$

$$= 1.10100... \times 10^{-7}$$

$$= 1.1 \times 10^{-7}\ C \text{ (to 2 s.f.)}$$

This is another case of the inverse square law — $E \propto \frac{1}{r^2}$. Field strength decreases as you go further away from Q.

On a diagram, the field lines of a radial field get further apart, and if you plot the electric field strength against r you get the same shape as for gravitational field strength on page 76. If the charge isn't a point charge (e.g. a charged metal sphere), then the electric field strength inside the object doesn't have the same $E \propto \frac{1}{r^2}$ relation (you don't need to worry what the electric field is inside a charged object, just that it doesn't follow an inverse square law).

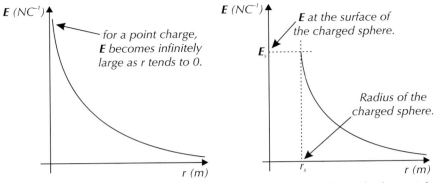

Tip: Graphs of **E** against r usually only show the magnitude of **E** (so they're always positive). But if **E** was negative, the graph would just be reflected in the horizontal axis.

Figure 5: Graphs of E against r for a point charge (left) and a charged sphere (right).

Electric field strength in uniform fields

A uniform field can be produced by connecting two parallel plates to the opposite poles of a battery — see Figure 6.

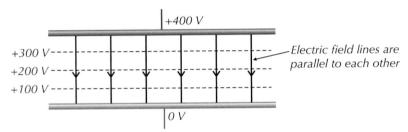

Figure 6: Electric field lines between parallel plates.

The field strength **E** is the same at all points between the two plates and is given by:

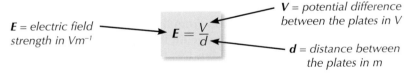

E = electric field strength in Vm⁻¹

$$E = \frac{V}{d}$$

V = potential difference between the plates in V

d = distance between the plates in m

Tip: Electric field strength can be measured in Vm⁻¹ as well as NC⁻¹.

Note that here the potential difference between the plates is the same as the potential of the top plate. This is because for this example the potential of the bottom plate is 0 V. This won't always be the case, so you might find it easier to think of it as ΔV.

Example — **Maths Skills**

What's the electric field strength between two parallel plates 0.15 m apart with a potential of +650 V and +200 V respectively?

$$E = \frac{\Delta V}{d} = \frac{650 - 200}{0.15} = 3000 \text{ Vm}^{-1}$$

Charged particles in a uniform electric field

A uniform electric field can be used to determine whether a particle is charged or not. The path of a charged particle moving through an electric field (at an angle to the field lines) will bend — the direction depends on whether it's a positive or negative charge.

A charged particle that enters an electric field at right angles to the field (as in Figure 7) feels a constant force parallel to the electric field lines. If the particle is positively charged then the force acts on it in the same direction as the field lines. If it's negatively charged, the force is in the opposite direction to the field lines. This causes the particle to accelerate at right angles to the particle's original motion — and so it follows a curved path (a parabola). In a 3D situation, the motion is the same (a parabola), as there are no other significant forces acting on the charged particle.

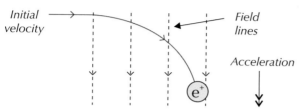

Figure 7: *Motion of a positron (which is positively charged) in a uniform electric field.*

> **Tip:** An electron (which is negatively charged) would accelerate upwards rather than downwards in the electric field in Figure 7.

Practice Questions — Application

Q1 Two electrons are fired towards each other and reach a separation of 5.22×10^{-13} m. What's the force on each electron at this point?

Q2 Copy the diagram on the right and complete the path of the electron as it travels through the electric field.

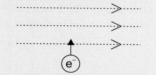

> **Tip:** The charge on an electron, $-e$, is -1.60×10^{-19} C.

Q3 The electric field generated by a charged sphere with charge 4.15 µC is measured as 15 000 NC^{-1}. How far from the centre of the sphere must the measuring instrument be?

Q4 A particle with charge 5.0×10^{-5} C is in a uniform electric field and experiences a force of 0.080 N. Calculate the magnitude of the electric field strength.

Q5 An alpha particle with a charge of $+2e$ and a mass of 6.64×10^{-27} kg is suspended freely between two parallel plates. The top plate has no charge and the bottom plate is charged to +5.00 nV. How far apart are the plates if the alpha particle isn't moving?

> **Tip:** For Q5, if the particle isn't moving, the upwards force from the electric field must equal the downwards force from gravity.

Q6 Initially, a sphere has radius r and charge Q. A point charge q is a distance of $3r$ away from the surface of the sphere. The force between the sphere and the point charge is F. The radius and charge of the sphere are then changed to $2r$ and $9Q$ respectively. Find the distance at which the point charge must be from the surface of the sphere in order for the force between the two to remain as F.

> **Tip:** 1 nV = 1×10^{-9} V.

Practice Questions — Fact Recall

Q1 Give the equation of Coulomb's law.

Q2 What's **E** a measure of?

Q3 Draw the electric field generated by a positive point charge.

Q4 Give two examples of pieces of equipment that can be used to investigate and map out electric field lines.

Q5 What kind of electric field is created by charged parallel plates?

6. Electric Potential

Learning Objectives:

- Understand the definition of absolute electric potential, V, including its value of zero at infinity.
- Be able to calculate the electric potential of a charged object in a radial electric field using $V = \frac{1}{4\pi\varepsilon_0}\frac{Q}{r}$.
- Be able to sketch graphs of V against r for positive and negative point charges, and know that the gradient of these graphs gives the electric field at a point r, where $E = \frac{\Delta V}{\Delta r}$.
- Understand the definition of electric potential difference.
- Know that ΔV can be found from the area under an E-r graph.
- Be able to calculate the work done in moving a charged object through an electric potential difference using $\Delta W = Q\Delta V$.
- Be able to derive the work done in moving a charge between plates from $Fd = Q\Delta V$.
- Know what equipotential surfaces are, and understand that no work is done on a charge moving along an equipotential surface.

Specification References 3.7.3.2 and 3.7.3.3

Tip: You can see the link between E and V from the formula for the field strength between parallel plates (see page 90).

As you might have guessed, you can also find a charged object's electric potential. Because electric forces can be both attractive and repulsive, electric potential can be both positive and negative.

What is electric potential?

All points in an electric field have an **absolute electric potential**, V. This is the **electric potential energy** that a unit positive charge (+1 C) would have at that point. The absolute electric potential of a point depends on how far it is from the charge creating the electric field and the size of that charge. In a radial field, absolute electric potential is given by:

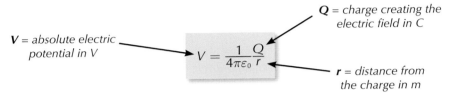

V = absolute electric potential in V

Q = charge creating the electric field in C

$$V = \frac{1}{4\pi\varepsilon_0}\frac{Q}{r}$$

r = distance from the charge in m

The sign of V depends on the charge Q — V is positive when Q is positive and the force is repulsive, and negative when Q is negative and the force is attractive. The absolute magnitude of V is greatest on the surface of the charge, and decreases as the distance from the charge increases — V will be zero at an infinite distance from the charge.

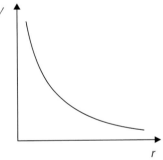

Figure 1: *V changing with r for a positive charge (repulsive force). V is initially positive and tends to zero as r increases towards infinity.*

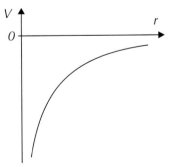

Figure 2: *V changing with r for a negative charge (attractive force). V is initially negative and tends to zero as r increases towards infinity.*

The gradient of a tangent to either graph gives the field strength E at that point:

$$E = \frac{\Delta V}{\Delta r}$$

Example — [Maths Skills]

A positively charged particle is placed 0.035 m from the centre of a sphere with a charge of +3.1 μC and radius 0.02 m. If the particle is then repelled by 0.19 m, what change in potential does it experience?

Change in potential = final potential − initial potential

$$= \frac{1}{4\pi\varepsilon_0}\frac{(3.1 \times 10^{-6})}{(0.035 + 0.19)} - \frac{1}{4\pi\varepsilon_0}\frac{(3.1 \times 10^{-6})}{0.035}$$

$$= -6.72529... \times 10^5$$

$$= -6.7 \times 10^5 \, V \ \ (\text{to 2 s.f.})$$

Electric potential difference

If two points in an electric field have a different absolute electric potential, then there is an **electric potential difference** between them. This is the energy needed to move a unit charge between those points.

You saw on the previous page that the absolute electric potential in a radial electric field changes with distance. The electric potential difference between two points in a radial electric field, ΔV, can be found from the area under a graph of electric field strength, E, against radial distance, r.

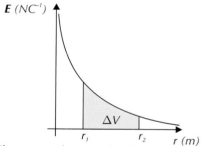

E (NC⁻¹)

ΔV

r_1　r_2　r (m)

Figure 3: The area under a graph of E against r is used to find the value of ΔV between r_1 and r_2.

Tip: If you're asked to estimate ΔV from an E−r graph, you'll probably have to work out the area under the graph by counting the number of squares on the graph paper or by splitting the area up into trapeziums.

To move a charge across a potential difference (i.e. from one electric potential to another) you need to use energy. The amount of energy you need (or the work done) depends on the size of the charge you're moving and the size of the potential difference you want to move it across:

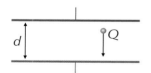

Δ W = work done in moving a charge in J ⟶

$$\Delta W = Q\Delta V$$

Q = the charge being moved in C

ΔV = electric potential difference in V

To see how this formula is derived, consider two parallel plates with a potential difference of ΔV across them, creating a uniform electric field.

d ↕　　↓ Q

The field strength is given by: $E = \dfrac{\Delta V}{d} = \dfrac{F}{Q}$　⟵ **E** defined as force per unit charge, page 88.

E in a uniform field, page 90.

Figure 4: Static charge on a comb doing work against gravity to lift a trickle of water.

This rearranges to give:

$$Q\Delta V = Fd$$

To move a charge Q from A to B:

work done = force × distance moved = Fd

So the work done in moving a charge Q through a potential difference ΔV is given by:

$$\Delta W = Q\Delta V$$

Tip: Ignore the work done against gravity — assume the strip of paper is massless.

Example — **Maths Skills**

A strip of paper with a charge of –0.053 µC is resting on the lower of two horizontal parallel plates 2.8 cm apart. A potential difference of 350 V is applied across the plates and the strip of paper is lifted so that it's touching the top plate. Find the work done by the plates in lifting the strip of paper.

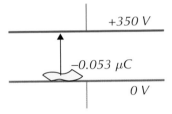

The paper moves through a potential difference of 350 V, so just put the numbers into the equation above:

$$\Delta W = Q\Delta V = (0.053 \times 10^{-6}) \times 350 = 1.855 \times 10^{-5} = 1.9 \times 10^{-5}\ \text{J (to 2 s.f.)}$$

Tip: Work is always positive, so ignore any minus signs here.

Equipotentials

Just like in gravitational fields, you find **equipotentials** (page 81) in electric fields too. And just like in gravitational fields, the field lines are always perpendicular to the equipotential lines. This means that for a point charge, the equipotentials are spherical surfaces, and between parallel plates, the equipotentials are flat planes. Remember, no work is done when you travel along an equipotential — an electric charge can travel along an equipotential without any energy being transferred.

Tip: The equipotential lines for a radial field are actually spherical shells. The diagram in Figure 5 is a 2D representation of this.

Figure 5: *Field lines (solid lines) and equipotential lines (dotted lines) for a radial field (left) and a uniform field (right).*

Practice Questions — Application

Q1 A small metal sphere is being held stationary on a platform directly next to a charged metal sphere, as shown below.

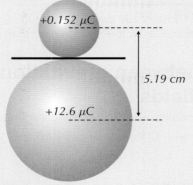

+0.152 µC

5.19 cm

+12.6 µC

Tip: Assume the smaller (green) sphere has a negligible radius.

a) What's the absolute electric potential due to the larger sphere at the centre of the smaller metal sphere?

b) The smaller sphere is released so that it can move freely, and is repelled by 12.9 cm before being stopped again. How much work is done in moving the sphere by this distance?

Tip: The value of ε_0 is 8.85×10^{-12} Fm^{-1}.

Q2 The graph of E against r for a point charge is shown below.

a) Use the graph to estimate the change in absolute electric potential between a distance of 3.0 m and a distance of 6.0 m from the charge.

b) Use the graph to estimate the charge of the point charge.

Tip: If you know the electric field E at a distance r, you can rearrange the equation $E = \frac{1}{4\pi\varepsilon_0}\frac{Q}{r^2}$ to find the charge.

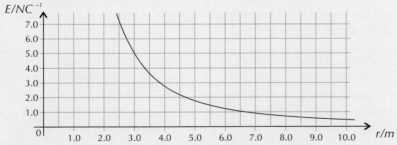

Q3 5.14 µJ of work is done to move a sphere with a charge of −83.1 nC from one parallel plate to another. Assuming no other forces are involved, what's the potential difference across the parallel plates?

Practice Questions — Fact Recall

Q1 What's absolute electric potential?

Q2 Sketch the graph of absolute electric potential V against distance r in a radial field for:

a) a positive point charge.

b) a negative point charge.

Q3 What's electric potential difference?

Q4 Say what each symbol represents in the equation $\Delta W = Q\Delta V$.

Q5 What shape do the electric equipotentials take for:

a) a radial field? b) a uniform field?

Learning Objectives:
- Know and understand the similarities and differences between gravitational and electrostatic forces.
- Be able to compare the magnitude of the gravitational and electrostatic forces between subatomic particles.

Specification References 3.7.1 and 3.7.3.1

Tip: The graphs of g against r (p. 76) and V against r for gravitational fields (p. 78) are also very similar to the graphs of E against r (p. 90) and V against r for electric fields (p. 92).

7. Comparing Electric and Gravitational Fields

You might have thought a lot of the formulas from the last topic looked familiar — electric and gravitational fields are more similar than you might think...

Similarities between gravitational and electric fields

A lot of the formulas used for electric fields are the same as those used for gravitational fields but with Q instead of m (or M) and $\frac{1}{4\pi\varepsilon_0}$ instead of G:

Newton's law of gravitation:
$$F = \frac{Gm_1m_2}{r^2}$$

Gravitational field strength for a radial field:
$$g = \frac{GM}{r^2}$$

Gravitational potential:
$$V = -\frac{GM}{r}$$

Coulomb's law:
$$F = \frac{1}{4\pi\varepsilon_0}\frac{Q_1Q_2}{r^2}$$

Electric field strength for a radial field:
$$E = \frac{1}{4\pi\varepsilon_0}\frac{Q}{r^2}$$

Absolute electric potential:
$$V = \frac{1}{4\pi\varepsilon_0}\frac{Q}{r}$$

There are several similarities between electric and gravitational fields that are useful to know:

Tip: Don't forget — a spherical mass acts as a point mass and a spherical charge acts as a point charge.

Gravitational field strength, g, is force per unit mass.	Electric field strength, E, is force per unit positive charge.
Newton's law of gravitation for the force between two point masses is an inverse square law. $F \propto \frac{1}{r^2}$	Coulomb's law for the electric force between two point charges is also an inverse square law. $F \propto \frac{1}{r^2}$
The gravitational field lines for a point mass...	The electric field lines for a negative point charge...
Gravitational potential, V, is potential energy per unit mass and is zero at infinity.	Absolute electric potential, V, is potential energy per unit positive charge and is zero at infinity.
The equipotential for a uniform spherical mass forms a spherical surface.	The equipotential for a point charge forms a spherical surface.
The work done to move a unit mass through a gravitational potential is $\Delta W = m\Delta V$.	The work done to move a unit charge through an electric potential is $\Delta W = Q\Delta V$.

The important difference

Although gravitational and electric fields are similar, they're not the same — there's an important difference between them too. Gravitational forces are always attractive, whereas electrostatic forces can be attractive or repulsive.

Forces at subatomic levels

When you get down to the **subatomic level** of electrons, protons and neutrons, the distances between particles become tiny. As both the gravitational and electrostatic forces have an inverse square relationship with distance, you'd expect these forces to be huge.

However, gravity at this level can pretty much be ignored. This is because although they're close together, all of the particles have incredibly small masses — the gravitational force at these distances is much weaker than the electrostatic force.

Thankfully, the nucleus doesn't break apart from all of this electrostatic repulsion — there are other forces at work (you met these in year 1 of A-level physics).

Figure 1: One of the reasons that the golfer doesn't fall to the centre of the Earth due to gravity is the electrostatic forces between the atoms in his shoes and the atoms at the surface of the ground.

Example — **Maths Skills**

Two protons in a nucleus are 3.00 fm apart. Calculate the gravitational and electrostatic forces between them.
m_{proton} = 1.67 × 10⁻²⁷ kg, Q_{proton} = 1.60 × 10⁻¹⁹ C,
ε_0 = 8.85 × 10⁻¹² Fm⁻¹ and G = 6.67 × 10⁻¹¹ Nm²kg⁻².

Gravitational:

$$F = \frac{Gm_1m_2}{r^2}$$
$$= \frac{6.67 \times 10^{-11} \times (1.67 \times 10^{-27})^2}{(3.00 \times 10^{-15})^2}$$
$$= 2.06688... \times 10^{-35}$$
$$= 2.07 \times 10^{-35} \text{ N (to 3 s.f.)}$$

Electrostatic:

$$F = \frac{1}{4\pi\varepsilon_0}\frac{Q_1Q_2}{r^2}$$
$$= \frac{1}{4 \times \pi \times 8.85 \times 10^{-12}} \times \frac{(1.60 \times 10^{-19})^2}{(3.00 \times 10^{-15})^2}$$
$$= 25.5766...$$
$$= 25.6 \text{ N (to 3 s.f.)}$$

By comparing these two forces, it's clear to see how much larger the electrostatic force is than the gravitational force at subatomic levels.

Tip: Remember — 1 fm = 1 × 10⁻¹⁵ m.

Tip: So here, the force on the protons due to electrostatic repulsion is over 10³⁶ times bigger than the force on the protons due to gravity.

Practice Question — Application

Q1 For this question, you can assume an electron has mass 9.11 × 10⁻³¹ kg and charge –1.60 × 10⁻¹⁹ C.

a) Calculate the gravitational and electrostatic forces between two electrons that are 6.00 × 10⁻¹⁰ m apart.

b) Each electron is replaced with a particle with the same charge as an electron but a different mass. Assuming the two particles are identical to each other, what mass would each particle need to have in order for the gravitational force to be equal to the electrostatic force?

Practice Questions — Fact Recall

Q1 List four similarities between electric and gravitational fields.

Q2 Give one major difference between electric and gravitational fields.

Section Summary

Make sure you know...

- That a force field (represented as a vector) is a region in which a body experiences a non-contact force.
- That a force field is present in interactions between masses or moving charges.
- That gravity acts as an attractive force between all matter (with mass).
- How to use Newton's law of gravitation to find the force between two point masses.
- That the gravitational law is an inverse square law, where G is the gravitational constant.
- How to represent radial and uniform gravitational fields using field lines.
- That the gravitational field strength, g, is the force per unit mass in a gravitational field.
- How to find the value of g in radial gravitational fields, and how to plot this as a graph of g against r.
- That gravitational potential, V, is the gravitational potential energy that a unit mass would have at a specific point, and that it is zero at an infinite distance from a point mass.
- How to calculate the gravitational potential at a given point in a gravitational field.
- Why there is a negative sign in the equation for calculating the gravitational potential.
- How to draw a graph of V against r, and that the gradient of the graph gives you g at that point.
- That the area under a graph of g against r gives you the gravitational potential difference, ΔV.
- That gravitational potential difference is the energy needed to move a unit mass between two points.
- How to calculate the work done in moving a mass through a gravitational potential difference.
- That gravitational equipotential surfaces are surfaces on which V is the same everywhere.
- How the orbital period and speed of an object are related to the radius of its orbit.
- How to derive the relation $T^2 \propto r^3$.
- That the combined potential and kinetic energy of an orbiting object is a constant.
- That escape velocity is the minimum speed an unpowered object needs to escape a gravitational field.
- The applications of geostationary and low orbiting satellites, and their planes of orbit and radii.
- How to use Coulomb's law to find the force between two point charges in a vacuum.
- That ε_o is the permittivity of free space.
- To treat air as a vacuum when using Coulomb's law.
- That a charged sphere can be treated as if all of its charge is concentrated at the centre of the sphere.
- That the electric field strength, E, is defined as the force per unit positive charge in an electric field.
- How to represent radial and uniform electric fields using field lines.
- How to calculate the magnitude of E in radial and uniform electric fields.
- What a graph of E against r for a radial field looks like.
- That a charged particle follows a parabolic trajectory when entering an electric field at right angles.
- That absolute electric potential, V, is the potential energy that a unit positive charge would have at a specific point, and that it becomes zero at an infinite distance from a point charge.
- How to calculate the absolute electric potential at a given point in a radial electric field.
- What a graph of V against r looks like, and that the gradient gives you E at that point.
- That electric potential difference is the energy needed to move a unit charge between two points.
- That ΔV can be found from the area under a graph of E against r.
- How to find the work done in moving a charge through an electric potential difference.
- How to derive the equation for the work done in moving a charge between charged plates.
- That electric equipotential surfaces are surfaces on which V is the same everywhere.
- The similarities and differences between gravitational and electric fields, including at subatomic levels.

Exam-style Questions

1 Which of the following statements is incorrect?

    A Gravitational forces are always attractive.

    B Gravitational potential increases with distance.

    C Electric forces are always attractive.

    D The electric field between charged parallel plates is uniform.

(1 mark)

2 The diagram shows a (positively charged) alpha particle being fired in a straight line through charged parallel plates, travelling parallel to the plates. Which of the following statements is correct?

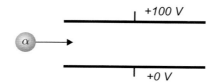

    A The particle's path will be deflected downwards.

    B The particle's path will be deflected upwards.

    C The particle's path will not be deflected.

    D The particle will be forced back in the direction it came.

(1 mark)

3 Which of the following graphs shows how the magnitude of the gravitational field strength varies with distance from a spherical mass with a radius R?

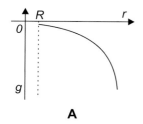
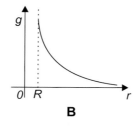
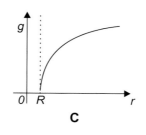
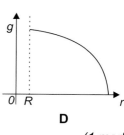

        A   **B**   **C**   **D**

(1 mark)

4 Which of the following gives the value of the change in gravitational potential, ΔV?

    A The gradient of the graph of g against r.

    B The area under the graph of g against r.

    C The gradient of the graph of T against r.

    D The area under the graph of T against r.

(1 mark)

5.1 Define a force field.

(1 mark)

5.2 Sketch the field lines and equipotential lines for the gravitational field around Earth.

(2 marks)

A student is conducting an experiment to determine the local value of g.
He finds g to be 9.83 Nkg^{-1}, to 3 significant figures.

Another student conducts the same experiment and finds the value to be 6.31 Nkg^{-1}.
He claims his value is about 1/3 lower because his work bench is about 1/3 taller
than the other student's.

5.3 Explain whether the student's claim is true or not.

(2 marks)

5.4 If the Earth has radius 6370 km and mass 5.98×10^{24} kg, calculate the altitude at
which the local value of g is 6.31 Nkg^{-1}. (Altitude is measured from sea level.)

(3 marks)

A satellite is in orbit around Earth at a gravitational potential of -20.6 $MJkg^{-1}$.

5.5 Calculate the value of g at this point.

(3 marks)

5.6 If the satellite has a mass of 2.53×10^3 kg, calculate the gravitational force acting on it.

(1 mark)

5.7 Calculate the speed that the satellite would need in order to escape the
gravitational field of Earth from its current position.

(2 marks)

5.8 Another satellite orbits the Earth once every 24 hours in the same plane as the
equator, and in the same direction as the Earth's rotation. Name this type of orbit.

(1 mark)

6.1 Describe the field produced between two charged parallel plates with
a potential difference across them.

(2 marks)

6.2 Describe the field produced by a (positively charged) alpha particle.

(2 marks)

In an experiment to demonstrate Rutherford scattering, alpha particles
are fired at thin gold foil.

One alpha particle is fired directly at a gold nucleus so that it comes to a full
stop 2.08×10^{-12} m from the nucleus, before being deflected backwards along
its initial path.

Alpha particles carry a charge of $+2e$ ($e = 1.60 \times 10^{-19}$) and have a mass
of 6.64×10^{-27} kg. A gold nucleus carries a charge of $+79.0e$.

6.3 Assuming the alpha particle started with zero electric potential, calculate the electric
potential difference between its start position and where it came to a stop.

(2 marks)

6.4 Calculate the speed at which the alpha particle is fired.

(4 marks)

7 A charged sphere carrying a charge of –34.7 µC is fixed to the bottom of a plastic tube. Another sphere, with a charge of –92.5 µC and a mass of 203 g, is dropped into the tube and allowed to move freely until it comes to a rest above the fixed sphere. This is shown in **Figure 1**.

Figure 1

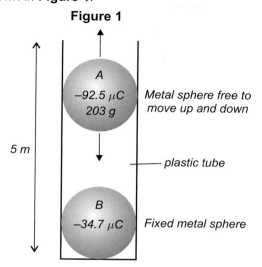

7.1 Calculate the force due to gravity on sphere *A*. (Use *g* = 9.81 Nkg⁻¹.)

(1 mark)

7.2 Assume the spheres have a negligible radius. Calculate the distance between the centres of the spheres when sphere *A* comes to rest.

(4 marks)

The top sphere is now placed on a smooth, horizontal plane between two parallel plates, as shown (from above) in **Figure 2**. For this question, assume the sphere has a negligible radius.

Figure 2

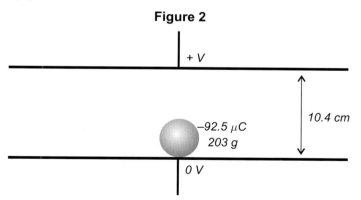

7.3 The electric field causes the sphere to roll from one plate to the other. If 18.5 mJ of work is done in moving the sphere, find the potential difference across the plates.

(2 marks)

7.4 Calculate the electric field strength between the plates.

(1 mark)

1. Capacitors

Capacitors are like big buckets. You can fill them up and empty them when you feel like it. The capacitance of a capacitor tells you how much charge it can hold. Almost all electrical products will contain one somewhere...

Capacitance

> The **capacitance** of an object is the amount of charge it is able to store per unit potential difference (p.d.) across it.

Capacitance is measured in **farads** — 1 farad (F) = 1 coulomb per volt (CV^{-1}).

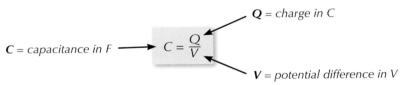

C = capacitance in F $\longrightarrow$ $C = \dfrac{Q}{V}$ $\longleftarrow$ Q = charge in C

V = potential difference in V

Capacitors

A **capacitor** is an electrical component that can store electrical charge. They are made up of two electrical conducting plates separated by an electrical insulator (a **dielectric**, p.107). The circuit symbol is two parallel lines:

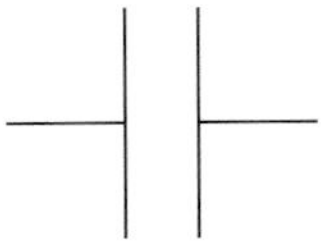

Figure 2: Circuit symbol for a capacitor.

Figure 1: A disassembled capacitor, showing two sheets of aluminium foil with paper between them. The foil and paper are rolled up like a big Swiss roll.

Tip: You can make your own simple capacitor using two pieces of aluminium separated by a piece of paper.

Tip: Remember, the potential difference between two points is the work done (energy transferred) in moving a unit charge between them.

When a capacitor is connected to a direct current (d.c.) power source, charge builds up on its plates — one plate becomes negatively charged and one becomes positively charged (there's more on this on p. 109). The plates are separated by an electrical insulator, so no charge can move between them. This means that a potential difference builds up between the plates of the capacitor.

The capacitance of a capacitor is the charge that the capacitor can store per unit potential difference across it. The voltage rating of a capacitor is the maximum potential difference that can be safely put across it. A capacitor will only charge up to the voltage of the power source it is connected to.

A good way to think of a capacitor is as a bucket that can hold electrical charge. The capacitance (C) is the area of the bucket's base, and the height is equal to its voltage rating (V). Then the charge (Q) that it can store is $C \times V$ (a rearrangement of the equation above).

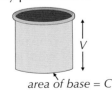

area of base = C

Figure 3: A capacitor can be thought of as a charge bucket.

A 100 μF capacitor is charged to a potential difference of 12 V.
How much charge is stored by the capacitor?

$C = \dfrac{Q}{V}$ so by rearranging, $Q = C \times V = (100 \times 10^{-6}) \times 12 = 1.2 \times 10^{-3}$ C

The voltage rating of the capacitor is 50 V.
What is the maximum charge that the capacitor can store?

$Q = C \times V = (100 \times 10^{-6}) \times 50 = 5 \times 10^{-3}$ C

Tip: In this formula, Q is the charge stored on each capacitor plate — the charge is just negative on one plate and positive on the other.

Investigating *V* and *Q*

$C = \dfrac{Q}{V}$ rearranged gives $Q = CV$, and since the capacitance of a capacitor is fixed, this means Q is directly proportional to V. You can investigate the relationship between the potential difference across, and the charge stored on, a capacitor experimentally by charging a capacitor using a constant current. First, set up a test circuit to measure current and potential difference:

Tip: A farad is a huge unit so you'll usually see capacitances expressed in terms of:

μF — microfarads (× 10^{-6})
nF — nanofarads (× 10^{-9})
pF — picofarads (× 10^{-12})

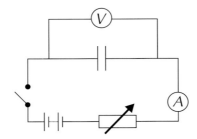

Figure 4: *Test circuit for a capacitor.*

Tip: This is the standard test circuit for testing the current through, and voltage across, a component. You should've seen it before when you learnt about electricity. Remember what each of the circuit symbols stands for:

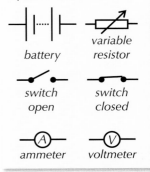

battery   variable resistor

switch open   switch closed

ammeter   voltmeter

After closing the switch, constantly adjust the variable resistor to keep the charging current constant for as long as you can (it's impossible when the capacitor is nearly fully charged). Record the p.d. at regular time intervals until it equals the battery p.d. Using the fixed charging current and the time taken to charge the capacitor you can plot the following graph:

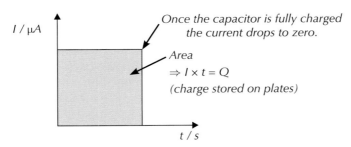

I / μA

Once the capacitor is fully charged the current drops to zero.

Area
$\Rightarrow I \times t = Q$
(charge stored on plates)

t / s

Figure 5: *Current-time graph for a capacitor.*

Tip: You should remember the formula for charge: $Q = It$.

The area under the graph in Figure 5 is the charge stored on the capacitor. Using $Q = It$ you can calculate the charge stored on the capacitor at a given time. Plotting a graph of charge stored against potential difference across the capacitor at each recorded time interval gives the graph in Figure 6.

Tip: There's more on why the current drops off like this on p.109-110.

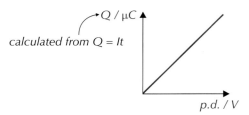

calculated from $Q = It$

Figure 6: *Charge-p.d. graph for a capacitor.*

Exam Tip
They love asking questions about this graph in the exam. You can find the capacitance by calculating $Q \div V$. Because the capacitance remains constant here (Q and V are proportional to each other), the gradient of the Q-V graph in Figure 6 is effectively the capacitance. But be careful — sometimes they flip the axes...

The graph in Figure 6 is a straight line through the origin — so this shows Q and V are directly proportional. You can use the same experiment to calculate the unknown capacitance of a capacitor, because the gradient of the graph is the capacitance, C.

Uses of capacitors

Capacitors can only store relatively small amounts of charge, so they aren't used instead of batteries. To store the same energy as an AA battery, you'd need around 6000 farads. The capacitor would be absolutely massive.

Capacitors are usually only used to provide power for a short amount of time — it is tricky to prolong the discharge time and the voltage through the circuit decreases as the capacitor discharges.

But capacitors are still very useful because they can store charge until it's needed, and then discharge all of their charge in a fraction of a second, whereas a battery might take several minutes. For this reason, charged capacitors can be very dangerous. Capacitors in electronics in your home could contain enough charge to kill you.

One example of capacitor use is in a camera flash — the camera battery charges the capacitor over a few seconds, and then the entire charge of the capacitor is dumped into the flash almost instantly. This allows the camera flash to be very bright for a very short time. Some other applications of capacitors are:

- 'Ultracapacitors' (really big capacitors) can be used in back-up power supplies to provide reliable power for short periods of time.

- To smooth out variations in d.c. voltage supplies — a capacitor absorbs the peaks and fills in the troughs.

Practice Questions — Application

Q1 A 0.10 F capacitor is used in a circuit as a back-up in case of a short interruption in the mains power supply. The p.d. supplied to the circuit is 230 V. How much charge can the capacitor store?

Q2 Explain why a capacitor would not be a good source for powering a portable media player.

Exam Tip
The formula for capacitance is given in the data and formulae booklet if you need a reminder in the exam.

Practice Questions — Fact Recall

Q1 Write down the definition of capacitance.

Q2 Explain how you could keep the current charging a capacitor constant.

2. Energy Stored by Capacitors

It takes energy to build up charge on the plates of a capacitor. This energy is stored by the capacitor and released when the charges are released.

Energy stored

Remember that when a capacitor charges, one plate becomes negatively charged while the other becomes positively charged (you'll see on p. 109 exactly how). Like charges repel, so when each plate of a capacitor becomes charged, the charges on that plate are being forced together 'against their will'. This requires energy, which is supplied by the power source and stored as electric potential energy for as long as the charges are held. When the charges are released, the electric potential energy is released.

Deriving the energy stored equations

You can find the energy stored by a capacitor by using the graph of potential difference against charge for the capacitor (see Figure 1).

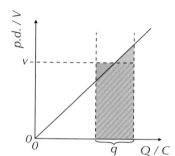

The p.d. across a capacitor is directly proportional to the charge stored on it, so the graph is a straight line through the origin (see page 104).

Consider a tiny increase in the charge on the plates during the charging process. The electric potential energy stored is the work done to move the extra

Figure 1: *A V-Q graph showing the energy stored by a small increase in charge on the plates of a capacitor.*

charge onto the plates against the potential difference across the plates, given by $E = \Delta W = Q\Delta V$ (page 94). Let the small charge being moved be q. The average p.d. over that step is v. So in that small step, the energy stored is $E = qv$, which is given by the area of the red rectangle in Figure 1. The area of the green parallelogram in Figure 1 is the area under the graph over the charge difference q, and it is the same as the area of the red rectangle.

The total energy stored by the capacitor is the sum of all the energies stored in each small step increase in charge, until the capacitor is fully charged. So it's just the area under the graph of V against Q. As it is just a triangle, the area under the graph would be the same if you swapped the axes, so the area under a Q-V graph also gives the energy stored by a capacitor.

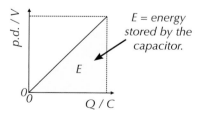
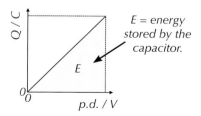

Figure 3: *The area under a V-Q graph (left) or the area under a Q-V graph (right) is the energy stored by a capacitor.*

The area of a triangle is given by ½ × base × height, so the energy stored by the capacitor is:

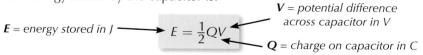

E = energy stored in J ⟶ $E = \frac{1}{2}QV$

V = potential difference across capacitor in V

Q = charge on capacitor in C

Learning Objectives:

- Know how to calculate the energy stored in a capacitor using $E = \frac{1}{2}QV$, $E = \frac{1}{2}CV^2$, $E = \frac{1}{2}\frac{Q^2}{C}$.
- Know that the area under the graph of charge against p.d. is the energy stored by a capacitor.

Specification Reference 3.7.4.3

Figure 2: *Storm clouds store massive amounts of energy — forming huge natural capacitor plates. Particle collisions cause electrons to be knocked off atoms, giving the bottom of the cloud a charge. This induces a charge in the surface of the Earth. The air in between acts as the dielectric, and if the potential difference gets big enough, the dielectric breaks down and lightning bolts travel between the cloud and the Earth.*

Exam Tip

This formula is given in the data and formulae booklet, but you need to understand where it comes from — remember it's the area under a V-Q graph or a Q-V graph.

The energy supplied by the power source in charging a capacitor is $\Delta W = V \Delta Q$, where V is the p.d. of the power source. So $E = QV$, which is exactly double the energy stored by the capacitor. In other words, the energy stored by the capacitor is half the energy supplied by the power source (the rest is lost to the resistance in the circuit and internal resistance of the battery).

Example ── **Maths Skills**

A capacitor is charged to a potential difference of 12 V using a constant current of 5 mA over 30 seconds. Calculate the energy stored by it.

Using $Q = It$, $Q = 5 \times 10^{-3} \times 30 = 0.15$ C

So $E = \frac{1}{2}QV = \frac{1}{2} \times 0.15 \times 12 = 0.9$ J

There are three expressions you need to know for the energy stored by capacitors:

- You know the first one already: $E = \frac{1}{2}QV$

- You know that $C = \frac{Q}{V}$, so $Q = CV$.
  Substitute that into the energy equation above: $E = \frac{1}{2}QV = \frac{1}{2}CV \times V$

$$E = \frac{1}{2}CV^2$$

- $C = \frac{Q}{V}$, so $V = \frac{Q}{C}$.
  Substituting into the first energy equation: $E = \frac{1}{2}QV = \frac{1}{2}Q \times \frac{Q}{C}$

$$E = \frac{1}{2}\frac{Q^2}{C}$$

Example ── **Maths Skills**

A 900 μF capacitor is charged up to a potential difference of 240 V. Calculate the energy stored by the capacitor.

First, choose the best equation to use — you've been given V and C, so you need $E = \frac{1}{2}CV^2$.

Substitute the values in:
$E = \frac{1}{2}CV^2 = \frac{1}{2} \times (900 \times 10^{-6}) \times 240^2 = 25.9$ J (to 3 s.f.)

Figure 4: *Benjamin Franklin was one of the first people to store electrical energy, in the mid 1700s. He invented what he called a 'battery' by grouping together many Leyden jars (simple capacitors).*

Tip: Remember:
$\Delta Q = I\Delta t$.

Practice Questions — Application

Q1 A 40 mF capacitor is connected to a 230 V power source. When fully charged, how much energy will be stored by the capacitor?

Q2 A 5 μF capacitor is charged with a constant current of 0.025 mA for 45 s. Calculate the energy stored by the capacitor in this time.

Practice Questions — Fact Recall

Q1 Explain why a build-up of charge on a capacitor results in a build-up of energy, and state what type of energy is stored in a capacitor.

Q2 How would you find the energy stored in a capacitor from a graph of charge against potential difference?

Q3 Write down three equations that can be used to calculate the energy stored in a capacitor.

3. Dielectrics

Dielectrics are insulators, plain and simple. They are a vital part of how capacitors work and are one of the factors which affect overall capacitance.

Permittivity

Altering the properties of a capacitor changes how much charge it can store at a given voltage (its capacitance). One of the things you can change is the dielectric material separating the two conducting plates. This changes the capacitance because different materials have different **relative permittivities**.

Permittivity is a measure of how difficult it is to generate an electric field in a medium. The higher the permittivity of a material, the more charge is needed to generate an electric field of a given size.

Relative permittivity is the ratio of the permittivity of a material to the permittivity of free space:

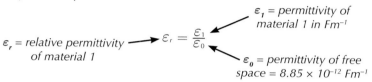

ε_r = relative permittivity of material 1 ⟶ $\varepsilon_r = \dfrac{\varepsilon_1}{\varepsilon_0}$

ε_1 = permittivity of material 1 in Fm^{-1}

ε_0 = permittivity of free space = $8.85 \times 10^{-12}\ Fm^{-1}$

Relative permittivity is sometimes also called the **dielectric constant**.

Polar molecules

Permittivity can be explained by the motion (or action) of the molecules inside a dielectric. Imagine that a dielectric is made up of lots of polar molecules — this means they have a positive end and a negative end. When no charge is being stored by a capacitor, no electric field is being generated. The molecules are aligned randomly, as shown in Figure 1.

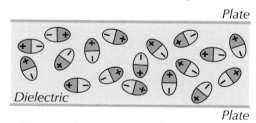

Figure 1: The alignment of molecules in a dielectric when no electric field is present.

When charge is applied to the plates of a capacitor an electric field is generated between them. The negative ends of the molecules are attracted to the positively charged plate, and vice versa. This causes the molecules to rotate (see Figure 2) and align themselves anti-parallel to the electric field generated between the plates.

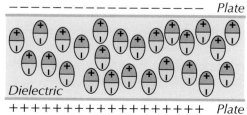

Figure 2: The alignment of molecules in a dielectric when an electric field is present.

Learning Objectives:
- Understand the terms relative permittivity and dielectric constant.
- Be able to describe the action of a simple polar molecule that rotates in the presence of an electric field.
- Know that the capacitance of a capacitor can be found using $C = \dfrac{A\varepsilon_0\varepsilon_r}{d}$.

Specification Reference 3.7.4.2

Tip: You can also think of the relative permittivity as the ratio of the size of an electric field generated in a vacuum (free space) to the size of the field generated in the medium by the same conditions.

Tip: A capacitor made from parallel plates separated by a dielectric is called a parallel-plate capacitor.

Tip: Anti-parallel means that the molecules are aligned with the electric field lines (see p.88), but are orientated in the opposite direction. So the positive poles are pointing to the negative plate, and vice versa.

The molecules each have their own electric field, which in this alignment now opposes the applied electric field of the capacitor. The larger the permittivity, the larger this opposing field is. This reduces the overall electric field between the parallel plates, which reduces the potential difference needed to transfer a given charge to the capacitor — so the capacitance increases ($Q = CV$).

Calculating capacitance

The capacitance of a capacitor depends on the dimensions of the capacitor as well as the dielectric inside it. Capacitance can be calculated using:

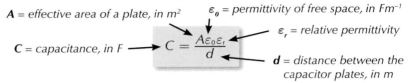

A = effective area of a plate, in m^2
ε_0 = permittivity of free space, in Fm^{-1}
ε_r = relative permittivity
C = capacitance, in F

$$C = \frac{A\varepsilon_0\varepsilon_r}{d}$$

d = distance between the capacitor plates, in m

Investigating capacitance

You can investigate how capacitance changes by setting up two parallel plates separated by a dielectric and connecting the plates to a capacitance meter. You can then alter how much the two plates overlap to change the effective area of the capacitor, or use different materials as the dielectric to vary the relative permittivity. Stacking multiple layers of the same material allows you to test how plate separation affects the capacitance. If you know the capacitance of a capacitor, the area of its plates and the distance between plates, you could also calculate the relative permittivity of a range of different dielectrics.

Figure 3: Some digital multimeters can also be used as capacitance meters.

Practice Questions — Application

Q1 A material has a permittivity of 1.99×10^{-11} Fm^{-1}. Calculate its relative permittivity.

Q2 A parallel-plate capacitor has plates that are 6.0 mm in length and 4.5 mm wide. The plates are 0.5 mm apart, and the gap between them is fully filled by a dielectric with a relative permittivity of 86. What is the capacitance of the capacitor?

Q3 A capacitor whose plates are separated by paper ($\varepsilon_r = 3.00$) has a capacitance of 16.6 nF. The paper is replaced with a different material with a dielectric constant of 2.55, and everything else remains constant. Calculate the new capacitance of the capacitor.

Practice Questions — Fact Recall

Q1 What is a dielectric material?

Q2 What is meant by the term permittivity?

Q3 What is the dielectric constant also known as?

Q4 What does the dielectric constant represent?

Q5 Describe the orientation of polar molecules in the dielectric of a capacitor if no charge is being stored by the capacitor.

Q6 Describe what happens to the molecules in the dielectric of a capacitor if charge is applied to the plates of the capacitor.

Q7 Give the equation for calculating the capacitance of a capacitor with plate area A, plate separation d and which contains a dielectric with a relative permittivity of ε_r.

4. Charging and Discharging

You've seen how capacitors are used to store charge and energy, and how to calculate the energy stored by them — but now it's time to look at the physics of how they actually charge and discharge a little closer.

Charging

When a capacitor is connected to a d.c. power supply (e.g. a battery), a current flows in the circuit until the capacitor is fully charged, then stops.

The electrons flow from the negative terminal of the supply onto the plate connected to it, so a negative charge builds up on that plate.

At the same time, electrons flow from the other plate to the positive terminal of the supply, making that plate positive. These electrons are repelled by the negative charge on the negative plate and attracted to the positive terminal of the supply.

The same number of electrons are repelled from the positive plate as are built up on the negative plate. This means an equal but opposite charge builds up on each plate, causing the potential difference between the plates. Remember that no charge can flow directly between the plates because they're separated by an insulator (dielectric).

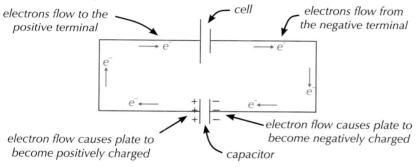

electrons flow to the positive terminal — *cell* — *electrons flow from the negative terminal*

electron flow causes plate to become positively charged — *capacitor* — *electron flow causes plate to become negatively charged*

Figure 1: *Flow of electrons as a capacitor charges.*

Initially the current through the circuit is high. But, as charge builds up on the plates, electrostatic repulsion makes it harder and harder for more electrons to be deposited. When the p.d. across the capacitor is equal to the p.d. across the supply, the current falls to zero. The capacitor is fully charged.

Charging through a fixed resistor

If you charge a capacitor through a fixed resistor, as in Figure 2, the resistance of the resistor will affect the time taken to charge the capacitor.

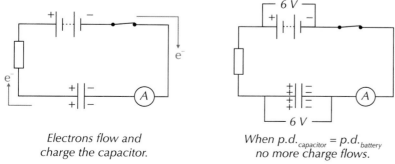

Electrons flow and charge the capacitor.

When p.d.$_{capacitor}$ = p.d.$_{battery}$ no more charge flows.

Figure 2: *(Left) A capacitor charging through a resistor. (Right) A charged capacitor.*

Learning Objectives:

- Be able to represent the charging and discharging of capacitors through resistors graphically.

- Be able to interpret the gradients of and areas under these graphs where appropriate.

- Know that the charge on a charging capacitor is given by: $Q = Q_0(1 - e^{-\frac{t}{RC}})$.

- Know that the charge, potential difference and current for a discharging capacitor are given by: $Q = Q_0 e^{-\frac{t}{RC}}$, $V = V_0 e^{-\frac{t}{RC}}$ and $I = I_0 e^{-\frac{t}{RC}}$.

- Be able to investigate the charge and discharge of capacitors and analyse the results by plotting a log-linear graph (Required Practical 9).

Specification Reference 3.7.4.4

Tip: The same is true for discharging a capacitor — see page 111.

As soon as the switch is closed, a current starts to flow. The potential difference across the capacitor is zero at first, so there is no p.d. opposing the current. The potential difference of the battery causes an initial relatively high current of $\frac{V}{R}$ to flow (where V is the voltage of the power supply and R is the resistance of the resistor). As the capacitor charges, the p.d. across the capacitor gets bigger, the p.d. across the resistor gets smaller, and the current drops. The charge (Q) on the capacitor is proportional to the potential difference across it, so the Q-t graph is the same shape as the V-t graph. This results in the following graphs for charging a capacitor through a resistor:

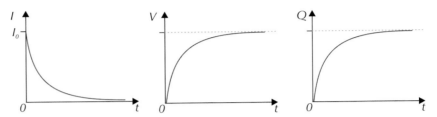

Figure 3: Graphs to show charging current against time (left), potential difference against time (middle) and charge against time (right) for a capacitor being charged through a fixed resistor.

The charge on a charging capacitor after a given time t is given by:

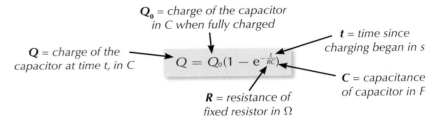

Q_0 = charge of the capacitor in C when fully charged

Q = charge of the capacitor at time t, in C

t = time since charging began in s

$$Q = Q_0(1 - e^{-\frac{t}{RC}})$$

C = capacitance of capacitor in F

R = resistance of fixed resistor in Ω

The formula for calculating the voltage across a charging capacitor is of the same form, but the formula for the charging current is different — it decreases exponentially. The formulas for both are:

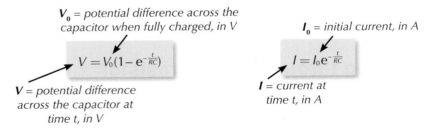

V_0 = potential difference across the capacitor when fully charged, in V

$$V = V_0(1 - e^{-\frac{t}{RC}})$$

V = potential difference across the capacitor at time t, in V

I_0 = initial current, in A

$$I = I_0 e^{-\frac{t}{RC}}$$

I = current at time t, in A

Example — Maths Skills

When fully charged, a 3.00 µF capacitor holds 36.0 µC of charge. It is connected in series with a 56.0 kΩ resistor. Calculate the charge on the capacitor 0.500 seconds after it begins charging.

$$Q = Q_0(1 - e^{-\frac{t}{RC}}) = 36.0 \times 10^{-6}(1 - e^{\frac{-0.500}{56.0 \times 10^3 \times 3.00 \times 10^{-6}}})$$
$$= 3.416... \times 10^{-5} = 34.2\,\mu C \text{ (to 3 s.f.)}$$

Discharging through a fixed resistor

To discharge a capacitor, take out the battery and reconnect the circuit. When a charged capacitor is connected across a resistor, the p.d. drives a current through the circuit. This current flows in the opposite direction from the charging current. The capacitor is fully discharged when the p.d. across the plates and the current in the circuit are both zero.

Tip: You need to reconnect the circuit to discharge a capacitor. If the capacitor isn't connected in a full circuit it will hold its charge.

Investigating capacitors discharging

REQUIRED PRACTICAL **9**

To investigate the behaviour of a discharging capacitor, you must first charge your capacitor using the method outlined on pages 109-110. Then open the switch, remove the power source and add a voltage sensor and data logger to create a circuit similar to the one shown in Figure 4.

Tip: Data loggers can be connected to a computer which will collect all your data for you and plot all sorts of graphs of it.

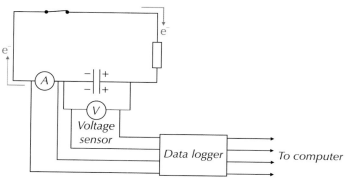

Figure 4: Test circuit for a discharging capacitor.

- Close the switch and allow the capacitor to discharge through the resistor.
- When the reading through the ammeter reaches zero, use the computer to calculate the charge on the capacitor over time.
- The computer can then plot a variety of graphs showing how the current, potential difference and charge vary over time.
- You can also create a log-linear plot of your results in order to draw a straight-line graph. This is explained in detail on the next page.

Tip: This experiment could be repeated with an assortment of capacitors to see how the choice of capacitor affects the results.

Graphs for discharging capacitors

Using the data collected by the data logger, you can plot discharge curves for the capacitor (as shown in Figure 5).

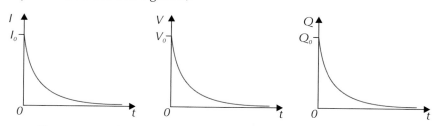

Figure 5: Graphs to show discharging current against time (left), potential difference against time (middle) and charge against time (right) for a capacitor discharging through a fixed resistor.

The I-t graph is the same as the one for charging (although the current is now flowing in the opposite direction). The current starts off relatively high and gradually decreases to zero. This is because the potential difference across the capacitor (and hence across the resistor) decreases as the charge on the capacitor decreases.

Tip: You could do a similar experiment to investigate a charging capacitor. Add a data logger and a voltage sensor across the capacitor in the circuit shown in Figure 2 on page 109. Then use a computer to draw the graphs on page 110.

Tip: V is proportional to Q (see pages 102), so you always get the same shape graph whether you use V or Q on the vertical axis.

As shown by the Q-t graph, when a capacitor is discharging, the amount of charge left on the plates falls exponentially with time. That means that, for a given capacitor, it always takes the same length of time for the charge to halve (see p.115), no matter how much charge you start with — like radioactive decay (p.180).

The charge left on the plates of a capacitor discharging from full is given by the equation:

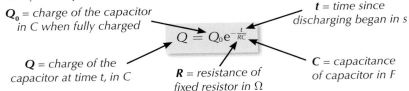

Q_0 = charge of the capacitor in C when fully charged

Q = charge of the capacitor at time t, in C

R = resistance of fixed resistor in Ω

t = time since discharging began in s

C = capacitance of capacitor in F

$$Q = Q_0 e^{-\frac{t}{RC}}$$

The same is true for the potential difference and current:

$$V = V_0 e^{-\frac{t}{RC}}$$

$$I = I_0 e^{-\frac{t}{RC}}$$

Example — **Maths Skills**

A 0.20 mF capacitor was charged to a potential difference of 12 V and then discharged through a fixed 50 kΩ resistor. Calculate the charge on the capacitor after 1 second.

The question tells you $C = 0.20 \times 10^{-3}$ F, $V = 12$ V, $R = 50 \times 10^3 \ \Omega$ and $t = 1$.

First you need to calculate the initial charge.

$C = \dfrac{Q}{V}$ so $Q = CV$, so the initial charge is:

$Q_0 = CV = 0.20 \times 10^{-3} \times 12 = 2.4 \times 10^{-3}$ C

Then use the equation:

$$Q = Q_0 e^{-\frac{t}{RC}}$$
$$= 2.4 \times 10^{-3} \times e^{-\frac{1}{(50 \times 10^3)(0.20 \times 10^{-3})}}$$
$$= 2.4 \times 10^{-3} \times e^{-0.1}$$
$$= 2.171... \times 10^{-3} = 2 \times 10^{-3} \text{ C (to 1 s.f.)}$$

Log-linear graphs

Log-linear graphs are plots where one of the axes is logarithmic. Log-linear and log-log plots are useful as they can often be used to produce a graph which is a straight line when linear axes would give a curve. So they're another good way of graphically displaying how charge, potential difference and current vary over time for a discharging capacitor.

Starting from the equation for charge on a discharging capacitor, the natural logarithm is taken of both sides:

$$Q = Q_0 e^{-\frac{t}{RC}}$$

$$\ln(Q) = \ln(Q_0 e^{-\frac{t}{RC}})$$

$\ln(A \times B) = \ln(A) + \ln(B)$, so this can be written as:

$$\ln(Q) = \ln(Q_0) + \ln(e^{-\frac{t}{RC}})$$

Another log rule is $\ln(e^A) = A$, so:

$$\ln(Q) = \left(-\frac{1}{RC}\right)t + \ln(Q_0)$$

From results gathered from the experiment outlined on page 111, you can calculate the charge, Q, from $\Delta Q = I\Delta t$ or from the area under an I-t graph. If you then plot a graph of $\ln(Q)$ against t, you will see it is a straight line (Figure 6). By comparing the equation for $\ln(Q)$ to the equation for a straight line, you can see that the gradient is equal to $-\dfrac{1}{RC}$.

Tip: The graph's intercept with the vertical axis is $\ln(Q_0)$ — it's the c in $y = mx + c$.

The value of RC is called the **time constant**. There's plenty more on the time constant on page 114, but for now all you need to know is that you can find it from the log-linear graph by dividing -1 by the gradient of the line. An example of this is shown on page 115.

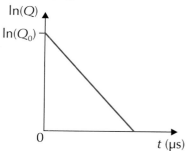

Figure 6: *A plot of ln(Q) against t.*

You could also plot $\ln(V)$ or $\ln(I)$ against time and find the time constant in the same way, as V and I both have the same dependence on R and C. You can do a similar experiment to investigate a charging capacitor — you could plot a graph of $\ln(I)$ against t which has the same shape as for a discharging capacitor.

Practice Questions — Application

Q1 A capacitor with capacitance 0.60 μF is initially charged to 15.0 V. It is then connected across a resistor with resistance 2.6 kΩ. Calculate the potential difference across the capacitor 1.5 ms after it has begun to discharge.

Q2 Calculate the charge stored on a 15 nF capacitor, originally charged to a potential difference of 230 V and then discharged for 0.01 seconds through a 50 kΩ resistor.

Tip: You'll need $Q = CV$ for this, which is given in the data and formulae book.

Q3 a) Sketch a graph of $\ln(V)$ against t for a discharging capacitor.

　　b) What does the vertical axis intercept of the graph represent?

Practice Questions — Fact Recall

Q1 Explain how negative charge is built up on the plate of a capacitor.

Q2 Draw curves showing how the p.d. across the capacitor, current through the circuit and charge on the capacitor vary with time when a capacitor is a) charging and b) discharging through a fixed resistor.

Q3 What is the equation for calculating the charge on a capacitor charging through a fixed resistor R, after a certain time t?

Q4 How can you alter the circuit used to charge a capacitor so that the capacitor discharges?

Q5 Write down the equation for the p.d. across a discharging capacitor.

Q6 Describe an experiment you could do to investigate a discharging capacitor.

Learning Objectives:

- Know what the time constant of a charging or discharging capacitor is, and that it is given by $\tau = RC$.
- Be able to calculate the time constant of a charging or discharging capacitor, including from graphical data.
- Be able to calculate the time to halve, knowing that $T_{\frac{1}{2}} = 0.69RC$.

Specification Reference 3.7.4.4

Tip: Remember that the charge remaining on a discharging capacitor is given by $Q = Q_0 e^{-\frac{t}{RC}}$.

Tip: τ is the Greek letter 'tau'.

Tip: Q is proportional to V, so the time constant is also the time taken for the voltage to decrease to about 37% of the source voltage (or increase to about 63% of the source voltage while charging).

5. Time Constant and Time to Halve

The time constant and the time to halve are two key values to calculate when you're talking about the time taken to charge or discharge a capacitor.

Charging and discharging times

The time taken for a capacitor to charge and discharge depends on two factors:

- The capacitance of the capacitor (C). This affects the amount of charge that can be transferred at a given voltage.
- The resistance of the circuit (R). This affects the current in the circuit.

Time constant

When the discharge time t is equal to RC the equation for the charge left on a discharging capacitor becomes:

$$Q = Q_0 e^{-1}.$$

So when $t = RC$:

$$\frac{Q}{Q_0} = \frac{1}{e}, \text{ where } \frac{1}{e} \approx \frac{1}{2.718} \approx 0.37$$

The time $t = RC$ is known as the **time constant**, τ, and is the time taken for the charge on a discharging capacitor (Q) to fall to about 37% of Q_0. It's also the time taken for the charge of a charging capacitor to rise to about 63% of Q_0 (Figure 1).

The larger the resistance in series with the capacitor, the longer it takes to charge or discharge. In practice, the time taken for a capacitor to charge or discharge fully is taken to be about $5RC$ or 5τ.

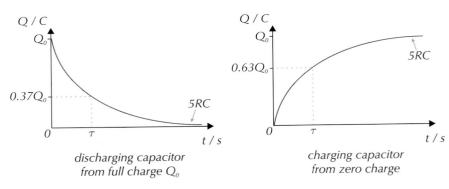

discharging capacitor from full charge Q_0

charging capacitor from zero charge

Figure 1: *Q-t graphs showing the time constant for discharging (left) and charging (right) a capacitor through a fixed resistor.*

You can calculate the time constant directly using $\tau = RC$ or you can find it from a graph. These may be graphs of Q against t (as shown above and on the next page) or log-plots of $\ln(Q)$ against t (which you saw on page 113, and can see again on the next page).

A capacitor was discharged from full through a 10 kΩ resistor (to 2 s.f.). The graph below shows how the charge on the capacitor changed over the first 2.0 seconds. Calculate the time constant and the capacitance.

The initial charge is 12×10^{-4} C, and you want to find the time taken for the charge to decrease to 37% of that.

So $0.37 \times (12 \times 10^{-4}) = 4.4 \times 10^{-4}$ C.

Using the graph, the charge is 4.4×10^{-4} C when the time is 1.0 s, so the time constant, τ, is 1.0 s.

$\tau = RC$,

so, $C = \dfrac{\tau}{R}$

$= \dfrac{1.0}{10 \times 10^{3}}$

$= 1.0 \times 10^{-4}$ F $= 0.10$ mF (to 2 s.f.)

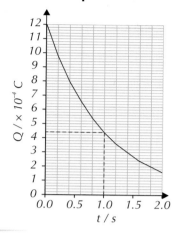

Tip: You might be asked to find the time it takes for a capacitor to discharge (or charge) to a certain percentage of its total charge.

Tip: Make sure that you remember that Q / Q_0 is the proportion of the charge <u>on the plates</u>, not the proportion of the charge that's been lost.

— Example — Maths Skills —

A student has done an experiment to find out how the charge of a discharging capacitor varies with time. She plots her results on a $\ln Q$-t graph and draws a line of best fit. The graph is shown below. Use the graph to find the time constant of the student's capacitor circuit.

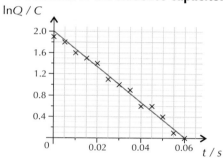

Tip: You met this experiment on page 111. Have a look at the experiment again to remind yourself of the log graphs.

The gradient of a $\ln Q$-t graph is equal to $-\dfrac{1}{RC}$. To find the time constant, $\tau = RC$, you divide -1 by the gradient.

Gradient $= -\dfrac{2.0 - 0}{0.06 - 0} = -33.333...$

$\tau = \dfrac{-1}{-33.333...} = 0.030$ s (to 2 s.f.)

Time to halve

The time to halve is the time taken for the charge, current or potential difference of a discharging capacitor to decrease to half of the initial value. So it's the time when $Q = \frac{1}{2}Q_0$.

Using the formula for charge remaining on a discharging capacitor:

$$Q = \tfrac{1}{2}Q_0 = Q_0 e^{-\frac{t}{RC}}$$

and then cancelling Q_0 gives:

$$\tfrac{1}{2} = e^{-\frac{t}{RC}}.$$

Take the natural log of both sides:

$$\ln\left(\frac{1}{2}\right) = \ln(e^{-\frac{t}{RC}})$$

$\frac{1}{2}$ can be written as 2^{-1} and $\ln(A^B) = B\ln A$, so

$$\ln\left(\frac{1}{2}\right) = \ln(2^{-1}) = -\ln 2.$$

Using the log rule $\ln(e^A) = A$, you know that $\ln(e^{-\frac{t}{RC}}) = -\frac{t}{RC}$. So you can write the above equation as:

$$-\ln(2) = -\frac{t}{RC}$$

Then rearrange to get:

$$t = \ln(2)RC.$$

$\ln(2) = 0.693...$, so the time to halve is given by:

Tip: Time to halve is not given in the data and formulae booklet, so make sure you know how to calculate it.

$T_{\frac{1}{2}}$ = 'time to halve', in s ⟶ $T_{\frac{1}{2}} = 0.69RC$

C = capacitance of the capacitor, in F

R = resistance of fixed resistor in Ω

Tip: As shown above, the "0.69" bit in $T_{\frac{1}{2}} = 0.69RC$ comes from $\ln(2) = 0.6931...$, and is rounded to 2 s.f. This means answers to questions that use this formula shouldn't be given to more than 2 s.f.

─ **Example** ─ **Maths Skills** ─────────

A fully charged 72 μF capacitor is put in series with an 18 kΩ resistor. Calculate the time taken for the charge to halve as the capacitor is discharged.

The time taken for the charge to halve is given by:

$T_{\frac{1}{2}} = 0.69RC = 0.69 \times (18 \times 10^3) \times (72 \times 10^{-6}) = 0.89424$
$= 0.89$ s (to 2 s.f.)

You can also find the time to halve from a graph of charge, potential difference or current against time. You read off the value of t at which the charge, potential difference or current reaches half of its initial value.

─ **Example** ─ **Maths Skills** ─────────

A capacitor is fully charged and then discharged by placing it in series with a resistor. The resistor has a resistance of 840 Ω. The graph below shows the potential difference across the discharging capacitor over time.
Use the graph to find the time taken for the potential difference to halve, and hence the capacitance of the capacitor.

Tip: This is similar to the time constant example on page 115. The difference is that you're looking for 50% of the original value on the vertical axis, rather than 37%.

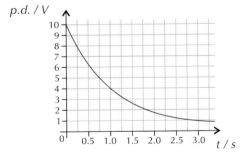

From the graph, you can see that the time taken for the voltage to decrease from 10 V to 5 V is 0.75 s.

Now rearrange the equation $T_{\frac{1}{2}} = 0.69RC$ to find the capacitance:

$$C = \frac{T_{\frac{1}{2}}}{0.69R} = \frac{0.75}{0.69 \times 840} = 1.2939... \times 10^{-3} = 1.3 \times 10^{-3} \text{ F (to 2 s.f.)}$$

Practice Questions — Application

Q1 A circuit has a time constant of 4.2 µs. What is its 'time to halve'?

Q2 A capacitor discharged in series with a resistor can be used to create a time delay function in electronics. In a particular burglar alarm, the alarm goes off when a 15 µF capacitor discharged through a 400 kΩ resistor has lost 63% of its initial charge.

a) Calculate the time delay on the alarm.

The company want to increase the capacitance of the capacitor in order to make the time delay 60 seconds.

b) What capacitance would be needed for this time delay?

c) Explain why this might not be practical, and suggest what other component they could replace instead.

Q3 A discharging capacitor loses 70% of its initial charge, Q_0, in 20.0 seconds. Find the time constant of the capacitor-resistor circuit.

Q4 A plot of ln(V) against t is created for a discharging capacitor.

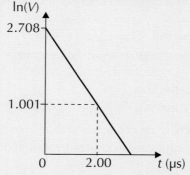

a) What does the vertical intercept of the graph represent?

b) Calculate the initial voltage.

c) Calculate the time constant for the capacitor.

d) Calculate the time taken for the voltage to halve.

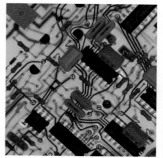

Figure 2: *A circuit from a burglar alarm containing capacitors. They can be used with big resistors to create a time delay. When the door is opened the capacitor begins to discharge and when it has lost a certain amount of its charge the alarm goes off, giving you time beforehand to enter the code.*

Tip: Look back at pages 112-113 for a reminder on this type of graph.

Practice Questions — Fact Recall

Q1 What two factors do the charging and discharging time of a capacitor depend on?

Q2 What is meant by the time constant when discharging a capacitor?

Q3 What is meant by the 'time to halve', and how can it be calculated?

Q4 Briefly describe how you would find the 'time to halve' for a discharging capacitor from a graph of current against time.

Section Summary

Make sure you know...

- That capacitance is defined as the amount of charge stored per unit potential difference, $C = \frac{Q}{V}$.
- That the charge on a capacitor and the potential difference across it are proportional.
- That the gradient of a graph of charge against p.d. is capacitance.
- That the area underneath a graph of charge against p.d. is the energy stored by the capacitor.
- The equations for energy stored: $E = \frac{1}{2}QV = \frac{1}{2}CV^2 = \frac{1}{2}\frac{Q^2}{C}$.
- What the term dielectric means.
- What the terms relative permittivity and dielectric constant represent.
- How to describe the rotation of particles in a dielectric material when charge is stored on a capacitor.
- Why dielectric materials increase the capacitance of a capacitor.
- That capacitance also depends on the dimensions of a capacitor, and is given by $C = \frac{A\varepsilon_0\varepsilon_r}{d}$.
- How charge is built up on the plates of a capacitor.
- How voltage, current and charge vary with time as a capacitor charges and discharges, and how to represent these relationships graphically.
- The equation for charge stored on a charging capacitor: $Q = Q_0(1 - e^{-\frac{t}{RC}})$.
- That, for a discharging capacitor, the equations for charge, potential difference and current are:
  $Q = Q_0 e^{-\frac{t}{RC}}$, $V = V_0 e^{-\frac{t}{RC}}$, $I = I_0 e^{-\frac{t}{RC}}$.
- How to investigate the charge and discharge of capacitors.
- How to plot a log-linear graph for a discharging capacitor.
- That the time constant is the time taken for a discharging capacitor to discharge to $\frac{1}{e}$ (about 37%) of its initial charge, or the time taken for a charging capacitor to charge to about 63% of its full charge.
- That the time constant τ is equal to RC.
- How to determine the time constant from graphs.
- That the time to halve is the time taken for the charge, potential difference or current of a discharging capacitor to fall to half of its initial value.
- That time to halve is given approximately by $T_{\frac{1}{2}} = 0.69RC$.

Exam-style Questions

1. When the switch S in the circuit shown below is in position 1, the capacitor C is fully charged by the battery through resistor R. The switch is then moved to position 2 and the capacitor is allowed to discharge fully through the resistor.

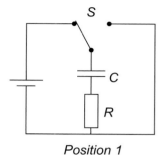

Position 1

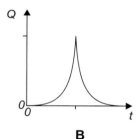

Position 2

Which graph correctly shows how the charge, Q, on the capacitor varies with time, t, during this process?

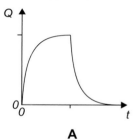

A

B

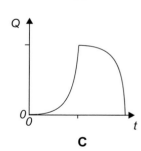

C

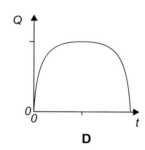

D

(1 mark)

2. A capacitor discharging through a fixed 100 kΩ resistor loses 35% of its charge in 1 s. What is the approximate capacitance of the capacitor?

A 9.5×10^{-6} F

B 5.3×10^{-5} F

C 2.3×10^{-5} F

D 2.2×10^{-2} F

(1 mark)

3   A 5 mF capacitor is charged to 24 V using a constant current for half a minute.
What current is used to charge the capacitor?

    A   4 A

    B   0.24 A

    C   4 mA

    D   0.24 mA

(1 mark)

4   A 20 μF capacitor is fully charged to a potential difference of 24 V through a variable
resistor R using the circuit below. The charging current is constant and charging takes
10 s. Which of the following statements is **correct**?

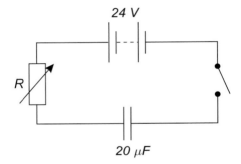

    A   The potential difference across the resistor is 20 V.

    B   The energy stored by the capacitor is 5.76 mJ.

    C   The total energy taken from the battery during charging is 5.76 mJ.

    D   The current supplied was 0.48 mA.

(1 mark)

5   A student creates a capacitor out of two square aluminium sheets, where each sheet
has a side length of 25 cm, separated by a 1.0 mm thick piece of paper. The relative
permittivity of paper is 3.3.

    5.1   Calculate the capacitance of the capacitor.

(2 marks)

    5.2   Explain how using a dielectric increases the capacitance of the capacitor, making
reference to the molecules in the dielectric.

(4 marks)

    5.3   The student wishes to increase the capacitance of her capacitor. She only has
aluminium and paper to work with. Suggest one change to the design of her
capacitor which would increase its capacitance.

(1 mark)

6   A student investigating capacitance charges a 1.0 μF capacitor from a 12 V d.c. supply and uses a voltage sensor and a charge sensor connected to a data logger to measure the potential difference across the capacitor at regular intervals of charge stored by the capacitor. A computer is used to plot a graph of p.d. against charge.

6.1   Copy the axes below and sketch the graph obtained by the computer on the axes.

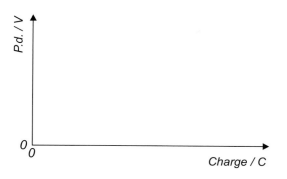

(1 mark)

6.2   State what is represented by the gradient of the graph.

(1 mark)

6.3   State what quantity the area enclosed by the line and the horizontal axis represents.

(1 mark)

The student then discharges the capacitor through a 29 kΩ resistor.

6.4   What is the potential difference across the capacitor plates after 0.030 s?

(2 marks)

6.5   Draw a graph showing how the current supplied to the resistor varies over time as the capacitor discharges.

(1 mark)

6.6   Calculate the time taken for the current through the resistor to halve.

(1 mark)

7   A 3.0×10^{-3} F capacitor was fully charged to 50.0 V through a 2.0 kΩ resistor.

7.1   During the charging, 7.5 J of energy was taken from the battery in total. Calculate how much energy was stored by the capacitor.

(1 mark)

7.2   Show that the time constant of the circuit is 6 seconds.

(1 mark)

7.3   Calculate the charge gained by the capacitor in 14 seconds.

(2 marks)

7.4   Sketch a graph of charge against time for the first 14 seconds of charging.

(1 mark)

Learning Objectives:

- Know that a force can act on a current-carrying wire in a magnetic field.
- Know Fleming's left-hand rule.
- Know what is meant by magnetic flux density, **B**, and know it's measured in teslas.
- Know the definition of the tesla.
- Be able to use **F** = **B**I*l* to find the force on a current-carrying wire when field is perpendicular to current.

Specification Reference 3.7.5.1

1. Magnetic Flux Density

You've met gravitational fields, you've met electric fields, and now it's time for the final sort — magnetic fields. Magnetic fields, as you might expect, have a field strength, except in this case it's known as magnetic flux density.

Magnetic fields

A **magnetic field** is similar to a gravitational field (page 72) and an electric field (page 87) — it's a region in which a force acts. In a magnetic field, a force is exerted on magnetic or magnetically susceptible materials (e.g. iron).

Magnetic fields can be represented by **field lines** (also called flux lines). Field lines go from the north to the south pole of a magnet, and the closer together the lines are, the stronger the field is (see Figure 1).

At a neutral point magnetic fields cancel out.

Figure 1: *The magnetic fields created by bar magnets.*

Magnetic fields around a wire

When current flows in a wire or any other long straight conductor, a magnetic field is induced around the wire. The field lines are concentric circles centred on the wire. The direction of a magnetic field around a current-carrying wire can be worked out with the **right-hand rule**:

- Curl your right hand into a fist and stick your thumb up.
- Point your thumb in the direction of the current through the wire.
- Your curled fingers will then show the direction of the field.

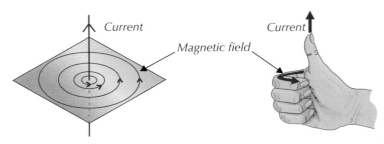

Current   *Current*

Magnetic field

Figure 3: *Diagram to show how your right hand can be used to show the direction of magnetic field lines around a current-carrying wire.*

Solenoids

If you loop a current-carrying wire into a coil in one plane, the surrounding magnetic field is doughnut shaped, while a coil with length (a **solenoid**) forms a field like a bar magnet (see Figure 4).

Figure 2: *A current-carrying wire induces a circular magnetic field around it — the needles of the small compasses follow a circle around the wire.*

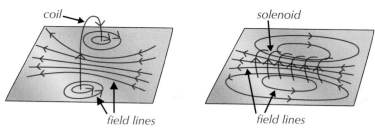

Figure 4: *The magnetic fields created by a current-carrying wire in a coil (left) and a solenoid (right).*

Figure 5: *A simple solenoid.*

Force on a current-carrying wire

If you put a current-carrying wire into an external magnetic field (e.g. between two magnets), the field around the wire and the field from the magnets are added together. This causes a resultant field — lines closer together show where the magnetic field is stronger. These bunched lines cause a 'pushing' force on the wire.

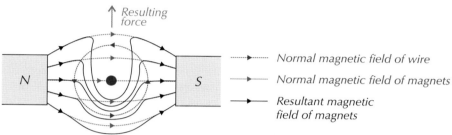

Figure 6: *A current-carrying wire in a magnetic field experiences a force.*

The size of the force depends on the component of the magnetic field that is perpendicular to the current. The direction of the force is always perpendicular to both the current direction and the magnetic field — it's given by Fleming's left-hand rule (see below). If the current is parallel to the field lines the size of the force is 0 N — there is no component of the magnetic field perpendicular to the current.

Fleming's left-hand rule

You can use your left hand to find the direction of the current, the direction of the external magnetic field or the direction of the force on the wire (as long as you know the other two). Stretch your thumb, forefinger and middle finger out, as shown in Figure 7, and use the following rules:

- The **F**irst finger points in the direction of the uniform magnetic **F**ield.
- The se**C**ond finger points in the direction of the conventional **C**urrent.
- The thu**M**b points in the direction of the force (the direction of **M**otion).

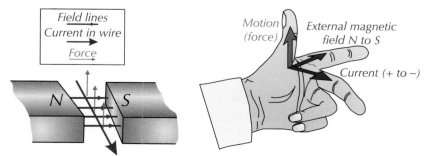

Figure 7: *Fleming's left-hand rule for a current-carrying wire in a magnetic field.*

Tip: A solid dot like in Figure 6 (●) shows current flowing out of the page (or towards the reader). A circle with a cross (⊗) shows current going into the page (or away from the reader). This notation can also be used to show field lines going into or coming out of the page.

Tip: The external magnetic field doesn't take into account the field around the wire as in Figure 6. It's just the field before the current was there.

Tip: Conventional current flows in the direction that a positive charge would flow, i.e. from positive to negative.

Tip: As long as the magnetic poles are wide enough, you can assume the magnetic field between them is uniform.

A current-carrying wire runs between two magnets, as shown below. What direction will the force on the wire be?

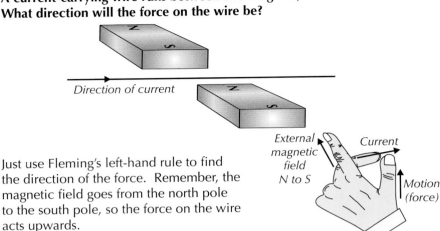

Direction of current

Just use Fleming's left-hand rule to find the direction of the force. Remember, the magnetic field goes from the north pole to the south pole, so the force on the wire acts upwards.

External magnetic field N to S *Current* *Motion (force)*

Figure 8: *British physicist John Ambrose Fleming.*

Tip: There's more on alternating current (ac) on pages 143-145.

By passing an alternating current through a wire in a magnetic field, the wire can be made to vibrate. The direction of the force acting on the wire is perpendicular to the direction of the current — so when the current is reversed, the direction of the force is also reversed. The constant reversal of an alternating current means the force is constantly alternated too, resulting in vibration of the wire.

Magnetic flux density

The force on a current-carrying wire at a right angle to a magnetic field is proportional to the **magnetic flux density**, **B**. Magnetic flux density is sometimes called the strength of the magnetic field. It is defined as:

> The force on one metre of wire carrying a current of one amp at right angles to the magnetic field.

Tip: One tesla is also equivalent to 1 weber per square metre. Webers are the unit of magnetic flux (page 133).

Magnetic flux density is a vector quantity with both a direction and a magnitude. It is measured in **teslas**, T. One tesla is equal to one newton per amp per metre:

$$1 \text{ tesla} = 1 \; \frac{\text{N}}{\text{Am}}$$

Tip: It helps to think of flux density as the number of magnetic field lines (measured in webers, Wb) per unit area.

When a current-carrying wire is at 90° to a magnetic field, the size of the force on the wire, *F* is proportional to the current, *I*, the length of wire in the field, *l*, as well as the flux density, *B*, of the external magnetic field. This gives the equation:

Tip: The derivation of this is pretty tricky, but luckily you don't need to know it.

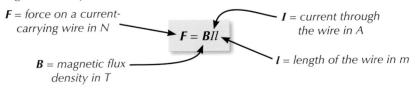

F = force on a current-carrying wire in N

I = current through the wire in A

$$F = BIl$$

B = magnetic flux density in T

l = length of the wire in m

This equation $F = BIl$ gives the maximum force the wire could experience. A force still acts on the wire if it's not at right angles to the magnetic field (as long as they're not parallel), but it will be smaller.

Tip: There's more about this on page 135.

Example — Maths Skills

A section of wire carrying a current of 5.2 A is placed at right angles to a uniform magnetic field with a flux density of 19 mT. If the wire experiences a force of 1.2×10^{-2} N, what length of wire is inside the magnetic field?

Tip: 1 mT (millitesla) is equal to 1×10^{-3} T.

Just rearrange the formula $F = BIl$ to make l the subject and then put in the numbers:

$$F = BIl \Rightarrow l = \frac{F}{BI}$$

$$= \frac{1.2 \times 10^{-2}}{(19 \times 10^{-3}) \times 5.2}$$

$$= 0.121...$$

$$= 0.12 \text{ m (to 2 s.f.)}$$

Practice Question — Application

Q1 A student runs a steady 1.44 A current through a wire from a dc supply. A 2.51 cm section of the wire is fixed at two points and then placed in a uniform magnetic field with a flux density of 9.21 mT, as shown.

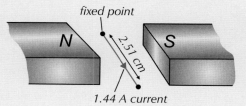

fixed point

N   S

2.51 cm

1.44 A current

a) In what direction will the force on the section of wire act?

b) Calculate the size of the force that will act on the section of wire.

c) The student wants to make the wire vibrate. What part of the experimental set-up could he alter to achieve this?

Practice Questions — Fact Recall

Q1 a) Which hand can you use to find the direction of the force acting on a current-carrying wire at a right angle to a magnetic field?

b) Describe what each finger represents on this hand when using it to find the direction of the force.

Q2 Give the condition needed to use the equation $F = BIl$ to calculate the force on a current-carrying wire in a magnetic field.

Q3 What does B represent in the equation $F = BIl$, and what are its units?

Tip: Make sure you do a full risk assessment before starting this experiment. Take care not to touch the circuit when current is flowing and always turn it off before making any changes.

Tip: The 'slab' magnets used in this experiment have poles on their largest faces.

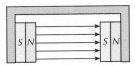

This sort of magnet is sometimes called a 'magnadur magnet'.

Tip: You only need to measure the length of wire shown on the diagram. Even though the vertical parts of the hoop are perpendicular to the field, the forces they feel act horizontally, and so don't affect the mass reading on the balance.

Tip: You need to be able to use the circuit diagram in Figure 2 to set up the circuit used in this experiment.

2. Investigating Force on a Current-Carrying Wire

You need to know how to use a top pan balance to investigate how the force on a wire varies with flux density, current and length of wire.

Investigating force on a wire

REQUIRED PRACTICAL **10**

You can use a top pan balance and the set-up shown in Figure 1 to investigate the relationship between the force on a wire, the length of wire perpendicular to a magnetic field, the current through it and flux density ($F = BIl$).

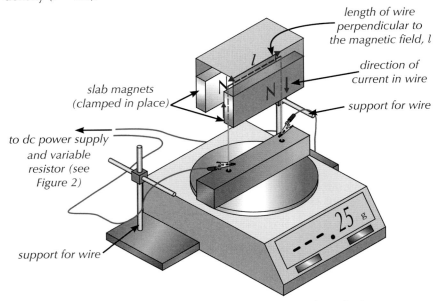

Figure 1: Set-up for an experiment to investigate how the force on a wire varies with flux density, current or length of wire.

Set up the experiment shown in Figure 1. A square hoop of metal wire is positioned so that the top of the hoop, length *l*, passes through the magnetic field, and is perpendicular to it. When a current flows, the length of wire in the magnetic field will experience a downwards force (Fleming's left-hand rule).

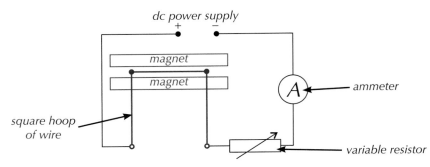

Figure 2: Circuit diagram for investigating how the force on a wire varies with flux density, current or length of wire. The section in red is shown in Figure 1.

The dc power supply should be connected to a variable resistor so that you can alter the current. Zero the digital balance when there is no current through the wire. Turn on the dc power supply — if the mass reading is negative, turn off the dc power supply and swap the crocodile clips over so that the mass is positive.

Note the mass showing on the digital balance and the current. Then use the variable resistor to change the current and record the new mass reading. Repeat this until you have tested a large range of currents, then do the whole thing twice more so that you have 3 mass readings for each current. Calculate the mean for each mass reading to improve the repeatability (see p.15).

Convert your mass readings into force using $F = mg$, then plot your data on a graph of force F against current I. Draw a line of best fit. You should find that you get a graph through the origin showing that force is proportional to current. Because $F = BIl$, the gradient of the line of best fit is equal to Bl. By keeping the length constant, you can divide the gradient by your value for l to get the magnetic flux density, B.

Tip: It's important to zero the balance when no current is flowing so that the mass reading is only due to the force caused by the current in the magnetic field (and not due to the mass of the equipment).

Tip: A dc power supply is used so that the direction of the force is constant — if an ac supply was used, the direction of the force would keep changing.

Example — Maths Skills

A student carries out the experiment described above to find the magnetic flux density, B. Complete the data table below and plot the results on a graph. Describe the relationship between force and current, and estimate a value for B.

Convert the masses in g to kg by dividing them by 1000.
Then use $W = mg$ to turn the masses into forces and draw the graph:

Tip: Remember $g = 9.81$ N Kg⁻¹.

current / A	mean mass recorded on top pan balance / g	force acting on current-carrying wire / N
1.0	0.20	0.0020
2.0	0.41	0.0040
3.0	0.61	0.0060
4.0	0.82	0.0080

Tip: Give the forces to the least number of s.f. that the data is given to.

Length l = 0.050 m throughout.

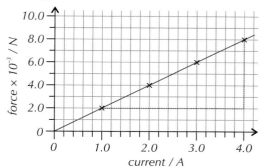

The graph is a straight line through the origin, so force is directly proportional to current.

$$Bl = \text{gradient} = \frac{\Delta y}{\Delta x} = \frac{(8.0 \times 10^{-3}) - (2.0 \times 10^{-3})}{4.0 - 1.0}$$
$$= 2.0 \times 10^{-3}$$

$$B = \text{gradient} \div l = 2.0 \times 10^{-3} \div 0.050$$
$$= 0.040 \text{ T}$$

The effect of length and flux density on force

You could also investigate how the length of the wire perpendicular to the magnetic field affects force by using different sized metal hoops so that the value of l changes while current I and the magnetic flux density B remain constant (you'll need to use the variable resistor to keep the current constant). Plotting a graph of force against length should give another straight line through the origin. In this case, the gradient of the line of best fit will give you BI. Dividing the gradient by the current I will give you the magnetic flux density, B, of the magnetic field.

Alternatively, you could keep the current and wire length the same and instead vary the magnetic field by changing the strength of the magnets used. Whichever variable you change, you should find the force is directly proportional to B, I, and l.

Tip: If you decide to change l, you'll need to make sure the length of the top of the largest metal hoop isn't longer than the length of the magnets — the whole length must be within the uniform magnetic field between the magnets.

Practice Question — Application

Q1 The diagram below shows an experiment set-up to determine the magnetic flux density of a uniform magnetic field between two slab magnets. A square wire hoop is placed such that the horizontal length l is perpendicular to the field between the magnets. The wire carries a current of 3.0 A. The horizontal length of the wire, l, is varied by using a range of different sizes of metal hoops. The hoop is connected to a dc power supply, a variable resistor and ammeter.

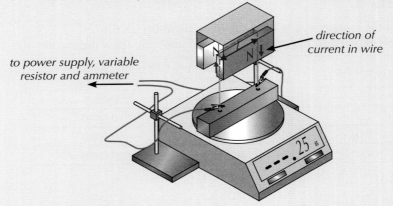

to power supply, variable resistor and ammeter

direction of current in wire

a) Draw a circuit diagram of the set-up described above.

b) Explain why a variable resistor is used in the circuit.

c) Explain why the vertical lengths of wire are unimportant and only length l is measured.

d) The mass shown on the balance is recorded for a range of currents. Explain how this information can be used to determine the magnetic flux density.

Tip: If you need a reminder of circuit symbols flick back to p.103 and p.126, or have a look over your year 1 of A-Level physics notes for a full reminder.

3. Forces on Charged Particles

*Any charged particle in a magnetic field feels a force as long as its moving —
you need to know how to calculate the size and direction of the force.*

Charged particles in a magnetic field

A force acts on a charged particle moving in a magnetic field. This is why a
current-carrying wire experiences a force in a magnetic field (page 123)
— electric current in a wire is the flow of negatively charged electrons.

- The force on a current-carrying wire in a magnetic field perpendicular
  to the current is given by $F = BIl$ (page 124).
- Electric current, I, is the flow of charge, Q, per unit time, t. So $I = \frac{Q}{t}$.
- A charged particle which moves a distance l in time t has
  a velocity, $v = \frac{l}{t}$. So $l = vt$.

Putting all these equations together gives the force acting on a single
charged particle moving through a magnetic field, where its velocity is
perpendicular to the magnetic field:

$$F = BIl = B\frac{Q}{t}\,v\,t$$

F = force in N
Q = charge on the particle in C

$$F = BQv$$

B = magnetic flux density in T
v = velocity of the particle in ms^{-1}

Learning Objectives:

- Understand that a charged particle moving through a magnetic field experiences a force.
- Be able to calculate the force acting on a charged particle whose velocity is perpendicular to a uniform magnetic field with $F = BQv$.
- Know how to work out the direction a force will act in for positive and negative charged particles.
- Know that a charged particle follows a circular path in a magnetic field perpendicular to its velocity.
- Know how the circular path of charged particles can be applied in devices such as the cyclotron.

Specification Reference 3.7.5.2

Example — **Maths Skills**

**An electron travels at a velocity of 2.00×10^4 ms^{-1} perpendicular to a
uniform magnetic field of strength 2.00 T. What is the magnitude of the
force acting on the electron? (The magnitude of the charge on an electron
is 1.60×10^{-19} C.)**

Just use the equation $F = BQv$ and put the correct numbers in:

$$F = BQv$$
$$= 2.00 \times (1.60 \times 10^{-19}) \times (2.00 \times 10^4)$$
$$= 6.40 \times 10^{-15} \text{ N}$$

The circular path of particles

By Fleming's left-hand rule the force on a moving charge travelling
perpendicular to a magnetic field is always perpendicular to its direction of
travel. Mathematically, that is the condition for circular motion (page 24).

To use Fleming's left-hand rule (page 123) for charged particles, use
your second finger (normally current) as the direction of motion for a positive
charge. If the particle carries a negative charge (i.e. an electron), point your
second finger in the opposite direction to its motion.

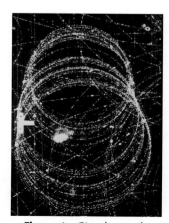

Figure 1: *Circular tracks
made by charged particles
in a cloud chamber with an
applied magnetic field.*

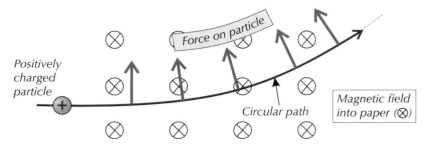

Figure 2: *A charged particle moving perpendicular to a magnetic field follows a circular path.*

Tip: The acceleration is towards the centre of the circle — it is centripetal (see page 23).

Tip: You'll often need to draw 3D situations in 2D like this (or interpret 2D diagrams as a 3D situation). Just remember that a cross always means "into the page".

Tip: Newton's laws of motion were covered in year 1 of A-Level physics — if you need a reminder, have a look at your year 1 notes.

Tip: A 'v' is cancelled from the top and bottom in the final rearrangement here.

The force due to the magnetic field ($F = BQv$) experienced by a particle travelling through a magnetic field is independent of the particle's mass, but the centripetal acceleration it experiences will depend on the mass — from Newton's 2nd law of motion.

- The particle's acceleration will be $a = \frac{v^2}{r}$ (see page 24).
- Combining this with Newton's 2nd law, $F = ma$, gives the force on a particle in a circular orbit $F = \frac{mv^2}{r}$.

The radius of the circular path followed by charged particles in a magnetic field can be found by combining the equations for the force on a charged particle in a magnetic field and for the force on a particle in a circular orbit.

$$F = \frac{mv^2}{r} \quad \text{and} \quad F = BQv \quad \Rightarrow \quad \frac{mv^2}{r} = BQv$$

$$\Rightarrow \quad r = \frac{mv}{BQ}$$

This means that:

- The radius of curvature increases (i.e. the particle is deflected less) if the mass or velocity of the particle increase.
- The radius of curvature decreases (i.e. the particle is deflected more) if the strength of the magnetic field or the charge on the particle increase.

Examples — **Maths Skills**

For each of the following, say which of the two particles would follow a circular path with the smaller radius in a magnetic field of flux density B.

a) **A carbon-12 nucleus with velocity v, relative mass of 12 and relative charge of +6, and a carbon-14 nucleus with velocity v, relative mass of 14 and relative charge of +6.**

The radius of the circular path followed by the particles is given by:

$$r = \frac{mv}{BQ}$$

In this case v, B and Q are identical for both particles but m is larger for the carbon-14 nucleus. As r is directly proportional to m, the carbon-12 nucleus will follow the circular path with the smaller radius.

b) **A carbon-14 nucleus with velocity v, relative mass of 14 and relative charge of +6, and a nitrogen-14 nucleus with velocity v, relative mass of 14 and relative charge of +7.**

m, v and B are identical for both particles but Q is larger for the nitrogen nucleus. As r is inversely proportional to Q, the nitrogen nucleus will follow the circular path with the smaller radius.

- The frequency of rotation for an object in circular motion is given by its velocity (*v*) divided by the distance it travels in each rotation ($2\pi r$):

$$f = \frac{v}{2\pi r}$$

Tip: $2\pi r$ is the circumference of a circle with radius *r*. There's more on the properties of circles on page 411.

- You can combine this with the formula for *r* on the previous page to get an expression for the frequency of rotation in terms of **B**, Q and *m*:

$$f = \frac{v}{2\pi r} \quad \text{and} \quad r = \frac{mv}{BQ} \quad \Rightarrow \quad f = \frac{\cancel{v}}{2\pi\left(\frac{m\cancel{v}}{BQ}\right)} = \frac{BQ}{2\pi m}$$

So the frequency of rotation of a charged particle in a magnetic field is independent of its velocity. The time it takes a particle to complete a full circle depends only on the magnetic flux density and its mass and charge. Increasing a particle's velocity will make it follow a circular path with a larger radius, but it will take the same amount of time to complete it. This is particularly relevant for particle accelerators.

Tip: For more on circular motion, see pages 19-25.

Particle accelerators

This effect is used in particle accelerators such as cyclotrons. Cyclotrons have many uses, for example in medicine. Cyclotrons can be used to produce radioactive tracers or high-energy beams of radiation for use in radiotherapy.

A cyclotron is made up of two hollow semicircular electrodes with a uniform magnetic field applied perpendicular to the plane of the electrodes, and an alternating potential difference applied between the electrodes.

Tip: The electrodes are sometimes called "dees" because of their D-shape.

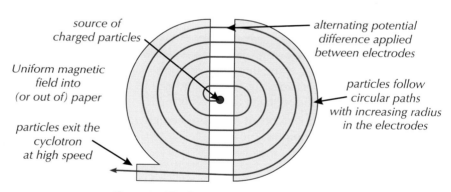

source of charged particles

alternating potential difference applied between electrodes

Uniform magnetic field into (or out of) paper

particles follow circular paths with increasing radius in the electrodes

particles exit the cyclotron at high speed

Figure 3: *The basic structure of a cyclotron.*

Figure 4: *A medical cyclotron used to produce radioactive tracers.*

Charged particles are produced and fired into one of the electrodes, where the magnetic field makes them follow a (semi)circular path and then leave the electrode. An applied potential difference between the electrodes then accelerates the particles across the gap until they enter the next electrode.

Because the particle's speed is slightly higher, it will follow a circular path with a larger radius (see previous page) before leaving the electrode again. At this point the direction of the potential difference will have been reversed and so the particle is accelerated again before entering the next electrode. This process repeats as the particle spirals outwards, increasing in speed, before eventually exiting the cyclotron.

Tip: If the potential difference wasn't alternating, the particle would slow down after leaving the second electrode.

Since the frequency of the circular motion doesn't depend on radius, the particle will always spend the same amount of time in each electrode (if **B**, Q and *m* are constant) so the alternating p.d. will have a fixed frequency.

Practice Questions — Application

Tip: Neutrons, along with protons, are one of the nucleons found in atoms. They have a mass of 1.67×10^{-27} kg and carry no charge.

Q1 Why would a neutron moving through a magnetic field perpendicular to its direction of motion not experience a force?

Q2 In which direction will the force due to the magnetic field act on the electron in the diagram below?

Electron

Magnetic field out of paper (●)

Q3 Find the force that acts on a particle with a charge of 3.2×10^{-19} C travelling at a velocity of 5.5×10^3 ms⁻¹ perpendicular to a magnetic field with a flux density of 640 mT.

Q4 a) The force needed to keep an object in circular motion is given by $\boldsymbol{F} = \frac{mv^2}{r}$. Combine this with $\boldsymbol{F} = \boldsymbol{BQv}$ to find an expression for the magnetic flux density $\boldsymbol{B}$ needed to keep a charged particle in circular motion for a given radius r and orbital speed $\boldsymbol{v}$.

b) In a circular particle accelerator with a radius of 5.49 m, protons are accelerated to 1.99×10^7 ms⁻¹. Find the magnetic flux density $\boldsymbol{B}$ that's required to keep the protons following the circular path of the accelerator. $m_\text{p} = 1.67 \times 10^{-27}$ kg and $Q_\text{p} = 1.60 \times 10^{-19}$ J.

Practice Questions — Fact Recall

Q1 If an electron is travelling through a uniform magnetic field perpendicular to its velocity, what shape will its path take? (Assume it has infinite space to move into.)

Q2 Briefly describe how a beam of high-speed charged particles is produced in a cyclotron.

4. Electromagnetic Induction

You might have come across electromagnetic induction before — it's the process at work in electromagnets, power generators, transformers, etc. It happens because of the force on charged particles in magnetic fields.

Magnetic flux

Magnetic flux density, **B**, is a measure of the strength of a magnetic field (or you can think of it as the number of field lines per unit area). The total **magnetic flux**, Φ, passing through an area, A, perpendicular to a magnetic field, **B**, is defined as:

Φ = magnetic flux in Wb (webers) $\longrightarrow$ $\boxed{\Phi = \boldsymbol{B}A}$ $\longleftarrow$ **B** = *magnetic flux density in T*

A = *area in m²*

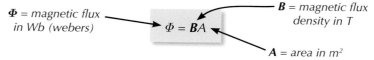

Example — **Maths Skills**

A square with sides of length 4.5 cm is placed in a magnetic field, normal to the field's direction. Find the magnetic flux passing through the square if the magnetic flux density is 0.92 T.

Start by finding the area of the square, remembering to convert the units:
$$A = (4.5 \times 10^{-2}) \times (4.5 \times 10^{-2}) = 2.025 \times 10^{-3} \text{ m}^2$$

Then just put the numbers into the equation above:
$$\Phi = \boldsymbol{B}A = 0.92 \times (2.025 \times 10^{-3}) = 1.863 \times 10^{-3}$$
$$= 1.9 \times 10^{-3} \text{ Wb (to 2 s.f.)}$$

Electromagnetic induction

If there is relative motion between a conducting rod and a magnetic field, the electrons in the rod will experience a force (see p.123), which causes them to accumulate at one end of the rod.

    This induces an **electromotive force (e.m.f.)** across the ends of the rod — this is called **electromagnetic induction**.

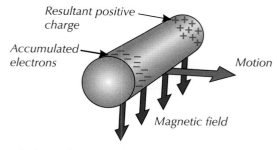

Resultant positive charge

Accumulated electrons

Motion

Magnetic field

Figure 1: *A conducting rod moving through a magnetic field.*

    An e.m.f. is induced by a conductor and a magnet when the conductor cuts the magnetic flux. The conductor can move and the magnetic field stay still or the other way round — you get an e.m.f. either way (see Figure 2).

    You can induce an e.m.f. in a flat coil or solenoid in the same way. In either case, the e.m.f. is caused by the magnetic field (or 'magnetic flux') that passes through the coil changing. If the coil is part of a complete circuit, an induced current will flow through it.

Learning Objectives:

- Know that magnetic flux is given by $\Phi = \boldsymbol{B}A$, where **B** is normal to A.

- Understand that the magnetic flux linkage is $N\Phi$, where N is the number of turns in a coil cutting the flux.

- Be able to calculate the magnetic flux and the magnetic flux linkage of a rectangular coil rotated in a magnetic field, e.g. using $N\Phi = \boldsymbol{B}AN\cos\theta$

Specification Reference 3.7.5.3

Tip: You can only use the equation $\Phi = \boldsymbol{B}A$ if **B** is normal to A (see below) — otherwise there's an extra term in the equation, which you'll see on page 135.

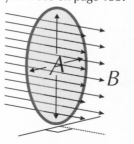

Tip: Think of magnetic flux as the total number of field lines.

Tip: E.m.f. is another way of saying voltage — you saw it in year 1 of A-Level physics if you need a reminder. See pages 139-142 for more on induced e.m.f.

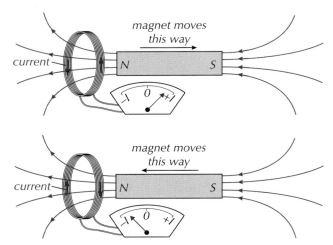

Figure 2: *Current and e.m.f. induced when a magnet is moved towards and away from a coil.*

Tip: The diagram shows a side-on view of a rectangular coil normal to a magnetic flux density **B**.

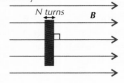

Tip: The unit of both flux linkage and Φ is the weber, Wb. Even though N has no units, the word "turns" is sometimes added to the units for magnetic flux, so you might see "weber-turns" used.

Tip: Don't be tempted to cancel down the 'N's here — consider the flux linkage, $N\Phi$, as a stand-alone term.

Flux linkage

When a wire coil is moved in a magnetic field, the size of the e.m.f. induced depends on the magnetic flux passing through the coil, Φ, and the number of turns on the coil cutting the flux. The product of these is called the flux linkage, $N\Phi$. For a coil of N turns normal to **B**, the **flux linkage** is given by:

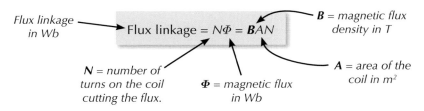

Flux linkage in Wb → Flux linkage $= N\Phi = BAN$ ← **B** = magnetic flux density in T

N = number of turns on the coil cutting the flux.

Φ = magnetic flux in Wb

A = area of the coil in m^2

The rate of change in flux linkage tells you how strong the electromotive force will be in volts:

> A change in flux linkage of one weber per second will induce an electromotive force of 1 volt in a loop of wire.

Example — Maths Skills

The flux linkage of a coil with a cross-sectional area of 0.33 m^2 normal to a magnetic field of flux density 0.15 T is 4.0 Wb. How many turns are in the coil?

Just rearrange the equation for flux linkage to make N the subject and then put the numbers in:

$$[N\Phi] = BAN \Rightarrow N = \frac{[N\Phi]}{BA}$$

$$= \frac{4.0}{0.15 \times 0.33} = 80.808...$$

$$= 81 \text{ turns (to 2 s.f.)}$$

Flux linkage at an angle

When the magnetic flux isn't perpendicular to the area you're interested in (e.g. Figure 3), you need to use trigonometry to resolve the magnetic field vector into components that are parallel and perpendicular to the area (see page 412).

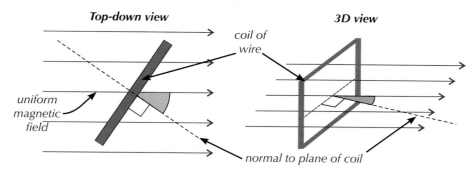

Top-down view   **3D view**

coil of wire

uniform magnetic field

normal to plane of coil

Figure 3: *A coil at an angle to a uniform magnetic field.*

Tip: Pulling the coil in Figure 3 out of the magnetic field would induce an e.m.f. — see page 139.

To find the magnetic flux passing through a coil, you're interested in the component of the magnetic field perpendicular to the area of the coil. By trigonometry, this is equal to $B\cos\theta$ — see Figure 4.

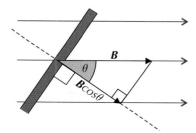

B

θ

$B\cos\theta$

Tip: Remember SOC CAH TOA:
$\cos\theta$ = adjacent ÷ hypotenuse

See page 412 for more.

Figure 4: *For a coil at an angle to a magnetic field, the component of the field perpendicular to the area of the coil is $B\cos\theta$, where θ is the angle between the field and the normal to the plane of the coil.*

So for a single loop of wire when B is not perpendicular to area, you can find the magnetic flux using this equation:

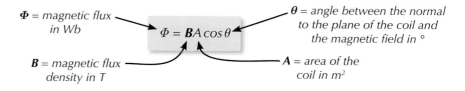

Φ = magnetic flux in Wb

θ = angle between the normal to the plane of the coil and the magnetic field in °

$$\Phi = BA\cos\theta$$

B = magnetic flux density in T

A = area of the coil in m^2

And for a coil with N turns, you can find the flux linkage with the equation:

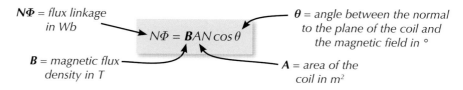

$N\Phi$ = flux linkage in Wb

θ = angle between the normal to the plane of the coil and the magnetic field in °

$$N\Phi = BAN\cos\theta$$

B = magnetic flux density in T

A = area of the coil in m^2

Example ─ Maths Skills ─────────────────────────────

A rectangular coil of wire with exactly 200 turns and sides of length
5.00 cm and 6.51 cm is rotating in a magnetic field with $B = 8.56 \times 10^{-3}$ T.
Find the flux linkage of the coil when the normal to the area of the coil is
at 12.6° to the magnetic field, as shown below.

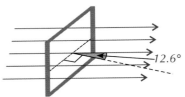

First find the area of the coil:

Area = $(5.00 \times 10^{-2}) \times (6.51 \times 10^{-2}) = 3.255 \times 10^{-3}$ m²

Then just put the numbers into the equation:

$N\Phi = BAN\cos\theta$
$= (8.56 \times 10^{-3}) \times (3.255 \times 10^{-3}) \times 200 \times \cos 12.6° = 5.438... \times 10^{-3}$
$= 5.44 \times 10^{-3}$ Wb (to 3 s.f.)

Tip: You could also give
your answer in Wb turns
— see page 134.

Practice Question — Application

Q1 A student sets up an experiment to measure the strength of the Earth's
magnetic field, **B**. He does this by measuring the flux linkage of a
rectangular wire coil.

a) Explain why the flux linkage of the coil changes when he rotates
the coil.

b) The student finds the highest value of flux linkage to be
1.3×10^{-6} Wb. If he used a coil with an area of
25 cm² and exactly 10 turns, find the local value of **B**.

c) Suggest two changes he could make to the coil to reduce the
uncertainty in his measurement of **B**.

Tip: See page 10 for
more on reducing
uncertainty.

Practice Questions — Fact Recall

Q1 What is induced when a magnetic field through a conductor changes?

Q2 Write down the equation that links magnetic flux, magnetic flux
density and the area A that's perpendicular to the field B that the
magnetic flux passes through.

Q3 What are the units of flux linkage?

Q4 Write down the formula for flux linkage through a coil that is not
perpendicular to the magnetic field, and briefly explain what each
term represents.

5. Investigating Flux Linkage

You need to know how to use a search coil and an oscilloscope to investigate how the e.m.f. induced in the search coil varies with the angle of the plane of the search coil to the solenoid.

Investigating flux linkage with a search coil

You can investigate the effect of angle to the flux lines on effective magnetic flux linkage in a search coil using the apparatus shown in Figure 1.

REQUIRED PRACTICAL **11**

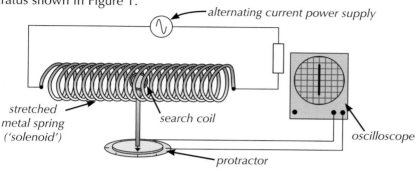

Figure 1: An experiment to investigate the effect on magnetic flux linkage of varying the angle between a search coil and magnetic field direction.

The stretched metal spring acts as a solenoid when connected to an alternating power supply. The alternating supply means the magnitude field of the solenoid is constantly changing — this means the flux through the search coil is changing which can induce an e.m.f. Make sure the peak of the ac voltage from the ac power supply stays the same throughout the experiment .

The search coil should have a known area and a set number of loops of fine wire. It is connected to an oscilloscope (see page 143) to record the induced e.m.f. in the coil. Set up the oscilloscope so that it only shows the amplitude of the e.m.f. as a vertical line (you'll need to turn off the time base).

A protractor is used to measure the orientation of the normal to the area of the search coil as an angle from the line of the magnetic field (see Figure 2).

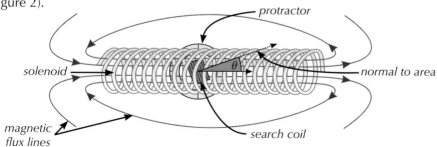

Figure 2: Overhead view showing position of search coil and protractor relative to the solenoid and its magnetic field.

Position the search coil so that it is about halfway along the solenoid and within the inside of the solenoid (but not touching it). The magnetic field of a solenoid is strongest inside the solenoid (see page 123), and can be assumed to be uniform and parallel to the side of the solenoid. Orientate the search coil so that it is parallel to the solenoid ($\theta = 0°$) and its area is perpendicular to the field. Then record the induced e.m.f. in the search coil from the amplitude of the oscilloscope trace.

Tip: You can use any metal spring as the solenoid in this experiment as long as it can be stretched — it must be easy to insert the search coil.

Tip: You must make sure you do a full risk assessment before carrying out this experiment. Make sure the current in the solenoid is not too high and take care not to touch the wires

Tip: If you need a reminder of the shape of the magnetic field surrounding a solenoid, see page 123.

Tip: See pages 143-144 for more information on how to use oscilloscopes.

Rotate the search coil so its angle to the solenoid and the magnetic flux lines (θ in Figure 2) changes by 10°. Record the induced e.m.f. and repeat until you have rotated the search coil by 90°. You'll find that as you turn the search coil, the induced e.m.f. decreases. This is because the search coil is cutting fewer flux lines as the component of the magnetic field perpendicular to the area of the coil gets lower, so the total magnetic flux passing through the search coil is lower. This means that the magnetic flux linkage experienced by the coil is lower.

Plot a graph of induced e.m.f. against θ. You should find that the induced e.m.f. is a maximum at 0°, and a zero at 90°.

Practice Question — Application

Q1 An experiment is set up as shown below. The search coil is rotated so that the angle of the search coil as recorded using the protractor changes from 90° to 0°.

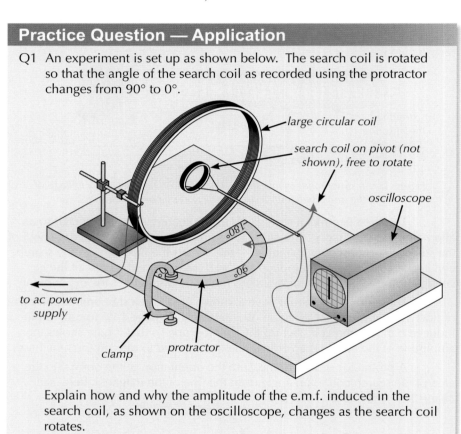

Explain how and why the amplitude of the e.m.f. induced in the search coil, as shown on the oscilloscope, changes as the search coil rotates.

6. Faraday's Law and Lenz's Law

There's more electromagnetic induction coming up for you to sink your teeth into. In this topic, you'll see how Faraday and Lenz tried to explain the phenomenon better with the inventively named Faraday's law and Lenz's law.

Faraday's law

The results from the last topic can be summed up in **Faraday's law**:

> Induced e.m.f. is directly proportional to the rate of change of flux linkage.

Faraday's law also gives us the magnitude of the induced e.m.f.:

> $$\frac{\text{Magnitude of}}{\text{induced e.m.f.}} = \frac{\text{Rate of change}}{\text{of flux linkage}}$$

Or, in symbols:

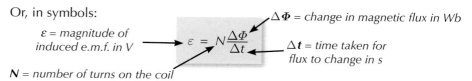

ε = magnitude of induced e.m.f. in V

$\Delta\varPhi$ = change in magnetic flux in Wb

$$\varepsilon = N\frac{\Delta\phi}{\Delta t}$$

Δt = time taken for flux to change in s

N = number of turns on the coil

The magnitude of the e.m.f. is shown by the gradient of a graph of flux linkage ($N\varPhi$) against time. The area under the graph of the magnitude of e.m.f. against time gives the flux linkage change.

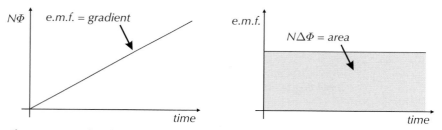

Figure 1: *Graphs of flux linkage against time (left) and magnitude of e.m.f. against time (right) for a conductor moving through a magnetic field at a constant speed.*

Example — Maths Skills

The diagram below shows a conducting rod of length *l* moving through a perpendicular uniform magnetic field, *B*, at a constant velocity, *v*. Show that the magnitude of the e.m.f. induced in the rod is equal to *Blv*.

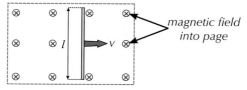

magnetic field into page

Displacement, $\mathbf{s} = \mathbf{v}\Delta t$ (displacement = velocity × time). Area of flux it cuts, $A = lv\Delta t$. Total magnetic flux cut through, $\Delta\varPhi = \mathbf{B}A = \mathbf{B}lv\Delta t$

Faraday's law gives $\varepsilon = N\dfrac{\Delta\varPhi}{\Delta t} = \dfrac{\Delta\varPhi}{\Delta t}$ (since $N = 1$)

So the magnitude of the induced e.m.f., $\varepsilon = \dfrac{\Delta\varPhi}{\Delta t} = \dfrac{\mathbf{B}lv\Delta t}{\Delta t} = \mathbf{B}lv$

Tip: Remember, a change in flux linkage of 1 Wb per second will induce an e.m.f. of 1 V in a loop of wire.

Figure 2: *British physicist Michael Faraday.*

Exam Tip
You might get asked to find the e.m.f. induced by the Earth's magnetic field across the wingspan of a plane — think of it as a moving straight conducting rod.

Tip: There's only 1 'turn' in a conducting rod, so you can ignore the N in Faraday's law.

Lenz's law

The direction of an induced e.m.f. (and current) are given by **Lenz's law**:

> The induced e.m.f. is always in such a direction as to oppose the change that caused it.

Tip: Remember, a current is only induced if the conductor is part of a complete circuit — if it isn't an e.m.f. will be induced but no current will flow.

The idea that an induced e.m.f. will oppose the change that caused it agrees with the principle of the conservation of energy — the energy used to pull a conductor though a magnetic field, against the resistance caused by magnetic attraction, is what produces the induced current.

Lenz's law can be used to find the direction of an induced e.m.f. and current in a conductor travelling at right angles to a magnetic field.

Example

Lenz's law says that the induced e.m.f. will produce a force that opposes the motion of the conductor — in other words a resistance. Picture a straight conductor being moved down through a perpendicular magnetic field:

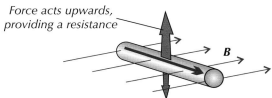

Motion of conductor

Direction of induced e.m.f. (and current)

Using Fleming's left-hand rule (see p.123), point your thumb in the direction of the force of resistance — which is in the opposite direction to the motion of the conductor. Point your first finger in the direction of the field. Your second finger will now give you the direction of the induced e.m.f.

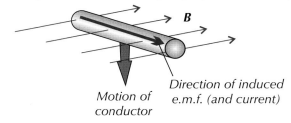

Force acts upwards, providing a resistance

Tip: If you've forgotten which finger is which, here's a reminder:

Force

Field

Current (+ to −)

If the conductor is connected as part of a circuit, a current will be induced in the same direction as the induced e.m.f.

Induced e.m.f. in a rotating coil

Tip: Note the axis of rotation is perpendicular to the magnetic field.

When a coil such as that in Figure 3 rotates uniformly (at a steady speed) in a magnetic field, the coil cuts the flux and an alternating e.m.f. is induced.

The amount of flux cut by the coil (flux linkage) is given by $N\Phi = BAN\cos\theta$ (see page 135). As the coil rotates, θ changes so the flux linkage varies sinusoidally between $+BAN$ and $-BAN$.

Tip: 'Sinusoidally' means it follows the same pattern as a sin (or cos) curve — see the next page for more.

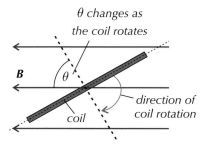

θ changes as the coil rotates

B

direction of coil rotation

coil

Figure 3: *The value of θ changes as the coil rotates.*

How fast θ changes depends on the angular speed, ω, of the coil (see page 20), $\theta = \omega t$. So you can write:

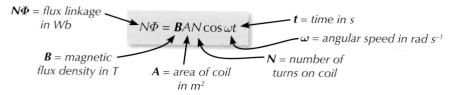

$N\Phi$ = flux linkage in Wb

B = magnetic flux density in T

A = area of coil in m^2

$$N\Phi = BAN\cos\omega t$$

t = time in s

ω = angular speed in rad s^{-1}

N = number of turns on coil

Tip: 'rad' (short for radians) is just another unit for measuring angles. It's the one that's normally used in circular motion (page 19).

The induced e.m.f., ε, depends on the rate of change of flux linkage (Faraday's law), so it also varies sinusoidally. The equation for the e.m.f. at time t is:

ε = induced e.m.f. in V

$$\varepsilon = BAN\omega\sin\omega t$$

Tip: You might see induced e.m.f. given as E instead of ε.

Tip: Flux linkage and induced e.m.f. are 90° out of phase.

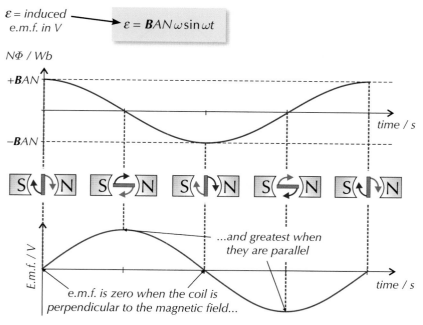

$N\Phi$ / Wb

$+BAN$

$-BAN$

time / s

...and greatest when they are parallel

E.m.f. / V

time / s

e.m.f. is zero when the coil is perpendicular to the magnetic field...

Tip: Flux linkage is at a maximum at 0° and 180° and zero at 90° and 270°. E.m.f. is at a maximum at 90° and 270° and zero at 0° and 180°.

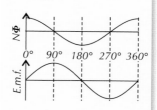

Figure 4: The sinusoidal graphs produced by plotting flux linkage and induced e.m.f. against time for a rotating coil in a uniform magnetic field.

Example — Maths Skills

A rectangular coil with 30.0 turns, each with an area of 0.200 m^2, is rotated as shown in the diagram at 20.0 rad s^{-1} in a uniform 1.50 mT magnetic field. Calculate the maximum e.m.f. induced in the coil.

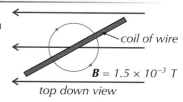

coil of wire

$B = 1.5 \times 10^{-3}$ T

top down view

$\varepsilon = BAN\omega\sin\omega t$, so ε will be greatest when $\sin\omega t = \pm1$. This gives:

$\varepsilon = BAN\omega\sin\omega t$

$\quad = 1.50 \times 10^{-3} \times 0.200 \times 30.0 \times 20.0 \times \pm1$

$\quad = \pm0.180$ V

Tip: $\sin\theta = \pm1$ when $\theta = \frac{\pi}{2}, \frac{3\pi}{2}, \frac{5\pi}{2}$ etc.

The shape of the graph of induced e.m.f. can be altered by changing the speed of rotation or the size of the magnetic field:

▪ Increasing the speed of rotation will increase the frequency and increase the maximum e.m.f.

▪ Increasing the magnetic flux density B will increase the maximum e.m.f., but will have no effect on the frequency.

Tip: The changes are directly proportional — doubling the speed of rotation doubles the maximum e.m.f. and halving the speed halves the maximum e.m.f. etc.

Below is a graph (in green) of induced e.m.f. against time for a coil, rotating with angular speed ω in a magnetic field with a flux density of B. Sketch, on the same axes, a graph of induced e.m.f. against time for the same coil rotating with angular speed 2ω in a magnetic field with a flux density of $0.25B$.

Doubling the speed will double the maximum e.m.f. and double the frequency. Dividing the flux density by 4 will have no effect on the frequency but will divide the maximum e.m.f. by 4. So the second graph (blue) will have double the frequency but half the amplitude of the first graph.

Tip: The amplitude is halved because it's doubled and then divided by 4.

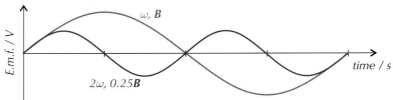

Generators

Generators, or dynamos, convert kinetic energy into electrical energy — they induce an electric current by rotating a coil in a magnetic field.

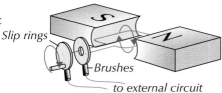

Figure 5: A simple alternator.

Figure 5 shows a simple alternator — a generator of alternating current. It has slip rings and brushes to connect the coil to an external circuit. The output voltage and current change direction with every half rotation of the coil, producing an alternating current.

Practice Questions — Application

Q1 Why is an e.m.f. often induced across the wings of a moving plane?

Q2 A wire coil with exactly 50 turns and an area of 0.24 m² is placed perpendicular to a magnetic field with flux density 1.5 T and rotated with a steady speed of 4.0π rad s⁻¹.

   a) Find the maximum e.m.f. induced across the wire coil.

   b) Sketch a graph of induced e.m.f. against time for the wire coil between 0 and 0.5 seconds.

   c) The speed of the coil's rotation is doubled. Sketch a new graph of induced e.m.f. against time for the coil between 0 and 0.5 s.

Tip: There are 2π rad in 360°.

Practice Questions — Fact Recall

Q1 What's Faraday's law?

Q2 Write down an expression for rate of change of flux linkage.

Q3 What's Lenz's law?

Q4 Sketch a graph of flux linkage against time for:

   a) A conductor moving in a straight line at a steady speed perpendicular to a uniform magnetic field.

   b) A wire coil rotating at a steady speed in a uniform magnetic field.

7. Alternating Current

Compared to induction this is a nice simple topic, thank goodness. You might remember this stuff from GCSE, but here it is again in a bit more detail.

What is alternating current?

An **alternating current** is one that changes direction with time. This means the voltage across a resistance goes up and down in a regular pattern — some of the time it's positive and some of the time it's negative.

Measuring alternating voltage with an oscilloscope

You can use an oscilloscope to display the voltage of an alternating current (and of a direct current too). The trace you see is made by an electron beam moving across a screen. The time base controls how fast the beam is moved across the screen. You can set this using a dial on the front of the oscilloscope.

Oscilloscopes are basically just voltmeters. The vertical height of the trace at any point shows the input voltage at that point. The oscilloscope screen has a grid on it — you can select how many volts per division you want the *y*-axis scale to represent using the *Y*-gain control dial, e.g. 5 V per division. On some oscilloscopes, the height of each square on the grid is 1 cm, so the scale may be set in terms of V per cm (Vcm^{-1}).

— **Example** —————————————

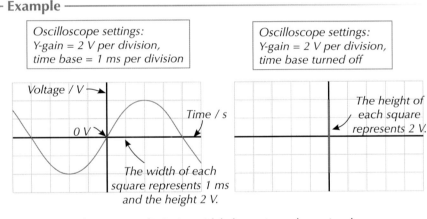

Oscilloscope settings:
Y-gain = 2 V per division,
time base = 1 ms per division

Oscilloscope settings:
Y-gain = 2 V per division,
time base turned off

The height of each square represents 2 V.

The width of each square represents 1 ms and the height 2 V.

Figure 1: *(Left) A sinusoidal alternating voltage signal.*
(Right) A sinusoidal alternating voltage with the time base turned off.

An alternating current (ac) source gives a regularly repeating sinusoidal waveform (see Figure 1). A direct current (dc) source is always at the same voltage, so you get a horizontal line (see Figure 3). Oscilloscopes can display ac voltage as a vertical line and dc voltage as a dot if you turn off the time base.

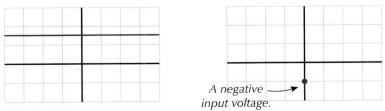

A negative input voltage.

Figure 3: *(Left) A dc supply shown on an oscilloscope.*
(Right) A different dc supply shown on an oscilloscope with the time base turned off.

Learning Objectives:

- Be familiar with the operation of the controls of an oscilloscope.
- Know how an oscilloscope can be used as a dc and ac voltmeter, and how to use it to measure time intervals and frequencies, and to display ac waveforms.
- Know how to calculate the peak, peak-to-peak and rms values for sinusoidal voltages and currents.
- Be able to use the equations $V_{rms} = \dfrac{V_0}{\sqrt{2}}$ and $I_{rms} = \dfrac{I_0}{\sqrt{2}}$.
- Be able to calculate the mains electricity peak voltage and peak-to-peak voltage values.

Specification Reference 3.7.5.5

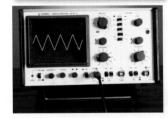

Figure 2: *An oscilloscope showing a wave form. The control dials can be used to change how the wave is displayed.*

Tip: If something is sinusoidal it means it varies like a sine curve. A sine curve is a smooth, repetitive wave, like the one shown on the left in Figure 1.

Tip: There's more coming up on how to read an oscilloscope on the next page.

Analysing oscilloscopes

Tip: The peak voltage is useful to know, though it's often easier to measure the peak-to-peak voltage and halve it.

There are three basic pieces of information you can get from an ac oscilloscope trace — the time period, T, the peak voltage, V_0, and the peak-to-peak voltage (see Figure 4).

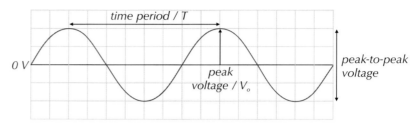

Figure 4: A diagram showing the three basic pieces of information you can get from an oscilloscope trace.

Tip: Remember, you can use this formula to calculate the frequency of any type of wave.

Measuring the distance between successive peaks along the time axis (the horizontal axis) gives you the time period (as long as you know the time base setting). You can use this to calculate the frequency:

$$\text{frequency} = \frac{1}{\text{time period}} \qquad f = \frac{1}{T}$$

Examples — Maths Skills

The diagram below shows the output on an oscilloscope of an alternating current.

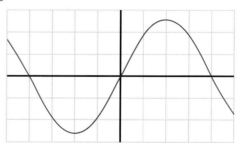

The Y-gain dial is set to 0.50 V per division and the time base is set to 1.0 s per division.

Calculate: a) the peak voltage of the wave,

b) the frequency of the wave.

a) The peak voltage is the height from the 0 V line to the top of the peak. The peak is 2.5 squares high.
So the peak voltage = 2.5 × 0.50 = 1.25 = 1.3 V (to 2 s.f.)

b) First you have to work out the time period of the wave. The time base is set to 1.0 s, so every square on the grid represents 1.0 s. The time period of the wave is 8 squares = 8 seconds.

$$\text{frequency} = \frac{1}{\text{time period}} = \frac{1}{8} = 0.125\,\text{Hz} = 0.13\ \text{Hz (to 2 s.f.)}$$

Figure 5: A sine wave shown on an oscilloscope.

Tip: A sine wave produces a sine curve on an oscilloscope (see previous page for more on sinusoidal curves).

Describing an alternating current

An ac supply with a peak voltage of 2 V will be below 2 V most of the time. That means it won't have as high a power output as a 2 V dc supply. To compare them properly, you need to average the ac voltage somehow. A normal average won't work, because the positive and negative bits cancel out. It turns out that something called the **root mean square (rms) voltage** does the trick. For a sine wave, you get the rms voltage, V_{rms}, by dividing the peak voltage, V_0, by $\sqrt{2}$. You do the same to calculate the **rms current** I_{rms}:

Tip: There's more on root mean square values on page 62.

$$V_{rms} = \frac{V_0}{\sqrt{2}}$$
V_0 = peak voltage in volts (V)

$$I_{rms} = \frac{I_0}{\sqrt{2}}$$
I_0 = peak current in amperes (A)

Tip: Even though this is only strictly true if the ac signal is a sine wave, it's also the only type of alternating signal that will come up in the exam, so I wouldn't worry too much about it.

If you want to work out the average power for an ac supply, just replace I and V in the power formula, $P = IV$ with the rms values:

$$\text{average power} = I_{rms} \times V_{rms}$$

Tip: You should remember the equation for power, $P = IV$, from year 1 of A-Level physics.

─ Example ─ **Maths Skills** ─────────

A light is powered by a sinusoidal ac power supply with a peak voltage of 2.12 V and a root mean square current of 0.40 A.

a) **Calculate the root mean square voltage of the power supply.**

$V_{rms} = \dfrac{V_0}{\sqrt{2}} = \dfrac{2.12}{\sqrt{2}} = 1.499... = 1.50\,\text{V}$ (to 3 s.f.)

b) **Calculate the average power of the power supply.**

$\text{average power} = I_{rms} \times V_{rms} = 0.40 \times 1.499... = 0.5996...$
$\qquad\qquad\qquad\qquad\qquad = 0.60\,\text{W}$ (to 2 s.f.)

Tip: You'll get given the equations for V_{rms} and I_{rms} in your exam data and formulae booklet.

Mains electricity

It's usually the rms voltage that's stated on a power supply. For example, the value of 230 V stated for the UK mains electricity supply is the rms value.

─ Example ─ **Maths Skills** ─────────

To calculate the peak voltage or peak-to-peak voltage of the UK mains electricity supply, just rearrange $V_{rms} = \dfrac{V_0}{\sqrt{2}}$ into $V_0 = \sqrt{2}\,V_{rms}$:

$V_0 = \sqrt{2} \times V_{rms} = \sqrt{2} \times 230 = 325.26... = 330\,\text{V}$ (to 2 s.f.)

$V_{\text{peak-to-peak}} = 2 \times V_0 = 2 \times 325.26... = 650.53...$
$\qquad\qquad\quad = 650\,\text{V}$ (to 2 s.f.)

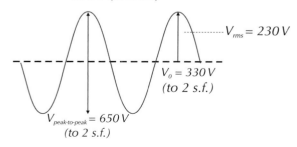

$V_{rms} = 230\,V$

$V_0 = 330\,V$ (to 2 s.f.)

$V_{\text{peak-to-peak}} = 650\,V$ (to 2 s.f.)

Practice Questions — Application

Q1 A sinusoidal ac supply provides a peak current of 8.2 A and a peak voltage of 9.0 V to a circuit.

   a) Calculate the root mean square current provided by the supply.

   b) Calculate the root mean square voltage of the supply.

Q2 The USA mains power supply has an rms voltage of 120 V. It is used to power a heater. The heater has an rms current flowing through it of 20.0 A.

   a) Calculate the peak voltage of the USA mains supply.

   b) Calculate the peak-to-peak voltage of the USA mains supply.

   c) Calculate the average power of the heater.

Q3 The diagram below shows an oscilloscope trace of a sinusoidal ac on a centimetre square grid.

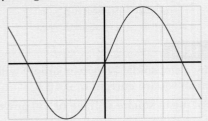

The time base of the oscilloscope is set to 20.0 ms cm^{-1} and the Y-gain is set at 0.50 V cm^{-1}.

   a) Calculate the frequency of the wave.

   b) Calculate the root mean square voltage of the ac current.

Practice Questions — Fact Recall

Q1 What is the trace on an oscilloscope created by?

Q2 What would an alternating current look like on an oscilloscope if the time base was turned off?

Q3 Why is it necessary to calculate the V_{rms} of an alternating power supply to be able to compare it to a dc power supply?

Q4 Write down the equation for calculating the root mean square current.

8. Transformers

Transformers are important in the transfer of electricity from power stations to the nation and are really useful in all sorts of electrical equipment. They're responsible for the buzzing you might have heard in chargers for laptops, speakers, and many more devices.

What's a transformer?

Transformers are devices that make use of electromagnetic induction to change the size of the voltage for an alternating current — see Figure 1.

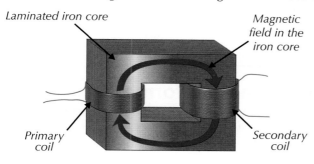

Figure 1: The basic structure of a (step-up) transformer.

An alternating current flowing in the primary (or input) coil causes the core to magnetise, demagnetise and remagnetise continuously in opposite directions. This produces a rapidly changing magnetic flux across the core. Because of this, a magnetically soft material is needed — usually iron or a special alloy.

The rapidly changing magnetic flux in the iron core passes through the secondary (or output) coil, where it induces an alternating voltage of the same frequency (but different voltage, assuming the number of turns is different). From Faraday's law (page 139), the voltage in both the primary and secondary coils can be calculated:

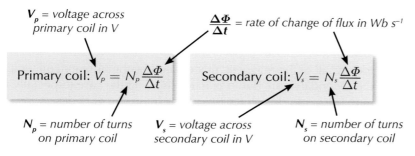

V_p = voltage across primary coil in V

$\dfrac{\Delta\Phi}{\Delta t}$ = rate of change of flux in Wb s⁻¹

Primary coil: $V_p = N_p \dfrac{\Delta\Phi}{\Delta t}$ Secondary coil: $V_s = N_s \dfrac{\Delta\Phi}{\Delta t}$

N_p = number of turns on primary coil V_s = voltage across secondary coil in V N_s = number of turns on secondary coil

These can be combined to give the equation for an ideal transformer:

$$V_p = N_p \frac{\Delta\Phi}{\Delta t} \Rightarrow \frac{V_p}{N_p} = \frac{\Delta\Phi}{\Delta t} \qquad V_s = N_s \frac{\Delta\Phi}{\Delta t} \Rightarrow \frac{V_s}{N_s} = \frac{\Delta\Phi}{\Delta t}$$

$$\Rightarrow \frac{V_p}{N_p} = \frac{V_s}{N_s} \quad \left(= \frac{\Delta\Phi}{\Delta t}\right)$$

Rearranging this gives:

$$\frac{N_s}{N_p} = \frac{V_s}{V_p}$$

Learning Objectives:

- Be able to use the equation for an ideal transformer, $\dfrac{N_s}{N_p} = \dfrac{V_s}{V_p}$, to find the number of turns or voltage on either coil.
- Know what can cause inefficiencies in a transformer.
- Know how eddy currents can be produced.
- Be able to calculate the efficiency of a transformer using efficiency $= \dfrac{I_s V_s}{I_p V_p}$
- Know how and why electrical power is transmitted at high voltage and be able to do calculations of power loss in transmission lines.

Specification Reference 3.7.5.6

Tip: A magnetically soft material is one whose magnetisation disappears after the current is removed.

Tip: The flux linkage will be higher in the coil with the most turns — it's proportional to the voltage in the coil.

Tip: In an ideal transformer, the flux through the secondary coil is the same as the flux through the primary coil and no energy is lost in the transfer.

Exam Tip
This is how the equation appears in your data and formulae booklet.

Step-up transformers increase the voltage by having more turns on the secondary coil than the primary. **Step-down transformers** reduce the voltage by having fewer turns on the secondary coil.

Figure 2: An early version of a transformer.

Figure 3: A transformer with both coils wrapped around the same core.

Tip: Resistivity is just a measure of how strongly a material opposes the flow of electric current.

Tip: Using low-resistance wires is particularly important in the secondary coil of a step-down transformer, or the primary coil of a step-up transformer, as they carry a higher current (see next page).

Tip: You can combine the two ideal transformer equations (this one and the one from the bottom of the last page) to give:
$$\frac{V_p}{V_s} = \frac{N_p}{N_s} = \frac{I_s}{I_p}$$

┌─ Example ── **Maths Skills** ──────────────────

What is the output voltage for a transformer with a primary coil of 120 turns, a secondary coil of 350 turns and an input voltage of 230 V?

Just use the equation for an ideal transformer and rearrange it to make the output voltage (V_s) the subject:

$$\frac{N_s}{N_p} = \frac{V_s}{V_p} \;\Rightarrow\; V_s = \frac{V_p N_s}{N_p}$$
$$= \frac{230 \times 350}{120} = 670.83...$$
$$= 670 \text{ V (to 2 s.f.)}$$

Inefficiency in a transformer

If a transformer was 100% efficient the power in would equal the power out. However, in practice there will be small losses of power from the transformer, mostly in the form of heat.

The metallic core is being cut by the continuously changing flux, which induces an e.m.f. in the core. In a continuous core this causes currents called **eddy currents**, which cause it to heat up and energy to be lost.

Eddy currents are looping currents induced by the changing magnetic flux in the core. They create a magnetic field that acts against the field that induced them, reducing the field strength. They also dissipate energy by generating heat. The effect of eddy currents can be reduced by laminating the core — this involves having layers of the core separated out by thin layers of insulator, so a current can't flow.

Heat is also generated by resistance in the coils. To reduce this, wires with a low resistance can be used. Thick copper wire is used for this, as copper has a low resistivity and a larger diameter means smaller resistance.

Energy is needed to magnetise and demagnetise the core, and this energy is wasted as it heats the core. To reduce this effect, a magnetically soft material that magnetises and demagnetises easily should be used (see previous page).

Ideally, all of the magnetic flux created by the primary coil would cut through the secondary coil, but in practice this isn't the case (especially if the two coils are far apart). To reduce this magnetic loss, a core design in which the coils are as close as possible can be used — this can include winding the coils on top of each other around the same part of the core (see Figure 3), rather than around different parts of the core (see Figure 1).

Calculating the efficiency of a transformer

Remember from year 1 of A-Level physics that the power transferred in a circuit is given by:

P = power in W ⟶ $P = IV$ ⟵ **I** = current in A
V = voltage in V

So for an ideal transformer, where power in = power out:

$$I_p V_p = I_s V_s \qquad \text{or} \qquad \frac{V_p}{V_s} = \frac{I_s}{I_p}$$

However, because not all the power is transferred, you can find the efficiency of a transformer by calculating the ratio of power out to power in:

$$\text{efficiency} = \frac{I_s V_s}{I_p V_p}$$

- I_s = current in secondary coil in A
- V_s = voltage in secondary coil in V
- V_p = voltage in primary coil in V
- I_p = current in primary coil in A

Tip: This gives the efficiency as a decimal — to make it a percentage, just multiply it by 100.

Example — Maths Skills

125 W of power is transferred to a transformer with 80.0% efficiency. If the current in the secondary coil is measured as 242 mA, what will the voltage be across the secondary coil?

Just rearrange the formula for efficiency to make V_s the subject, then put the numbers in:

$$\text{efficiency} = \frac{I_s V_s}{I_p V_p} \Rightarrow V_s = \frac{\text{efficiency} \times [I_p V_p]}{I_s}$$

$$= \frac{\left(\frac{80.0}{100}\right) \times 125}{(242 \times 10^{-3})} = 413.2...$$

$$= 413 \text{ V (to 3 s.f.)}$$

Tip: $P_p = I_p V_p$ is used here.

If a transformer isn't 100% efficient, it will lose energy (mostly through heat — see previous page). The power that isn't transferred to the secondary coil must be transferred to something else. You can find the energy 'lost' using the equation:

$$E = Pt$$

- E = energy in J
- P = power in W
- t = time in s

Exam Tip
$E = Pt$ is given to you in the Mechanics section of the data and formulae booklet in the form $P = \frac{\Delta W}{\Delta t}$ (remember, work done, W, and energy transferred are the same thing). You'll also find $P = IV$ in the data and formulae booklet.

Example — Maths Skills

A device charger contains a transformer which is 91.0% efficient. The supply voltage is 120 V and the supply current is 150 mA. If the device takes 2.0 hours to fully charge, how much energy is lost when the device is charged?

First find the power 'lost' when the charger is in operation.
If it's 91% efficient, (100 − 91 =) 9% of the power input is wasted:

$$\text{power wasted} = \frac{9.0}{100} \times I_p V_p$$

$$= 0.090 \times (150 \times 10^{-3}) \times 120 = 1.62 \text{ W}$$

Then find the energy wasted over 2 hours:

$$\text{energy wasted} = Pt = 1.62 \times (2.0 \times 60 \times 60) = 11\,664$$

$$= 12 \text{ kJ (to 2 s.f.)}$$

Tip: Don't forget to convert the time to seconds.

Transformers in the National Grid

Transformers are an important part of the National Grid. Electricity from power stations is sent round the country in the National Grid at the lowest possible current. This is because a high current causes greater energy losses due to heating in the cables. The power losses due to the resistance of the cables is equal to $P = I^2 R$ — so if you double the transmitted current, you quadruple the power lost. Using cables with the lowest possible resistance can also reduce energy loss.

Tip: Remember from year 1 of A-Level physics that $P = I^2 R$.

Tip: Although some energy is lost inside the transformers at each end, it's nowhere near as much as the energy that would be lost if the electricity were transmitted at 230 V.

Tip: Transformer efficiency was covered on pages 148-149.

Tip: Cables between pylons aren't insulated. This is because the air gap between the cables and the earth or other cables is big enough to avoid sparks.

Since power = current × voltage, a low current means a high voltage for the same amount of power transmitted. Transformers allow us to step up the voltage to around 400 000 V for transmission through the National Grid.

High voltage raises safety and insulation issues, and has to be stepped back down to a safer 230 V before it can be used in homes. This is done in stages, with power transferred from overhead lines to underground wires.

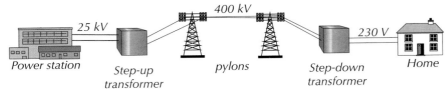

Power station   25 kV   *Step-up transformer*   400 kV   *pylons*   *Step-down transformer*   230 V   *Home*

Figure 4: *The power supply from a power station is stepped up to 400 kV before being stepped back down to 230 V for domestic use.*

Example — **Maths Skills**

A current of 1330 A is used to transmit 1340 MW of power through 147 km of cables. The resistance of the transmission wire is 0.130 Ω per kilometre. Calculate the power wasted.

Total resistance = 0.130 × 147 = 19.11 Ω

Power lost = I^2R = 1330^2 × 19.11 = 3.3803... × 10^7
= 3.38 × 10^7 W (to 3 s.f.)

Figure 5: *An electricity substation where the voltage is stepped down for domestic use.*

Tip: If you do this experiment, don't forget to do a full risk assessment before you start. Take care to keep the voltages and currents you use at a safe level.

Investigating transformers

You can investigate the relationship between the number of turns and the voltages across the coils of a transformer by setting up the equipment as shown in Figure 6.

Put two C-cores together and wrap wire around each to make the coils. Begin with 5 turns in the primary coil and 10 in the secondary coil (a ratio of 1:2).

Turn on the ac supply to the primary coil. Use a low voltage — remember transformers increase voltage, so make sure you keep it at a safe level. Record the voltage across each coil.

Keeping V_p the same so it's a fair test, repeat the experiment with different ratios of turns. Try 1:1 and 2:1. Divide N_s by N_p and V_s by V_p. You should find that for each ratio of turns, $\frac{N_s}{N_p} = \frac{V_s}{V_p}$.

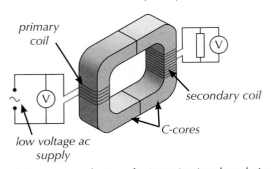

primary coil   *secondary coil*   *C-cores*   *low voltage ac supply*

Figure 6: *Experimental set-up for investigating the relationship between the number of turns and voltages in transformers.*

To investigate the relationship between current and voltage of the transformer coils for a given number of turns in the coil you can use the same equipment as above, but add a variable resistor to the primary coil circuit and an ammeter to both circuits as shown in Figure 7.

Turn on the power supply and record the current through and voltage across each coil. Leaving the number of turns constant, adjust the variable resistor to change the input current. Record the current and voltage for each coil, then repeat this process for a range of input currents.

You should find that for each current, $\dfrac{N_s}{N_p} = \dfrac{V_s}{V_p} = \dfrac{I_p}{I_s}$.

Tip: These formulas won't quite work in your investigation because real transformers aren't 100% efficient.

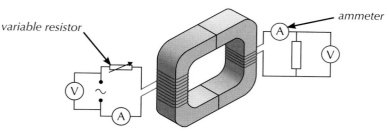

Figure 7: *Set-up for investigating the relationship between current and voltage in a transformer.*

Practice Questions — Application

Q1 a) The primary coil of a transformer has 250 turns. The secondary coil has 420 turns. If the voltage across the primary coil is 190 V, calculate the voltage across the secondary coil.

b) There is a current of 13 A in the secondary coil of a different transformer. The voltage across the primary coil is 120 V and the voltage across the secondary coil is 75 V. Calculate the current in the primary coil.

Q2 What's the efficiency of a transformer with a power input of 65 W and a power output of 62 W?

Q3 Laptop chargers contain a transformer that reduces the voltage from the mains supply.

a) When in operation, the transformer heats up. Why does this happen? Give two reasons.

b) The laptop requires a voltage of 19 V and is supplied with 120 W of power.

(i) Does the charger use a step-up or step-down transformer?

(ii) If the transformer has an efficiency of 0.85, how much current runs through the primary coil?

Tip: Mains electricity in the UK supplies 230 V.

Q4 12.3 MW of power is wasted when power is transmitted through wire with a resistance of 0.0985 Ω km^{-1} at a current of 1250 A. Over what distance is the power transmitted?

Practice Questions — Fact Recall

Q1 How does a transformer change the voltage of an electricity supply?

Q2 What is an eddy current? How do they affect transformer efficiency?

Q3 Give three ways of reducing energy loss in a transformer.

Q4 Why is a high voltage used to transmit power in the National Grid?

Section Summary

Make sure you know...

- That a current-carrying wire in a magnetic field will experience a force.
- That the direction of the force acting on a current-carrying wire perpendicular to a magnetic field can be found using Fleming's left-hand rule.
- What magnetic flux density (B) is, and that it's measured in teslas (T).
- That one tesla is equal to one newton per amp per metre.
- How to calculate the force acting on a current-carrying wire perpendicular to a magnetic field.
- How to investigate the force on a wire in a magnetic field using a top pan balance.
- That a charged particle moving through a magnetic field will experience a force.
- How to calculate the force on a charged particle moving in a magnetic field when the magnetic field is perpendicular to the particle's velocity.
- How to work out the direction of the force acting on both positive and negative charged particles in a magnetic field.
- That a charged particle moving perpendicular to a uniform magnetic field will follow a circular path.
- How particle accelerators such as cyclotrons make use of the force that acts on charged particles moving perpendicular to a magnetic field.
- What magnetic flux is, and how to calculate the value of magnetic flux when B is perpendicular to A.
- That an e.m.f. is induced across a conductor moving through a magnetic field, or a conductor inside a changing magnetic field. If the conductor is connected to a complete circuit, a current will flow.
- What flux linkage is, and how to calculate the flux linkage for a coil normal to B with N turns cutting the flux.
- How to calculate the magnetic flux in a rectangular coil when the coil is not normal to B.
- How to calculate the flux linkage of a rectangular coil when the coil is not normal to B.
- How to use an oscilloscope to investigate the effect of varying the angle between a search coil and the direction of a magnetic field.
- That Faraday's law shows that the magnitude of the induced e.m.f. is equal to the rate of change of flux linkage in a conductor.
- What Lenz's law is.
- How to apply Faraday's and Lenz's laws to a straight conductor moving in a magnetic field.
- That the e.m.f. induced in a coil rotating uniformly in a magnetic field varies sinusoidally.
- How to calculate the maximum induced e.m.f. for a coil rotating uniformly in a magnetic field.
- Simple experimental applications of Faraday's law, such as generators and dynamos.
- How to operate an oscilloscope, use it as an ac or dc voltmeter, measure time intervals and frequencies, and display ac waveforms.
- How to calculate peak and peak-to-peak voltages and currents for alternating currents varying sinusoidally, including mains electricity.
- How to calculate root mean square currents and potential differences.
- What a transformer is, how it works and the different types of transformer (step-up and step-down).
- How to use the ideal transformer equation that links the number of turns and the voltage of the primary and secondary coils.
- What causes energy loss and inefficiency in a transformer, and ways to reduce this energy loss.
- How to calculate the efficiency of a transformer.
- How electrical power is transmitted in the National Grid and why transformers are useful in transmitting power in the National Grid, including how to calculate power losses in transmission.

Exam-style Questions

1   Which of the following statements best describes Faraday's law?

    A   The induced e.m.f. is always in such a direction as to oppose the change
        that caused it.

    B   The induced e.m.f. is always in such a direction as to increase the change
        that caused it.

    C   The induced e.m.f. is directly proportional to the rate of change of flux linkage.

    D   The induced e.m.f. is inversely proportional to the rate of change of flux linkage.

(1 mark)

2   The diagram shows a straight conductor moving perpendicular to a magnetic field.
    Which of the following statements is NOT true about the e.m.f. induced?

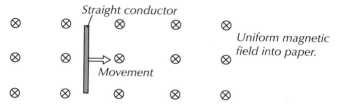

    A   The faster the conductor moves, the greater the induced e.m.f.

    B   The slower the conductor moves, the greater the induced e.m.f.

    C   The longer the conductor, the greater the induced e.m.f.

    D   The stronger the magnetic field, the greater the induced e.m.f.

(1 mark)

3   Which of the following statements is true for a rectangular coil rotating in a uniform
    magnetic field?

    A   The magnitude of the induced e.m.f. is greatest when the plane of the area of the
        coil is parallel to the direction of the magnetic field.

    B   Increasing the frequency of rotation increases the maximum induced e.m.f.

    C   The induced e.m.f. is independent of the area of the coil.

    D   A graph of induced e.m.f. against time would be linear.

(1 mark)

4   Which of the following is NOT a way of improving transformer efficiency?

    A   Decreasing the thickness of the wires in the coils.

    B   Laminating the core.

    C   Reducing the distance between the two coils.

    D   Using a more magnetically soft material for the core.

(1 mark)

5 A straight wire carrying a current of 190 mA is perpendicular to a magnetic field with a flux density of 1.7 T. The wire is fixed in position at the ends of the wire.

5.1 The wire experiences a force of 4.5 × 10⁻² N. Calculate the length of the wire.

(2 marks)

5.2 The experimental set-up is changed so that the ends of the current-carrying wire are still fixed in position, but the whole wire vibrates with a steady frequency.
State two ways this could have been achieved.

(2 marks)

5.3 The wire is replaced by a rectangular coil with an area of 10.0 cm².
The coil is free to rotate and is connected to a dc power supply. It experiences a force that causes it to rotate in the clockwise direction, as shown:

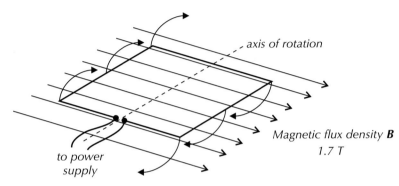

State in which direction current flows around the coil.

(1 mark)

5.4 Calculate the flux linkage of the coil when the normal to the area of the coil is at 41° to the magnetic field.

(2 marks)

5.5 The rest of the circuit is removed and the coil is attached to slip rings. The coil is rotated with ω = 10.0π rad s⁻¹ so that an alternating e.m.f. is induced across the coil.
Calculate the maximum e.m.f. induced across the coil.

(1 mark)

5.6 On the axes below, show how the induced e.m.f. varies with angle θ during one complete rotation of the coil, starting at θ = 0, where θ is the angle the normal to the coil makes to the field.

(2 marks)

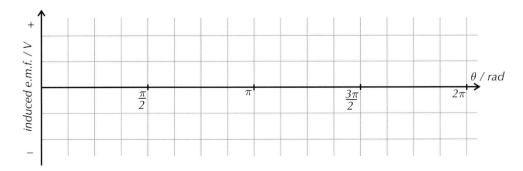

6 This question is about moving charged particles in uniform magnetic fields.

6.1 If a moving charged particle enters a uniform magnetic field perpendicular to its velocity, what shape will the particle's path take?

(1 mark)

An electron is fired into a uniform magnetic field with a flux density of 0.93 T at a speed of 8.1×10^7 ms^{-1}, as shown below.

Uniform magnetic field into paper.

Electron

6.2 Calculate the magnitude of the force the electron will experience, and state its direction.

(2 marks)

6.3 An alpha particle of charge $+2e$ is fired into the same magnetic field as the electron at the same speed. State the magnitude and direction of the force it will experience.

(2 marks)

7 A wire coil is rotated at a steady rate in a uniform magnetic field, with the axis of rotation perpendicular to the field, so that an e.m.f. is induced across the coil.

7.1 State the values of θ at which the induced e.m.f. across the coil will be greatest, where θ is the angle the normal to the area of the coil makes with the magnetic field.

(1 mark)

7.2 The coil is rotated with a frequency of 5 Hz. Sketch a graph of the induced e.m.f. against time for $0 \leq t \leq 0.4$ s.

(3 marks)

8 This question is about transformers.

8.1 State which kind of current supply is required for a transformer.

(1 mark)

8.2 Explain why it is important to use wires with as low a resistance as possible in the coils of a transformer.

(1 mark)

8.3 A transformer is 91% efficient and has a power input of 1.2 kW. Calculate the power output of the transformer.

(1 mark)

8.4 If the transformer is in operation for 3.0 hours, calculate how much energy will be wasted.

(1 mark)

9 An experiment is set up as shown in Figure 1 to investigate how the force on a wire varies with flux density, current, and length of wire. A stiff metal wire is clamped in place so that is passes through a metal cradle resting on a top pan balance. A magnadur magnet is attached to either side of the metal cradle, with opposite sides facing each other so that there is a uniform magnetic field between the two. Crocodile clips connect the stiff metal wire to a circuit containing an ammeter and a variable dc power supply. The balance is zeroed before the power supply is switched on.

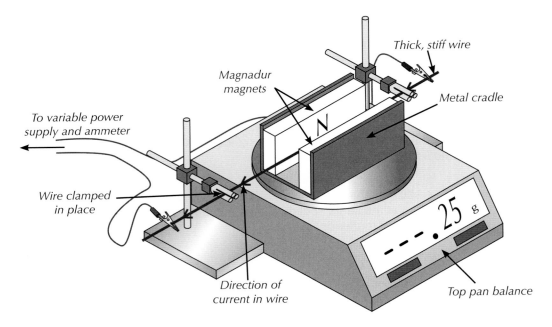

Figure 1

9.1 State the direction of the force acting on the metal cradle due to the current in the wire.

(1 mark)

9.2 The length of the wire in the magnetic field is 9.0 cm. Explain how the set-up shown in Figure 1 could be used to determine the magnetic flux density of the magnetic field due to the magnadur magnets without altering the length of the wire in the field.

(4 marks)

9.3 The magnets are replaced with magnets of different size and strength such that the magnetic flux density of the resultant magnetic field is now 25 mT. When the power is switched on, the ammeter reads 1.6 A and the top pan balance reads 0.65 g. Calculate the length of the wire in the magnetic field.

(2 marks)

9.4 The crocodile clips are swapped over. Explain what will be observed.

(2 marks)

1. Rutherford Scattering

You won't be very surprised to hear that nuclear physics is all about atomic nuclei. What is a bit surprising, is that we've only known nuclei exist for just over 100 years...

Learning Objectives:

- Understand how Rutherford scattering of alpha particles demonstrates the existence of the atomic nucleus.

- Appreciate how the knowledge and understanding of the structure of the nucleus has changed over time.

Specification Reference 3.8.1.1

The history of the atom

The idea of atoms has been around since the time of the Ancient Greeks in the 5th Century BC. A man called Democritus proposed that all matter was made up of little, identical lumps called 'atomos'.

Much later, in 1804, a scientist called John Dalton put forward a hypothesis that agreed with Democritus — that matter was made up of tiny spheres ('atoms') that couldn't be broken up. He reckoned that each element was made up of a different type of 'atom'.

Nearly 100 years later, J. J. Thomson discovered that electrons could be removed from atoms. So Dalton's theory wasn't quite right (atoms could be broken up). Thomson suggested that atoms were spheres of positive charge with tiny negative electrons stuck in them like fruit in a plum pudding. This "plum pudding" model of the atom was known as the Thomson Model.

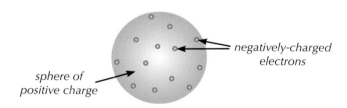

negatively-charged electrons

sphere of positive charge

Figure 1: *The "plum pudding" model of the atom.*

Until this point though, nobody had proposed the idea of the **nucleus**. Rutherford was the first to suggest atoms did not have uniformly distributed charge and density.

The discovery of the atomic nucleus

In 1909, in Ernest Rutherford's laboratory, Hans Geiger and Ernest Marsden studied the scattering of **alpha particles** by thin metal foils.

A stream of alpha particles from a **radioactive source** are fired at very thin gold foil. When alpha particles from a radioactive source strike a fluorescent screen a tiny visible flash of light is produced (see Figure 3). The fluorescent screen is circular and surrounds the experiment so that alpha particles scattered by any angle can be detected. This is now known as the **Rutherford scattering experiment**.

Tip: Alpha particles are formed of two protons and two neutrons. They're positively charged, so they're repelled by other positive charges.

Figure 2: *Ernest Rutherford, the New Zealand-born physicist who proposed the existence of the atomic nucleus.*

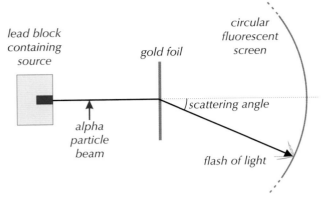

Figure 3: *The Rutherford scattering experiment.*

Geiger and Marsden recorded the number of alpha particles scattered at different angles. If the Thomson model was right, all the flashes should have been seen within a small angle of the beam, because the positively-charged alpha particles would be deflected by a small amount by the electrons.

In fact, most alpha particles passed straight through the gold film, and Geiger and Marsden observed that occasionally some alpha particles scattered at angles greater than 90°.

Conclusions from Rutherford scattering

Rutherford scattering shows that atoms must have a small, positively charged nucleus that contains most of the atom's mass — shown in Figure 4.

- The atom must be mostly empty space because most of the alpha particles just pass straight through.

- The nucleus must have a large positive charge, as some of the positively-charged alpha particles are repelled and deflected by a large angle.

- The nucleus must be tiny as very few alpha particles are deflected by an angle greater than 90°.

- Most of the mass must be in the nucleus, since the fast alpha particles (with high momentum) are deflected by the nucleus.

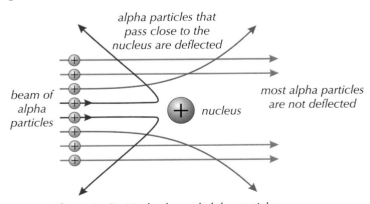

Figure 4: *Positively charged alpha particles deflected by an atomic nucleus.*

Tip: The conclusions from Rutherford scattering are really important — make sure you know them and understand how Rutherford arrived at them from the experimental evidence.

Tip: In order for the fast alpha particles to be deflected by the nucleus, the alpha particles must be striking something more massive than themselves.

The structure of the nucleus — protons and neutrons

The proton was discovered in 1919 by Rutherford and William Kay when firing high-energy alpha particles at different gases. Scientists thought the nucleus must be made up of protons, but there was a problem. If the nucleus was only made up of protons, you'd expect massive nuclei to have very high charges compared to lower mass nuclei — but the charges observed were much lower than expected.

Rutherford proposed the idea of a "proton-electron doublet" being part of the nucleus in 1920 to help explain this observation. He didn't realise it at the time, but he was referring to the neutron. It took until 1932 for James Chadwick to provide experimental evidence for the neutron.

Practice Questions — Fact Recall

Q1 Describe the Rutherford scattering experiment.

Q2 Explain how the Rutherford scattering experiment shows that atoms contain a small, positive nucleus containing most of the atom's mass.

2. Measuring Nuclear Radius

Learning Objectives:

- Estimate the radius of a nucleus by calculating the distance of closest approach of an alpha particle using Coulomb's law.

- Be able to determine the radius of a nucleus using electron diffraction.

- Be familiar with the graph of intensity against angle for electron diffraction by a nucleus.

Specification Reference 3.8.1.5

Rutherford's scattering experiment gives a good estimate of nuclear radii, but to get a really good measurement, you'll need to diffract electrons...

Closest approach of a scattered particle

You can estimate the radius of an atomic nucleus by using Rutherford's scattering experiment (see p.157–158). An alpha particle that 'bounces back' and is deflected through 180° will have stopped a short distance from the nucleus (see Figure 1).

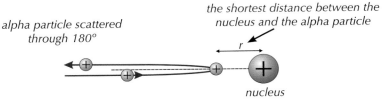

alpha particle scattered through 180°

the shortest distance between the nucleus and the alpha particle

r

nucleus

Figure 1: *The closest approach of a scattered alpha particle.*

The alpha particle does this at the point where its electric potential energy (see page 92) equals its initial kinetic energy. This combined with Coulomb's law (see page 87) gives the equation:

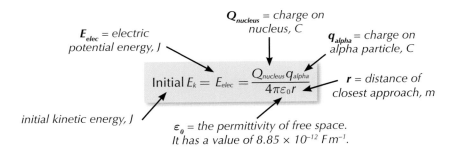

E_{elec} = electric potential energy, J

$Q_{nucleus}$ = charge on nucleus, C

q_{alpha} = charge on alpha particle, C

$$\text{Initial } E_k = E_{elec} = \frac{Q_{nucleus}q_{alpha}}{4\pi\varepsilon_0 r}$$

initial kinetic energy, J

r = distance of closest approach, m

ε_0 = the permittivity of free space. It has a value of 8.85×10^{-12} F m^{-1}.

Exam Tip
The value of ε_0 will be in the data and formulae booklet you'll be given in the exam, so don't worry about memorising it.

This is just conservation of energy — and if you know the initial kinetic energy of the alpha particle you've fired, you can use it to find how close the particle can get to the nucleus.

To find the charge of a nucleus you need to know the atom's **proton number**, Z — that tells you how many protons are in the nucleus (surprisingly). A proton has a charge of $+e$ (where e is the magnitude of the charge on an electron), so the charge of a nucleus must be $+Ze$.

The distance of closest approach is an estimate of nuclear radius — it gives a maximum value for it. However electron diffraction (which you'll meet on the next page) gives much more accurate values for nuclear radii.

Tip: The magnitude of the charge on an electron, e, is 1.60×10^{-19} C, but an electron has a negative charge of $-e$.

Tip: Alpha particles contain 2 protons, so they have a charge of $+2e$.

Tip: To convert from electronvolts (eV) to joules (J), just multiply by the charge on an electron, e.

┌─ **Example** ── **Maths Skills** ──────────────

An alpha particle with an initial kinetic energy of 6.0 MeV is fired at a gold nucleus. Estimate the radius of the nucleus by finding the closest approach of the alpha particle to the nucleus. (Z_{gold} = 79)

Initial particle energy = 6.0 MeV = 6.0×10^6 eV

Convert energy into joules: $6.0 \times 10^6 \times 1.60 \times 10^{-19} = 9.6 \times 10^{-13}$ J

So initial $E_k = E_{elec} = \dfrac{Q_{gold}\,q_{alpha}}{4\pi\varepsilon_0 r}$

$\qquad\qquad\qquad = 9.6 \times 10^{-13}$ J at closest approach.

Rearrange to get $r = \dfrac{(+79e)(+2e)}{4\pi\varepsilon_0(9.6\times 10^{-13})}$

$\qquad\qquad = \dfrac{2\times 79 \times (1.60 \times 10^{-19})^2}{4\pi \times 8.85\times 10^{-12}\times 9.6 \times 10^{-13}}$

$\qquad\qquad = 3.788... \times 10^{-14}$

$\qquad\qquad = 3.8 \times 10^{-14}$ m (to 2 s.f.)

Therefore nuclear radius $\approx 3.8 \times 10^{-14}$ m

Exam Tip
Exam questions will normally give you particle energies in eV. Make sure you always convert from eV to joules before dropping the energy into this equation.

Tip: The values of 2 and 79 are exact here, since proton number has to be a whole number — so these numbers don't need to be taken into consideration when deciding how many significant figures you should give your answer to.

Electron diffraction

Electrons are a type of particle called a lepton. Leptons don't interact with the strong nuclear force (whereas neutrons and alpha particles do). Because of this, electron diffraction is an accurate method for measuring the nuclear radius.

Like other particles, electrons show wave-particle duality (see p.383) — so electron beams can be diffracted. A beam of moving electrons has an associated **de Broglie wavelength**, λ, which at high speeds (where you have to take into account relativistic effects (see p.390-395)) is approximately:

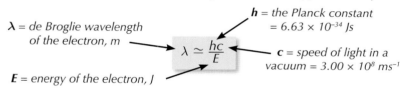

$\lambda = $ de Broglie wavelength of the electron, m

$h = $ the Planck constant $= 6.63 \times 10^{-34}$ Js

$$\lambda \simeq \dfrac{hc}{E}$$

$c = $ speed of light in a vacuum $= 3.00 \times 10^8$ ms^{-1}

$E = $ energy of the electron, J

Exam Tip
Don't worry about learning the value of the Planck constant — it's in the data and formulae booklet you get in the exam. The speed of light is also in there, but you should really know that by now.

The wavelength must be tiny (~10^{-15} m) to investigate the nuclear radius — so the electrons will have a very high energy. If a beam of high-energy electrons is directed onto a thin film of material in front of a screen, a diffraction pattern will be seen on the screen (see Figure 2).

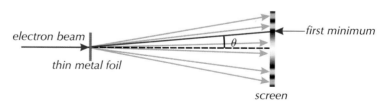

electron beam

thin metal foil

θ

first minimum

screen

Figure 2: *Electron diffraction.*

Tip: Each high-energy electron is diffracted by an individual nucleus. This is <u>not</u> the same as the electron diffraction covered in year 1 of A-level, where the electrons diffracted through the gaps between the atoms, a bit like a diffraction grating. This is why the equation is for a minimum, not a maximum.

The first minimum appears where:

$\theta = $ scattering angle

$$\sin\theta \simeq \dfrac{1.22\lambda}{2R}$$

$R = $ the radius of the nucleus the electrons have been scattered by

Tip: The derivation of this equation is quite hard, so you don't need to worry about it.

Using measurements from this diffraction pattern, you can rearrange the above equation to find the radius of the nucleus.

Figure 3: *Louis de Broglie's wave-particle duality theory correctly predicted that electrons would form diffraction patterns.*

Example ⸺ Maths Skills

A beam of 300 MeV electrons is fired at a piece of thin foil, and produces a diffraction pattern on a fluorescent screen. The first minimum of the diffraction pattern is at an angle of 30° from the straight-through position. Estimate the radius of the nuclei the electrons were diffracted by.

$E = 300 \text{ MeV} = 3.00 \times 10^8 \times 1.60 \times 10^{-19} = 4.80 \times 10^{-11} \text{ J}$

$\lambda \simeq \dfrac{hc}{E} = \dfrac{6.63 \times 10^{-34} \times 3.00 \times 10^8}{4.80 \times 10^{-11}} = 4.143... \times 10^{-15} \text{m}$

So $R \simeq \dfrac{1.22\lambda}{2\sin\theta} = \dfrac{1.22 \times 4.143... \times 10^{-15}}{2\sin 30°} = 5.055... \times 10^{-15}$

$= 5.06 \times 10^{-15} \text{ m (to 3 s.f.)}$

Variation of intensity with diffraction angle

The diffraction pattern is very similar to that of a light source shining through a circular aperture — a central bright maximum (circle) containing the majority of the incident electrons, surrounded by other dimmer rings (maxima).

The intensity of the maxima decreases as the angle of diffraction increases. The graph in Figure 4 shows the relative intensity of electrons in each maximum.

Tip: You might also see a logarithmic plot of this graph — it's almost exactly the same shape, but the peak heights are less pronounced.

Tip: The intensity never actually hits zero, it just gets very close.

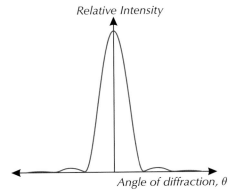

Figure 4: *A graph to show the variation of intensity of electrons in an electron diffraction pattern.*

Practice Questions — Application

Tip: The proton number of lead is 82.

Tip: The proton number of zinc is 30.

Q1 An alpha particle has an electric potential energy of 1.0×10^{-13} J at its closest approach to a zinc nucleus. Use this information to estimate the size of the nuclear radius.

Q2 An alpha particle with an initial kinetic energy of 4.0 MeV is fired towards a lead nucleus. Calculate the distance of closest approach of the alpha particle to the nucleus.

Q3 What is the approximate de Broglie wavelength of an electron with an energy of 50 MeV (to 2 s.f.)?

Q4 A beam of electrons is fired towards a thin sheet of gold foil. The electrons in the beam have an energy of 200 MeV (to 2 s.f.). If the first diffraction minimum is observed on a screen at an angle of 33° to the beam's original direction, calculate the radius of a gold nucleus.

Practice Questions — Fact Recall

Q1 Write down the equation you would use to calculate the distance of closest approach for an alpha particle scattered by an atomic nucleus.

Q2 What is the relationship between the de Broglie wavelength of a fast-moving particle and its energy?

Q3 Give an equation for where the first minimum appears for the diffraction of an electron beam of wavelength λ around a nucleus of radius R.

Q4 Sketch a graph to show how relative electron intensity varies with the angle of diffraction in the diffraction pattern formed when electrons are diffracted by a thin film of material.

3. Nuclear Radius and Density

Now you know how nuclear radii can be measured, it's time to see how they vary with nucleon number...

The size of the atom

By probing atoms using scattering and diffraction methods, we know that the radius of an atom is about 0.05 nm (5×10^{-11} m) and the radius of the smallest nucleus is about 1 fm (1×10^{-15} m — pronounced "femtometres"). Basically, nuclei are really, really tiny compared with the size of the whole atom. To make this easier to visualise, try imagining a large Ferris wheel (which is pretty darn big) as the size of an atom. If you then put a grain of rice (which is rather small) in the centre, this would be the size of the atom's nucleus.

Molecules are just a number of atoms joined together. As a rough guide, the size of a molecule equals the number of atoms in it multiplied by the size of one atom.

Nucleons

The particles that make up the nucleus (protons and neutrons) are called **nucleons**. The number of nucleons in an atom is called the **nucleon (mass) number**, A. As more nucleons are added to the nucleus, it gets bigger.

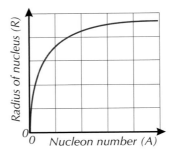

Figure 1: *A graph to show how nuclear radius increases with increasing nucleon number.*

When data from nuclear radii experiments is plotted on a graph of nuclear radius R against the cube root of the nucleon number $A^{1/3}$, the line of best fit gives a straight line. This shows a linear relationship between R and $A^{1/3}$. As the nucleon number increases, the radius of the nucleus increases proportionally to the cube root of A.

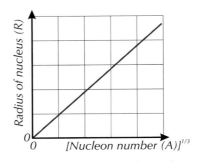

Figure 2: *A graph to show how nuclear radius is directly proportional to the cube root of the nucleon number.*

This relationship can be written as: $R \propto A^{1/3}$. You can make this into an equation by introducing a constant, R_0, which is equal to about 1.4 fm (= 1.4×10^{-15} m). This gives:

$$R = R_0 A^{1/3}$$

Example — Maths Skills

Calculate the radius of an oxygen nucleus containing 16 nucleons.

$R = R_0 A^{1/3} = (1.4 \times 10^{-15}) \times (16)^{1/3} = 3.527... \times 10^{-15}$ m = 3.5 fm (to 2 s.f.)

Tip: Don't forget — something to the power of 1/3 is just the same as taking the cube root of that something.

Nuclear density

The equation $R = R_0 A^{1/3}$ shows that the volume that each nucleon takes up in the nucleus is about the same, since $R^3 \propto A$ and so volume $\propto A$. Combining this with the fact that the mass of each nucleon is also about the same, the equation $R = R_0 A^{1/3}$ provides evidence that the density of nuclear matter is constant, regardless of how many nucleons make up the nucleus.

- Protons and neutrons have nearly the same mass (we'll call it $m_{nucleon}$). A nucleus with a nucleon number A has mass $A \times m_{nucleon}$.

- Assuming nuclei are spherical, the volume of a nucleus is given by $V = \frac{4}{3}\pi R^3$.

- Density = mass ÷ volume, so:

$$\rho = \frac{\text{mass}}{\text{volume}} = \frac{A \times m_{nucleon}}{\frac{4}{3}\pi R^3} = \frac{A \times m_{nucleon}}{\frac{4}{3}\pi \left(R_0 A^{\frac{1}{3}}\right)^3} = \frac{3m_{nucleon}}{4\pi R_0^3} = \text{constant}$$

Tip: The difference between the masses of a proton and a neutron is so small that they are both equal to 1.7×10^{-27} kg to 2 s.f. Their values will be given in the data and formulae booklet in your exam.

Tip: The volume of a sphere is $V = \frac{4}{3}\pi r^3$.

If you substitute the constants into this formula, you'll get that nuclear density is around 1.45×10^{17} kg m^{-3}. A teaspoon of pure nuclear matter would have a mass of about five hundred million tonnes.

Example — Maths Skills

Show that a carbon nucleus (mass = 2.00×10^{-26} kg, A = 12) and a gold nucleus (mass = 3.27×10^{-25} kg, A = 197) have roughly the same density.

Carbon:

Radius = $R = R_0 A^{1/3} = (1.4 \times 10^{-15}) \times (12)^{1/3}$
$= 3.205... \times 10^{-15}$ m

Volume = $V = \frac{4}{3}\pi R^3 = 1.379... \times 10^{-43}$ m^3

Density = $\rho = \frac{m}{V} = \frac{2.00 \times 10^{-26}}{1.379... \times 10^{-43}} = 1.450... \times 10^{17}$
$= 1.5 \times 10^{17}$ kg m^{-3} (to 2 s.f.)

Gold:

Radius = $R = R_0 A^{1/3} = (1.4 \times 10^{-15}) \times (197)^{1/3}$
$= 8.146... \times 10^{-15}$ m

Volume = $V = \frac{4}{3}\pi R^3 = 2.264... \times 10^{-42}$ m^3

Density = $\rho = \frac{m}{V} = \frac{3.27 \times 10^{-25}}{2.264... \times 10^{-42}} = 1.444... \times 10^{17}$
$= 1.4 \times 10^{17}$ kg m^{-3} (to 2 s.f.)

Tip: Just to make you gasp in awe and wonder, out in space nuclear matter makes up neutron stars, which are several kilometres in diameter.

Tip: Remember that density, ρ, is just mass, m, divided by volume, V.

Tip: Remember, you need to assume that the nucleus is spherical.

Nuclear density is significantly greater than atomic density (which is approximately between 10^3 and 10^5 kg m^3) — this suggests three important facts about the structure of an atom:

- Most of an atom's mass is in its nucleus.

- The nucleus is small compared to the atom.

- An atom must contain a lot of empty space.

Practice Questions — Application

Q1 A nucleus is made up of 21 protons and 24 neutrons. Given that each nucleon has a mass of 1.7×10^{-27} kg, calculate the mass of the nucleus.

Q2 Phosphorus has a nucleon number of 31. Taking $R_0 = 1.4 \times 10^{-15}$ m, find its nuclear radius.

Q3 An unknown element has a nuclear radius of 4.2 fm. Estimate the element's nucleon number.

Q4 Calculate the volume of a nucleus containing 23 nucleons.

Q5 What is the density of a lead nucleus, given that it has a nucleon number of 207, and that each nucleon has a mass of 1.7×10^{-27} kg?

Practice Questions — Fact Recall

Q1 What is the typical radius of:
   a) The smallest atomic nucleus?
   b) An atom?

Q2 What are nucleons?

Q3 a) How are nuclear radius and nucleon number related? Write down an equation to show this relation.

   b) Describe how this relationship could be shown using experimental data of nucleon number and nuclear radius.

   c) Explain briefly what this equation tells you about nuclear density.

4. Properties of Nuclear Radiation

There are four different types of nuclear radiation — each has different properties, and you can use these to identify what type a source is emitting.

Radioactive decay

If an atomic nucleus is unstable, it will 'break down' to become more stable. The nucleus decays by releasing energy and/or particles, until it reaches a stable form — this is called **radioactive decay**. An individual radioactive decay is random and can't be predicted.

Types of nuclear radiation

There are four types of **nuclear radiation** — **alpha**, **beta-minus**, **beta-plus** and **gamma** — and each is made up of different constituents. They are listed in Figure 1. The masses here have been given in atomic mass units (u) — one atomic mass unit = 1.661×10^{-27} kg, and is about the same as the mass of a proton or neutron (see page 165).

Radiation	Symbol	Constituent	Relative Charge	Mass (u)
Alpha	α	A helium nucleus — 2 protons & 2 neutrons	+2	4
Beta-minus (Beta)	β⁻ or β	Electron	−1	(negligible)
Beta-plus	β⁺	Positron	+1	(negligible)
Gamma	γ	Short-wavelength, high-frequency electromagnetic wave.	0	0

Figure 1: *Types of nuclear radiation.*

Penetration of nuclear radiation

Different types of radiation have different penetrating powers. The more penetrating the type of radiation, the thicker or denser a material needs to be to absorb it.

- Alpha — absorbed by paper, skin or a few centimetres of air.
- Beta-minus — absorbed by about 3 millimetres of aluminium.
- Gamma — absorbed by many centimetres of lead, or several metres of concrete.

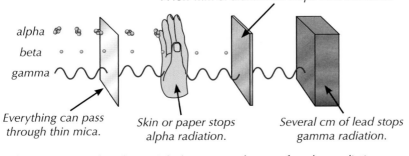

A few mm of aluminium stops beta radiation.

Everything can pass through thin mica.

Skin or paper stops alpha radiation.

Several cm of lead stops gamma radiation.

Figure 2: *Examples of materials that stop each type of nuclear radiation.*

Learning Objectives:

- Know the properties of each type of nuclear radiation.

- Know how to identify alpha, beta and gamma radiation using simple absorption experiments.

- Know some applications of each type of nuclear radiation, including how radiation can be used to measure the thickness of paper, aluminium foil and steel.

- Know the relative hazards of exposure to humans of each type of nuclear radiation.

- Appreciate the balance between the risk and benefits in the use of radiation in medicine.

Specification Reference 3.8.1.2

Tip: You met positrons in year 1 of A-level — they're the antiparticles of electrons.

Tip: Alpha radiation is stopped by the outer layer of dead skin cells on a person's body — the radiation is stopped before it has time to reach the live skin cells, which would otherwise be ionised.

Tip: Both the thickness and the density of a material affect whether radiation will penetrate it — a very thick piece of aluminium <u>could</u> stop gamma radiation.

Beta-plus particles almost immediately annihilate with electrons, so they effectively have zero range.

Identifying nuclear radiation

You can identify the type(s) of radiation emitted by a source by testing to see if they pass through various materials using the apparatus in Figure 3.

radiation   Geiger-Müller tube

radioactive source   absorber   Geiger counter

Figure 3: *Using different absorbers to experimentally identify radiation.*

Tip: A Geiger-Müller tube produces a 'count' in the form of an electrical pulse each time radiation enters it. The counter shows how many counts there are each second — this is called the count rate.

Tip: For more on background radiation and how to measure it, see page 171.

Tip: Any experiments involving radiation can be dangerous if done incorrectly. As always, a full risk assessment must be carried out beforehand.

- Record the background radiation count rate when no source is present.
- Place an unknown source near to a Geiger-Müller tube and record the count rate.
- Place a sheet of paper between the source and the Geiger-Müller tube. Record the count rate.
- Replace the paper with a 3 mm thick sheet of aluminium and record the count rate.
- For each count rate that is recorded, take away the count rate of the background radiation to find the actual count rate.

Depending on when the count rate significantly decreased, you can deduce what kind of radiation the source was emitting. For example, if paper has no effect and aluminium causes a significant (but not complete) reduction in count rate, the source must be emitting beta and gamma radiation.

Magnetic fields

Tip: Gamma radiation isn't made up of charged particles, so it won't be deflected by a magnetic field.

You can also identify types of radiation by looking at how they travel through magnetic fields. Charged particles moving perpendicular to a uniform magnetic field are deflected in a circular path. The direction in which a particle curves depends on the charge — if it's positive the particle will curve one way, if it's negative it'll curve the other way. The radius of curvature of its path can also tell you about its charge and mass — see pages 129–131 for more.

Applications of nuclear radiation

What a radioactive source can be used for often depends on its **ionising** properties.

Alpha radiation

Alpha particles are strongly positive — so they can easily pull electrons off (ionise) atoms. Ionising an atom transfers some of the energy from the alpha particle to the atom. The alpha particle quickly ionises many atoms (about 10 000 ionisations per mm in air for each alpha particle) and loses all its energy. This makes alpha-sources suitable for use in smoke alarms because they allow current to flow, but won't travel very far. When smoke is present, the alpha particles can't reach the detector and this sets the alarm off.

Figure 4: *The components inside a smoke alarm, including a source of alpha radiation.*

Although alpha particles can't penetrate your skin, sources of alpha particles are dangerous if they are ingested. They quickly ionise body tissue in a small area, causing lots of damage.

Beta radiation

The beta-minus particle has lower mass and charge than the alpha particle, but a higher speed. This means it can still knock electrons off atoms. Each beta particle will ionise about 100 atoms per mm in air, losing energy with each interaction. This lower number of interactions means that beta radiation causes much less damage to body tissue.

When creating sheets of material, such as paper, aluminium foil or steel, beta radiation can be used to control its thickness (see Figure 5).

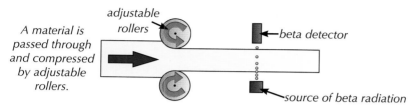

A material is passed through and compressed by adjustable rollers.

adjustable rollers

beta detector

source of beta radiation

Figure 5: *A diagram to show beta radiation being used to measure the thickness of a material.*

The material is flattened as it is fed through rollers. A radioactive source is placed on one side of the material, and a radioactive detector on the other. The thicker the material, the more radiation it absorbs and prevents from reaching the detector. If too much radiation is being absorbed, the rollers move closer together to make the material thinner. If too little radiation is being absorbed, they move further apart.

Tip: Alpha radiation can't be used in this process because the particles would not pass through the material. Gamma radiation can't be used because the thicknesses used are too thin to stop gamma radiation and so it would always pass through the sheet of material.

Gamma radiation

Gamma radiation is even more weakly ionising than beta radiation, so will do even less damage to body tissue. This means it can be used in medicine.

Radioactive tracers are used to help diagnose patients without the need for surgery. A radioactive source with a short half-life to prevent prolonged radiation exposure is either eaten or injected into the patient. A detector, e.g. a PET scanner, is then used to detect the emitted gamma rays.

Gamma rays can be used in the treatment of cancerous tumours — damaging cells and sometimes curing patients of cancer. Radiation damages all cells though — cancerous or not, and so sometimes a rotating beam of gamma rays is used. This lessens the damage done to surrounding tissue, whilst giving a high dose of radiation to the tumour at the centre of rotation.

Damage to other, healthy cells is not completely prevented however and treatment can cause patients to suffer side effects — such as tiredness and reddening or soreness of the skin. Some forms of gamma ray treatments can also cause long term side effects like infertility.

As well as patients, the risks towards medical staff giving these treatments must be kept as low as possible. Exposure time to radioactive sources is kept to a minimum, and generally staff leave the room (which is itself shielded) during treatment.

Simply put, radiation use in medicine has benefits and risks. The key is trying to use methods which reduce the risks (shielding, rotating beams etc.) while still giving you the results you want. It's all one big balancing act.

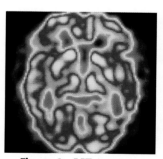

Figure 6: *PET (positron emission tomography) scanning uses a radioactive tracer injected into the bloodstream. Positrons emitted by the tracer annihilate with electrons in the body, producing gamma rays that are then detected by the scanner.*

Summary of nuclear radiation properties

Tip: Learn this table and the one at the start of the topic — they're packed full of useful information you need to know.

Here's a handy table of some of the properties of nuclear radiation:

Radiation	Symbol	Ionising power	Speed	Penetrating power	Affected by magnetic field
Alpha	α	Strong	Slow	Absorbed by paper or a few cm of air	Yes
Beta-minus (Beta)	β^- or β	Weak	Fast	Absorbed by ~3 mm of aluminium	Yes
Beta-plus	β^+	Annihilated by electron, so virtually zero range			
Gamma	γ	Very weak	Speed of light	Absorbed by many cm of lead, or several m of concrete	No

Figure 7: *Properties of nuclear radiation.*

Practice Question — Application

Q1 A radioactive source emits an unknown type of radiation. Using a Geiger-Müller tube placed in front of the source, the count rate from the source was measured. There was no decrease in the count rate recorded when a few sheets of paper were placed between the source and Geiger-Müller tube. The count rate dropped to almost zero when the sheets of paper were replaced with a thin sheet of aluminium. Identify the type of radiation being emitted by the source.

Q2 Some types of nuclear radiation are used in medical treatments. Briefly outline one such treatment, and the benefits and risks that are involved in the patient being exposed to the radiation used in this treatment.

Practice Questions — Fact Recall

Q1 What is radioactive decay?

Q2 List the four types of nuclear radiation and their constituents.

Q3 Give the relative ionising powers of alpha, beta and gamma radiation, and use this to describe the relative danger of human exposure to each.

Q4 Give two methods of identifying types of nuclear radiation.

Q5 Describe a practical application for each of alpha, beta and gamma radiation, and explain why that form of radiation is used.

5. Background Radiation and Intensity

Background radiation is nuclear radiation found everywhere and that comes from lots of different sources. The inverse square law describes how the intensity of gamma radiation falls off the further away you get from a source.

Background radiation

Wherever you are, there's always a low level of radiation — this is called **background radiation**. When you take a reading of the count rate from a radioactive source, you need to measure the background radiation count rate separately and subtract it from your measurement. To do this accurately you should:

- Take three readings of the count rate using a Geiger counter without a radioactive source present.

- Average these three readings (see p.5) and subtract this average from each measurement you take of a radioactive source's count rate.

Sources of background radiation

There are many sources of background radiation, including:

- The air — radioactive radon gas is released from rocks. It emits alpha radiation. The concentration of this gas in the atmosphere varies a lot from place to place, but it's usually the largest contributor to the background radiation.

- The ground and buildings — nearly all rock contains radioactive materials.

- Cosmic radiation — cosmic rays are particles (mostly high-energy protons) from space. When they collide with particles in the upper atmosphere, they produce nuclear radiation.

- Living things — all plants and animals contain carbon, and some of this will be radioactive carbon-14.

- Man-made radiation — in most areas, radiation from medical or industrial sources makes up a tiny, tiny fraction of the background radiation.

The inverse square law

A gamma source will emit gamma radiation in all directions. This radiation spreads out as you get further away from the source. The **intensity** of radiation is the amount of radiation per unit area — it will decrease the further you get from the source. If you took readings of intensity, I, at a distance, x, from the source you would find that it decreases by the square of the distance from the source (see Figure 2).

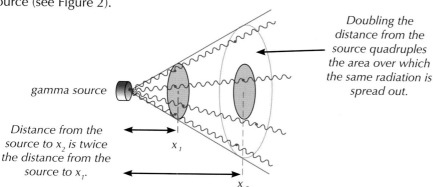

gamma source

Doubling the distance from the source quadruples the area over which the same radiation is spread out.

Distance from the source to x_2 is twice the distance from the source to x_1.

x_1

x_2

Figure 2: *How the intensity of radiation varies with distance from a source.*

Learning Objectives:

- Know examples of the origins of background radiation.

- Be able to accurately eliminate background radiation from experiment results.

- Know that the intensity of gamma radiation decreases with distance from a source according to the inverse square law, $I = \dfrac{k}{x^2}$.

- Be able to investigate and experimentally verify the inverse square law for gamma radiation. (Required Practical 12)

- Know how the inverse square law can be applied to the safe handling of radioactive sources.

Specification Reference 3.8.1.2

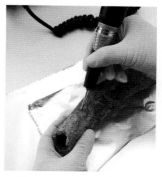

Figure 1: *The carbon-14 in living tissue decays after an organism dies. By measuring how much carbon-14 remains in dead tissue, the age of a sample can be estimated (see pages 181–182).*

Tip: The $\frac{1}{x^2}$ term comes about because the gamma radiation spreads out over the surface of an imaginary sphere of area $A = 4\pi x^2$ (see pages 73–74).

This can be written as the equation:

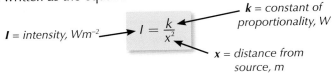

k = constant of proportionality, W

I = intensity, Wm^{-2} → $I = \dfrac{k}{x^2}$

x = distance from source, m

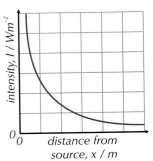

Figure 3: A graph of how the intensity varies with distance from a source.

Example — Maths Skills

The intensity of radiation 0.50 m from a gamma source was measured as 2.5×10^{-10} Wm^{-2}. What intensity would be measured 1.5 m from the source?

Intensity at 0.50 m = $\dfrac{k}{0.50^2}$ = 2.5×10^{-10}

Therefore $k = 2.5 \times 10^{-10} \times 0.50^2 = 6.25 \times 10^{-11}$

So intensity at 1.5 m = $\dfrac{6.25 \times 10^{-11}}{1.5^2}$ = $2.777... \times 10^{-11}$
$= 2.8 \times 10^{-11}$ Wm^{-2} (to 2 s.f.)

Investigating the inverse square law

This relationship can be proved by taking measurements of intensity at different distances from a gamma source, using a Geiger counter — see Figure 4.

REQUIRED PRACTICAL **12**

In this experiment, the collection area of the Geiger-Müller tube stays constant, so the intensity is proportional to the count rate (in counts per second) detected by the tube.

Tip: You need to do a risk assessment before carrying out this practical. You should be especially careful when dealing with the radioactive source in this experiment. See the next page for some safety advice when working with radioactive sources. You must follow the local rules set by your school or college's designated Radiation Protection Supervisor.

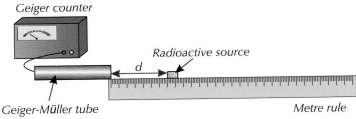

Geiger counter

Radioactive source

d

Geiger-Müller tube

Metre rule

Figure 4: An experiment to investigate the inverse square law for gamma radiation.

1. Set up the equipment as shown in Figure 4 but without the radioactive source. Make sure the Geiger-Müller tube is lined up with the start of the measuring scale on the ruler.

2. Turn on the Geiger counter. Take three readings of the background radiation count rate and average them (see the previous page).

3. Carefully place the radioactive source at a distance d from the tube.

4. Record the count rate at that distance. Take three readings of the count rate at this distance.

5. Move the source so the distance between it and the tube doubles ($2d$) and record the count rate. Again, take three readings of the count rate at this distance.

6. Repeat step 5 for distances of $3d$, $4d$ etc.

7. Once the experiment is finished, put away the radioactive source immediately — you don't want to be exposed to more radiation than you need to be.

8. Average the count rates recorded for each distance. Eliminate the counts at each distance due to background radiation by subtracting the average background radiation count rate (see p.171) from the average count rate at each distance. Then plot a graph of the corrected count rate against distance of the tube from the source — see Figure 5.

Tip: You could label the y-axis as relative intensity, since the intensity is proportional to count rate in this experiment (see previous page).

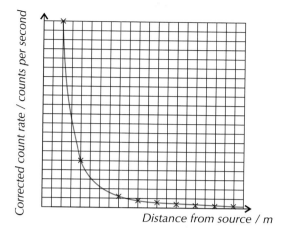

Figure 5: *A graph showing the inverse square relationship between count rate detected against the distance of the detector from the source of gamma radiation.*

Tip: You can also plot a graph of the different count rate (or intensity) readings against the inverse of the distance squared. If the data fits a straight line through the origin, then the inverse square law is verified.

You should see that as the distance doubles, the corrected count rate will drop to a quarter of its value, supporting the inverse square law.

Safe handling of radioactive sources

From the inverse square law, you know that using a radioactive source becomes significantly more dangerous the closer you get to the source (as the intensity of the radiation increases). This is why you should always hold a source away from your body when transporting it through the lab. Long handling tongs should also be used to minimise the radiation absorbed by the body. For those not working directly with radioactive sources, it's best to just keep as far away as possible.

Sources of gamma radiation should always be stored in a lead box (as this will absorb the radiation, see p.167). Make sure you only have radioactive sources out of their storages boxes for the shortest time possible — don't get them out until you need to use them and put them away as soon as you're done.

Practice Questions — Application

Q1 A Geiger-Müller tube placed 20.0 cm from a gamma source measures a count rate of 54 counts per second. Ignoring background radiation, calculate the expected count rate if the tube was placed 45 cm from the source.

Q2 The intensity of radiation from a gamma source is found to be 4.0×10^{-10} Wm^{-2} at a distance of 0.50 m from the source, and 1.8×10^{-10} Wm^{-2} at a distance of 0.75 m from the source. Use these measurements to show that the intensity of gamma radiation obeys the inverse square law.

Q1 Describe how you would eliminate background radiation from counts of a radioactive source.

Q2 Give three sources of background radiation.

Q3 What is the relationship between the intensity of gamma radiation and the distance from the source?

Q4 Describe an experiment you could use to demonstrate the relationship between the intensity of gamma radiation and the distance from the source.

Q5 Explain, in terms of the inverse square law, why you should hold gamma radiation sources at arm's length when you are handling them.

6. Exponential Law of Decay

The number of unstable nuclei that decay each second in a radioactive sample depends on how many unstable nuclei are left in the sample.

The rate of radioactive decay

Radioactive decay is completely random — you can't predict which atom's nucleus will decay when. But although you can't predict the decay of an individual nucleus, if you take a very large number of nuclei, their overall behaviour shows a pattern. **Isotopes** of an element have the same number of protons, but different numbers of neutrons in their nuclei. Any sample of a particular isotope has the same rate of decay — i.e. the same proportion of atomic nuclei will decay in a given time. Each unstable nucleus within the isotope will also have a constant decay probability.

The decay constant and activity

The **activity**, A, of a sample is the number of nuclei that decay each second. It is proportional to the number of unstable nuclei in the sample, N. For a given isotope, a sample twice as big would give twice the number of decays per second. The **decay constant**, λ, is the constant of proportionality. It is the probability of a specific nucleus decaying per unit time, and is a measure of how quickly an isotope will decay — the bigger the value of λ, the faster the rate of decay. The decay constant has units s^{-1} and the activity is measured in becquerels (1 Bq = 1 decay per second).

A = activity, Bq ⟶ $A = \lambda N$ ⟵ N = number of unstable nuclei in sample

λ = decay constant, s^{-1}

Because the activity, A, is the number of nuclei that decay each second, you can write it as the change in the number of unstable nuclei, ΔN, during a given time (in seconds) Δt:

$$A = -\frac{\Delta N}{\Delta t}$$

There's a minus sign in this equation because ΔN is always a decrease. Combining these two equations for the activity then gives the rate of change of the number of unstable nuclei:

$\frac{\Delta N}{\Delta t}$ = rate of change of number of unstable nuclei, s^{-1} ⟶ $\frac{\Delta N}{\Delta t} = -\lambda N$ ⟵ N = number of unstable nuclei in sample

λ = decay constant, s^{-1}

Example — Maths Skills

A sample of a radioactive isotope contains 3.0×10^{19} nuclei. Its activity is measured to be 2.4×10^{12} Bq. Calculate the isotope's decay constant.

Rearrange $A = \lambda N$ to give $\lambda = \frac{A}{N} = \frac{2.4 \times 10^{12}}{3.0 \times 10^{19}} = 8.0 \times 10^{-8}\, s^{-1}$

Learning Objectives:

- Know that radioactive decay is random and that a given nucleus has a constant probability of decay.

- Know and be able to use the term activity, A, given by $A = \lambda N$, where N is the number of unstable nuclei and λ is the decay constant.

- Be able to use the equation for the rate of change of N: $\frac{\Delta N}{\Delta t} = -\lambda N$.

- Be able to model the rate of change of N, e.g. using spreadsheet modelling.

- Understand that N varies exponentially with time: $N = N_0 e^{-\lambda t}$.

- Be able to answer questions on radioactive decay involving molar mass or the Avogadro constant.

- Understand that A varies exponentially with time: $A = A_0 e^{-\lambda t}$.

Specification Reference 3.8.1.3

Tip: You might see exam questions talking about A or N in terms of atoms — don't worry, it's just the same as the number of nuclei.

Tip: This is how you'll get this equation in the exam — the minus sign has just been taken over to the other side by multiplying both sides by −1.

Tip: Don't get the decay constant confused with wavelength — they both use the symbol λ.

Tip: So N decreases by 9.0×10^{16} nuclei every second.

A radioactive isotope has a decay constant of 1.2×10^{-4} s⁻¹.
What is the rate of change of N for a sample containing 7.5×10^{20} nuclei?

$$\frac{\Delta N}{\Delta t} = -\lambda N = -(1.2 \times 10^{-4}) \times (7.5 \times 10^{20}) = -9.0 \times 10^{16} \text{ s}^{-1}$$

Radioactive decay is an iterative process (the number of nuclei that decay in one time period controls the number that are available to decay in the next). This means that you can model the decay of a sample with time using an iterative spreadsheet. The spreadsheet can be used to produce a graph of N against t for a sample. There's more on this on pages 406–407.

The decay equation

Tip: If you modelled $\frac{\Delta N}{\Delta t} = -\lambda N$ with a spreadsheet to get a graph of N against t, the graph would have this equation, and the shape shown in Figure 1.

The number of unstable nuclei remaining, N, depends on the number originally present, N_0. The number remaining can be calculated using the equation:

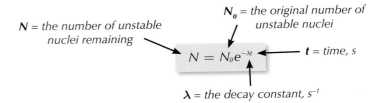

N = the number of unstable nuclei remaining

N_0 = the original number of unstable nuclei

$$N = N_0 e^{-\lambda t}$$

t = time, s

λ = the decay constant, s⁻¹

Tip: Remember to always round your final answer to the same no. of significant figures (s.f.) as the least precise data you use in your calculation. Here the 800 seconds is an exact value, and the other pieces of data are to 3.s.f. — so the answer to this calculation should be given to 3 s.f.

A sample of the radioactive isotope ¹³N contains 5.00×10^6 nuclei.
The decay constant for this isotope is 1.16×10^{-3} s⁻¹.
How many nuclei of ¹³N will remain after exactly 800 seconds?

$$N = N_0 e^{-\lambda t} = (5.00 \times 10^6) \times e^{-(1.16 \times 10^{-3}) \times 800}$$
$$= 1.976... \times 10^6$$
$$= 1.98 \times 10^6 \text{ nuclei (to 3 s.f.)}$$

The number of unstable nuclei decreases **exponentially** with time (see Figure 1).

Tip: Exponential change is where the change in the amount of something is proportional to the amount of that something left — see page 409.

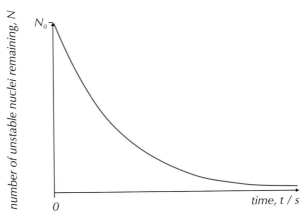

Figure 1: A graph showing the exponential decrease in the number of unstable nuclei of a radioactive isotope.

Tip: There's more on logs and log graphs on pages 402 and 409–410.

Plotting the natural log (ln) of the number of undecayed nuclei (or the activity) against time gives a straight-line graph (see Figure 2).

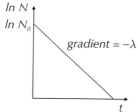

Figure 2: A graph showing the natural log of the number of unstable nuclei against time.

Tip: Taking natural log of both sides of $N = N_0 e^{-\lambda t}$ and using the log rules on p.402 gives $\ln N = \ln N_0 - \lambda t$. Comparing this to the straight line graph equation $y = mx + c$, you can see that a plot of $\ln N$ against t will be a straight line with a y-intercept of $\ln N_0$ and a gradient of $-\lambda$ (see p.410).

Dice simulation of radioactive decay

Radioactive decay is a random process where there is a constant probability that an unstable nucleus will decay. The probability is given by the decay constant. The result of rolling a fair 6-sided dice is also random with a constant probability — the probability of rolling any one number is 1/6. These similarities mean that you can simulate radioactive decay using dice, where each dice represents an unstable nucleus.

You need at least 100 dice for a good simulation of the unstable nuclei in a small radioactive sample. Roll all the dice and count how many of them landed on a 6 — these dice represent the nuclei that have decayed. Record the total number of dice rolled and the number of dice that have 'decayed' in a table. Remove the 'decayed dice' and roll the remaining dice again.

Repeat this process until all of the dice have 'decayed'. Each roll counts as 1 unit of time passing in the life span of the radioactive sample. If you plot a graph of the number of dice rolled each time (i.e. the number of unstable nuclei left in the sample, N) against time, then you'll see the same exponential relationship as shown in Figure 1 for radioactive decay.

Figure 3: Wooden cubes being used to simulate radioactive decay. Each cube represents an unstable nucleus that 'decays' when it lands on the black side facing upwards.

Calculations involving molar mass or the Avogadro constant

You might be asked to work out the number of atoms (or nuclei) in a sample from a **molar mass** (see page 57). The molar mass of a substance is the mass that 1 **mole** of the substance would have (usually in grams per mole, $g\,mol^{-1}$), and is equal to its relative atomic or relative molecular mass.

You can use the molar mass of a substance and the total mass of a sample of it to work out the number of moles in the sample. Then you just need to multiply the number of moles, n, by the Avogadro constant, N_A (see page) to calculate the number of atoms.

N = number of atoms in a sample ⟶ $N = n N_A$ ⟵ N_A = the Avogadro constant $(= 6.02 \times 10^{23}\ mol^{-1})$

n = the number of moles in a sample, mol

Tip: For a sample containing only one isotope, the value of the molar mass in grams is equal to the value of the mass number. E.g. for U-238 (which can also be written as ^{238}U), the molar mass is $238\ g\,mol^{-1}$.

Tip: It's no coincidence that the molar mass of ^{234}Pa is $234.0\ g\,mol^{-1}$ — remember, the molar mass of a sample is the same as the mass number if there are no other atoms in the sample.

── **Example** ── 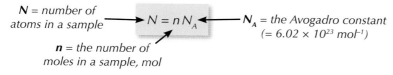 ────────────

A sample contains 12 g of the radioactive isotope ^{234}Pa. Calculate the number of atoms of ^{234}Pa that would have been present in the sample 3.0 minutes ago. The decay constant for this isotope is $2.87 \times 10^{-5}\ s^{-1}$ and its molar mass is $234.0\ g\,mol^{-1}$.

The number of moles in the sample = $12 \div 234.0 = 0.0512...$ mol

This means the number of ^{234}Pa atoms, N, in the sample now is
$N = nN_A$
$\quad = 0.0512... \times 6.02 \times 10^{23} = 3.087... \times 10^{22}$

Rearrange $N = N_0 e^{-\lambda t}$ to find N_0, the number of atoms of ^{234}Pa there were 3 minutes (180 s) ago:

$N_0 = N \div e^{-\lambda t} = 3.087... \times 10^{22} \div e^{-(2.87 \times 10^{-5}) \times 180}$
$\qquad\qquad\qquad\quad = 3.103... \times 10^{22} = 3.1 \times 10^{22}$ atoms (to 2 s.f.)

Activity and the decay equation

The number of unstable nuclei decaying per second (the activity) is proportional to the number of nuclei remaining. As a sample decays, its activity goes down — there's an equation for that too:

Figure 4: *Graph showing the exponential decay of the activity against time.*

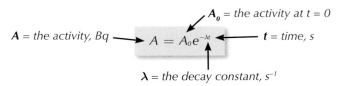

A = the activity, Bq ⟶ $A = A_0 e^{-\lambda t}$ ⟵ t = time, s

A_0 = the activity at $t = 0$

λ = the decay constant, s^{-1}

Example — **Maths Skills**

The isotope radon-220 has a decay constant of 1.25×10^{-2} s^{-1}. How long will it take for the activity of a sample of Radon-220 to fall from 85 Bq to 55 Bq?

First rearrange $A = A_0 e^{-\lambda t}$ to make t the subject.
Divide by A_0, take the natural log (ln) of both sides then divide by $-\lambda$.

$$\frac{A}{A_0} = e^{-\lambda t} \implies \ln\left(\frac{A}{A_0}\right) = -\lambda t \implies t = \frac{\ln\left(\frac{A}{A_0}\right)}{-\lambda}$$

Now just plug the numbers in and solve for t.

$$t = \frac{\ln\left(\frac{A}{A_0}\right)}{-\lambda} = \frac{\ln\left(\frac{55}{85}\right)}{-1.25 \times 10^{-2}} = 34.825... = 35 \text{ s (to 2 s.f.)}$$

Practice Questions — Application

Q1 A sample contains 4.5×10^{18} atoms of a radioactive isotope. If it has a decay constant of 1.1×10^{-13} s^{-1}, what is the activity of the sample?

Q2 A sample of a radioactive isotope has an activity of 3.2 kBq. If the isotope has a decay constant of 1.3×10^{-4} s^{-1}, what will the sample's activity be 6.5 hours later?

Q3 Protactinium-234m has a decay constant of 9.87×10^{-3} s^{-1}. If a sample initially contains 2.5×10^{15} unstable nuclei, how many will there be after 35 minutes?

Q4 The radioactive isotope ^{60}Co has a decay constant of 4.17×10^{-9} s^{-1} and molar mass 60.0 g mol^{-1}.
A sample initially contains 32.0 g of this isotope.

a) How many atoms of ^{60}Co are initially in the sample?

b) Calculate how many grams of ^{60}Co will remain after 3.50 years.

Tip: 1 kBq (kilobecquerel) is equal to 1000 Bq.

Q1 a) What is the decay constant of a radioactive isotope?

b) What is meant by the activity of a radioactive sample?

c) Write down an equation that relates activity, the decay constant and the number of unstable nuclei in a sample.

Q2 a) Write down an equation for the rate of change in the number of unstable nuclei in a radioactive sample.

b) State one method you could use to model the decay of an isotope over time using the formula for rate of change given in part a).

Q3 Give the equation for the number of undecayed radioactive nuclei as a function of time.

Q4 A student rolls 10 ordinary dice. She records how many dice show a 6, removes those showing 6, and rolls the remaining dice again. She repeats the process until she has removed all of the dice.

a) Explain why this process is often used to demonstrate radioactive decay.

b) The graph of the number of dice left against the number of rolls does not resemble an exponential relationship. Give a reason why this is the case.

Q5 Describe how to calculate the number of atoms in a radioactive substance given the mass present in a sample and the molar mass.

Q6 Give an equation that shows how the activity of a radioactive sample will change with time.

Tip: Make sure you check your units carefully when you're calculating half-life.

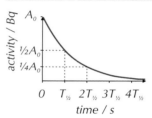

Figure 2: A graph of activity against time. The activity halves after each half-life $T_{1/2}$.

Tip: When measuring the activity and half-life of a source, you've got to remember background radiation — it needs to be subtracted from the activity readings to give the source activity (see page 171).

7. Half-life and its Applications

The time it takes the number of unstable nuclei in a sample to halve is constant for a given substance. This quantity is called half-life, and you need to be able to calculate it...

Half-life of radioactive isotopes

The **half-life** ($T_{1/2}$) of an isotope is the average time it takes for the number of unstable nuclei to halve. Of course, measuring the number of unstable nuclei isn't the easiest job in the world.

In practice, half-life isn't measured by counting nuclei, but by measuring the time it takes the activity to halve. The longer the half-life of an isotope, the longer it takes for the radioactivity level to fall.

Calculating half-life from decay curves

You can calculate the half life of an isotope from a decay curve — see Figure 1.

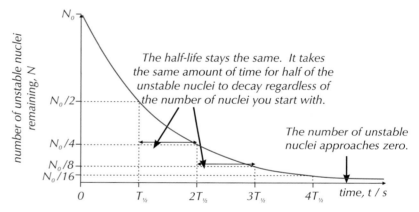

Figure 1: A graph showing that the time taken for the number of unstable nuclei remaining to halve is always equal to the half-life.

- Read off the value of unstable nuclei when $t = 0$.
- Go to half the original number of unstable nuclei on the y-axis.
- Draw a horizontal line to the curve, then a vertical line down to the x-axis. Read off the half-life where the line crosses the x-axis.
- It's always a good idea to check your answer — repeat these steps for a quarter of the original value and divide your answer by two. That will also give you the half-life. Check that you get the same answer both ways.

You can calculate half-life from an activity-time or count rate-time graph in exactly the same way, see Figure 2. N, A and count rate for an isotope will halve in the same amount of time, giving the half-life.

The half-life equation

You can derive an equation for half-life from the formula $N = N_0 e^{-\lambda t}$ on p.176.

- When $t = T_{1/2}$, the number of undecayed nuclei has halved, so $N = \frac{1}{2}N_0$.
- Substituting these values into the equation for N:

$$\tfrac{1}{2}N_0 = N_0 e^{-\lambda T_{1/2}}$$

- Cancelling N_0 and taking the natural log of both sides:

$$\ln\left(\tfrac{1}{2}\right) = -\lambda T_{1/2}$$

- $\tfrac{1}{2} = 2^{-1}$ so $\ln\left(\tfrac{1}{2}\right) = \ln(2^{-1}) = -\ln(2)$ using the rules on page 402. So:

$T_{1/2}$ = the half-life, s ⟶ $$T_{1/2} = \frac{\ln 2}{\lambda}$$

λ = the decay constant, s^{-1}

Tip: You'll need to use the power law $\frac{1}{a^n} = a^{-n}$ here.

Example — Maths Skills

A radioactive isotope has a decay constant of 1.16×10^{-3} s^{-1}.
What is the half-life for this isotope?

$$T_{1/2} = \frac{\ln 2}{\lambda} = \frac{\ln 2}{1.16 \times 10^{-3}} = 597.54... = 598 \text{ s (to 3 s.f.)}$$

Tip: Here it's fine to give the half-life in seconds. Some isotopes have much longer half-lives though, which are more sensible to give in years.

You can also use this equation to help you determine half-life from a graph of $\ln N$, where N is the number of unstable nuclei, against time, t.

Example — Maths Skills

The graph shows the radioactive decay of isotope X. Use the graph to find the half life of this isotope.

From page , you know that the gradient of this graph is equal to $-\lambda$.

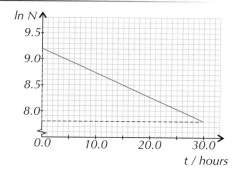

$$-\lambda = \frac{\Delta(\ln N)}{\Delta t} = \frac{7.8 - 9.2}{30.0 \times 60 \times 60}$$
$$= -1.296... \times 10^{-5} \text{ s}^{-1}$$

So $\lambda = 1.296... \times 10^{-5}$ s^{-1}

$$T_{1/2} = \frac{\ln 2}{\lambda} = \frac{\ln 2}{1.296... \times 10^{-5}} = 53\,471.3... \text{ s}$$
$$= 53\,000 \text{ s (to 2 s.f.)}$$

Tip: Make sure the time you plug into this equation is in the correct units — it should be in seconds.

Applications of radioactive isotopes

Radioactive substances are extremely useful. You can use them for all sorts — to date organic material, diagnose medical problems, sterilise food, and in smoke alarms. The best isotope for a particular application and what we do with any radioactive waste products usually depends a lot on half-life.

Radioactive dating

The radioactive isotope carbon-14 is used in **radioactive dating**. Living plants take in carbon dioxide from the atmosphere as part of photosynthesis, including the radioactive isotope carbon-14. When they die, the activity of carbon-14 in the plant starts to fall, with a half-life of around 5730 years. Archaeological finds made from once-living material (like wood) can be tested to find the current amount of carbon-14 in them, and date them.

Tip: You'll need to use the equations for N back on pages 175–176 to answer this question.

A sample of 6.5×10^{23} carbon atoms is taken from a spear, and found to have an activity of 0.052 Bq. The ratio of radioactive carbon-14 to stable carbon-12 in living wood is $1 : 1.4 \times 10^{12}$, and the half-life of carbon-14 is 5730 years. How old is the wood the spear is made from?

The half-life of carbon-14 in seconds is

$$T_{1/2} = 5730 \times (365 \times 24 \times 3600) = 1.8070... \times 10^{11} \text{ s}$$

So the decay constant, $\lambda = \dfrac{\ln 2}{T_{1/2}} = \dfrac{\ln 2}{1.8070... \times 10^{11}} = 3.8358... \times 10^{-12} \text{ s}^{-1}$

Rearrange and use the equation for activity to find the number of carbon-14 nuclei in the wood:

$$N = \frac{A}{\lambda} = 0.052 \div (3.8358... \times 10^{-12}) = 1.3556... \times 10^{10}$$

Use the ratio given in the question to calculate the expected number of carbon-14 nuclei in a sample of 6.5×10^{23} carbon atoms from living wood:

$$N_0 = (1 \div 1.4 \times 10^{12}) \times 6.5 \times 10^{23} = 4.6428... \times 10^{11}$$

N and N_0 are related by $N = N_0 e^{-\lambda t}$. You can rearrange this by dividing by N_0 and taking the natural log (ln) of both sides to make t the subject and find the age of the wood:

Tip: $\ln(e^a) = a$, which is why $\ln(e^{-\lambda t})$ becomes $-\lambda t$ when you're rearranging to find t. For more help on logs, skip on over to page 402.

$$N = N_0 e^{-\lambda t} \rightarrow \frac{N}{N_0} = e^{-\lambda t} \rightarrow \ln\left(\frac{N}{N_0}\right) = \ln(e^{-\lambda t}) \rightarrow \ln\left(\frac{N}{N_0}\right) = -\lambda t$$

$$\text{So } t = -\frac{1}{\lambda} \times \ln\left(\frac{N}{N_0}\right) = -\frac{1}{3.8358... \times 10^{-12}} \times \ln\left(\frac{1.3556... \times 10^{10}}{4.6428... \times 10^{11}}\right)$$

$$= 9.2121... \times 10^{11} \text{ s}$$
$$= 29\,000 \text{ years (to 2 s.f.)}$$

However, it can be difficult to get a reliable age from radioactive dating as:

- For man-made objects crafted from natural materials like wood, you can only find the age of the material used — not the object itself.
- The object may have been contaminated by other radioactive sources.
- There may be a high background count that obscures the object's count.
- There may be uncertainty in the amount of carbon-14 that existed thousands of years ago.
- The sample size or count rate may be small, and so might be statistically unreliable.

Tip: If a house was built from wood from a tree cut down years ago, carbon dating would only give the length of time since the tree was cut down, not since the house was built.

Medical diagnosis

Technetium-99m is widely used in medical tracers — radioactive substances that are used to show tissue or organ function. The tracer is injected into or swallowed by the patient and then moves through the body to the region of interest. The radiation emitted is recorded and an image of inside the patient produced.

Technetium-99m is suitable for this use because it emits γ-radiation, has a half-life of 6 hours (long enough for data to be recorded, but short enough to limit the radiation to an acceptable level) and decays to a much more stable isotope.

Storage of radioactive waste

Nuclear fission reactors use uranium-235 to generate electricity. The uranium decays into several different radioactive isotopes with different half-lives. These isotopes emit alpha, beta and gamma radiation, and must be stored carefully for hundreds of years until their activity has fallen to safe levels.

This radioactive waste is a serious problem as it has a very long half-life and so stays highly radioactive for a really long time. See pages 190–195 for more about nuclear fission, nuclear fission reactors and their waste products.

Figure 3: *Spent nuclear fuel rods are stored in water for several years while the activity decreases.*

Practice Questions — Application

Q1 The activity of a radioactive sample fell from 2400 Bq to 75 Bq over a period of 24 hours. How long is the isotope's half-life?

Q2 An isotope has a half-life of 483 seconds. What is its decay constant?

Q3 The graph below shows the natural log of the number of unstable nuclei in a sample, N, against time. Use this graph to calculate the half-life of the sample.

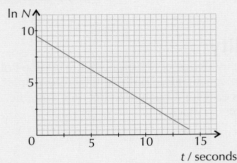

Q4 A sample of carbon atoms from an ancient bone has an activity of 0.45 Bq. The same number of carbon atoms from a piece of living tissue has an activity of 1.2 Bq. Given that the half-life of carbon-14 is 5730 years, how old is the bone?

Practice Questions — Fact Recall

Q1 What is the half-life of an isotope?

Q2 Describe how you would find an isotope's half-life from a graph showing its activity against time.

Q3 Give two applications of radioactive isotopes where the half-life is important. Explain why the half-life is important in each case.

Learning Objectives:

- Be able to sketch the graph of N (number of neutrons) against Z (proton number) for stable nuclei, and identify regions of particles that will undergo α and β decay.

- Know the possible decay modes of unstable nuclei, including: α, β^-, β^+ and electron capture.

- Know how N and Z are changed by radioactive decay, and be able to represent these decays using simple decay equations.

- Know that γ-rays can be emitted by nuclei in excited states and its applications, e.g. technetium-99m as a gamma source in medical diagnosis.

- Understand and be able to use nuclear energy level diagrams.

Specification Reference 3.8.1.4

8. Nuclear Decay

Nuclei are radioactive either because they have too few or too many neutrons, or too many nucleons altogether, making them unstable because they have too much energy.

Nuclide notation

The nuclide notation of an element summarises all the information about its atomic structure. Figure 1 shows the nuclide notation for carbon-12, with nucleon number $A = 12$ and proton number $Z = 6$:

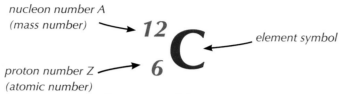

nucleon number A (mass number)

proton number Z (atomic number)

$$^{12}_{6}\text{C}$$

element symbol

Figure 1: *Nuclide notation.*

Nuclear stability

The nucleus is under the influence of the strong nuclear force holding it together and the electromagnetic force pushing the protons apart. It's a very delicate balance, and it's easy for a nucleus to become unstable.
A nucleus will be unstable if it has:

- too many neutrons
- too few neutrons
- too many nucleons altogether — i.e. it's too heavy
- too much energy

There are several types of decay or 'decay modes' that an unstable nucleus can undergo to make itself more stable. You can get a stability graph by plotting N (number of neutrons) against Z (proton number) — see Figure 2.

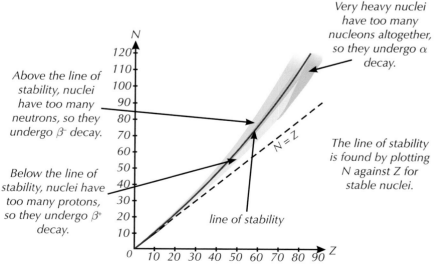

Figure 2: *A graph showing stable and unstable nuclei.*

Alpha emission

Alpha emission only happens in very heavy atoms, like uranium and radium. The nuclei of these atoms are too massive to be stable. When an alpha particle is emitted, the proton number decreases by two, and the nucleon number decreases by four.

Example

Uranium-238 decays to thorium-234 by emitting an alpha particle:

nucleon number decreases by 4

$$^{238}_{92}U \longrightarrow \, ^{234}_{90}Th + \, ^{4}_{2}\alpha$$

proton number decreases by 2

Tip: Isotopes that will undergo α decay have a large number of neutrons and protons. Emitting an alpha particle decreases Z by 2 and decreases N by 2, which moves the nucleus towards a more stable region on a graph of N against Z.

Beta emission

Beta-minus (usually just called beta) decay is the emission of an electron from the nucleus along with an antineutrino. Beta-minus decay happens in isotopes that are **neutron rich** (i.e. have many more neutrons than protons in their nucleus). When a nucleus ejects a beta-minus particle, one of the neutrons in the nucleus is changed into a proton — the proton number increases by one, and the nucleon number stays the same.

Example

Rhenium-188 decays to osmium-188 by emitting a beta-minus particle:

nucleon number stays the same

$$^{188}_{75}Re \longrightarrow \, ^{188}_{76}Os + \, ^{0}_{-1}\beta + \, ^{0}_{0}\overline{\nu}_e$$

proton number increases by 1

Tip: Remember that isotopes that will undergo β^- decay are above the line of stability on a graph of N against Z. Converting a neutron into a proton increases Z and decreases N, which brings the nucleus closer to the line of stability.

Beta-plus decay happens in isotopes that are **proton rich** (i.e. have a high proton to neutron ratio). When a nucleus ejects a beta-plus particle, a proton gets changed into a neutron — the proton number decreases by one, and the nucleon number stays the same. A neutrino is also emitted.

Example

Sodium-22 decays to neon-22 by emitting a beta-plus particle:

nucleon number stays the same

$$^{22}_{11}Na \longrightarrow \, ^{22}_{10}Ne + \, ^{0}_{+1}\beta + \, ^{0}_{0}\nu_e$$

proton number decreases by 1

Tip: A beta-plus β^+ particle is a positron.

Tip: Isotopes that will undergo β^+ decay are below the line of stability on a graph of N against Z. Converting a proton into a neutron increases N and decreases Z, which brings the nucleus closer to the line of stability.

Gamma emission

After alpha or beta decay, the nucleus often has excess energy — it's in an excited state. This energy is lost by emitting a gamma ray. During gamma emission, there is no change to the nuclear constituents — the nucleus just loses excess energy.

Another way that gamma radiation is produced is **electron capture**. This is when a nucleus captures and absorbs one of its own orbiting electrons, which causes a proton to change into a neutron. A neutrino is also released.

Exam Tip
You'll need to be able to write out (and complete) nuclear equations — so make sure you learn what changes with each decay. There's more coming up on how to tackle decay equations on the next page.

Electron capture has the same effect on the nucleon and proton numbers of the nucleus as beta-plus decay — both cause N to increase by 1 and Z to decrease by 1 in nuclei that are below the line of stability. This makes the nucleus unstable and it emits gamma radiation.

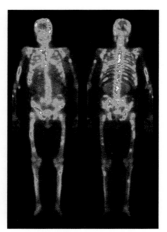

Figure 3: The artificial isotope technetium-99m is formed in an excited state from the decay of another element. It is used as a tracer in medical imaging (see page 299).

> **Example**
>
> Beryllium-7 decays to lithium-7 by electron capture:
>
> nucleon number stays the same
>
> $$^{7}_{4}Be + ^{0}_{-1}\beta \longrightarrow ^{7}_{3}Li + ^{0}_{0}\nu_e + \gamma$$
>
> proton number decreases by 1

Conservation rules in nuclear reactions

In every nuclear reaction energy, momentum, charge, nucleon number and lepton number must be conserved. As you might remember from year 1 of A-level, electrons and neutrinos have a lepton number of +1. Positrons and antineutrinos have a lepton number of –1.

Tip: When balancing equations, writing the charge of particles where the proton number would usually go helps you to make sure the charges are balanced.

> **Example**
>
> Here's the nuclear equation for the beta-minus decay of rhenium-188, showing how nucleon number and charge are conserved:
>
> $188 = 188 + 0 + 0$ — nucleon numbers balance
>
> $$^{188}_{75}Re \longrightarrow ^{188}_{76}Os + ^{0}_{-1}\beta + ^{0}_{0}\overline{\nu}_e$$
>
> $75 = 76 - 1 + 0$ — charges balance

Energy level diagrams for nuclear reactions

Just like electron energy transitions, energy transitions of nuclei during radioactive decays can be shown using **energy level diagrams**.

Often energy level diagrams have a vertical energy axis to show the relative energy of each level — the lower down the axis the lower the energy of the nucleus. The energy levels are drawn as horizontal lines.

Tip: For more on the energy that a nucleus has, see pages 188–189.

The element is usually written in nuclide notation on the energy levels. The decay and energy change is shown by an arrow between the energy levels. This arrow is labelled with the type of decay and the change in energy of the nucleus, ΔE. This change in energy is equal to the amount of energy that is released by the decay.

> **Example**
>
> The energy diagram shown on the right represents the alpha decay of a uranium-238 nucleus into thorium-234. The nucleus loses 4.3 MeV of energy through this decay.
>
>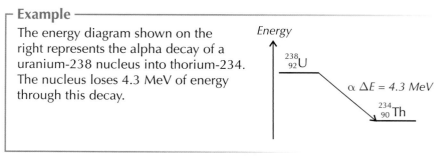

Example

The energy diagram below represents one possible decay of cobalt-60 into nickel-60 via a beta-minus and a gamma decay. The order of the decays on the diagram shows that the beta-minus decay occurred before the gamma decay.

The cobalt is transformed into nickel by the beta-minus decay — but the nucleus is excited and unstable at this energy level. It becomes de-excited once it has emitted a gamma ray. The nuclide notation for the nickel isotope only needs to be written on one of the levels.

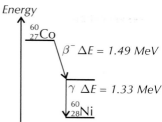

Tip: From this diagram, you can work out that the total decay energy to get from cobalt-60 to nickel-60 is:
1.49 + 1.33 = 2.82 MeV.

Practice Questions — Application

Q1 Complete the following nuclear equation for the beta-minus decay of caesium-137: $^{137}_{55}Cs \longrightarrow ^{?}_{?}Ba + ? + ?$

Q2 Write out the nuclear equation for the alpha decay of At-211 ($Z = 85$) to an isotope of bismuth. (The chemical symbol of bismuth is Bi.)

Q3 Rubidium-83 decays via electron capture to krypton. Write out the nuclear equation for this process. (Rubidium has 37 protons and the chemical symbol Rb, krypton has 36 protons and the chemical symbol Kr.)

Q4 A simple energy level diagram for the decay of an iodine nucleus to a xenon nucleus is shown below.

a) Describe the decay process that is represented by this diagram.

b) What does ΔE represent in these types of diagram?

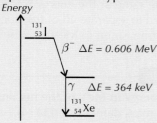

Tip: Remember that the number in the name of an isotope is the nucleon number.

Practice Questions — Fact Recall

Q1 Give four reasons why a nucleus might be unstable.

Q2 Sketch a graph of neutron number N against proton number Z, showing the line of stable nuclei, and indicating the regions of nuclei which will decay by α, β⁻ and β⁺ decay.

Q3 Describe the particle(s) emitted and the change that occurs to the decayed nucleus for each of the following types of decay:

a) alpha   b) beta-minus   c) beta-plus

Q4 Technetium-99m is a radioactive isotope that emits gamma radiation.

a) Describe what causes this isotope to emit gamma radiation.

b) Give one application of this isotope.

Q5 Give three things that are always conserved in a nuclear reaction.

9. Mass Defect and Binding Energy

Learning Objectives:

- Appreciate that $E = mc^2$ applies to all energy changes.
- Know that a mass defect of 1 u (where u is an atomic mass unit) is equivalent to a binding energy of 931.5 MeV.
- Be able to do simple calculations involving mass difference and binding energy.
- Be able to sketch the graph of average binding energy per nucleon against nucleon number.

Specification Reference 3.8.1.6

The binding energy is a measure of how strongly a nucleus is held together — the greater the binding energy per nucleon, the more stable the nucleus.

Mass defect and binding energy

The mass of a nucleus is less than the mass of its constituent nucleons — the difference is called the **mass defect**. Einstein's equation says that mass and energy are equivalent:

E = energy in J ⟶ $E = mc^2$ ⟵ c = the speed of light in a vacuum in ms^{-1}

m = mass in kg

This equation applies to all energy changes. As nucleons join together, the total mass decreases — this 'lost' mass is converted into energy and released. The amount of energy released is equivalent to the mass defect.

Example — **Maths Skills**

The mass of a nucleus of potassium, $^{40}_{19}$K, is 39.9536 u. The mass of a proton is 1.00728 u and the mass of a neutron is 1.00867 u. Calculate the mass defect of the nucleus in u.

Number of protons = 19, number of neutrons = (40 − 19) = 21
Mass of nucleons = (19 × 1.00728 u) + (21 × 1.00867 u) = 40.32039 u

So mass defect = mass of nucleons − mass of nucleus
= 40.32039 u − 39.9536 u = 0.36679 u

If you pulled the nucleus completely apart, the energy you'd have to use to do it would be the same as the energy released when the nucleus formed. The energy needed to separate all of the nucleons in a nucleus is called the **binding energy** (measured in MeV), and it is equivalent to the mass defect.

Tip: Don't forget — 1 u = 1.661 × 10^{-27} kg.

Example — **Maths Skills**

Calculate the binding energy in MeV of the nucleus of a lithium-6 atom, $^{6}_{3}$Li, given that its mass defect is 0.0343 u.

Convert the mass defect into kg:
Mass defect = 0.0343 × (1.661 × 10^{-27}) = 5.697... × 10^{-29} kg

Use $E = mc^2$ to calculate the binding energy:
E = (5.697... × 10^{-29}) × (3.00 × 10^8)^2 = 5.127... × 10^{-12} J
Convert to MeV:
E = ((5.127... × 10^{-12}) ÷ (1.60 × 10^{-19})) ÷ 10^6 = 32.046... = 32.0 MeV (to 3 s.f.)

Tip: To convert from J to eV, you divide by the magnitude of the charge on an electron, e = 1.60 × 10^{-19} C. Then to convert to MeV you just divide by 10^6.

The binding energy per unit of mass defect is the same for all nuclei:

$$1 \text{ u} \approx 931.5 \text{ MeV}$$

Example — **Maths Skills**

Using the mass defect given above, the binding energy in MeV of lithium-6 is:
$E \approx$ 0.0343 u × 931.5 MeV = 31.950... = 32.0 MeV (to 3 s.f.)

Exam Tip
Remember that 1 u ≈ 931.5 MeV — it'll be really useful for your exam.

Average binding energy per nucleon

A useful way of comparing the binding energies of different nuclei is to look at the average binding energy per nucleon.

$$\text{Average binding energy per nucleon} = \frac{\text{Binding energy } (B)}{\text{Nucleon number } (A)}$$

Example — Maths Skills

What is the average binding energy per nucleon for a ^6_3Li nucleus?

You know from the second example on the previous page that binding energy = 32.046... MeV. Nucleon number = A = 6

Binding energy per nucleon = $\frac{B}{A}$ = 32.046... ÷ 6 = 5.341...

$\qquad\qquad\qquad\qquad\qquad\qquad\qquad$ = 5.34 MeV (to 3 s.f.)

A graph of average binding energy per nucleon against nucleon number, for all elements, shows a curve (Figure 1). Higher average binding energy per nucleon means more energy is needed to remove nucleons from the nucleus. Therefore, the higher the average binding energy per nucleon, the more stable the nucleus. This means that the most stable nuclei occur around the maximum point on the graph — which is at nucleon number 56 (i.e. iron, Fe).

Tip: Make sure you remember the average nucleon number of iron.

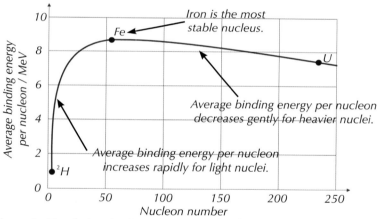

Iron is the most stable nucleus.

Average binding energy per nucleon decreases gently for heavier nuclei.

Average binding energy per nucleon increases rapidly for light nuclei.

Figure 1: *Graph showing how the average binding energy per nucleon varies with nucleon number. The red line shows the line of best fit.*

Exam Tip
Make sure you know this graph really well, including the units and axes — you could be asked to sketch it in an exam.

Practice Questions — Application

Q1 The mass defect of a carbon-12 nucleus is 0.0989 u.
Calculate the binding energy of the nucleus.

Q2 Calculate the mass defect of a nucleus of $^{16}_8\text{O}$ (mass = 15.994915 u).

Q3 The binding energy per nucleon of a nucleus of iron-56 is 8.79 MeV.
What is its mass defect in u?

Tip: The mass of a proton is 1.00728 u, and the mass of a neutron is 1.00867 u.

Practice Questions — Fact Recall

Q1 What type of energy changes does the equation $E = mc^2$ apply to?

Q2 What is meant by the binding energy of a nucleus?

Q3 How many MeV of binding energy are equivalent to a 1 u mass defect?

Q4 Sketch the graph of the average binding energy per nucleon against nucleon number. Label the first element and the most stable element on the graph.

10. Nuclear Fission and Fusion

Learning Objectives:

- Know that nuclear fission is the splitting of larger nuclei into smaller nuclei.

- Know that nuclear fusion is the combining of two smaller nuclei into one larger nucleus.

- Know that energy can be released during fission or fusion due to an increase in binding energy per nucleon.

- Identify, on a plot of average binding energy per nucleon against nucleon number, the regions where nuclei will release energy when undergoing fission/fusion.

- Be able to calculate the energy released during fission or fusion from nuclear masses.

Specification Reference 3.8.1.6

Tip: Heavy nuclei don't always fission to form the same daughter nuclei — ^{235}U can split into lots of different pairs of nuclei (with nucleon numbers around 90 and 140).

Radioactive decay isn't the only way that nuclei can change — they can also split into two smaller nuclei or fuse with other nuclei to form larger ones.

Fission

Large nuclei (e.g. uranium), are unstable and some can randomly split into two smaller nuclei — this is called **nuclear fission**. This process is called spontaneous if it just happens by itself, or induced if we encourage it to happen.

Energy is released during nuclear fission because the new, smaller nuclei have a higher average binding energy per nucleon (see the previous page). The larger the nucleus, the more unstable it will be — so large nuclei are more likely to spontaneously fission. This means that spontaneous fission limits the number of nucleons that a nucleus can contain — in other words, it limits the number of possible elements.

Example

Fission can be induced by making a neutron enter a ^{235}U nucleus, causing it to become very unstable. Only low-energy neutrons can be captured in this way. A low-energy neutron is called a **thermal neutron**.

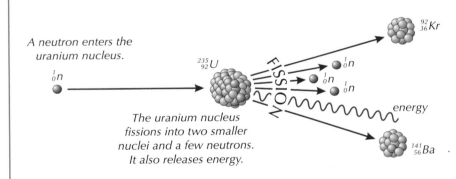

A neutron enters the uranium nucleus.

$^{1}_{0}n$

$^{235}_{92}U$

$^{1}_{0}n$

$^{1}_{0}n$
$^{1}_{0}n$

energy

$^{92}_{36}Kr$

$^{141}_{56}Ba$

The uranium nucleus fissions into two smaller nuclei and a few neutrons. It also releases energy.

Figure 1: *A possible fission of a uranium-235 nucleus.*

Fusion

Two light nuclei can combine to create a larger nucleus — this is called **nuclear fusion**. A lot of energy is released during nuclear fusion because the new, heavier nuclei have a much higher average binding energy per nucleon (as long as the lighter nuclei are light enough — see Figure 1 on the previous page).

Example

In the Sun, hydrogen nuclei fuse in a series of reactions to form helium. One of the reactions is: $^{2}_{1}H + ^{1}_{1}H \longrightarrow ^{3}_{2}He +$ energy.

$^{2}_{1}H$

$^{1}_{1}H$

fusion

energy

$^{3}_{2}He$

Figure 2: *Two isotopes of hydrogen fuse to form helium.*

All nuclei are positively charged — so there will be an electrostatic (or Coulomb) force of repulsion between them (see Figure 3). Nuclei can only fuse if they overcome this electrostatic force and get close enough for the attractive force of the strong interaction to hold them both together. About 1 MeV of kinetic energy is needed to make nuclei fuse together — that's a lot of energy.

Low-energy nuclei are deflected by electrostatic repulsion.   *High-energy nuclei overcome electrostatic repulsion and are attracted by the strong interaction.*

Figure 3: *Nuclei must overcome their mutual electrostatic repulsion to fuse together.*

Energy released by fission and fusion

You can tell whether it is energetically favourable for an element to undergo fission or fusion by looking at the graph of average binding energy per nucleon against nucleon number. Only elements to the right of ^{56}Fe can release energy through nuclear fission. Similarly only elements to the left of ^{56}Fe can release energy through nuclear fusion. This is because energy is only released when the average binding energy per nucleon increases.

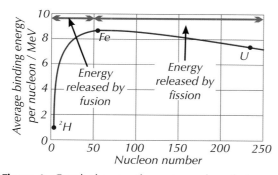

Figure 4: *Graph showing the regions where fusion and fission reactions are energetically favourable.*

The change in binding energy gives the energy released. The average binding energy per nucleon graph can be used to estimate the energy released in nuclear reactions.

┌ **Example** ── **Maths Skills** ──────────────────

Energy released through the nuclear fusion of ^{2}H and ^{3}H:

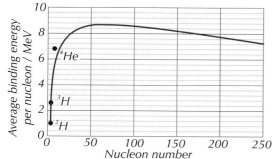

Binding energy of ^{4}He = 4 × 6.8 = 27.2 MeV

Binding energy of ^{2}H and ^{3}H = (2 × 1.1) + (3 × 2.6) = 10.0 MeV

Therefore total energy released = 27.2 − 10.0 = 17.2 MeV

Figure 5: *Graph showing the energy released by the fusion of ^{2}H and ^{3}H.*

Energy released in the induced nuclear fission of ^{235}U:

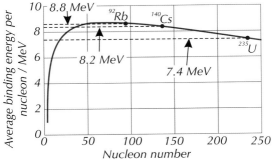

Figure 6: *Graph showing the energy released by the fission of ^{235}U.*

Binding energy of ^{235}U
= 235 × 7.4 = 1739 MeV

Binding energy of ^{92}Rb and ^{140}Cs
= (92 × 8.8) + (140 × 8.2)
= 1957.6 MeV

So the total energy released = 1957.6 − 1739
= 218.6
= 220 MeV (to 2 s.f.)

You can find the number of neutrons released per fission, x, by balancing the nucleon number on both sides of the reaction equation:

$$^{235}_{92}\text{U} + ^{1}_{0}\text{n} \longrightarrow ^{92}_{37}\text{Rb} + ^{140}_{55}\text{Cs} + x^{1}_{0}\text{n}$$
$$235 + 1 = 92 + 140 + x$$

So number of neutrons released per fission = x = 236 − 232 = 4

The energy released during a fission or fusion reaction can also be calculated using the equation $E = \Delta m \times c^2$, where Δm is the total difference in mass between the initial and final nuclei (as long as you take the neutrons released into account too).

Example — **Maths Skills**

Calculate the energy released by the following fission reaction using the data in Figure 7: $^{235}_{92}\text{U} + ^{1}_{0}\text{n} \longrightarrow ^{89}_{36}\text{Kr} + ^{144}_{56}\text{Ba} + 3^{1}_{0}\text{n}$

$\Delta m = m_{U-235} + m_n - m_{Kr-89} - m_{Ba-144} - 3m_n$
$\quad = 234.99333 + 1.00867 - 88.89783 - 143.89215 - (3 \times 1.00867)$
$\quad = 0.18601$ u

In MeV: 0.18601 × 931.5 = 173.26... = 173.3 MeV (to 4 s.f.)

Tip: Here the mass difference has been converted from u to MeV by multiplying by 931.5 MeVu^{-1} (see p.188).

Isotope	Mass (u)
^{235}U	234.99333
^{89}Kr	88.89783
^{144}Ba	143.89215
^{139}Te	138.90613
^{94}Zr	93.88431
^{2}H	2.01355
proton	1.00728
neutron	1.00867
positron	0.00055
neutrino	0

Figure 7: *The masses of some nuclei and particles.*

Practice Questions — Application

Q1 Using the data in Figure 7, calculate the energy released during the fusion of two hydrogen nuclei: $^{1}_{1}\text{p} + ^{1}_{1}\text{p} \longrightarrow ^{2}_{1}\text{H} + ^{0}_{1}\text{e} + \nu$.

Q2 Calculate the energy released when a ^{235}U nucleus is hit by a neutron and fissions into ^{94}Zr and ^{139}Te and a number of neutrons.

Practice Questions — Fact Recall

Q1 What is nuclear fission?

Q2 What is the difference between spontaneous and induced fission?

Q3 What is nuclear fusion?

Q4 Why is fission/fusion only energetically favourable for certain nuclei?

Q5 Explain why light nuclei must have a lot of energy to fuse together.

Q6 Sketch the graph of average binding energy per nucleon against nucleon number. Show on it what nuclear reaction is energetically favourable for the elements on either side of the peak.

11. Nuclear Fission Reactors

You need to know all about the different bits and pieces that make up a nuclear reactor, as well as the potential problems that come from releasing energy in this way.

How reactors work

We can harness the energy released during nuclear fission reactions in a thermal nuclear reactor (see Figure 1), but it's important that these reactions are very carefully controlled.

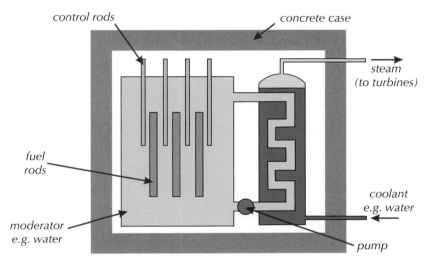

Figure 1: *The key features of a thermal nuclear reactor.*

Chain reactions

Nuclear reactors use rods of uranium that are rich in ^{235}U (or sometimes plutonium rods rich in ^{239}Pu) as 'fuel' for fission reactions. (The rods also contain other isotopes, but they don't undergo fission.) These are placed into the reactor remotely, which keeps workers as far away from the radiation as possible (see p.173).

These fission reactions produce more neutrons which then induce other nuclei to fission — this is called a **chain reaction**. The neutrons will only cause a chain reaction if they are slowed down, which allows them to be captured by the uranium nuclei — these slowed down neutrons are called thermal neutrons (see page 190).

Moderator

Fuel rods need to be placed in a **moderator** (for example, water) to slow down and/or absorb neutrons. You need to choose a moderator that will slow down some neutrons enough so they can cause further fission, keeping the reaction going at a steady rate.

The moderator slows down neutrons through **elastic collisions** (kinetic energy is conserved) with nuclei of the moderator material. When neutrons collide with particles that are of a similar mass, they are slowed down more efficiently (see next page). Water is often used as a moderator since it contains hydrogen, which is of a similar mass to a neutron.

Learning Objectives:

- Know that thermal neutrons can induce fission in uranium nuclei and start a fission chain reaction.
- Understand the term critical mass.
- Know functions of the moderator, control rods and coolant in a thermal nuclear reactor, be able to give examples of materials used for these functions and explain why certain materials are chosen for each of these.
- Know ways in which the risks of nuclear power are minimised, e.g. remote handling of fuel, shielding, emergency shut down mechanisms.
- Know why fission waste products are dangerous, and how they are handled and stored.
- Appreciate the balance between the benefits and risks of nuclear power, and that scientific knowledge of nuclear energy allows society to make informed decisions.

Specification References 3.8.1.6, 3.8.1.7 and 3.8.1.8

Figure 2: *Fuel rods being lowered into the reactor at a nuclear power station. The core is surrounded by water which acts as a moderator.*

Moderation by elastic collisions

Assuming that the collision of a moderator particle and a neutron is perfectly elastic, you know that kinetic energy and momentum are both conserved. If you assume the moderator particle is stationary before the collision (see Figure 3), then you can write the equations below for the conservation of kinetic energy and momentum of the particles.

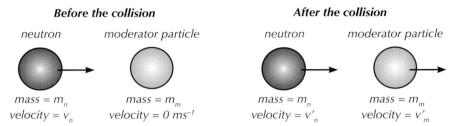

Before the collision   *After the collision*

neutron   *moderator particle*   *neutron*   *moderator particle*

$mass = m_n$ $mass = m_m$ $mass = m_n$ $mass = m_m$
$velocity = v_n$ $velocity = 0\ ms^{-1}$ $velocity = v'_n$ $velocity = v'_m$

Figure 3: *A diagram to show the mass and velocity of a moderator particle and a neutron before and after an elastic collision.*

Conservation of momentum: $m_n v_n = m_n v'_n + m_m v'_m$

Conservation of kinetic energy: $\frac{1}{2} m_n v_n^2 = \frac{1}{2} m_n v'^2_n + \frac{1}{2} m_m v'^2_m$

Tip: You should have met conservation of momentum in year 1 of A-level.

You can use these equations to find the following equations for v_n' and v_m' in terms of v_n.

$$v_n' = \frac{(m_n - m_m)}{(m_n + m_m)} v_n \qquad v_m' = \frac{2m_n}{(m_n + m_m)} v_n$$

Tip: You don't need to worry about how the above equations were rearranged to find the equations for v_n' and v_m'.

If the mass of the moderator particle is roughly equal to the mass of a neutron, $m_n = m_m = m$, then the final velocity of the neutron would be 0 ms⁻¹:

$$v_n' = \frac{(m - m)}{(m + m)} v_n = 0 \qquad v_m' = \frac{2m}{(m + m)} v_n \Rightarrow v_m' = v_n$$

All of the kinetic energy and momentum would be transferred to the moderator particle.

The more similar the masses m_n and m_m are, the more kinetic energy and momentum will be transferred from the neutron to the moderator particle. In a thermal nuclear reactor, you don't want to stop the neutrons, but you do need to slow them down a lot (to around 2200 ms⁻¹) — so you need the mass of the moderator particles to be around the same size as the mass of a neutron.

Control rods

You want the chain reaction to continue on its own at a steady rate, where one fission follows another. The amount of 'fuel' you need to do this is called the **critical mass** — any less than the critical mass (sub-critical mass) and the reaction will just peter out. Nuclear reactors use a supercritical mass of fuel (where several new fissions normally follow each fission) and control the rate of fission using **control rods**.

Control rods control the chain reaction by limiting the number of neutrons in the reactor. They absorb neutrons so that the rate of fission is controlled. Control rods are made up of a material that absorbs neutrons (e.g. boron), and they can be inserted by varying amounts to control the reaction rate.

Coolant

Coolant is sent around the reactor to remove heat produced by fission. The material used should be a liquid or gas at room temperature, and be efficient at transferring heat. Often the coolant is the same water that is being used in the reactor as a moderator. The heat from the reactor can then be used to make steam for powering electricity-generating turbines.

Safety

Reactor shielding

The nuclear reactor is surrounded by a thick concrete case, which acts as shielding. This prevents radiation escaping and reaching the people working in the power station.

Emergency shut-down

In an emergency, the reactor can be shut down automatically by the release of control rods into the reactor. The control rods are lowered fully into the reactor, which slows down the reaction as quickly as possible.

Handling and storing fission waste products

Unused uranium fuel rods emit only alpha radiation, which is weakly penetrating and so is easily contained. Spent fuel rods are more dangerous, since fission waste products usually have a larger proportion of neutrons than nuclei of a similar atomic number — this makes them unstable and radioactive. The fission waste products emit beta and gamma radiation, which are strongly penetrating.

The products can be used for practical applications such as tracers in medical diagnosis (see page 169). However, their handling and storage needs great care since they are highly radioactive. When material is removed from the reactor, it is initially very hot, so it is placed in cooling ponds until the temperature falls to a safe level. This is done remotely — just like the handling of fuel — to limit the radiation that workers are exposed to. The radioactive waste should then be stored in sealed containers until its activity has fallen sufficiently (see page 183).

Society and nuclear power

Many countries, including the UK, use nuclear power to generate electricity. There are some huge benefits to generating electricity using nuclear power. Unlike with burning fossil fuels such as coal, oil and gas, there is enough fuel for us to keep generating electricity using nuclear power for centuries to come. The process also doesn't release greenhouse gases which can affect our atmosphere. The biggest benefit of nuclear power is its efficiency (the energy produced per unit mass of fuel) — it generates many thousands times more electrical energy per kg of nuclear fuel than you would get per kg of fossil fuel.

However, all these benefits need to be weighed up against the risks of nuclear power. Nuclear reactors have to be designed and built extremely carefully to minimise the danger of a nuclear disaster. One of the biggest risks comes from how we deal with the waste produced and making sure it doesn't endanger people or the environment.

It's important that our society properly understands the science, benefits and risks of nuclear power to be able to make informed decisions, e.g. on whether to build and develop more nuclear power stations, where to build power stations and store waste etc. It will never be a risk-free way of generating electricity, but there are lots of measures in place to make it as safe as possible.

Tip: If the chain reaction in a nuclear reactor is left to continue unchecked, large amounts of energy are released in a very short time. Many new fissions will follow each fission, causing a runaway reaction which could lead to reactor meltdown and the release of radioactive material into the atmosphere.

Tip: See page 167 for more about materials that will block the different types of radiation.

Tip: Fission doesn't produce any greenhouse gases, but parts of the process do, e.g. transporting uranium fuel rods to a power station.

Practice Questions — Fact Recall

Q1 Describe a fission chain reaction. Explain what is meant by a 'critical mass' of fuel and why it is needed in a thermal nuclear reactor.

Q2 What properties do the materials used for each of the moderator and coolant need to have? Give an example material that could be used for each.

Q3 Describe how the moderator slows down neutrons within a nuclear reactor. What is the name given to these neutrons and why do they need to be slowed down?

Q4 How are the control rods used to control the rate of reaction? Give an example of a material used to make control rods.

Q5 Give two safety features of a nuclear reactor, and explain how they are used.

Q6 Why are used fuel rods more dangerous than unused ones?

Q7 Why is it important that the fuel and waste should be handled remotely?

Q8 How are used fuel rods disposed of?

Q9 Explain why it's important that society understands the risks and benefits of generating electricity using nuclear power. Give an example of both a risk and a benefit.

Section Summary

Make sure you know...

- What the Rutherford scattering experiment is and that it provided evidence for an atomic nucleus.
- How our knowledge of the structure of the nucleus has changed over time.
- A typical value for nuclear radius and that the closest approach of an alpha particle is an estimate for it.
- How to determine a nuclear radius using electron diffraction, and be familiar with the graph of relative intensity against angle for electron diffraction by a nucleus.
- That experimental data shows that nuclear radius is proportional to the cube root of the nucleon number, and that this is evidence that nuclear material has a constant density.
- How to calculate nuclear density.
- The four types of nuclear radiation: alpha, beta-plus, beta-minus and gamma, and their constituents.
- The relative ionising strengths and penetrating powers of the different types of nuclear radiation, and how they affect what the radiation can be used for.
- How to identify nuclear radiation from its penetrating power and its behaviour in a magnetic field.
- Some applications of each type of radiation, and know that there's a balance between the risk and the benefits of using radiation in medical diagnosis and treatment.
- Where background radiation comes from, and how to remove it from readings of radioactive sources.
- That the intensity of gamma radiation decreases with distance from a source by the inverse square law, and the application of this to the safe handling of radioactive sources.
- How to verify the inverse square law by taking readings at different distances from a radioactive source.
- That radioactive decay is random, but that an unstable nucleus has a constant probability of decay.
- The equations $A = \lambda N$ and $\frac{\Delta N}{\Delta t} = -\lambda N$ and how to model the second equation using a spreadsheet.
- That N and A for a radioactive sample both decrease exponentially, and the equations showing this.
- How to answer radioactive decay questions using the molar mass and the Avagadro constant.
- What half-life is and how it affects applications of nuclear radiation, and how to calculate half-life from either the decay constant of a material, a graph of A or N against t or a graph of $\ln N$ against t.
- The graph of neutron number against proton number for stable and unstable nuclei.
- How proton and nucleon numbers change for each of α, β^- and β^+ decay, and electron capture.
- How to write balanced nuclear equations for each of these decay modes.
- That excited nuclear states can emit gamma radiation, and that gamma sources can be used as tracers.
- How to interpret energy level diagrams showing the radioactive decay of a nucleus.
- The equation $E = mc^2$ and that it applies to all energy changes.
- That a mass defect of 1 u is equivalent to a binding energy of 931.5 MeV.
- How to do simple calculations with mass defect and binding energy.
- How to sketch the graph of average binding energy per nucleon against nucleon number and identify the regions where energy will be released by fission and fusion.
- How nuclear fission and fusion both release energy due to an increase in binding energy per nucleon.
- How to calculate the energy released during fission and fusion reactions using nuclear masses.
- What is meant by a fission chain reaction and a critical mass.
- The functions of the moderator, coolant, control rods and shielding in a nuclear fission reactor.
- Examples of materials used for the moderator, coolant and control rods in a nuclear fission reactor.
- The safety aspects of nuclear power, e.g. remote handling of fuel, storage and disposal of waste etc.
- That there is a balance between the risk and benefits of nuclear power, and knowledge of nuclear physics allows society to be able to make informed decisions concerning nuclear power.

Exam-style Questions

1 Which of the following graphs correctly shows how the relative intensity of electrons detected on a screen varies with angle for electron diffraction by a nucleus?

A *Relative Intensity*

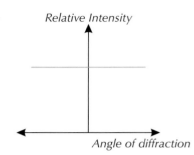

Angle of diffraction

B *Relative Intensity*

Angle of diffraction

C *Relative Intensity*

Angle of diffraction

D *Relative Intensity*

Angle of diffraction

(1 mark)

2 The energy level diagram shows the decay of an argon-41 nucleus into potassium-41. What is the total energy lost by the nucleus in this decay?

A 1295.198 keV

B 1.198 MeV

C 2492 keV

D 13 274 keV

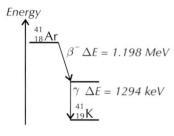

Energy

$^{41}_{18}Ar$

$\beta^{-}\ \Delta E = 1.198\ MeV$

$\gamma\ \Delta E = 1294\ keV$

$^{41}_{19}K$

(1 mark)

3 0.60 g of the radioisotope ^{60}Co is present in a sample. ^{60}Co has a half life of 5.2 years and a molar mass of 60 g mol^{-1}. How long will it take for there to be 3.01×10^{21} atoms of ^{60}Co in the sample?

A 5.2 years

B 10.4 years

C 7.8 years

D 15.6 years

(1 mark)

4 A beam of electrons each with an energy of 250 MeV (to 3 s.f.) is fired at a thin sheet of aluminium foil. Concentric bright and dark rings are seen on a luminescent screen behind the foil. The first minimum is at an angle of 57.4° to the beam's original path.

4.1 Estimate the de Broglie wavelength of the electron beam.

(3 marks)

4.2 Use the results of the above experiment to calculate the radius of an aluminium nucleus.

(2 marks)

^{23}Na has a nuclear mass of 3.8×10^{-26} kg.

4.3 What is its nuclear radius? (You can assume the value of $R_0 = 1.4 \times 10^{-15}$ m.)

(2 marks)

4.4 Calculate the density of a nucleus of ^{23}Na.

(3 marks)

4.5 Explain how observations from Rutherford's alpha scattering experiment led him to conclude that there is a very small, positively charged nucleus at the centre of an atom that contains most of the atom's mass.

(2 marks)

5 The diagram below shows the main features of a nuclear reactor.

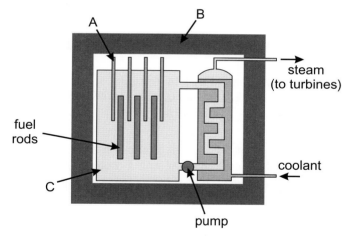

5.1 Identify and explain the function of components *A*, *B* and *C* labelled in the diagram.

(6 marks)

5.2 The mass of uranium used as fuel in the reactor is greater than the critical mass. Explain what is meant by a critical mass and why this amount of fuel is used.

(3 marks)

5.3 Coolant is passed through a nuclear reactor to remove heat produced by fission. Describe the properties needed for a substance to be suitable to use as a coolant, and give an example of a suitable coolant material.

(3 marks)

5.4 Describe what happens in an emergency shut-down of a nuclear reactor.

(2 marks)

6   This question is about binding energy.

6.1   What is meant by the term *binding energy*?

(1 mark)

6.2   Sketch the graph of average binding energy per nucleon against nucleon number. Indicate which nucleus is found at the peak of the graph, and give an approximate value for its average binding energy per nucleon.

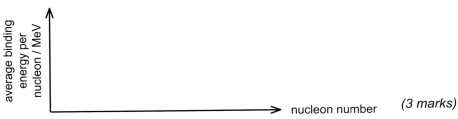

(3 marks)

6.3   Calculate the binding energy per nucleon of zinc-66, given that it has a mass defect of 0.62065 u.

(1 mark)

After absorbing a neutron, ^{235}U can fission into ^{94}Sr and ^{140}Xe, along with a number of neutrons:

$$^{235}_{92}U + \,^1_0n \longrightarrow \,^{140}_{54}Xe + \,^{94}_aSr + b\,^1_0n$$

6.4   Calculate the proton number of ^{94}Sr, a, and the number of neutrons produced, b.

(2 marks)

6.5   Explain how the fission of heavy nuclei releases energy.

(1 mark)

6.6   Calculate the energy released by this reaction.
(Nuclear masses: ^{235}U = 234.99333 u, ^{94}Sr = 93.89446 u, ^{140}Xe = 139.89194 u, ^1n = 1.00867 u.)

(3 marks)

7   An isotope of phosphorus, ^{33}P, emits β^- radiation as it decays to an isotope of sulfur. ^{33}P has a half-life of 25.4 days, and a proton number of 15.

7.1   Write the nuclear equation for the decay of ^{33}P. (The chemical symbol of sulfur is S.)

(3 marks)

7.2   Calculate the decay constant of ^{33}P.

(2 marks)

7.3   A sample of ^{33}P contains 1.6×10^{15} atoms. How long will it take for the size of the sample to fall to 7.0×10^{13} atoms?

(2 marks)

Technetium-99m is a radioactive isotope that only emits γ radiation.

7.4   A Geiger-Müller tube records an intensity of 3.6×10^{-10} Wm^{-2} when 0.2 m from a sample of Te-99m. What would you expect the intensity to be 0.5 m from the source?

(2 marks)

7.5   Technetium-99m is commonly used as a tracer for medical imaging. Describe how its properties make it useful for this purpose.

(2 marks)

1. Lenses

A lot of what we know about astrophysics was discovered by looking at objects in space. One way of doing this is to use optical telescopes to collect light from space. Some optical telescopes use lenses, so you need to start with a bit of lens theory before we get onto the really cool bits.

Converging lenses

Lenses change the direction of light rays by **refraction**. **Converging lenses** are convex (thicker across the middle than at the edges) and cause rays of light to bend towards each other. In Figure 1, the horizontal axis through the centre of a lens is called the **principal axis** and the vertical axis is called the **lens axis**.

Rays parallel to the principal axis of the lens are known as **axial rays**. Axial rays passing through the lens will converge on a point called the **principal focus** (see Figures 1 and 3). Rays that aren't parallel to the principal axis are called **non-axial rays**. A converging lens will cause parallel non-axial rays to converge somewhere else on the **focal plane** — the plane perpendicular to the principal axis that contains the principal focus (see Figure 2).

The **focal length** of the lens, f, is the perpendicular distance between the lens axis and the focal plane.

> **Tip:** Mirrors also have a principle axis — it's perpendicular to the surface of the mirror and passes through its centre.

> **Tip:** You should hopefully remember a bit of this lens theory from GCSE.

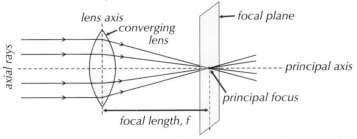

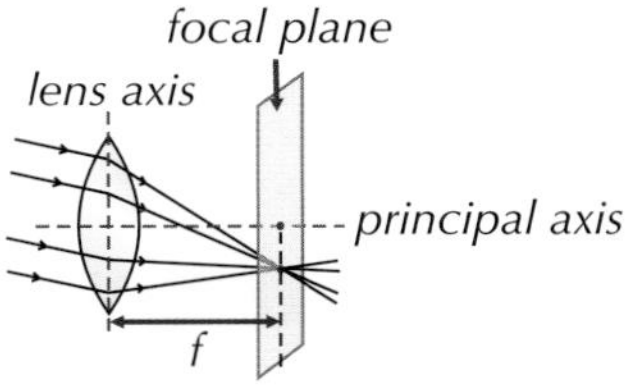

Figure 1: Axial rays passing through a lens converging at the principal focus.

Figure 2: Parallel non-axial rays will converge on the focal plane.

Images

When light rays from an object pass through a lens, an image is formed where the rays meet. To work out where an image will appear, you can draw a **ray diagram**.

Ray diagrams

A ray diagram shows how light rays travel, e.g. from an object through a lens. Draw rays coming from the top of the object — where they meet each other is where the top of the image will be formed.

You only need to draw two rays from the top of the object to work out where the top of the image will be: one parallel to the principal axis (an axial ray) that passes through the principal focus, and one passing through the centre of the lens that doesn't get refracted at all — see Figure 4. If an object sits on the principal axis, the bottom of the image will be on the principal axis.

If an object doesn't sit on the principal axis you'll also need to draw two rays from the bottom of the object to find where the bottom of the image will be formed.

Figure 3: A converging lens focusing axial rays onto the principal focus.

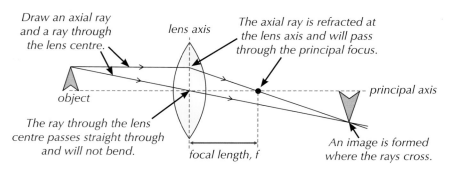

Draw an axial ray and a ray through the lens centre.

lens axis

The axial ray is refracted at the lens axis and will pass through the principal focus.

object

principal axis

The ray through the lens centre passes straight through and will not bend.

focal length, f

An image is formed where the rays cross.

Figure 4: *How to draw a ray diagram for light passing through a converging lens.*

A ray diagram can also tell you whether the image is diminished (smaller than the object), the same size, or magnified. If the image is further from the lens axis than the object, then it's magnified, if it's closer it's diminished, and if it's the same distance away, then it's the same size. (See page 205 for more on magnification.)

Real and virtual images

Images can be real or virtual. A **real image** is formed when light rays from a point on an object are made to pass through another point in space. The light rays are actually there, and the image can be captured on a screen. A **virtual image** is formed when light rays from a point on an object appear to have come from another point in space. The light rays aren't really where the image appears to be, so the image can't be captured on a screen.

Figure 5: *When you look into a mirror, you see a virtual image. Surprisingly, there's not really a bird behind the mirror — it just looks like the light rays are coming from behind the mirror.*

Converging lenses can form both real and virtual images, depending on where the object is. If the object is further than the focal length away from the lens, the image is real and inverted. If the object's closer, the image is virtual. The ray diagrams in Figures 7 and 8 show how real and virtual images are formed by a converging lens.

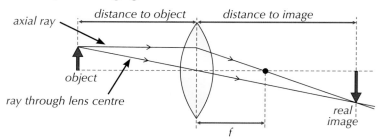

axial ray

distance to object

distance to image

object

ray through lens centre

f

real image

Figure 7: *A real image of an object is formed by a converging lens when the object is further from the lens than the focal length.*

Figure 6: *The garden is further from the converging lens than the focal point of the lens, so the image formed is real and inverted (upside down).*

The image is formed where the rays meet. Here you have to extend the rays back to where they appear to meet — this is where the virtual image is formed.

distance to image

distance to object

virtual image

object

Dotted lines show the virtual rays.

f

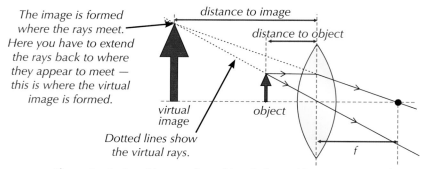

Figure 8: *A virtual image of an object is formed by a converging lens when the object is closer to the lens than the focal length.*

The lens equation

The distance between the object and the lens axis is known as u, and the distance between image and the lens axis is known as v (positive if image is real, negative if image is virtual). The values u, v and f are related by the lens equation:

$$\frac{1}{f} = \frac{1}{u} + \frac{1}{v}$$

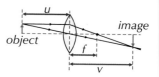

Figure 9: *The positions of u and v for a real image formed by a lens with a focal length f.*

Example — **Maths Skills**

An object is placed 3.0 metres away from a converging lens with a focal length of 1.2 m. At what distance from the lens will an image be produced? Will the image be real or virtual?

$u = 3.0$ and $f = 1.2$ m, substituting into the lens equation gives

$$\frac{1}{1.2} = \frac{1}{3.0} + \frac{1}{v}$$

$$\Rightarrow \frac{1}{v} = \frac{1}{1.2} - \frac{1}{3.0} = 0.50$$

$$\Rightarrow \text{so } v = \frac{1}{0.50} = 2.0 \text{ m.}$$

The value of v is positive, so the image is real. (The object is further than the focal length away from the lens, so this also tells you it'll be real.)

You don't need to learn the lens equation for your exams, but knowing how the distances u and v depend on one another and on the focal length, f, will help you get to grips with astronomical refracting telescopes (which are coming up on the next few pages).

2. Optical Telescopes

This topic's all about the two main types of optical telescope — refracting telescopes and reflecting telescopes. You need to know their set-up and how they're used with CCDs to detect and record images of objects in space.

Astronomical refracting telescopes

Ray diagram in normal adjustment

An **astronomical refracting telescope** is usually made up of two converging lenses. The **objective lens** converges the rays from the object to form a real image inside the telescope. The eye lens (confusingly this is part of the telescope — not the lens inside your eye) acts as a magnifying glass on this real image to form a magnified virtual image, which the observer can then view.

When you're viewing an object from space, the object is so far away that you can assume it is at infinity and the rays from each point of it are parallel to each other. When viewed with an astronomical refracting telescope, a real image is formed on the focal plane of the objective lens.

A telescope that is in normal adjustment is set up so that the principal focus of the objective lens is in the same position as the principal focus of the eye lens (see Figure 1). This means the rays from the real image come out of the eye lens parallel and the final magnified image appears to be at infinity. The length of the telescope is the focal length of the objective lens, f_o, added to the focal length of the eye lens, f_e (see Figure 2).

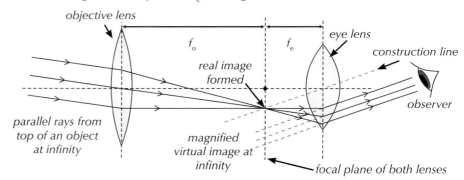

Figure 1: Light rays from a distant object passing through an astronomical refracting telescope in normal adjustment.

You need to know how to draw a ray diagram for an astronomical telescope in normal adjustment (see Figure 1). Here's how to go about it:

- Start by drawing a straight non-axial ray that passes through the centre of the objective lens and ends at the eye lens axis.

- Draw a non-axial ray on either side of the ray you've just drawn, making sure both rays are parallel to the original ray and end at the objective lens axis. Draw a straight line from where each of these rays meet the objective lens axis, so that all of the rays cross at the same point on the focal plane and reach the eye lens axis. A real image is formed where these three rays intersect.

- Draw a dotted line that passes through the point the rays cross and the centre of the eye lens — this is a construction line. Continue the three rays drawn so they are refracted at the eye lens axis and leave the lens parallel with the construction line (and each other). You can show a virtual image is formed at infinity by extending these lines backwards using dotted lines.

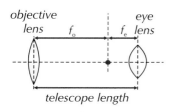

Figure 2: The length of a telescope is the sum of the focal lengths of the lenses.

Angular magnification

The magnification, M, of the telescope can be calculated in terms of angles, or the focal lengths. The **angular magnification** is the angle subtended by the image at the eye, θ_i, divided by the angle subtended by the object at the unaided eye, θ_o:

Tip: It doesn't matter what the units are for the angles used in this formula, as long as both angles are in the same units.

$$M = \frac{\text{angle subtended by image at eye}}{\text{angle subtended by object at unaided eye}} = \frac{\theta_i}{\theta_o}$$

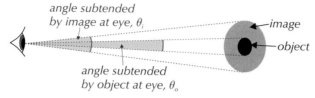

angle subtended by image at eye, θ_i

image

object

angle subtended by object at eye, θ_o

Figure 3: *Calculating the angular magnification of a refracting telescope.*

Example — Maths Skills

A galaxy subtends an angle of 9.5×10^{-3} rad at the eye when viewed from Earth. The image seen through an astronomical refracting telescope subtends an angle of 0.71 rad at the eye.
Calculate the angular magnification of the telescope.

$$M = \frac{\text{angle subtended by image at eye}}{\text{angle subtended by object at unaided eye}}$$

$$= \frac{0.71}{9.5 \times 10^{-3}} = 74.736... = 75 \text{ (to 2 s.f.)}$$

The angular magnification of an astronomical refracting telescope in normal adjustment can also be calculated from the focal lengths of the objective and eye lenses (f_o and f_e respectively):

$$M = \frac{f_o}{f_e}$$

Since a large magnification M is required to view objects from space, f_o is typically much greater than f_e.

Figure 4: *Refracting telescopes have to be very long to get a good magnification (see p.204). This one at the Yerkes Observatory is the largest one currently in operation.*

Example — Maths Skills

Calculate the focal length of the objective lens in an astronomical refracting telescope in normal adjustment with an angular magnification of 45 and a total length of 0.80 m.

- You know the angular magnification $M = \frac{f_o}{f_e} = 45$.
- The length of the telescope is the sum of the focal lengths of the lenses: $0.80 = f_o + f_e$, so $f_e = 0.80 - f_o$.
- Substituting this value for f_e into the magnification equation gives:
$$\frac{f_o}{0.80 - f_o} = 45$$
- Rearrange and solve to find: $f_o = 45(0.80 - f_o) = 36 - 45f_o$
$$\Rightarrow 46f_o = 36 \Rightarrow f_o = \frac{36}{46} = 0.782... = 0.78 \text{ m (to 2 s.f.)}$$

Tip: Notice that 0.78 m of the 0.80 m telescope is made up of the focal length of the objective lens — the focal length of the eye lens is tiny in comparison.

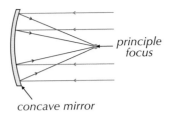

concave mirror

Figure 5: *The principle focus of a parabolic concave mirror.*

Figure 7: *Reflecting telescopes can be much shorter than refracting ones because the light 'doubles back' on itself inside.*

Tip: Light intensity is just how bright the light is.

Figure 8: *A CCD chip in a circuit.*

Reflecting telescopes

Reflecting telescopes use mirrors to reflect and focus light. A parabolic concave primary mirror converges axial rays from an object at its principle focus, forming a real image (see Figure 5). An eye lens magnifies this image in the same way as in a refracting telescope.

The principle focus of the primary mirror is in front of the mirror. If you tried to observe the light from that side of the mirror, you'd be in the way of the rays coming in. So an arrangement needs to be devised where the observer doesn't block out the incoming light. A set-up called the Cassegrain arrangement, which uses a convex secondary mirror, is a common solution to this problem — see Figure 6.

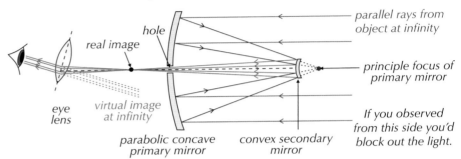

Figure 6: *The Cassegrain reflecting telescope arrangement.*

Charge-coupled devices (CCDs)

Astronomers use sensitive light detectors called **CCDs** to capture images digitally. CCDs are silicon chips about the size of a postage stamp, divided up into a grid of millions of identical picture elements (pixels).

When photons hit the silicon in a pixel, this creates free electrons that are confined to the pixel, causing a charge to accumulate on it. This charge can be measured and used to create a digital signal. This signal describes not only where the light hits, but its intensity too, as the charge on each pixel will vary depending on how many photons hit it. This allows a digital image of an object to be created.

CCDs vs. the human eye

CCDs and eyes are both light detectors. You need to be able to compare them:

- Not every photon that hits the silicon in a CCD causes an electron to be released. **Quantum efficiency** is the proportion of the incident photons that are detected by a light detector, and therefore the proportion of incident photons that release electrons for a CCD. The quantum efficiency of a CCD is typically 80% or more. In comparison, a cell in the human eye detects on average about 1 in every 100 photons and so has a quantum efficiency of about 1%. So CCDs detect far more of the light that falls on them than the eye does.

- CCDs can detect a wider spectrum of light than the human eye. The eye can only detect visible light, whereas CCDs can detect infrared, visible and U-V light.

- If you were to project the whole visual field of an eye onto a screen, you'd need over 500 megapixels for the eye to not see any pixelation. CCDs on the other hand have around 50 megapixels. The more megapixels a sensor has, the more detail it can capture, so you might think that the eye captures more detail than a CCD. However, what's also important is how far apart different parts of the object being viewed need to be in order for them to be distinguishable — this is called **spatial resolution**. The minimum resolvable distance of the human eye is around 100 μm, whereas CCDs can have a spatial resolution of around 10 μm. So CCDs are better for capturing fine detail.

Tip: See page 208 for more about the resolving power of a telescope.

- CCDs are generally less convenient to use — the human eye doesn't need any extra equipment, and looking down a telescope is simpler than setting up a CCD. However, CCDs produce digital images which can be stored, copied, and shared globally.

Practice Question — Application

Q1 The telescope shown in Figure 9 is used to observe distant galaxies. f_o = 0.52 m and f_e = 0.0010 m.

 a) Calculate the magnification of the telescope.

 b) The image formed of a distant galaxy by the telescope subtends an angle of 0.331 rad at the eye.
 Calculate the angle subtended by the galaxy at the unaided eye.

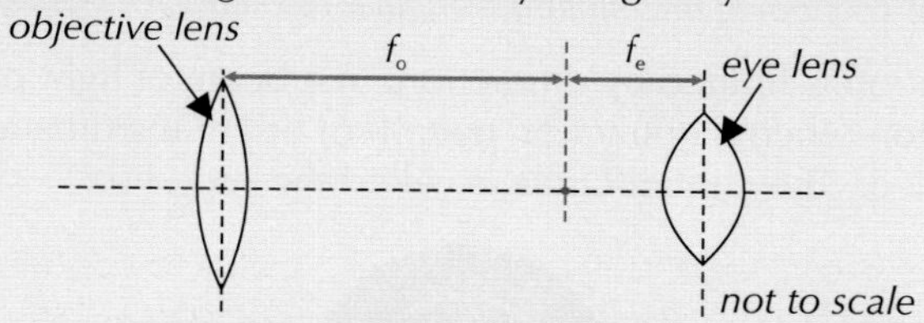

Figure 9: A refracting telescope.

Exam Tip
There are two formulas for calculating linear magnification — just look at the quantities you're given in the question to work out which one you should use.

Practice Questions — Fact Recall

Q1 a) What assumption can be made about the light rays reaching Earth from a distant object in space?

 b) Draw a ray diagram of an astronomical refracting telescope in normal adjustment showing the paths of three non-axial rays from a distant object passing through the telescope. Include any virtual rays in your diagram.

 c) What type of image is viewed through the eye lens?

Q2 Give an equation for the magnification of an astronomical refracting telescope in terms of the focal lengths of the lenses.

Q3 Explain why astronomical refracting telescopes have to be very long to produce a highly magnified image.

Q4 Draw a ray diagram of a Cassegrain reflecting telescope showing the path of rays as they pass through the telescope and form a virtual image at infinity.

Q5 a) What is meant by the quantum efficiency of a sensor?

 b) How does the quantum efficiency of a CCD differ from that of the eye?

Q6 Compare CCDs to the eye by considering the spatial resolution and convenience of use of each sensor.

Exam Tip
Make sure you can draw ray diagrams for the two types of optical telescopes to show how light passes through them.

Learning Objectives:

- Understand what the minimum angular resolution of a telescope is and be able to calculate it using the Rayleigh criterion: $\theta \approx \frac{\lambda}{D}$.
- Know that the radian is a unit of angle.
- Know the relative merits of using reflecting and refracting telescopes, including the problems of spherical and chromatic aberration.

Specification References 3.9.1.2 and 3.9.1.4

Tip: When light passes through a circular aperture, it spreads out. This is called diffraction. The light interferes constructively and destructively to produce a diffraction pattern, which you met in year 1 of A-Level Physics.

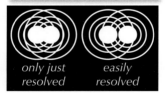

| only just resolved | easily resolved |

Figure 3: *Two light sources can be easily resolved if the Airy discs of their diffraction patterns do not touch.*

Tip: Remember, radians are just another unit of angle (see p.19).

Tip: $\theta \propto \lambda$, so the resolving power of a telescope varies with the wavelength of light — it's not a fixed value for the telescope. The lower the minimum angular resolution, the better the resolving power.

3. Comparing Telescopes

Telescopes come in all sorts of different sizes and set-ups. Different ones will perform better in different situations so it's important to choose the right one. Your decision may also be based on factors such as how much space you have, or how much money you have.

Resolving power of a telescope

The **resolving power** of a telescope is just a measure of how much detail you can see. It's very important when choosing a telescope — you could have a huge magnification, but if your resolving power's low, you'll just see a big blurry image.

The resolving power of an instrument is dependent on the **minimum angular resolution**, which is the smallest angular separation at which the instrument can distinguish two points. The smaller the minimum angular resolution, the better the resolving power of the telescope. There's more about the resolving power on page 213.

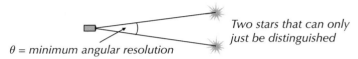

Two stars that can only just be distinguished

θ = minimum angular resolution

Figure 1: *Finding the minimum angular resolution of a telescope.*

Resolution is limited by diffraction. If a beam of light passes through a circular aperture, then a diffraction pattern of bright maxima and dark minima is formed. The central circle is called the Airy disc.

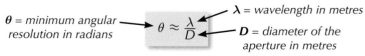

Airy disc — *First minimum*

Figure 2: *Diffraction pattern through a circular aperture.*

The Rayleigh criterion

Two light sources can just be distinguished if the centre of the Airy disc from one source is at least as far away as the first minimum of the other source — see Figure 3.

This observation led to the **Rayleigh criterion**:

$\boldsymbol{\theta}$ = minimum angular resolution in radians

$\theta \approx \frac{\lambda}{D}$

$\boldsymbol{\lambda}$ = wavelength in metres

$\boldsymbol{D}$ = diameter of the aperture in metres

For telescopes, D is the diameter of the objective lens or the objective mirror. So very large lenses or mirrors are needed to see fine detail.

Example ——— Maths Skills

Find the diameter of lens needed to resolve radiation with a wavelength of 910 nm from two objects that are 1.7 × 10⁻⁶ rad apart in the sky.

Rearrange the Rayleigh criterion:

$D \approx \frac{\lambda}{\theta} = \frac{910 \times 10^{-9}}{1.7 \times 10^{-6}} = 0.535... = 0.54$ m (to 2 s.f.)

Reflectors or refractors?

There are disadvantages and difficulties to using both refracting and reflecting telescopes.

Refracting telescopes

- Glass refracts different colours of light by different amounts and so the image for each colour is in a slightly different position. This blurs the image and is called **chromatic aberration**.

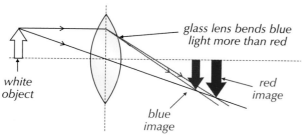

Figure 4: Chromatic aberration.

Tip: Different colours of light have different wavelengths, and different wavelengths of light are refracted different amounts.

- Any bubbles or impurities in the glass absorb and scatter some of the light, which means that very faint objects aren't seen. Building large lenses that are of a sufficiently good quality is difficult and expensive.

- Large lenses are very heavy and can only be supported from their edges, so their shape can become distorted.

- For a large magnification, the objective lens needs to have a very long focal length. This means that refracting telescopes have to be very long, leading to very large, expensive buildings being needed to house them.

Figure 5: Different wavelengths of light are refracted by different amounts by the lens, which causes the blurred fringe effect seen here.

Reflecting telescopes

Large mirrors of good quality are much cheaper to build than large lenses. They can also be supported from underneath so they don't distort as much as lenses.

Mirrors don't suffer from chromatic aberration but can have **spherical aberration**. If the shape of the mirror isn't quite parabolic, parallel rays reflecting off different parts of the mirror do not all converge onto the same point.

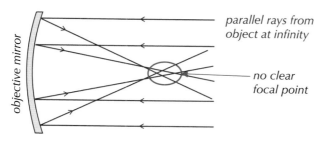

Figure 6: Spherical aberration.

Tip: When the Hubble Space Telescope was launched it suffered from spherical aberration. They had to find a way round the problem before it could be used.

The secondary mirror in a Cassegrain telescope can also cause problems. Some incoming light will be blocked by the secondary mirror and mirror supports, and some of the light reflected from the primary mirror will diffract round the secondary mirror — both leading to a decrease in image clarity.

Tip: See p.206 if you need a recap on the Cassegrain telescope.

Practice Questions — Application

Q1 Telescope A has a dish diameter of 3 m, and Telescope B has a dish diameter of 2.5 m. Which will have a better resolving power for a given wavelength of radiation?

Q2 A reflecting telescope with a mirror of diameter 3.2 m is used to detect light with a wavelength of 650 nm.

   a) Calculate the minimum angular resolution of the telescope at this wavelength.

   b) The light detected is emitted by two sources, separated by an angle of 1.99×10^{-7} rad. Calculate the maximum wavelength of light these sources would need to emit to be distinguished as two sources by the telescope.

Practice Questions — Fact Recall

Q1 What is the resolving power of a telescope?

Q2 What sort of pattern is produced when light passes through a circular aperture? What is the central part of this pattern called? What effect does this have on the resolution of a telescope?

Q3 Write down the Rayleigh criterion.

Q4 a) Describe the problem of chromatic aberration when using refracting telescopes.

   b) Give two more examples of the problems with using refracting telescopes.

Q5 Give two advantages of using a reflecting telescope rather than a refracting telescope with the same resolution power and magnification.

Q6 The mirror of an old reflecting telescope has become slightly distorted and now creates images that are slightly blurred.
Name this effect and describe how this problem occurs.

4. Non-optical Telescopes

Astronomers aren't only interested in the visible electromagnetic (EM) radiation coming from space — they also detect and analyse EM radiation that is not visible — like radio waves, I-R, U-V and X-rays.

Radio telescopes

Radio telescopes are similar to optical telescopes in some ways. The most obvious feature of a radio telescope is its parabolic dish. This works in exactly the same way as the objective mirror of an optical reflecting telescope. Instead of a polished mirror, a wire mesh can be used since the long wavelength radio waves don't notice the gaps. EM radiation is reflected and focused by the dish and an antenna is used as a detector at the principle focus. There is no equivalent to the eye lens of an optical telescope.

A preamplifier amplifies the weak radio signals without adding too much noise to the signal. The signal is then amplified further by a second amplifier before being passed through a tuner to filter out any unwanted wavelengths. A computer creates something called a false-colour image of the detected radio signals. Different colours are assigned to different wavelengths or intensities to produce false-colour images of non-visible EM radiation.

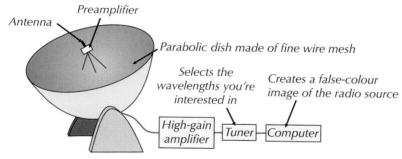

Antenna — *Preamplifier*

Parabolic dish made of fine wire mesh

Selects the wavelengths you're interested in — High-gain amplifier → Tuner → *Creates a false-colour image of the radio source* → Computer

Figure 1: *A radio telescope focuses a signal onto an antenna. The signal is amplified, filtered and analysed to produce a digital image.*

Most radio telescopes are manoeuvrable, allowing the source of the waves to be tracked (in the same way as by optical telescopes). The telescope moves with the source, stopping it 'slipping out of view' as the Earth rotates.

Resolving power

The wavelengths of radio waves are about a million times longer than the wavelengths of light. The resolving power of a telescope is dependent on the Rayleigh criterion (see p.208), which is $\theta \approx \frac{\lambda}{D}$.

So for a radio telescope to have the same resolving power as an optical telescope, its dish would need to be a million times bigger (about the size of the UK for a decent one). The resolving power of a radio telescope is worse than the unaided eye.

Radio astronomers get around this by linking lots of telescopes together. Using some nifty computer programming, their data can be combined to form a single image. This is equivalent to one huge dish the size of the separation of the telescopes. Resolutions thousands of times better than optical telescopes can be achieved this way.

Learning Objectives:

- Understand how the following telescopes work: single-dish radio telescopes, I-R telescopes, U-V telescopes and X-ray telescopes.

- Understand the similarities and differences between optical telescopes and the telescopes listed above — to include the structure, where they should be positioned, and how they are used.

- Know that the collecting power of a telescope is proportional to the square of the diameter of the mirror or dish.

- Be able to compare the resolving powers and collecting powers of optical, single-dish radio, I-R, U-V and X-ray telescopes.

Specification References 3.9.1.3 and 3.9.1.4

Figure 2: *A false-colour radio image of a galaxy.*

Tip: Remember — the <u>smaller</u> the minimum angular resolution, θ, the <u>better</u> the resolving power.

Figure 3: *The Very Large Array (VLA) in New Mexico consists of 27 radio dishes, each with a 25 m diameter.*

Benefits of radio telescopes

Radio telescopes are much easier to make than optical telescopes:

- Being able to make the radio telescope's dish using a wire mesh makes their construction much easier and cheaper than optical reflectors.
- The longer the wavelength of the radiation being detected, the less it's affected by imperfections in the shape of the dish or mirror collecting it. So for radio telescopes, the dish doesn't have to be anywhere near as perfect as the mirrors and lenses used in optical telescopes to avoid problems like spherical aberration (p.209).

Tip: A dish needs to have a precision of about $\lambda/20$ to avoid spherical aberration.

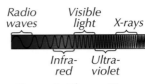

Figure 4: *The positions of types of radiation on the EM spectrum — the wavelength increases from right to left.*

Tip: Infrared is produced by anything that's hot (or even moderately warm) — so if an infrared telescope isn't cooled it'll produce its own infrared radiation which will be mixed up with the radiation being observed.

I-R and U-V telescopes

Infrared (I-R) and ultraviolet (U-V) telescopes are very similar to optical reflecting telescopes. They use the same parabolic mirror set-up to focus the radiation onto a detector. In both cases, CCDs (see p.206) or special photographic paper are used as the radiation detectors, just as in optical telescopes.

The longer the wavelength of the radiation, the less it's affected by imperfections in the mirror (see above). So the mirrors in infrared telescopes don't need to be as perfectly shaped as in optical telescopes. But because U-V waves have a shorter wavelength, the mirrors in U-V telescopes have to be even more precisely made (see Figure 4).

I-R telescopes have the added problem that they produce their own infrared radiation due to their temperature. They need to be cooled to very low temperatures using liquid helium, or refrigeration units.

X-ray telescopes

X-ray telescopes have a different structure from other telescopes. X-rays don't reflect off surfaces in the same way as most other EM radiation. Usually X-ray radiation is either absorbed by a material or it passes straight through it.

X-rays do reflect if they just graze a mirror's surface though. By having a series of nested mirrors, you can gradually alter the direction of X-rays enough to bring them to a focus on a detector. This type of telescope is called a grazing telescope.

Figure 5: *The XMM-Newton telescope actually contains three separate X-ray telescopes, each with 58 mirrors which gradually alter the direction of the X-rays.*

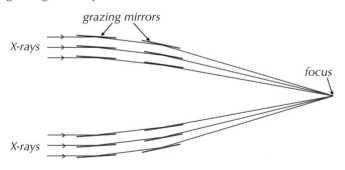

Figure 6: *Grazing mirrors focusing X-rays in an X-ray telescope.*

The X-rays can be detected using a modified Geiger counter or a fine wire mesh. Modern X-ray telescopes use highly sensitive X-ray CCD cameras.

Telescope positioning

One of the big problems with doing astronomy on Earth is trying to look through the atmosphere. Our atmosphere only lets certain wavelengths of electromagnetic radiation through and is opaque to all the others. Figure 7 shows how the transparency of the atmosphere varies with wavelength.

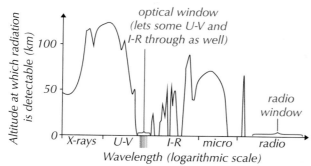

Figure 7: *The transparency of the Earth's atmosphere to different wavelengths of EM radiation.*

We can use optical and radio telescopes on the surface of the Earth because the atmosphere is transparent to these wavelengths. Observing other wavelengths can be a bit more tricky.

A few wavelengths of infrared radiation can reach the Earth's surface, but most are absorbed by water vapour in the atmosphere. On Earth, the best way to observe I-R radiation is to set up shop in high and dry places, like the Mauna Kea volcano in Hawaii.

Most ultraviolet, X-rays and the other wavelengths of infrared radiation are absorbed higher up in the atmosphere, so being on a mountain doesn't help.

One way to get round this problem is to strap U-V, X-ray and I-R telescopes to high-altitude weather balloons or aeroplanes. They can take the telescope high enough into the atmosphere to detect the radiation.

The ideal situation is to get your telescope above the atmosphere altogether, by launching it into space and setting it in orbit around the Earth (see Figure 5 on the previous page).

Figure 8: *The UK Infra-Red Telescope (UKIRT) is one of twelve telescopes around the summit of the Mauna Kea volcano.*

Figure 9: *The Stratospheric Observatory for Infrared Astronomy's (SOFIA) airborne observatory — a telescope embedded in a plane.*

Resolving power

You need to be able to compare resolving powers of various non-optical telescopes with that of an optical telescope. The resolving power of a telescope is limited by two main factors:

- The Rayleigh criterion (see page 208):
  This says that resolving power depends on the wavelength of the radiation and the diameter of the objective mirror or dish. So, for the same size of dish, a U-V telescope has a much better resolving power than a radio telescope, as the radiation it detects has a much shorter wavelength.

- The quality of the detector:
  The resolving power of a telescope is limited by the resolving power of the detector. That can be how many pixels there are on a CCD, or for a wire mesh X-ray detector, how fine the wire mesh is.

Tip: Remember, the Rayleigh criterion is:
$$\theta \approx \frac{\lambda}{D}.$$

A lower value of θ means a better resolving power.

Collecting power

The **collecting power** of a telescope is proportional to its collecting area. For a radio, optical, U-V or I-R telescope, this is the area of the objective mirror or dish. For X-ray telescopes, it's the size of the opening through which X-rays can enter the telescope. In general, X-ray telescopes have a much lower collecting power than other types of telescope.

A bigger dish or mirror collects more energy from an object in a given time. This gives a more intense image, so the telescope can observe fainter objects. The collecting power (energy collected per second) is proportional to the area:

$$collecting\ power \propto dish\ diameter^2$$

Practice Question — Application

Q1 The VISTA (Visible and Infrared Survey Telescope for Astronomy) is positioned at an altitude of 2500 m in the Atacama Desert in Chile. The telescope has a mirror diameter of 4.1 m.

a) How many times greater is the collecting power of VISTA than that of the largest astronomical refracting telescope with an objective lens of diameter 1.02 m?

b) The resolving power of this telescope is found to be better in the visible light region than in the infrared region.
Suggest a reason for this.

c) The Atacama desert is one of the driest places on Earth and the telescope is at an altitude of around 2500 m. Why is this a good location?

d) Give an example of how the resolving power of the telescope for a particular wavelength of light could be increased without changing the size of the dish.

Practice Questions — Fact Recall

Q1 Describe the structure of a radio telescope and how it is designed to detect radio waves.

Q2 Why is the resolving power of a radio telescope much lower than that of a similar sized optical telescope?

Q3 Explain how radio telescopes are used to get resolutions much higher than optical telescopes.

Q4 Give two reasons why building a large radio telescope is easier than building a large optical telescope.

Q5 Why do the mirrors of U-V telescopes have to be made more precisely than optical telescopes?

Q6 Explain why I-R telescopes must be cooled to very low temperatures.

Q7 Why do X-ray telescopes use grazing mirrors?

Q8 Suggest a suitable location for:
a) an X-ray telescope,
b) an I-R telescope.

Q9 What's the relationship between collecting power and diameter?

5. Parallax and Parsecs

The distances to stars can be difficult to calculate. Astronomers have several methods to work them out — but not all of the methods work for all distances. Astronomers also use a few different units to measure distance.

Parallax

Imagine you're in a moving car. You see that (stationary) objects in the foreground seem to be moving faster than objects in the distance. This apparent change in position is called **parallax**.

The distance to nearby stars can be calculated by observing how they move relative to stars that are so distant that they appear not to move at all — background stars. This is done by comparing the position of the nearby star in relation to the background stars at different parts of the Earth's orbit.

Parallax is measured in terms of the angle of parallax. If you observe the position of the star at either end of the Earth's orbit (6 months apart), the angle of parallax is half the angle that the star moves in relation to the background stars. The greater the angle, the nearer the object is to you.

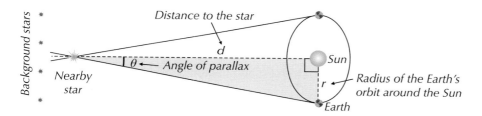

Figure 1: *The distance of a nearby star can be calculated by measuring the angle of parallax and knowing the diameter of the Earth's orbit.*

Using the triangle in Figure 1, you can calculate the distance to the nearby star, *d*, using trigonometry, if you know the angle of parallax and the radius of the Earth's orbit:

$$\tan \theta = \frac{r}{d}$$

$$\Rightarrow d = \frac{r}{\tan \theta}$$

For small angles $\tan \theta \approx \theta$, where θ is in radians. Because the angles used in astronomy are so tiny, you can assume this for calculations of parallax (as long as you're working in radians). So:

d = distance to the star ⟶ $d \approx \dfrac{r}{\theta}$ ⟵ ***r*** = radius of the Earth's orbit
⟵ **θ** = angle of parallax in radians

Remember, the angle in radians = angle in degrees × $\frac{\pi}{180}$ (see page 19).

Tip: Very distant stars appear not to move as we orbit the Sun because the angle by which they move is too tiny for us to measure.

Tip: You can see parallax at work by holding your hand out in front of your face and closing just your right eye, followed by just your left eye. Your hand will move in relation to the background. The closer your hand is to your face, the more it moves in relation to the background.

Tip: Remember, trigonometry says that $\tan \theta = \dfrac{\text{opposite side}}{\text{adjacent side}}$.

Tip: You've seen small angle approximations before on page 37.

Tip: If you use this formula in the exam, make sure you tell them that you're assuming that θ is small.

Tip: The units of *d* and *r* don't matter, as long as they have the same units.

Parsecs

Parallax gives a unit of distance called a **parsec** (pc).

> A star is exactly one parsec (pc) away if the angle of parallax,
> $\theta = 1$ arcsecond $= \left(\frac{1}{3600}\right)^{\circ}$

The distances measured in astronomy are usually huge — even the nearest large galaxy to the Milky Way is 780 000 parsecs away. Astronomers often use parsecs (pc) or megaparsecs (Mpc) to measure these large distances. You need to be able to use these conversions:

> 1 pc = 3.08×10^{16} m   1 Mpc = 1×10^{6} pc

Example — Maths Skills

Proxima Centauri has an angle of parallax of 0.77 arcseconds. Calculate the distance to Proxima Centauri in parsecs. The average radius of the Earth's orbit is 1.50×10^{11} m.

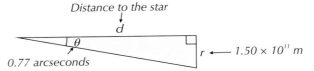

Distance to the star
d
0.77 arcseconds θ r ← 1.50×10^{11} m

- Convert the angle to degrees, and then into radians:

  0.77 arcseconds $= 0.77 \times \left(\frac{1}{3600}\right)^{\circ} = 2.138... \times 10^{-4 \, \circ}$
  $2.138... \times 10^{-4 \, \circ} \times \frac{\pi}{180} = 3.73... \times 10^{-6}$ rad

- $d \approx \frac{r}{\theta} = \frac{1.50 \times 10^{11}}{3.73... \times 10^{-6}} = 4.018... \times 10^{16}$ m

- Convert this into parsecs by dividing by 3.08×10^{16}:

  $d = 4.018... \times 10^{16}$ m
  $= (4.018... \times 10^{16}$ m$) \div (3.08 \times 10^{16})$
  $= 1.304...$
  $= 1.3$ pc (to 2 s.f.)

Light years (ly)

All electromagnetic waves travel at the speed of light, c, in a vacuum (where $c = 3.00 \times 10^{8}$ ms^{-1}). The distance that electromagnetic waves travel through a vacuum in one year is called a **light year** (ly).

If we see the light from a star that is, say, 10 light years away then we are actually seeing it as it was 10 years ago. The further away the object is, the further back in time we are actually seeing it.

> 1 light year = 9.46×10^{15} m

> 1 parsec = 3.26 light years

- Light from the Sun takes around 8 minutes to reach Earth, so the light that we see from the Sun actually left the Sun 8 minutes earlier.

- Proxima Centauri is 1.3 pc away from Earth (see previous example), so it is 1.304... × 3.26 = 4.252... = 4.3 (to 2 s.f.) light years away. So the light from Proxima Centauri will take just over 4 years and 3 months to reach us.

The light from very distant galaxies has taken billions of years to reach us. Astronomers search for distant galaxies so that they can 'look into the past' at what the Universe was like billions of years ago.

Measuring distances

Figure 2 shows the diameter of an object, the angle subtended by it in the sky, and the distance to the object.

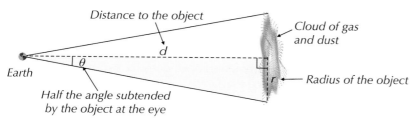

Figure 2: *The distance of an object can be calculated by measuring the angle it subtends in the sky and knowing the actual diameter of the object.*

As long as the angle is small, you can use the formula on page 215 to relate all three:

θ = half the angle subtended in radians ⟶ $\theta \approx \dfrac{r}{d}$ ⟵ r = radius of object

d = distance to object

Tip: You might need to use this method to find the magnification of a telescope (see p.205). If you're given the distance to and the size of an object, you can use this method to work out the angle the <u>object</u> subtends in the sky. If you know the angle the <u>image</u> subtends, you can work out the magnification.

Tip: Don't forget to work with angles in radians when using the small angle approximation.

Tip: $\frac{r}{d}$ is just a ratio, so the two distances just need to be in the same units — it doesn't matter which.

Example — **Maths Skills**

A satellite in orbit around the Earth has a radius of 6.6 m. Given that it is in orbit at 569 km from Earth, find the angle subtended by it as viewed from Earth.

Half the angle subtended = $\theta \approx \dfrac{r}{d} = \dfrac{6.6}{569 \times 10^3} = 1.159... \times 10^{-5}$ rad

So the angle subtended = $1.159... \times 10^{-5} \times 2 = 2.31... \times 10^{-5}$
$= 2.3 \times 10^{-5}$ rad (to 2 s.f.)

You can only use parallax to measure distance for nearby objects, and using the angle subtended only works if you already know the object's size, which is not very likely if you don't know its distance.

There are lots of other ways of measuring distance that astronomers can use, including standard candles (page 221), red shift (pages 241-244) and quasars (page 250).

It's important for astronomers to have several ways of measuring distance. Measuring the same distance using different methods helps our results to be more precise. Also, some methods only work for certain distances.

Q1 The star Alpha Tauri in the constellation Taurus is 20 pc away. Calculate this distance in light years.

Q2 An observatory is studying the star Sirius from Earth. One image of Sirius is captured, and another is captured 6 months later. The angle of parallax is recorded to be 0.37 arcseconds.

a) Why does the observatory take two images, 6 months apart, to measure the parallax of Sirius?

b) The average radius of the Earth's orbit is 1.50×10^{11} m. Calculate the distance to Sirius in metres.

c) How long will it take for light from Sirius to reach Earth in years and months?

Q3 An astronomer wants to calculate the distance to Mars from Earth. He knows the following information about Mars:

Radius: 3389.5 km

Mass: 6.4185×10^{23} kg

Volume: 1.6318×10^{11} km^3

a) He measures the angle subtended by Mars in the sky. Which of the measurements above does he also need to know to calculate the distance to Mars?

b) The angle subtended by Mars in the sky was measured as 5.0 arcseconds. Calculate the distance to Mars at the time the measurement was taken.

Tip: The distance from Earth to Mars isn't always the same — it varies depending on where the two planets are in their orbits of the Sun.

Practice Questions — Fact Recall

Q1 What is meant by the angle of parallax?

Q2 What is a parsec?

Q3 What is the definition of a light year?

Q4 Why does it take time for light emitted by the Sun to reach us? How long does it take?

Q5 What formula would you use to calculate the distance to an object if you knew the size of the object, and the angle subtended by the object in the sky. What assumption needs to be made?

Q6 Why is it important for astronomers to have several different methods of calculating distances to objects in space?

6. Magnitude

There are a few properties that are used to classify stars. Some stars in the sky look a lot brighter than others — this could be because they're actually a lot brighter, or just much closer. When classifying stars by their brightness you can either talk about how bright they appear, or how bright they actually are.

Power output

Stars can be classified according to their **luminosity** — that is, the total amount of energy emitted in the form of electromagnetic radiation each second (see p.231). Luminosity is also known as the **power output** of a star, as it's just a rate of energy transfer, and is measured in watts (W). The Sun's luminosity is about 4×10^{26} W. The most luminous stars have a luminosity about a million times that of the Sun.

The **intensity**, I, of an object that we observe is the power received from it per unit area at Earth. This is the effective **brightness** of an object.

Apparent magnitude

Brightness is a subjective scale of measurement — the brightness of an object will appear to vary, depending on how far you are from the object. The brightness of a star in the night sky depends on two things — its power output and its distance from us (if you ignore weather and light pollution, etc.). So the brightest stars will be close to us and have a high luminosity. The **apparent magnitude**, m, of an object is a measure of the brightness (or intensity) of the object.

An Ancient Greek called Hipparchus invented a system where the very brightest stars were given an apparent magnitude of 1 and the dimmest visible stars an apparent magnitude of 6, with other levels catering for the stars in between. In the 19th century, the scale was redefined using a strict logarithmic scale:

> A magnitude 1 star has an intensity 100 times greater than a magnitude 6 star.

In other words, 5 magnitudes difference corresponds to a difference in intensity of 100 times. This means a difference of 1 on the magnitude scale corresponds to a difference in intensity of $100^{1/5} \approx 2.51$ times.

Example — Maths Skills

A magnitude 1 star is about 2.51 times brighter than a magnitude 2 star and $2.51 \times 2.51 = 2.51^2$ times brighter than a magnitude 3 star.

You can also calculate the brightness (or intensity) ratio between two stars using:

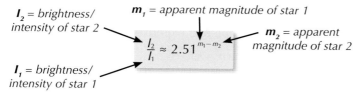

I_2 = brightness/intensity of star 2

m_1 = apparent magnitude of star 1

$$\frac{I_2}{I_1} \approx 2.51^{m_1 - m_2}$$

m_2 = apparent magnitude of star 2

I_1 = brightness/intensity of star 1

Learning Objectives:

- Understand how stars can be classified by their luminosity (or power output).
- Know that brightness is a subjective scale of measurement.
- Know what is meant by apparent magnitude, m, and that it's measured with a scale invented by Hipparchus, where the dimmest visible stars have an apparent magnitude of 6.
- Know the relationship between brightness and m.
- Know what is meant by absolute magnitude, M.
- Understand how M and m are related by the formula:
  $m - M = 5 \log \frac{d}{10}$
- Understand how type 1a supernovae can be used as standard candles to determine distances.

Specification References 3.9.2.1, 3.9.2.2 and 3.9.2.6

Tip: Don't get confused between brightness and luminosity. Luminosity is the power emitted by the star. Brightness is the power received at Earth per unit area.

Tip: It's important to know that the magnitude scale <u>isn't linear</u>. The intensity difference between a magnitude 1 and a magnitude 2 star isn't the same as it is between a magnitude 2 and a magnitude 3 star. Only the <u>ratios of the intensities</u> are the same.

The scale is continuous — an object can have an apparent magnitude of any value, including decimals.

Tip: Make sure you substitute in the right apparent magnitude values. If you get them the wrong way round you'll get the wrong ratio.

┌─ **Example** ── **Maths Skills** ─────────────────

Star A has an apparent magnitude of 3.44 and star B has an apparent magnitude of 5.72. Show that star A appears around 8 times brighter than star B.

- Find the difference in apparent magnitude:
$$m_B - m_A = 5.72 - 3.44 = 2.28$$

- So brightness/intensity ratio is approximately:
$$\frac{I_A}{I_B} \approx 2.51^{2.28} = 8.151... = 8.2 \text{ (to 2 s.f.)}$$

- So star A appears around 8 times brighter than star B.

Tip: The answer here has been given to the same number of s.f. as there are d.p. in the power, which is the convention when raising a number to a power (see page 402).

At the same time as the logarithmic scale was introduced, the range was extended in both directions, with the very brightest objects in the sky having negative apparent magnitude. The Sun has an apparent magnitude of −26.7, and the dimmest objects observed by the Hubble Space Telescope have an apparent magnitude of around 30. The whole scale is tied to a star called Vega, which is defined to have an apparent magnitude of 0.

Tip: The Moon doesn't have a very high luminosity, but it's very close. Its apparent magnitude during a full moon is about −12.7 and it's by far the brightest object in the night sky.
Have a look — it's lovely.

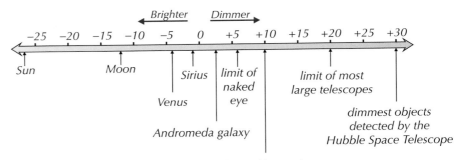

Figure 1: The apparent magnitude scale and the positions of some key features.

Tip: The absolute magnitude of the Sun is 4.83, so if we put it 10 parsecs away, we'd only just be able to see it with the naked eye.

Absolute magnitude

The **absolute magnitude** of an object, M, is based only on the power output of the object. It does not depend on its distance from Earth. It is defined as what its apparent magnitude would be if it were 10 parsecs away from Earth.

The relationship between M and m is given by the following formula:

Tip: The logarithm here is to the base 10 — don't get it confused with the natural logarithm 'ln'.

For more on logs, see pages 401-402.

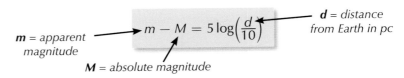

m = apparent magnitude

$$m - M = 5\log\left(\frac{d}{10}\right)$$

d = distance from Earth in pc

M = absolute magnitude

Tip: You can assume the 5 and 10 here are both exact when dealing with significant figures.

Example — Maths Skills

The star Aludra in the constellation Canis Major has an apparent magnitude of 2.5 and is 3200 light years from Earth. Calculate the absolute magnitude of Aludra.

- Convert the distance to parsecs:

  3200 light years = 3200 ÷ 3.26 = 981.5... pc

- You know the apparent magnitude so use the formula $m - M = 5\log\left(\frac{d}{10}\right)$ to find the absolute magnitude.

- So $M = m - 5\log\left(\frac{d}{10}\right)$

  $= 2.5 - 5\log\left(\frac{981.5...}{10}\right)$

  $= -7.459...$

  $= -7.5$ (to 2 s.f.)

> **Tip:** Just like with apparent magnitude, you can get negative absolute magnitudes. The more negative, the brighter the star.

If you know the absolute magnitude of an object, you can use this equation to calculate its distance from Earth. This is really handy, since the distance to most stars is too big to measure using parallax (see page 215).

Standard candles

You can also use the absolute magnitude formula to calculate the distance to distant objects using **standard candles**. Standard candles are objects that you can calculate the absolute magnitude of directly, e.g. **type 1a supernovae**. All type 1a supernovae have the same peak in absolute magnitude. If you find a type 1a supernova within a galaxy, you can work out how far that galaxy is from us by looking at how bright the supernova appears in comparison to how powerful you know it really is, then use the formula on page 220.

> **Tip:** Supernovae occur when stars much more massive than the Sun explode — see p.236 for more.

Figure 2: A type 1a supernova (the bright dot) within the galaxy NGC 4526. The absolute magnitude is known, so its distance can be calculated (around 55 million light years).

Example — Maths Skills

A type 1a supernova in a galaxy has a peak absolute magnitude of −19.3, and a peak apparent magnitude of 2.2. Calculate the distance to the galaxy in parsecs.

Rearrange the equation for absolute magnitude to get $\log\left(\frac{d}{10}\right)$ on its own on one side of the equation.

$m - M = 5\log\left(\frac{d}{10}\right)$

$\Rightarrow \frac{m - M}{5} = \log\left(\frac{d}{10}\right)$

You want to find d, so you need to get rid of the log in the equation. To do that, raise each side of the equation to the power 10:

$10^{\frac{m-M}{5}} = 10^{\log\left(\frac{d}{10}\right)} \Rightarrow 10^{\frac{m-M}{5}} = \frac{d}{10}$

Make d the subject and substitute in the values for m and M:

$d = 10 \times 10^{\frac{m-M}{5}}$

So $d = 10 \times 10^{\frac{2.2-(-19.3)}{5}} = 10 \times 10^{\frac{21.5}{5}}$

$= 199\ 526.2...$ pc

$= 200\ 000$ pc (to 1 s.f.)

> **Tip:** It's really handy to remember log rules like $10^{\log x} = x$ in calculation questions. For more on log rules, see page 402.

Practice Questions — Application

Q1 Albireo B and Deneb are stars in the constellation Cygnus.
Albireo B has an apparent magnitude of 5.1 and Deneb has an
apparent magnitude of 1.25.

   a) Approximately how much brighter is Deneb than Albireo B?

   b) Deneb is 975 parsecs from Earth.
      Calculate the absolute magnitude of Deneb.

Q2 The following table shows some information about two stars in the
constellation of Orion.

	Alnilam	Bellatrix
Apparent magnitude, m		1.64
Absolute magnitude, M	−6.39	
Distance from Earth, pc	413 pc	75 pc

   a) Calculate the apparent magnitude of Alnilam.

   b) Calculate the absolute magnitude of Bellatrix.

   c) Which star appears to be brighter?

   d) Which star would appear brighter from 10 pc away?

Q3 Type 1a supernovae all have the same peak in absolute magnitude
of −19.3. A type 1a supernova occurs in a galaxy 3.4 Mpc away.
Calculate the peak apparent magnitude of the supernova.

Practice Questions — Fact Recall

Q1 What is another name for the power output of a star?

Q2 a) Explain why brightness is a subjective scale of measurement.

   b) What two quantities affect the brightness of a star in the night sky?

Q3 a) Describe what is meant by the apparent magnitude of a star.

   b) When Hipparchus invented the scale for the brightness of stars,
      what apparent magnitude were the dimmest visible stars given?

Q4 Roughly how much brighter is a magnitude 1 star than a
magnitude 2 star?

Q5 Write down the equation that links the brightness and apparent
magnitudes of two stars, including the approximation.

Q6 What does it mean if a star has a negative apparent magnitude value?

Q7 Define the absolute magnitude of a star.

Q8 Write down the equation linking apparent magnitude and absolute
magnitude.

Q9 What is a standard candle? Why are type 1a supernovae used as
standard candles?

Tip: Remember (p.219),
brightness and intensity
are the same thing.

Tip: The Hubble
constant (see p.245)
was worked out using
standard candles.

7. Stars as Black Bodies

Astronomers can estimate a star's temperature by assuming that stars are perfect absorbers and emitters — something known as a black body...

Black body radiation

All objects that are hotter than **absolute zero** emit electromagnetic radiation due to their temperature. At room temperature, this radiation lies mostly in the infrared part of the spectrum (which we can't see) — but heat something up enough and it'll start to glow and emit visible light. This is because the wavelengths of radiation emitted depend on the temperature of the object.

A **black body** is an object with a pure black surface that emits radiation strongly and in a well-defined way. It absorbs all light incident on it — we call them black bodies because they don't reflect any light.

> A black body is a body that absorbs all electromagnetic radiation of all wavelengths and can emit all wavelengths of electromagnetic radiation.

Because they emit all wavelengths of electromagnetic radiation, they emit a continuous spectrum of electromagnetic radiation. We call it black body radiation. A graph of radiation power output against wavelength for a black body varies with temperature, but they all have the same general shape (as shown in Figure 1). They're known as black body curves:

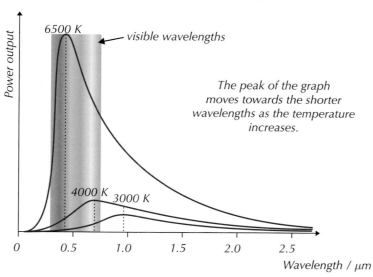

Figure 1: Black body radiation curves for black bodies at 3000 K, 4000 K and 6500 K.

There is no such thing as a perfect black body, but to a reasonably good approximation stars (and some other astronomical sources) behave as black bodies. We can use black body radiation curves to make estimations about an object's temperature and other properties.

Tip: Remember, power output is the energy emitted per second.

Tip: The Sun can be assumed to be a black body. It absorbs almost all of the light incident on it, and emits the most intense radiation in the yellow part of the visible region of the electromagnetic spectrum.

Figure 3: *Wilhelm Wien was a German physicist who was awarded the 1911 Nobel Prize in physics for his work on black bodies.*

Wien's displacement law

All black body spectra have a peak intensity. The wavelength that this peak occurs at is called the peak wavelength, λ_{max}. The higher the surface temperature of a star, the shorter the wavelength, λ_{max}.

λ_{max} is related to the temperature by **Wien's displacement law**:

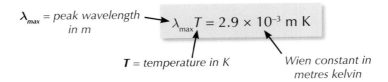

λ_{max} = peak wavelength in m

$$\lambda_{max} T = 2.9 \times 10^{-3} \text{ m K}$$

T = temperature in K

Wien constant in metres kelvin

A hotter black body will emit more radiation than a cooler one, and it's peak wavelength will be a shorter wavelength. It also has a higher total power output (assuming it has the same surface area) — see Figure 4.

The hotter star emits a lot more radiation near its peak wavelength than a cooler star...

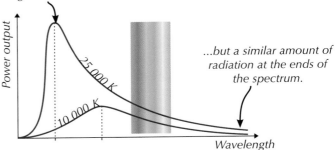

...but a similar amount of radiation at the ends of the spectrum.

Figure 4: *Black body curves to show the different range and intensity of wavelengths of radiation emitted by stars of different temperatures.*

A hotter star may not appear as bright as a cooler one, as it may mostly emit radiation that isn't in the visible part of the electromagnetic spectrum. If a cooler star emits more radiation in the visible region than a hotter one, it will appear brighter (see p.219 for more on brightness).

If you assume that a source acts as a black body, you can use Wien's displacement law to estimate the surface temperature of the source using measurements of its peak wavelength. The temperature you calculate is the source's black-body temperature.

┌─ **Example** ──── **Maths Skills** ────────────────────────

An astronomer observed radiation from a star and recorded the peak wavelength to be in the visible light region at 520 nm. Assuming that the star behaves as a black body, estimate the temperature of the star in kelvin.

▪ The peak wavelength in metres is $\lambda_{max} = 520 \times 10^{-9}$ m.

▪ Rearranging Wien's displacement law: $T = \dfrac{2.9 \times 10^{-3}}{\lambda_{max}}$

▪ So $T = \dfrac{2.9 \times 10^{-3}}{520 \times 10^{-9}} = 5576.9...$ K = 5600 K (to 2 s.f.)

Stefan's law

As you probably remember from p.219, the power output of a star is the total energy it emits per second. The power output is proportional to the fourth power of the star's surface temperature and is directly proportional to the surface area. This is **Stefan's law**:

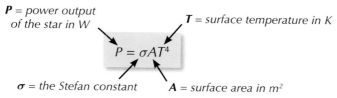

P = power output of the star in W

T = surface temperature in K

$$P = \sigma A T^4$$

σ = the Stefan constant

A = surface area in m²

Measurements give the Stefan constant as $\sigma = 5.67 \times 10^{-8} \, \text{W} \, \text{m}^{-2} \, \text{K}^{-4}$.

Exam Tip
Stefan's law and the Stefan constant are given in the data and formulae booklet.

Example — Maths Skills

The Sun has a black-body temperature of 5800 K. What surface area would a star have if it had a black-body temperature of 3000 K but the same power output as the Sun?

- Start by calculating the surface area of the Sun. Assuming the Sun is a sphere, the surface area is given by $4\pi r^2$, where r is the radius of the Sun:
$$A_{\text{Sun}} = 4 \times \pi \times (6.96 \times 10^8)^2 = 6.08... \times 10^{18} \, \text{m}^2$$

- $P = \sigma A T^4$. The power outputs of the star and the Sun are the same, so:
$$\sigma A_{\text{Sun}} (T_{\text{Sun}})^4 = \sigma A_{\text{star}} (T_{\text{star}})^4$$
$$\Rightarrow A_{\text{star}} = \frac{A_{\text{Sun}} (T_{\text{Sun}})^4}{(T_{\text{star}})^4} \quad \text{(cancelling } \sigma \text{ from both sides and rearranging)}$$
$$\Rightarrow A_{\text{star}} = \frac{(6.08... \times 10^{18})(5800)^4}{(3000)^4}$$
$$= 8.50... \times 10^{19} = 9 \times 10^{19} \, \text{m}^2 \text{ (to 1 s.f.)}$$

Exam Tip
The formula for the area of a sphere, $A = 4\pi r^2$, and the value of the mean radius of the Sun, 6.96×10^8 m, are both given in the data and formulae booklet in the exam.

Tip: A cooler star needs to be a lot bigger to have the same power output.

Inverse square law

From Earth, we can measure the intensity of a star. The intensity is the power of radiation per square metre, so as the radiation spreads out and becomes diluted, the intensity decreases. If the energy has been emitted from a point or a sphere (like a star, for example) then it obeys the inverse square law:

P = power output of the star in W

I = intensity in W m⁻²

$$I = \frac{P}{4\pi d^2}$$

d = distance from the star in m

To use the inverse square law, you need to assume that the star is spherical, and that it gives out an even amount of power in every direction.

You can use Wien's displacement law, Stefan's law and the inverse square law to work out all sorts of things about stars, as long as you assume that stars behave as black bodies.

Tip: There's a full explanation of the inverse square law on pages 73-74.

Exam Tip
You won't be given the inverse square law in the exam. You'll need to remember it.

Tip: The inverse square law means that if you moved a star 3 times further away, it'd be $3^2 = 9$ times less intense.

A star has a surface area of 4.1 × 10¹³ m² and produces a black body spectrum with a peak wavelength of 115 nm. The intensity of the light from the star when it reaches Earth is 1.12 × 10⁻¹¹ Wm⁻². How long does the light from the star take to reach Earth?

- First, find the temperature of the star using Wien's displacement law:

  $\lambda_{max}T = 2.9 \times 10^{-3}$ m K,

  so $T = 2.9 \times 10^{-3} \div \lambda_{max} = 2.9 \times 10^{-3} \div 115 \times 10^{-9} = 25\ 217.39...$ K.

- Now, you can use Stefan's law to find the power output:

  $P = \sigma A T^4 = (5.67 \times 10^{-8}) \times (4.1 \times 10^{13}) \times (25\ 217.39...)^4$

  $= 9.40... \times 10^{23}$ W

Tip: The Stefan constant:
$\sigma = 5.67 \times 10^{-8}$ Wm⁻²K⁻⁴

- Then use $I = \dfrac{P}{4\pi d^2}$ to find the distance of the star from Earth:

  $d = \sqrt{\dfrac{P}{4\pi I}} = \sqrt{\dfrac{9.40... \times 10^{23}}{4\pi \times 1.12 \times 10^{-11}}} = 8.17... \times 10^{16}$ m

Tip: Speed of light in a vacuum:
$c = 3.00 \times 10^8$ ms⁻¹

- Finally, use $c = d \div t$ to find the time taken

  $t = d \div c = 8.17... \times 10^{16} \div 3.00 \times 10^8 = 272\ 426\ 013.2$ s

  ≈ 8.6 years (to 2 s.f.)

Practice Questions — Application

Q1 Sunspots are cooler spots on the Sun's surface that have a temperature of around 3000-4500 K, compared with the average surface temperature of 5800 K on the rest of the Sun's surface.
Use this information to explain why calculations using the inverse square law for the Sun might be inaccurate.

Q2 The data in the table below gives some of the properties of the star Rigel in the constellation of Orion.

Surface temperature / K	11 000
Distance / ly	773
Intensity of light / W m⁻²	3.7 × 10⁻⁸

a) Calculate the wavelength of the peak in the black body radiation curve of Rigel, λ_{max}.

b) Sketch the black body radiation curve for Rigel, labelling λ_{max}.

c) Calculate the surface area of Rigel in metres squared.

Practice Questions — Fact Recall

Q1 What is a black body?

Q2 Draw the general shape of a graph of wavelength against intensity for a black body.

Q3 What does the peak wavelength of radiation emitted by a black body depend on?

Q4 Write down Wien's displacement law linking black-body temperature and peak wavelength.

Q5 Write down Stefan's law linking power, temperature and surface area.

Q6 Write down the law linking intensity and distance.

8. Stellar Spectral Classes

You can classify stars using their light spectra as well as by their power output. One way astronomers analyse the light emitted from stars is by splitting the light received from them using a prism or diffraction grating.

Line spectra

You saw line spectra in year 1 of your A-Level, but for year 2 you need to know about line absorption spectra in a bit more detail...

Energy levels in atoms

Electrons in an atom can only exist in certain well-defined energy levels. Each level is given a number. n = 1 represents the lowest energy level an electron can be in — the ground state. An atom is said to be excited when one or more of its electrons is in an energy level higher than the ground state.

Electrons can move up an energy level by absorbing a photon. Since these transitions are between definite energy levels, the energy of each photon absorbed can only take a certain allowed value. The energy of a photon is given by $E = hf$ (h = the Planck constant and f = frequency), so only certain frequencies, and so certain wavelengths, of light can be absorbed by electrons.

Line absorption spectra

If you split the light from a star using a prism or diffraction grating, you get a spectrum. Stars are approximately black bodies (see page 223), so they emit a continuous spectrum of electromagnetic radiation. You get **absorption lines** in the spectrum when radiation from the star passes through a cooler gas e.g. in the star's atmosphere.

At low temperatures, most of the electrons in the gas atoms will be in their ground states (n = 1). Photons of particular wavelengths are absorbed by the electrons to excite them to higher energy levels. When the electrons de-excite, the same wavelength of light is emitted and radiated in all directions. This means the intensity of radiation with this wavelength reaching Earth is reduced, which is shown by dark lines in the otherwise continuous spectrum, corresponding to the absorbed wavelengths.

The intensity of an absorption line is how dark it is — don't get this confused with the intensity of radiation. The more intense the absorption line at a particular wavelength, the more radiation of that wavelength has been absorbed.

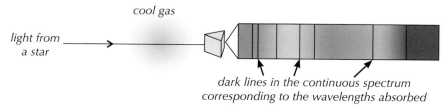

dark lines in the continuous spectrum corresponding to the wavelengths absorbed

Figure 1: *An absorption spectrum of a cool gas.*

Hydrogen Balmer absorption lines

The wavelengths corresponding to the visible part of hydrogen's line absorption spectrum are caused by the electrons in atomic hydrogen moving between the first excitation level (n = 2) and higher energy levels. This leads to a series of lines called the **Balmer series**.

Learning Objectives:

- Know that Hydrogen Balmer absorption lines are caused by the excitation of electrons in the n = 2 energy level.

- Understand how astronomers can use Hydrogen Balmer absorption lines to determine stellar surface temperature.

- Know how stars can be classified according to their spectra.

- Know the seven spectral classes used to classify stars: O, B, A, F, G, K, M.

- Be able to classify stars into one of the seven spectral classes from their colour, temperature, or prominent absorption lines.

Specification Reference 3.9.2.4

Tip: You learnt about diffraction gratings in year 1 of A-level physics — they are just thin slides that contain lots of equally spaced slits very close together. They are often used to diffract light.

These lines are seen in stellar spectra where light emitted by the star has been absorbed by hydrogen atoms in the stellar atmosphere as the light passes through it.

Determining stellar temperatures

For a hydrogen absorption line to occur in the visible part of a star's spectrum, electrons in the hydrogen atoms already need to be in the $n = 2$ state. This happens at high temperatures, where collisions between the atoms give the electrons extra energy.

If the temperature is too high, though, the majority of the electrons will reach the $n = 3$ level (or above) instead, which means there won't be so many Balmer transitions. So, the intensity of the Balmer lines depends on the temperature of the star (see Figure 3).

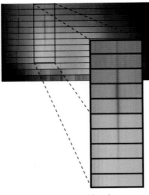

Figure 2: *The line absorption spectra of several stars of decreasing temperature (top to bottom). The close-up shows a Hydrogen Balmer absorption line, and how its intensity varies with temperature.*

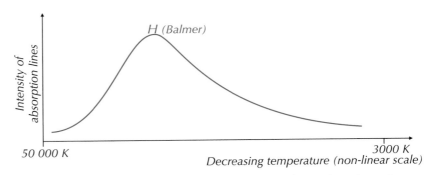

Figure 3: *The intensity of Balmer lines in the hydrogen line absorption spectrum of a star depends on the temperature of the star.*

You can use the intensity of Balmer lines to find the temperature of a star. For a particular intensity of the Balmer lines, two temperatures are possible. Astronomers get around this by looking at the absorption lines of other atoms and molecules as well.

Spectral classes

Stars are classified into groups called **spectral classes**. The spectral class of a star depends on the relative strength of certain absorption lines. There are seven main spectral classes that we classify stars into — here they are in order of decreasing temperature:

> **O, B, A, F, G, K and M**

O class stars are the hottest and appear blue in colour. M class stars are the coolest and appear red in colour. All the in-between classes cover the temperatures and colours in between. The Sun is a G class star.

Remember... the spectral classification system is all to do with the spectra of stars — properties like magnitude, distance and apparent brightness don't affect it.

Figure 4 shows how the intensity of the visible spectral lines changes with temperature, and how the spectral classes are split up:

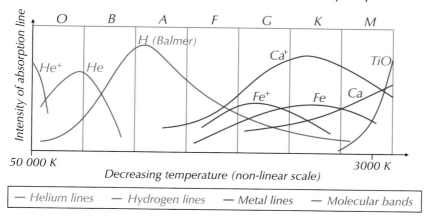

Figure 4: *A graph showing how the intensity of absorption lines in the visible region of stellar spectra changes with stellar temperature (and spectral class).*

The visible spectral characteristics of spectral classes

The seven spectral classes of stars all have different characteristics, e.g. colour, temperature and line absorption spectra, that you need to know...

Spectral Class	Colour	Temperature / K	Absorption lines
O	Blue	25 000 – 50 000	The strongest spectral lines are helium-plus ion (He^+) and helium atom (He) absorptions, since these need a really high temperature. They have weak hydrogen (H) Balmer lines too.
B	Blue	11 000 – 25 000	These spectra show strong helium atom (He) and hydrogen (H) absorptions.
A	Blue-white	7500 – 11 000	Visible spectra are governed by the strongest Hydrogen Balmer (H) lines, but there are also some metal ion absorptions.
F	White	6000 – 7500	These spectra have strong metal ion absorptions.
G	Yellow-white	5000 – 6000	These have both metal ion and metal atom absorptions.
K	Orange	3500 – 5000	Spectral lines are mostly from neutral metal atoms.
M	Red	< 3500	Spectral lines are from neutral atoms, as well as molecular band absorptions from compounds like titanium oxide (TiO), since these stars are cool enough for molecules to form.

Only cooler stars contain atoms with electrons in low enough energy levels to bond and form molecules, and therefore show molecular bands in their absorption line spectra.

Tip: Hot stars only have absorption lines corresponding to a few different elements (mostly hydrogen and helium), whereas cooler stars will have absorption lines corresponding to many more elements — see the table below.

Tip: This type of diagram, and the H-R diagram which you'll see on page 231, tend to be drawn with spectral class along the horizontal axis. The relationship between spectral class and temperature isn't linear or logarithmic.

Exam Tip
You'll need to be able to work out what spectral class a star belongs to given its characteristics, so make sure you know all the information in this table off by heart.

Tip: Spectral class A stars have the strongest absorption lines.

Tip: Spectral classes used to be ordered alphabetically according to the strength of their Balmer lines, but when astronomers realised that this didn't work, they just rearranged them into this funny order. Useful.

Tip: The colour goes from blue to red as the temperature decreases, because the peak wavelength increases ($\lambda_{max} T$ = constant) and so moves towards the red end of the spectrum.

Practice Question — Application

Q1 Figure 5 shows the line absorption spectra of two stars, with some absorption lines labelled with the elements or molecules that are responsible for them.

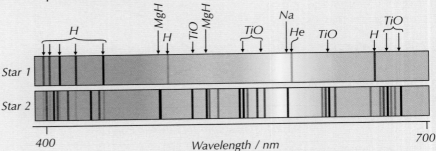

Figure 5: Line absorption spectra of two stars.

a) Which star has the strongest hydrogen Balmer lines?

b) Which star would you expect to have the lower temperature? Give one reason for your answer.

c) Which of the two stars would you expect to be in spectral class B? Give a reason for your answer.

d) Suggest a spectral class for the other star and explain your answer.

e) Give a range of temperatures in which you'd expect to find Star 1.

Practice Questions — Fact Recall

Q1 Why do absorption lines only appear at particular wavelengths in the spectra of stars?

Q2 Why do hydrogen atoms need to have electrons in the $n = 2$ energy level for hydrogen Balmer lines to be seen?

Q3 Sketch a graph showing how the intensity of hydrogen Balmer absorption lines varies with temperature.

Q4 Write down the seven spectral classes of stars in order of decreasing temperature.

Q5 Which would you expect to be hotter, a blue star or a white star?

Q6 Briefly state what ions, atoms or molecules cause the strongest absorption lines in the line absorption spectrum of a star in spectral class:

    a) G   b) M   c) O   d) K

Q7 What spectral class would a white star be in?

Q8 What spectral class would you expect a star with metal ion absorption lines and very strong hydrogen Balmer lines to be in?

9. The Hertzsprung-Russell Diagram

This whole section is about one diagram... but it's a pretty important one. It's also very pretty — so think of this part of the book as a bit of light relief.

The diagram

Independently, Hertzsprung and Russell noticed that a plot of absolute magnitude (see p.220) against temperature (or spectral class) didn't just throw up a random collection of stars but showed distinct areas.

This diagram ended up being really important for studying how stars evolve (p.233). A graph of absolute magnitude vs temperature/spectral class for stars became known as the **Hertzsprung-Russell** (H-R) **diagram** (Figure 1).

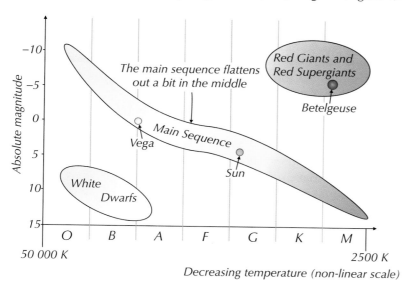

Figure 1: *The Hertzsprung-Russell diagram with the positions of the Sun and the stars Vega and Betelgeuse labelled.*

Distinct areas of the H-R diagram

The three distinct areas in which stars fall on the H-R diagram correspond to three main types of stars:

- The long, diagonal band is called the **main sequence**. Main sequence stars are in their long-lived stable phase where they are fusing hydrogen into helium. The Sun is a main sequence star.

- Stars that have a high luminosity and a relatively low surface temperature must have a huge surface area because of Stefan's law (page 225). These stars are called **red giants** and are found in the top-right corner of the H-R diagram. Red giants are stars that have moved off the main sequence, and fusion reactions other than hydrogen to helium are also happening in them (see page 233).

Learning Objectives:

- Know the general shape of the Hertzsprung-Russell (H-R) diagram, including the position of the main sequence, dwarf and giant stars.

- Know that the absolute magnitude axis on an H-R diagram ranges from −10 to 15.

- Know that the *x*-axis on an H-R diagram either shows temperature ranging from 50 000 K to 2500 K, or the spectral classes OBAFGKM.

Specification Reference 3.9.2.5

Tip: Remember... stars with negative values of absolute magnitude are brighter (page 220).

Tip: The temperature scale's a bit weird because of how the spectral classes (O-M) are defined. Remember, temperature goes the "wrong way" along the *x*-axis — from hotter to cooler.

Tip: In several billion years, the Sun will become a red giant (see page 234). It will expand to be around 20% larger than the Earth's current orbit and shine 3000 times brighter than it does now.

Tip: A white dwarf is a small hot blue or white star about the size of the Earth.

- Stars that have a low luminosity but a high temperature must be very small, again because of Stefan's law. These stars are called **white dwarfs** and are about the size of the Earth. They lie in the bottom-left corner of the H-R diagram. White dwarfs are stars at the end of their lives, where all of their fusion reactions have stopped and they are just slowly cooling down (see page 234).

Practice Question — Application

Q1 Star 1 is a spectral class B star and has an absolute magnitude of 10. Star 2 is also in spectral class B, but has an absolute magnitude of –4.

    a) Which star you would expect to be bigger, and why?

    b) What type of star is Star 1?

    c) What type of star is Star 2?

    d) Which star is further along in its evolution sequence?

Practice Questions — Fact Recall

Exam Tip
Make sure you've learnt the axis scales off by heart — you might need to draw an H-R diagram in your exam.

Q1 What quantity is shown by the vertical axis of an H-R diagram? What range of values does this axis have?

Q2 What two quantities can be plotted on the horizontal axis of an H-R diagram?

Q3 Sketch the H-R diagram. Mark on your diagram the main sequence, the red giants and red supergiants, the white dwarfs and the position of the Sun. Make sure you label both your axes.

Q4 What process is occurring in the core of a main sequence star?

10. Evolution of Sun-like Stars

Stars like our Sun go through several different stages in their lives and move around the H-R diagram as they go (see p.231). The Sun is on the main sequence right now, but it's still got a long way to go.

Formation

All stars are born in a cloud of dust and gas, most of which was left when previous stars blew themselves apart in supernovae (see page 236). The denser clumps of the cloud contract (very slowly) under the force of gravity.

When these clumps get dense enough, the cloud fragments into regions called protostars, that continue to contract and heat up. Eventually the temperature at the centre of the protostar reaches a few million degrees, and hydrogen nuclei start to fuse together to form helium (see page 190).

This releases an enormous amount of energy and creates enough pressure (radiation pressure) to stop the gravitational collapse. The star has now reached the main sequence (see page 231) and will stay there, relatively unchanged, while it fuses hydrogen into helium.

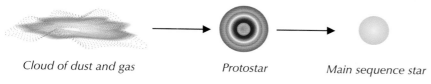

Cloud of dust and gas   Protostar   Main sequence star

Figure 2: *A star in the early stages of its stellar evolution.*

Core and shell burning sequence

Stars spend most of their lives as main-sequence stars. The pressure produced from hydrogen fusion in their core balances the gravitational force trying to compress them. This stage is called core hydrogen burning.

When all the hydrogen in the core has fused into helium, nuclear fusion stops, and with it the outward pressure stops. The helium core contracts and heats up under the weight of the star. As a result, the outer layers expand and cool, and the star becomes a red giant.

The material surrounding the core still has plenty of hydrogen. The heat from the contracting helium core raises the temperature of this material enough for the hydrogen to fuse. This is called shell hydrogen burning. (Very low-mass stars stop at this point. They use up their fuel and slowly fade away...)

The helium core continues to contract until, eventually, it gets hot enough and dense enough for helium to fuse into carbon and oxygen. This is called core helium burning. This releases a huge amount of energy, which pushes the outer layers of the star further outwards.

When the helium runs out, the carbon-oxygen core contracts again and heats a shell around it so that helium can fuse in this region — shell helium burning.

Figure 4 shows a summary of the stages a star goes through between the main sequence and becoming a red giant.

Learning Objectives:

- Understand the stellar evolution of a Sun-like star from formation to white dwarf.

- Understand how a Sun-like star moves around the H-R diagram as it evolves.

Specification Reference 3.9.2.5

Figure 1: *The 'Pillars of Creation' photographed by the Hubble Space Telescope. These pillars of gas and dust in the Eagle Nebula are thought to be areas of star formation. Nebulae are clouds of interstellar gas and dust and are often star-forming regions.*

Tip: As a star contracts, the temperature increases due to conservation of energy — gravitational potential energy is converted to thermal energy.

Tip: The cooling of the outer layers of the star makes the star's colour change to become redder — this is why we call them red giants.

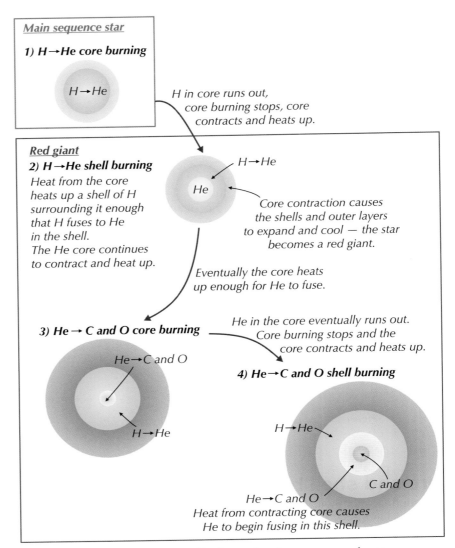

H in core runs out, core burning stops, core contracts and heats up.

Red giant

2) H→He shell burning
Heat from the core heats up a shell of H surrounding it enough that H fuses to He in the shell.
The He core continues to contract and heat up.

H→He

He

Core contraction causes the shells and outer layers to expand and cool — the star becomes a red giant.

Eventually the core heats up enough for He to fuse.

3) He → C and O core burning

He→C and O

H→He

He in the core eventually runs out. Core burning stops and the core contracts and heats up.

4) He→C and O shell burning

H→He

C and O

He→C and O
Heat from contracting core causes He to begin fusing in this shell.

Figure 3: A red giant with a radius of around 1 billion km.

Figure 4: The helium and hydrogen burning sequence of a star transitioning from a main sequence star to a red giant.

White dwarfs

In low-mass stars, like our Sun, the carbon-oxygen core won't get hot enough for any further fusion and so it continues to contract under its own weight. Once the core has shrunk to about Earth-size, electrons exert enough pressure (**electron degeneracy pressure**) to stop it collapsing any more (fret not — you don't have to know how).

The helium shell becomes more and more unstable as the core contracts. The star pulsates and ejects its outer layers into space as a **planetary nebula**, leaving behind the dense core.

The star is now a very hot, dense solid called a white dwarf, which will simply cool down and fade away.

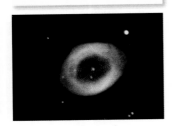

Figure 5: When a low-mass red giant cools down, the shells are ejected and create a beautiful planetary nebula, like this one, leaving a white dwarf at the centre.

Stellar evolution and the H-R diagram

As a Sun-like star evolves through its life, it moves in a certain path on the H-R diagram:

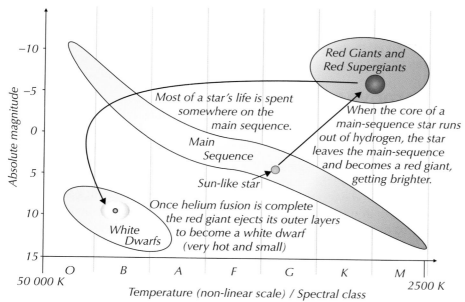

Figure 6: The movement of a Sun-like star on the H-R diagram as it evolves.

Tip: Stars begin as a cloud of dust and gas — as they contract and get hotter they appear in the M spectral class on the H-R diagram and move across to the main sequence as they get hotter. They reach the main sequence when fusion starts in their cores.

Tip: Most of the stars on the main sequence will eventually leave the main sequence and become a red giant. The path they take on the H-R diagram is actually a bit more wiggly than the one shown for the yellow star in Figure 6 — but you just need to know the basic shape.

Practice Question — Application

Q1 Five stars **a**, **b**, **c**, **d** and **e** are plotted on the graph below.

a) Which star is a red giant? Which is the hottest main-sequence star?

b) Which star could be the Sun?

c) Write down a sequence of letters that represents the evolution of the Sun from its current position until it becomes a white dwarf.

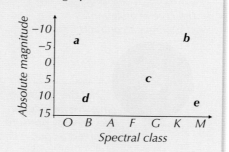

Practice Questions — Fact Recall

Q1 What causes a cloud of dust and gas to contract and form a star?

Q2 In what stage of evolution does a star spend most of its life?

Q3 Explain how the process of core hydrogen burning stops a main sequence star from compressing under the gravitational force.

Q4 What happens to the core of a star when it runs out of hydrogen? What effect does this have on the hydrogen layer surrounding the core of the star?

Q5 What causes a main-sequence star to become a red giant?

Q6 How is a white dwarf formed?

Q7 What is a planetary nebula?

- Know the defining properties of supernovae, including a rapid increase in absolute magnitude.

- Know that a burst of gamma rays can be emitted when a supergiant star collapses.

- Be able to compare the energy output of a supernova to that of the Sun.

- Be able to recognise the light curve for a type 1a supernova, and know how this relates to a type 1a supernova's use as a standard candle.

- Understand how neutron stars and black holes are formed.

- Know the composition and density of a neutron star.

- Know that a black hole is an object that has an escape velocity greater than the speed of light, c.

- Know that astronomers believe that there is a supermassive black hole at the centre of every galaxy.

- Know that the radius of the event horizon of a black hole is called the Schwarzschild radius (R_s), and be able to calculate it using $R_s = \dfrac{2GM}{c^2}$.

Specification Reference 3.9.2.6

Tip: This is a huge amount of energy — a nearby supernova could destroy the Earth's atmosphere.

11. Supernovae, Neutron Stars and Black Holes

Eventually the Sun will become a white dwarf and slowly fade away — but stars a lot more massive than our Sun end their lives in much more exciting and spectacular ways...

Supernovae

High-mass stars have a shorter life and a more exciting death than lower-mass stars like the Sun. Even though stars with a large mass have a lot of fuel, they use it up more quickly and don't spend so long as main-sequence stars.

When they are red giants the 'core burning to shell burning' process can continue beyond the fusion of helium, building up layers in an onion-like structure to become red supergiants. For really massive stars this can go all the way up to iron. Nuclear fusion beyond iron isn't energetically favourable though, so once an iron core is formed then very quickly it's goodbye star.

When the core of a star runs out of fuel, it starts to contract. If the star is massive enough, though, electron degeneracy can't stop the core contracting. This happens when the mass of the core is more than 1.4 times the mass of the Sun.

The core of the star continues to contract, and as it does, the outer layers of the star fall in and rebound off the core, setting up huge shockwaves. These shockwaves cause the star to explode cataclysmically in a **supernova**, leaving behind the core, which will either be a **neutron star** or (if the star was massive enough) a **black hole**.

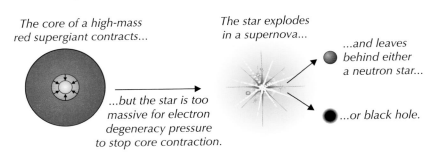

The core of a high-mass red supergiant contracts...

...but the star is too massive for electron degeneracy pressure to stop core contraction.

The star explodes in a supernova...

...and leaves behind either a neutron star...

...or black hole.

Figure 1: *The possible evolution paths of a high-mass star from its red supergiant phase.*

When the star explodes in a supernova, it will experience a brief and rapid increase in absolute magnitude (see page 220). The light from a supernova can briefly outshine an entire galaxy, before fading over the next few weeks or months.

For some very massive stars, bursts of high energy gamma rays are emitted. The gamma burst can go on for minutes or very rarely, hours.

The energy output of a type 1a supernova is around 10^{44} J. This is roughly the same as the energy output of the Sun over its entire lifetime. Other types of supernova may release much more energy than this.

Type 1a supernovae

The defining characteristic of a supernova is a rapid, massive increase in brightness. Different types of supernovae have different characteristics. These can be represented using **light curves**. A light curve is a graph of absolute magnitude, M, plotted against the time since the supernova reached peak magnitude.

A type I supernova is a supernova which has no hydrogen lines in its spectrum. If a supernova does have hydrogen lines (i.e. the Balmer series — see page 227), then it's a type II supernova.

Light curves for type I supernovae have two defining features — a sharp initial peak followed by a gradually decreasing curve (see Figure 2).

A subset of type I supernovae, called type 1a, are formed when a white dwarf core absorbs matter from a nearby binary partner. Because type 1a supernovae all have the same mass when they explode, they all have identical light curves, showing the same peak in absolute magnitude. This means they can be used as a standard candle (p.221). They are so bright that distances up to 1000 Mpc can be measured. Figure 2 shows what the light curve for a type 1a supernova looks like.

Tip: A light curve can sometimes be a graph of the brightness of the supernova against time, or the apparent magnitude of the supernova against time — these graphs have the same shaped curves as in Figure 2.

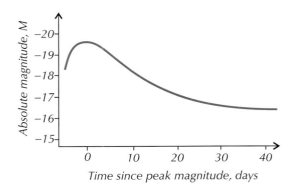

Figure 2: A characteristic light curve for a type 1a supernova.

Tip: You don't need to know the light curves for other kinds of supernova.

Neutron stars

As the core of a massive star contracts, the electrons in the core material get squashed onto the atomic nuclei and combine with protons to form neutrons and neutrinos.

If the star's core is between 1.4 and 3 solar masses, this is as far as the star can contract — the core suddenly collapses to become a neutron star made mostly of neutrons. The outer layers of the star fall onto the neutron star, which causes shockwaves in these layers and leads to a supernova. After the supernova, the neutron star is left behind.

Neutron stars are incredibly dense (about 4×10^{17} kg m^{-3}). They're very small, typically about 20 km across, and they can rotate very fast (up to 600 times a second).

They emit radio waves in two beams as they rotate. These beams sometimes sweep past the Earth and can be observed as radio pulses rather like the flashes of a lighthouse. These pulsing neutron stars are called pulsars.

Figure 3: The Crab Nebula — a supernova remnant with a pulsar (rotating neutron star) in the centre.

Tip: When I say 'incredibly' dense I mean it... if you squashed the entire population of the Earth into the size of an average sugar cube — it still wouldn't quite be as dense as a neutron star. Good luck getting your head around that.

Black holes

If the core of the star is more than 3 times the Sun's mass, the core will contract until neutrons are formed, but now the gravitational force on the core is greater — the neutrons can't withstand this gravitational force.
There's nothing left to stop the core collapsing to an infinitely dense point. At that point, the laws of physics break down completely.

Tip: Remember, the kinetic energy of an object is given by $E_k = \frac{1}{2}mv^2$.

The **escape velocity** is the velocity that an object would need to travel at to have enough kinetic energy to escape a gravitational field. When a massive star collapses into an infinitely dense point, a region around it has such a strong gravitational field that it becomes a black hole — an object whose escape velocity is greater than the speed of light, c. If you enter this region, there's absolutely no escape — not even light can escape it.

Tip: Nothing can travel faster than the speed of light, c.

The boundary of the region around the infinitely dense point in which the escape velocity is greater than c is called the **event horizon**. It is the distance at which the escape velocity is equal to c, so light has just enough kinetic energy to escape the gravitational pull of the black hole.

The radius of the event horizon of a black hole is called the **Schwarzschild radius** — it is thought of as being the radius of a black hole. Inside the Schwarzschild radius, everything, including light, can do nothing but travel further into the black hole.

Figure 4: Astronomers think the intense radiation seen at the centre of galaxies is caused by matter falling onto a supermassive black hole.

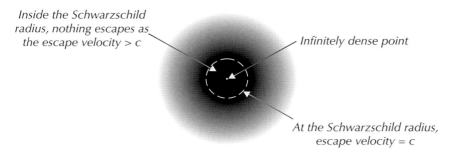

Inside the Schwarzschild radius, nothing escapes as the escape velocity > c

Infinitely dense point

At the Schwarzschild radius, escape velocity = c

Figure 5: The structure of a black hole.

Astronomers now believe that there is a **supermassive black hole** (more than 10^6 times more massive than our Sun) at the centre of every galaxy. As they consume stars close to them, they produce intense radiation, making the centre of galaxies very bright (see pages 250-251 for more).

Deriving the Schwarzschild radius

The size of the Schwarzschild radius for a black hole with a given mass can be found by thinking about the energy needed to move an object from a distance of r from the centre to infinity.

Consider a black hole with a mass of M. Assuming that its entire mass is at the centre, it is a point mass and the gravitational potential, V, at a distance of r from it is given by the formula (page 78):

Tip: The gravitational constant, G, is 6.67×10^{-11} Nm²kg⁻². It'll be given in the data and formulae booklet in the exam.

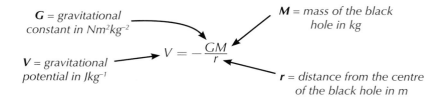

G = gravitational constant in Nm²kg⁻²

M = mass of the black hole in kg

V = gravitational potential in Jkg⁻¹

$$V = -\frac{GM}{r}$$

r = distance from the centre of the black hole in m

The gravitational potential energy (E_p) of an object with mass m is (page 80):

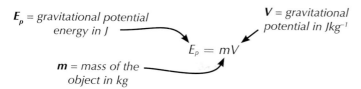

$$E_p = mV$$

E_p = gravitational potential energy in J

m = mass of the object in kg

V = gravitational potential in Jkg^{-1}

So the gravitational potential energy of an object at a distance of r from the centre of a black hole is:

$$E_p = -\frac{GMm}{r}$$

To escape the black hole, the object needs to gain enough gravitational potential energy so that $E_p = 0$. This means it needs to gain a gravitational potential energy of $\frac{GMm}{r}$.

As it travels away from the centre of the black hole, its kinetic energy is transferred into gravitational potential energy, so it just needs enough kinetic energy to escape.

If the object has a velocity v at a distance of r from the centre of the black hole, then it has kinetic energy of $E_k = \frac{1}{2}mv^2$.

The kinetic energy needed to escape is just equal to gravitational potential energy that it needs to gain:

$$\frac{1}{2}mv^2 = \frac{GMm}{r}$$

Dividing through by m and making r the subject gives:

$$r = \frac{2GM}{v^2}$$

The Schwarzschild radius is the distance at which the escape velocity is equal to the speed of light. By replacing v with the speed of light, c, you get the Schwarzschild radius, R_s:

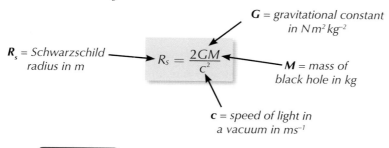

$$R_S = \frac{2GM}{c^2}$$

R_s = Schwarzschild radius in m

G = gravitational constant in N m^2 kg^{-2}

M = mass of black hole in kg

c = speed of light in a vacuum in ms^{-1}

Figure 6: Karl Schwarzschild, the German astronomer who derived the Schwarzschild radius.

Tip: This doesn't quite work in such strong gravitational fields. You really need to use Einstein's general relativity, but that's way too complicated for A-Level and it just so happens that this method still gives the right answer — hooray.

Tip: You don't need to know this derivation, but being able to follow it will help you to understand what the Schwarzschild radius actually is, and you need to be able to use the equation at the end.

Example — Maths Skills

a) **Calculate the Schwarzschild radius for a black hole that has a mass of 6.00×10^{30} kg.**

Use the formula above:

$$R_S = \frac{2GM}{c^2} = \frac{2(6.67 \times 10^{-11})(6.00 \times 10^{30})}{(3.00 \times 10^8)^2} = 8893.33...$$
$$= 8890 \text{ m (to 3 s.f.)}$$

Tip: Make sure you give your answer to the appropriate number of significant figures. You should give your answer to the lowest number of s.f. given in the question.

b) Calculate the average density of the matter within the event horizon of the black hole.

The radius of the event horizon is 8893.33... m, and the volume of a sphere is $\frac{4}{3}\pi r^3$. So the volume of the black hole is:

$$V = \frac{4}{3}\pi r^3 = \frac{4}{3}\pi(8893.33...)^3 = 2.946... \times 10^{12} \text{ m}^3$$

So density $\rho = \frac{m}{V} = \frac{6 \times 10^{30}}{2.946... \times 10^{12}} = 2.036... \times 10^{18}$

$$= 2.04 \times 10^{18} \text{ kg m}^{-3} \text{ (to 3 s.f.)}$$

Practice Questions — Application

Q1 A black hole has a Schwarzschild radius of 1.2 km. A photon is emitted from a source 950 m from the centre of the black hole. Will the photon be able to escape the black hole? Explain your answer.

Q2 There is thought to be a supermassive black hole 4.31 million times as massive as the Sun at the centre of the Milky Way.

a) Calculate the mass of this black hole.

b) What is its Schwarzschild radius?

c) How many times larger is the volume of the black hole than the volume of the Sun?

Practice Questions — Fact Recall

Q1 Why can stars with a core mass greater than 1.4 solar masses not form white dwarfs?

Q2 Describe how a contracting core of a star creates a supernova. What change in absolute magnitude occurs when a star explodes as a supernova?

Q3 How long would it take for the Sun to output as much energy as the energy of a gamma ray burst that can accompany a supernova?

Q4 What is a light curve of a supernova? Sketch the light curve of a type 1a supernova.

Q5 What is the typical density of a neutron star? What are they made of?

Q6 Explain what is meant by escape velocity.

Q7 What is a black hole? How is one formed?

Q8 a) What is meant by an event horizon?

b) What is meant by the Schwarzschild radius of a black hole?

c) Write down the formula for the Schwarzschild radius of a black hole.

Q9 What is thought to be at the centre of the every galaxy?

12. Doppler Effect and Red Shift

When a light source is moving, the wavelength of the light it gives out gets shifted. Depending on where you are standing in relation to its motion, it may look redder or bluer than it should.

The Doppler effect

You'll have experienced the **Doppler effect** loads of times with sound waves. Imagine a police car driving past you. As it moves towards you its siren sounds higher-pitched, but as it moves away, its pitch is lower. This change in frequency and wavelength is called the Doppler effect.

The frequency and the wavelength change because the waves bunch together in front of the source and stretch out behind it. The amount of stretching or bunching together depends on the velocity of the source.

Sound waves emitted at the same wavelength/ frequency in all directions.

Sound waves travelling in the opposite direction to the motion are spread out.

Sound waves travelling in the same direction as the motion are 'bunched up'.

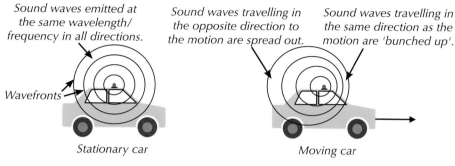

Wavefronts

Stationary car

Moving car

Figure 1: *Doppler shift of sound waves from a moving police car siren.*

Red shift

The Doppler effect happens with all waves, including electromagnetic radiation. When a light source moves away from us, the wavelength of the light reaching us becomes longer and the frequencies become lower. This shifts the light that we receive towards the red end of the electromagnetic spectrum and is called a **red shift**. The light that we receive from a star moving away from us is redder than the actual light emitted by the star.

Observer from Earth sees light with a longer wavelength than that emitted.

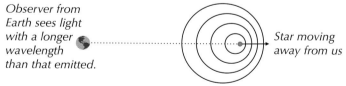

Star moving away from us

Figure 2: *The red shift of light emitted by a star moving away from us.*

When a light source moves towards us, the opposite happens and the light undergoes a **blue shift**. The light from the star looks bluer than it actually is.

Observer from Earth sees light with a shorter wavelength than that emitted.

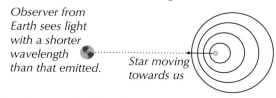

Star moving towards us

Figure 3: *The blue shift of light emitted by a star moving towards us.*

Learning Objectives:

- Understand the Doppler effect.
- Understand and be able to apply the formulas for red shift:
$$z = \frac{\Delta f}{f} = \frac{v}{c} \text{ and}$$
$$\frac{\Delta \lambda}{\lambda} = -\frac{v}{c} \text{ for } v \ll c.$$

Specification References 3.9.3.1 and 3.9.3.2

Tip: The distance between wavefronts is equal to λ, the wavelength.

Tip: The Doppler effect is used in weather radars to detect the movement of precipitation. A microwave signal is directed at the precipitation, which reflects it. The reflected signal is measured to see how the motion of the precipitation has affected the signal.

Tip: Remember, red light is the lowest frequency of light in the visible spectrum.

Tip: It's important to realise that the frequency of the source doesn't change, just the frequency of the radiation reaching us.

Tip: Blue light has the highest frequency of the visible light spectrum. In blue shift, light gets shunted towards (or beyond) the blue end of the visible spectrum.

The amount of red (or blue) shift depends on how fast the star is moving away from (or towards) us. The higher the velocity, the more the waves are shifted.

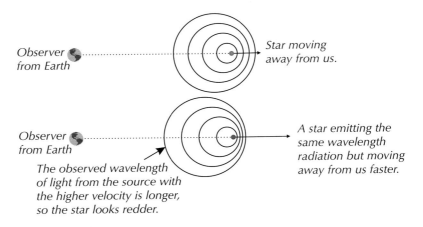

Figure 4: *Red (or blue) shift increases with the velocity of the source.*

The amount of red shift, z, is linked to the velocity that the object is moving away from us at, the **recessional velocity**, by the following formula:

Tip: $v \ll c$ means that v is much less than c. For velocities close to c, this doesn't work.

Tip: You can use any units for v and c in this equation, as long as they are the same.

v = recessional velocity of the source in line with the observer in ms^{-1}

z = red shift $\longrightarrow$ $z = \dfrac{v}{c}$ if $v \ll c$

c = speed of electromagnetic radiation in a vacuum in ms^{-1}

If the source is moving away, $v > 0$ and so the red shift is positive. If the source is moving towards the observer, $v < 0$ and the red shift is negative — the radiation is blue-shifted.

This equation works for all electromagnetic radiation, but you'll only need to use it to calculate red and blue shifts of optical and radio frequencies.

An increase in wavelength is always referred to as a red shift, even if it's a shift of wavelength taking place at a wavelength higher than that of red visible light, e.g. a shift from infrared to radio wavelengths. A decrease in wavelength is always referred to as a blue shift.

Tip: The speed of light, $c = 3.00 \times 10^8$ ms^{-1}.

Tip: Remember that red shift has no units.

┌─ **Example** ── **Maths Skills** ────────────

A star is receding from the Earth at a velocity of 525 kms^{-1}. Calculate the red shift of the star.

The recessional velocity is in kms^{-1}, so convert it into ms^{-1} to match the speed of light units.

$$z = \frac{v}{c} = \frac{525\,000}{3.00 \times 10^8} = 1.75 \times 10^{-3}$$

Example — Maths Skills

The red shift of the Andromeda galaxy is –3.70 × 10⁻⁴.
How fast is the Andromeda galaxy moving? Is it receding from or approaching the Milky Way?

- Rearranging the equation $z = \frac{v}{c}$ gives $v = zc$

  $v = zc = (-3.70 \times 10^{-4}) \times (3.00 \times 10^{8}) = -111\,000 = -1.11 \times 10^{5}\ \text{ms}^{-1}$

- The velocity is negative, so the Andromeda galaxy is approaching us.

Observing red shift

The entire radiation spectrum of a moving source will be shifted, depending on its motion — including any absorption lines in the spectrum.

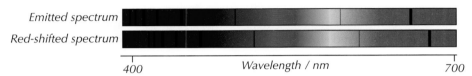

Emitted spectrum

Red-shifted spectrum

400   *Wavelength / nm*   700

Figure 5: *The absorption lines on the line absorption spectrum of a red-shifted star will all be shifted closer to (or beyond) the red end of the spectrum.*

Different atoms and molecules only absorb particular wavelengths of light that correspond to the differences between their energy levels (see page 227). You can identify the actual wavelengths of the absorption lines for a particular atom or molecule in the lab.

You can work out the red shift of a star by looking at the wavelength or frequency of known absorption lines (like the Balmer series — page 227) in the observed spectrum and comparing them to the wavelengths or frequencies they should be. You can calculate the red shift from the change in frequency and the frequency emitted by the source:

$$z = \frac{\Delta f}{f}$$

Δf = difference between emitted and observed frequencies

f = frequency of light emitted by the star

Δf is the difference between the emitted frequency (f) and the observed frequency (f_{obs}), $f - f_{obs}$.

- If the star is red-shifted, $f > f_{obs}$, so Δf and z will be positive.
- If the star is blue-shifted, $f < f_{obs}$, so Δf and z will be negative.

You can also find red shift by observing the change in wavelength of light from a source:

$$z = -\frac{\Delta \lambda}{\lambda}$$

$\Delta \lambda$ = difference between emitted and observed wavelengths

λ = wavelength of light emitted by the star

Combing these two equations with equation from the previous page gives, for v << c:

$$\frac{\Delta f}{f} = \frac{v}{c}$$

$$\frac{\Delta \lambda}{\lambda} = -\frac{v}{c}$$

Tip: Measuring a laboratory source will give the actual positions of the absorption lines of an atom or molecule without red shift.

Tip: The units don't matter — Δf and f (or $\Delta \lambda$ and λ) just need to be in the same units.

Tip: $\Delta \lambda$ is the difference between the emitted wavelength (λ) and the observed wavelength (λ_{obs}), $\lambda - \lambda_{obs}$.

Tip: Notice the negative sign in this equation. You'll get this formula and the one above in the data and formulae booklet in the exam.

Tip: Remember, the wavelength of the absorption line found from a laboratory source isn't red-shifted and is the actual wavelength of the absorption line.

Example — Maths Skills

A star is receding from Earth at an unknown velocity. The wavelength of a hydrogen absorption line in the spectrum of this star is 661 nm. The wavelength of the same absorption line observed in a laboratory source is 656 nm. Calculate the velocity at which the star is receding from Earth.

- Combining $z = -\frac{\Delta\lambda}{\lambda}$ and $z = \frac{v}{c}$, we get $\frac{v}{c} = -\frac{\Delta\lambda}{\lambda}$ so $v = -\frac{\Delta\lambda}{\lambda} \times c$

- So $v = -\frac{656 - 661}{656} \times (3.00 \times 10^8) = 2.286... \times 10^6 \text{ ms}^{-1}$.

- So the star is receding at $2.29 \times 10^6 \text{ ms}^{-1}$ (to 3 s.f.).

Cosmological red shift

All distant galaxies show red shift, and are moving away from us. The way astronomers tend to look at this stuff, the galaxies aren't actually moving through space away from us. Instead, space itself is expanding and the light waves are being stretched along with it. This is called cosmological red shift to distinguish it from red shift produced by sources that are moving through space.

The same formula works for both types of red shift as long as v is much less than c. If v is close to the speed of light, you need to use a nasty, relativistic formula instead (you don't need to know that one).

Tip: There's more on expanding space later — see pages 245-247.

Practice Questions — Application

Q1 A hydrogen absorption line is measured in the spectra of two stars. The line is at a frequency of 4.37×10^{14} Hz in the spectrum of star A and 5.2×10^{14} Hz in the spectrum of star B. The frequency of the hydrogen absorption line is measured in a laboratory to be 4.57×10^{14} Hz.
   a) Which star is moving away from us? Explain how you know.
   b) Calculate the red shift of this star.
   c) How fast is this star travelling?

Q2 The Whirlpool galaxy is receding from us at a velocity of 463 kms⁻¹.
   a) Will its red shift be positive or negative? Explain how you know.
   b) Calculate its red shift.
   c) A lab source of atomic hydrogen shows a strong absorption line at a wavelength of 0.21121 m. Calculate the wavelength this absorption line would be observed at if emitted from this galaxy.

Tip: Remember, $\Delta\lambda = \lambda - \lambda_{obs}$.

Practice Questions — Fact Recall

Q1 Explain how the Doppler effect makes police car sirens sound higher pitched as they travel towards us.

Q2 What is red shift? What is blue shift?

Q3 Write down a formula for red shift, z, in terms of the velocity of the radiation source. What assumption does this formula make?

Q4 Explain how you can tell that a star is moving away from or towards us by looking at its spectrum.

Q5 Write down two formulas for calculating red shift, from frequency and wavelength measurements.

Q6 What is meant by cosmological red shift?

13. The Big Bang Theory

As you've seen by now, astrophysics is full of loads of weird and wonderful objects and ideas. We can't explain them all, but astronomers are trying to use all the evidence we've got to discover more about the origin, evolution and fate of the universe. And right now, the best idea they've got is the Big Bang theory.

The cosmological principle

When you read that nearly all the galaxies in the universe are moving away from the Earth (see p.244 and below), it's easy to imagine that the Earth is at the centre of the universe, or that there's something really special about it. Earth is special to us because we live here — but on a universal scale, it's just like any other lump of rock.

The idea that no part of the universe is any more special than any other is summarised by the **cosmological principle**:

> On a large scale the universe is:
> - **homogeneous** (every part is the same as every other part) and
> - **isotropic** (everything looks the same in every direction)
>   — so it doesn't have a centre.

Until the 1930s, astronomers believed that the universe was infinite in both space and time (that is, it had always existed), and static. This seemed the only way that it could be stable using Newton's law of gravitation (page 73). Even Einstein modified his theory of general relativity to make it consistent with this Steady-State Universe.

Hubble's law

Edwin Hubble was the first scientist to realise that the universe is expanding. He used type 1a supernovae as standard candles (see page 221) to calculate the distances to galaxies, as well as measuring their red shift. The spectra from galaxies all show red shift (apart from a few very close ones). The amount of red shift gives the recessional velocity — how fast the galaxy is moving away (see page 242). Hubble realised that the speed that galaxies moved away from us depended on how far they were away.

A plot of recessional velocity against distance showed that they were proportional, which suggests that the universe is expanding. The further away a galaxy is, the faster it's travelling away from us. This gives rise to **Hubble's law**:

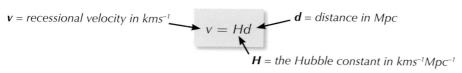

v = recessional velocity in kms^{-1} → $v = Hd$ ← d = distance in Mpc

H = the Hubble constant in kms^{-1}Mpc^{-1}

Since distance is very difficult to measure, astronomers used to disagree greatly on the value of H, with measurements ranging from 50 to 100 kms^{-1}Mpc^{-1}. It's now generally accepted that H lies between 65 and 80 kms^{-1}Mpc^{-1} and most agree it's in the mid to low 70s. You'll be given a value to use in the exam, either in the data and formulae booklet or in a question where you need it.

Learning Objectives:

- Know that Hubble's law, $v = Hd$, provides a simple interpretation of the expansion of the universe.

- Understand the Big Bang theory as the current consensus for the origin of the universe.

- Be able to estimate the age of the universe, assuming H is constant.

- Understand the evidence for the Big Bang theory, including the red shift of distant galaxies, CMBR and the relative abundance of H and He.

- Understand that dark energy is a possible explanation for an expanding universe, but that this is a controversial theory.

Specification References 3.9.2.6 and 3.9.3.2

Figure 1: *The relative brightness and distribution of over 1.6 million galaxies. On a large scale, the universe is homogeneous and isotropic.*

Tip: The SI unit (p.400) for H is s^{-1}. To get H in SI units, you need v in ms^{-1} and d in m (1 Mpc = 3.08×10^{22} m).

Tip: You can use $v = Hd$ to calculate the distance to a distant source if you know its red shift, by using $z = \frac{v}{c}$ to work out its recessional velocity.

Tip: If $v = Hd$ then $H = \frac{v}{d}$. So a graph of recessional velocity (v) against distance from Earth (d) can be used to estimate H. The gradient of the line of best fit is an estimate of H.

Tip: A closed system just means that energy and momentum are conserved. If there's only so much energy in the universe, as it expands, it must also be cooling.

Tip: Hubble's law says $v = Hd$ so $\frac{1}{H} = \frac{d}{v}$.

Tip: The universe is actually thought to be 13.75 billion years old to within 1%.

Example — Maths Skills

A galaxy is receding from us at 950 kms⁻¹. Assuming the Hubble constant is 65 kms⁻¹ Mpc⁻¹, how far away is the galaxy in Mpc?

$v = Hd$ so $d = \frac{v}{H} = \frac{950}{65} = 14.61... = 15$ Mpc (to 2 s.f.)

The Big Bang theory

The universe is expanding and cooling down (because it's a closed system). So further back in time it must have been denser and hotter. If you trace time back far enough, you get a hot Big Bang:

> **The Big Bang theory:**
>
> The universe started off very hot and very dense (perhaps as an infinitely hot, infinitely dense point) and has been expanding ever since.

The spectra from all galaxies (apart from a few very close ones that are moving towards us) show red shift. This shows they're all moving apart (see page 244). The red shift of distant galaxies is one piece of evidence for the Big Bang theory (see the next page for more).

Since the universe is expanding uniformly away from us it seems as though we're at the centre of the universe, but this is an illusion. You would observe the same thing at any point in the universe.

A good way to visualise this is to think of the universe as the surface of a balloon with lots of dots on it, representing galaxies. As you blow up the balloon, the space between all the galaxies (dots) gets bigger. Each galaxy sees all the other galaxies moving away from it, even though it's not at the centre of the motion.

Age and size

If the universe has been expanding at the same rate for its whole life, the age of the universe can be estimated by:

$$t = \frac{distance}{velocity} = \frac{1}{H}$$

(This is only an estimate though — see dark energy on the next page.)

Example — Maths Skills

If H = 75 kms⁻¹ Mpc⁻¹, calculate the age of the universe.

The Hubble constant is in kms⁻¹Mpc⁻¹, but you need it in s⁻¹ to get the time in s.

Multiplying by 10^3 gives it in ms⁻¹Mpc⁻¹ and dividing by 1 Mpc $= 3.08 \times 10^{22}$ m gives it in ms⁻¹m⁻¹, which is just s⁻¹.

75 kms⁻¹Mpc⁻¹ $= (75 \times 10^3) \div (3.08 \times 10^{22})$ s⁻¹ $= 2.4... \times 10^{-18}$ s⁻¹

So $t = \frac{1}{H} \approx \frac{1}{2.4... \times 10^{-18} \text{s}^{-1}} = 4.106... \times 10^{17}$
$= 4.11 \times 10^{17}$ s
$= 13$ billion years (to 2 s.f.).

The absolute size of the universe is unknown, but there is a limit on the size of the observable universe. This is simply a sphere (with the Earth at its centre) with a radius equal to the maximum distance that light can travel during its age, taking into account the expansion of space itself. So if H = 75 kms^{-1} Mpc^{-1}, then this sphere would have a radius of about 13 billion light years, but taking into account the expansion of the universe, it is thought to be more like 46-47 billion light years.

Tip: The universe is bigger than this sphere, but it isn't observable to us, because we can only see light that has had time to travel to us.

Dark energy

Even though we can estimate the age and observable size of the universe, it is just that — an estimate. This is because the rate of expansion of the universe hasn't been constant.

All the mass in the universe is attracted together by gravity. This attraction tends to slow down the rate of expansion of the universe. It's thought that the expansion was decelerating until about 5 billion years ago.

But in the late 90s, astronomers found evidence that the expansion is now accelerating. Astronomers are trying to explain this acceleration using **dark energy** — a type of energy that fills the whole of space. There are various theories of what this dark energy is, but it's really hard to test them.

Tip: There has been a lot of controversy surrounding dark energy in the astrophysics community — nobody really knows what it is, or if it's even there — but it is now generally accepted that it does exist.

Cosmic microwave background radiation

The Big Bang model predicts that loads of electromagnetic radiation was produced in the very early universe. This radiation should still be observed today (it hasn't had anywhere else to go).

Because the universe has expanded, the wavelengths of this cosmic background radiation have been stretched and are now in the microwave region — so it's known as **cosmic microwave background radiation** (**CMBR**). This was picked up accidentally by the astronomers Penzias and Wilson in the 1960s.

The Cosmic Background Explorer (COBE) and Wilkinson Microwave Anisotropy Probe (WMAP) satellites have been sent into space to have a detailed look at the radiation. They found the CMBR has a perfect black body spectrum corresponding to a temperature of 2.73 K (see page 224).

The CMBR is largely isotropic and homogeneous, which agrees with the cosmological principle (see page 245). There are very tiny fluctuations in temperature due to tiny energy-density variations in the early universe, which were needed for the initial 'seeding' of galaxy formation.

The background radiation also shows the Doppler effect, indicating the Earth's motion through space. It turns out that the Milky Way is rushing towards an unknown mass (the Great Attractor) at over a million miles an hour.

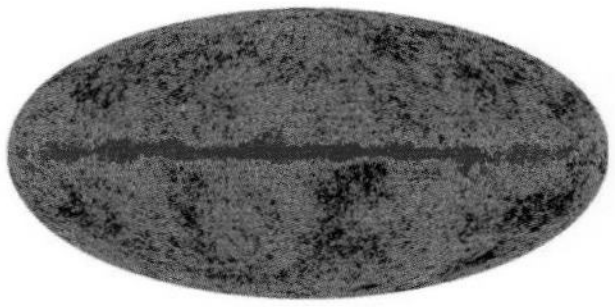

Figure 2: *A WMAP image of the CMBR. You can see that, except for the red horizontal band which shows Milky Way emissions, the CMBR is mostly homogeneous and isotropic, but shows small variations.*

Relative abundance of H and He

The Big Bang model also explained the large abundance of helium in the universe (which had puzzled astrophysicists for a while). The early universe had been very hot, so at some point it must have been hot enough for hydrogen fusion to happen. This means that, together with the theory of the synthesis of the heavier elements in stars, the relative abundances of all of the elements can be accounted for.

Practice Questions — Application

Q1 The Antennae galaxies are a pair of colliding galaxies at a distance of 1.4×10^7 pc. Calculate the maximum and minimum recessional velocities of the Antennae galaxies using $H = 70 \pm 5$ kms^{-1} Mpc^{-1}.

Q2 A star in a distant galaxy is receding from Earth. A physicist measures the wavelength of a hydrogen absorption line in a laboratory source to be 725 nm. The wavelength of the same absorption line in the spectrum of the star has been shifted by 35 nm. Calculate the star's distance from Earth. Use the value $H = 72$ kms^{-1} Mpc^{-1}.

Q3 Assuming that H has been constant since the universe began, the universe is approximately 13.8 billion years old.

    a) Why is this figure only an estimate?

    b) A student claims that this means he can calculate the absolute size of the universe. Why is he wrong?

Practice Questions — Fact Recall

Q1 State the cosmological principle.

Q2 What link did Hubble find between red shift and distance?

Q3 Write down Hubble's law, including the units of each quantity.

Q4 What is the Big Bang theory?

Q5 How can we estimate the age of the universe from the Hubble constant?

Q6 Why can't we measure the absolute size of the universe?

Q7 What might dark energy explain?

Q8 Explain what CMBR is and how it supports the Big Bang theory.

Q9 How does the relative abundance of H and He in the universe support the Big Bang theory?

14. Detection of Binary Stars, Quasars and Exoplanets

Since astronomers discovered the Doppler effect, it has helped to explain and observe quite a few strange things going on in space...

Spectroscopic binary stars

About half of the stars we observe are actually two stars that orbit each other. Many of them are too far away from us to be resolved with telescopes (see page 208 for more on resolving power), but the lines in their spectra show a binary star system. These are called **spectroscopic binary stars**. An eclipsing binary system is one whose orbital plane lies almost in our line of sight, so the stars eclipse each other as they orbit.

By observing how the absorption lines in the spectrum change with time the orbital period can be calculated. For simplicity, think about only one absorption line from the spectrum of an eclipsing binary system — see Figure 1.

Position of binary star system:

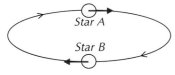

• *Observer*

Both stars are moving at right angles to our line of sight, so there's no Doppler effect (other than that of the binary system as a whole, as a result of its recessional velocity — which will be the same for both stars).

Absorption line spectrum:

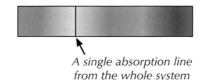

A single absorption line from the whole system

• *Observer*

Both stars are moving along our line of sight. Star A shows maximum blue shift. Star B shows maximum red shift.

Two separate absorption lines (one red-shifted and one blue-shifted) — one from each star.

Figure 1: *How the absorption lines in the spectrum of a spectroscopic binary star system change as the stars orbit each other.*

As you can see in Figure 1, as the stars orbit each other, the separation between the lines goes from zero up to the maximum separation and back to zero again in half a period. So the orbital period is double the time between the two positions in Figure 1.

Learning Objectives:

- Be able to apply the Doppler effect to binary stars.
- Know that quasars are bright radio sources and are the most distant objects that we can measure.
- Know that quasars show large optical red shifts, indicating that they are very far away.
- Be able to estimate the distance and power output of quasars.
- Know that quasars are thought to form around supermassive black holes.
- Know what exoplanets are and why they are hard to detect.
- Know that Doppler shift and the transit method can be used to detect exoplanets.
- Recognise the typical light curve for an exoplanet in transit across a star.

Specification References 3.9.3.1, 3.9.3.3 and 3.9.3.4

Tip: Two stars that look like one star are called optical doubles. Some optical doubles may not actually be binary systems. They might just be in the same line of sight but nowhere near each other.

Tip: The orbital period of a spectroscopic binary system can be anywhere from a few minutes to hundreds of thousands of years.

Apparent magnitude of an eclipsing binary system

If you plot a light curve of apparent magnitude against time for an eclipsing binary system, you'll get a graph like the one in Figure 3. As the stars eclipse each other, the apparent magnitude drops because some of the light is blocked out. It drops more as the dimmer star passes in front of the brighter star.

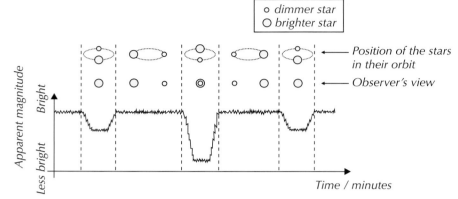

Figure 3: *Varying apparent magnitude of a spectroscopic binary star system over time.*

Figure 2: *The binary system Alpha Centauri is the third brightest star system in the sky. Alpha Centauri A and B orbit each other once every 80 years. Their companion Proxima Centauri (out of the frame of this image) is thought to orbit the pair with a period of millions of years.*

Quasars

Quasars (quasi-stellar objects) were discovered in the late 1950s and were first thought to be stars in our galaxy. The puzzling thing was that their spectra were nothing like normal stars. They sometimes shot out jets of material, and many of them were very active radio sources.

The 'stars' produced a continuous spectrum that was nothing like a black body radiation curve, and instead of absorption lines there were emission lines of elements that astronomers had not seen before.

However, these lines looked strangely familiar and in 1963 Maarten Schmidt realised that they were simply the Balmer series of hydrogen (see p.227) but red-shifted enormously.

This huge red shift suggests they're a huge distance away (see page 245) — in fact, the most distant objects seen. The measured red shifts give us distances of billions of light years. Using the inverse square law for intensity (see p.225) gives an idea of just how bright quasars are:

Example — Maths Skills

A quasar has the same intensity as a star. The star is 20 000 ly away and has the same power output as the Sun (4×10^{26} W). The quasar's red shift gives a distance of 1×10^{10} ly. Calculate its power output.

$$P \propto Id^2 \Rightarrow \frac{P_{quasar}}{P_{star}} = \left(\frac{d_{quasar}}{d_{star}}\right)^2 \quad (I_{quasar} = I_{star} \text{ so they cancel out of the equation}).$$

$$\Rightarrow P_{quasar} = P_{star} \times \left(\frac{d_{quasar}}{d_{star}}\right)^2 = 4 \times 10^{26} \times \left(\frac{1 \times 10^{10}}{20000}\right)^2 = 1 \times 10^{38} \text{ W}$$

That's bright — about 10 times the power output of the entire Milky Way galaxy.

There's also very good evidence to suggest that quasars are only about the size of the Solar System. Let me run that past you again. That's the power of a trillion Suns from something the size of the Solar System.

These numbers caused a lot of controversy in the astrophysics community — they seemed crazy. Many astrophysicists thought there must be a more reasonable explanation. But then evidence for the distance of quasars came when sensitive CCD equipment detected the fuzzy cloud of a galaxy around a quasar.

The current consensus is that a quasar is a very powerful galactic nucleus centred around a huge black hole more than 10^6 times the mass of the Sun. Almost all galaxies are thought to have these 'supermassive' black holes at their centres (see page 238), but only some eject huge amounts of material from their nuclei. Those that do are known as **active galactic nuclei** and a galaxy containing one is known as an active galaxy.

The black hole in an active galactic nucleus is surrounded by a doughnut shaped mass of whirling gas falling into it, which emits matter and radiation. In the same way as a pulsar (see p.237), magnetic fields produce jets of matter and radiation streaming out from the poles. The black hole must consume the mass of about 10 Suns per year to produce the energy observed.

Figure 4: *The active galaxy M87 — 50 million light years from Earth. The supermassive black hole at the centre is emitting a 5000-light-year-long jet of matter.*

Exoplanets

Doppler shift can also be used to discover other objects like **exoplanets**. An exoplanet (sometimes called an extrasolar planet) is any 'planet' not in our Solar System — this is because the word planet is usually reserved for only things within our Solar System, orbiting the Sun.

Exoplanets are pretty hard to find though, because:

- They're orbiting stars which are much brighter than them. Most exoplanets cannot be seen, as the bright light from the stars or other objects they're orbiting drowns out any light from the exoplanet.

- They're too small to distinguish from nearby stars (the subtended angle is too small for the resolving power of most telescopes — see p.208.)

- Only a few of the largest and hottest exoplanets that are furthest away from their stars can be seen directly using specially built telescopes.

Exoplanet detection — Doppler shift

As mentioned above, exoplanets can be rather tricky to find. One method that has been developed to tackle this involves the use of the Doppler effect.

Sometimes called the **radial velocity method**, the Doppler shift method measures how much the emissions from stars have been red or blue shifted (similar to binary stars on p.249).

- An exoplanet orbiting a star has a small effect on the star's orbit. It causes tiny variations (a wobble) in the star's orbit (see Figure 5).

- This is because the star and the exoplanet are actually orbiting around the centre of mass between them — but as the star is so much bigger than the exoplanet, the centre is much closer to the centre of the star.

- This wobble causes tiny red and blue shifts in the star's emissions which can be detected on Earth and can suggest the presence of an exoplanet.

- From this, the minimum mass of the exoplanet can also be calculated.

- There are problems with this method though, as the movement needs to be aligned with the observer's line of sight — if the planet orbits the star perpendicular to the line of sight then there won't be any detectable shift in the light from the star.

Tip: This method is also sometimes referred to as 'Doppler spectroscopy'.

Tip: The Doppler effect is commonly referred to as just red shift — but remember that light / EM radiation can be either red or blue shifted, so you could use Doppler shift as the generic term.

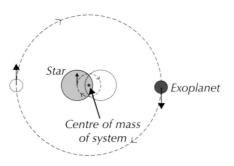

Figure 5: *An exoplanet orbiting a star. The exoplanet's gravitational field has a small effect on the motion of the star.*

Exoplanet detection — the transit method

An alternative to using Doppler shift is the **transit method**. This measures the change in apparent magnitude as an exoplanet travels ('transits') in front of a star.

- As an exoplanet orbits a star, if it passes in front of the star (as viewed by the observer), then it will block out some of the star's light, causing a change in apparent magnitude.

- This change can be viewed as a dip in the star's light curve (see Figure 6).

Tip: The dip in apparent magnitude must be periodic. If it doesn't occur at regular intervals, then it probably wasn't caused by an orbiting planet.

Tip: You first met light curves on page 237. Take a look back if you need a reminder.

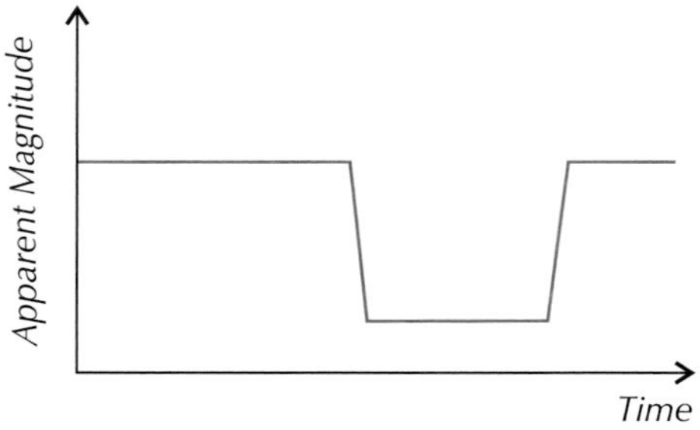

Figure 6: *A typical light curve for the transit of an exoplanet in front of a star.*

- The amount that the star's light is dimmed depends on the relative sizes of the star and the orbiting planet — the bigger the planet relative to the star, the greater the dip in apparent magnitude. The measurements from the transit method can therefore be used to find the radius of the exoplanet.

- However, the chances of the planet's path being perfectly aligned so that it passes directly between the star and the observer is incredibly low.

- Even if it is aligned with the observer, the transit is likely to last only a tiny fraction of its whole orbital period — so transits can be few and far between, making them easy to miss.

- This means you can only use this method to confirm already observed exoplanets. You can't use it to rule out any potential exoplanet locations — just because there's no dip in apparent magnitude doesn't necessarily mean there's no exoplanet there.

Tip: The Doppler shift and transit methods are often used together to find out as much information as possible about an exoplanet.

Practice Questions — Application

Q1 Algol is an eclipsing binary system in the constellation Perseus. The graph below shows how its apparent magnitude varies over a period of time.

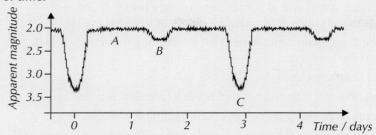

a) Do both the stars have the same apparent magnitude?

b) Describe the positions of the two stars at points A, B and C, as viewed by an observer in the plane of their orbit.

c) Write down the approximate period of the system in days.

Q2 Calculate the distance to a quasar with a red shift of 0.12.

Q3 An astrophysicist is trying to detect exoplanets in a nearby galaxy.

a) Give two reasons why it can be difficult to detect exoplanets.

b) The astrophysicist decides to use Doppler shift in order to help her detect any exoplanets. Explain why the measurement of a red or blue shift in emissions from distant stars can suggest the presence of an exoplanet.

c) The astrophysicist detects no red or blue shift in the emissions of the first star she observes. She concludes that the star has no exoplanets. Is she correct? Explain your answer.

> **Tip:** You can use Hubble's law for this question (page 245) and $H = 65$ kms^{-1} Mpc^{-1}.

Practice Questions — Fact Recall

Q1 Explain how spectroscopic binary stars can be identified by observing how their spectral lines change.

Q2 How would you find the orbital period of a spectroscopic binary star?

Q3 Why does the apparent magnitude of an eclipsing binary star change with time?

Q4 What were quasars first thought to be?

Q5 How do we know that quasars are very far away?

Q6 Describe what a quasar is and what is thought to cause them to emit such large amounts of radiation.

Q7 Sketch a typical light curve for an exoplanet as it transits across a star, as viewed by an observing astronomer.

Q8 Give one advantage and one disadvantage of using the transit method in the detection of exoplanets.

Section Summary

Make sure you know...

- How to draw a ray diagram of an astronomical refracting telescope in normal adjustment.
- How to calculate magnification of an astronomical refracting telescope using
  M = (angle subtended by image at eye) / (angle subtended by object at unaided eye) or $M = f_o / f_e$.
- How to draw a ray diagram for the Cassegrain arrangement for a reflecting telescope showing the path of rays through the telescope up to the eyepiece.
- How to compare the eye and CCDs as detectors.
- What the minimum angular resolution of a telescope is and how to calculate it using $\theta \approx \frac{\lambda}{D}$.
- The relative merits of, and problems with, using reflecting and refracting telescopes.
- The similarities and differences between optical telescopes and single-dish radio telescopes, I-R, U-V and X-ray telescopes (including their positioning, use, resolving power and collecting power).
- That the collecting power of a telescope is proportional to the square of the diameter of the mirror/dish.
- The definitions of a parsec (pc) and a light year (ly).
- How stars can be classified by their luminosity (or power output).
- What is meant by apparent magnitude, m, how it is related to intensity/brightness, and that Hipparchus gave the dimmest stars an apparent magnitude of 6.
- What is meant by absolute magnitude, M, and that $m - M = 5 \log \left(\frac{d}{10} \right)$.
- How type 1a supernovae can be used as standard candles to measure the distance to galaxies.
- The general shape of black body curves and how they vary with temperature.
- How stars can be assumed to be black bodies and classified according to their black-body temperature.
- Wien's displacement law for black bodies: $\lambda_{max} T = 2.9 \times 10^{-3}$ m K, and Stefan's law: $P = \sigma A T^4$
- The inverse square law $I = \frac{P}{4\pi d^2}$ and the assumptions needed to use it.
- That Balmer absorption lines are due to electron transitions in atomic hydrogen from the n = 2 state.
- The colour, temperature and prominent absorption lines of each spectral class of OBAFGKM.
- The Hertzsprung-Russell (H-R) diagram, including the path of a Sun-like star on it.
- That a supernova has a rapid increase in absolute magnitude and gamma ray bursts can be emitted.
- The defining features of a type 1a supernova light curve and how the energy output of a supernova compares with that of the Sun.
- That a neutron star is a very small and dense remnant of a high-mass star and is made of neutrons.
- That a black hole is produced by the collapse of the core of a very high-mass star to an infinitely dense point, and that it has an escape velocity $v > c$.
- What is meant by the terms event horizon and Schwarzchild radius (R_s), and be able to calculate R_s.
- That it is thought that there is a supermassive black hole at the centre of every galaxy.
- What is meant by the Doppler effect and red shift, z, and that $z = \frac{\Delta f}{f} = \frac{v}{c}$ and $\frac{\Delta \lambda}{\lambda} = -\frac{v}{c}$ for $v \ll c$.
- Hubble's law: $v = Hd$ and how to use it to estimate the age of the universe.
- How to calculate the red shift of distant galaxies.
- The Big Bang theory and how the red shift of galaxies, CMBR and the relative H and He abundances are evidence for it.
- That dark energy might explain the accelerating expansion of the universe.
- What the red shift tells you about the motion of objects and their distance.
- That quasars have large red shifts, and how to estimate their distance and power output.
- What exoplanets are, and how Doppler shift and the transit method can be used to detect them.
- What a typical light curve for the transit of an exoplanet in front of a star looks like.

Exam-style Questions

1   The Hubble Space Telescope (HST) is a Cassegrain telescope in orbit around the Earth.

   1.1   Draw a ray diagram to show two axial rays travelling through a typical Cassegrain arrangement as far as the eye lens.

(2 marks)

        When the HST was launched, it suffered from a problem called spherical aberration which had to be corrected in the first servicing mission of 1993.

   1.2   Explain the problem of spherical aberration.

(1 mark)

   1.3   Describe how spherical aberration can be avoided.

(1 mark)

   1.4   The HST can observe light with wavelengths in the infrared, ultraviolet and visible regions of the electromagnetic spectrum. Explain why HST would be unable to observe the same radiation if it was based on Earth.

(1 mark)

   1.5   The HST is one of many telescopes that records images using charge-coupled devices (CCDs). Discuss the advantages and disadvantages of using a CCD as a detector instead of the human eye.

(6 marks)

2   Mu Cephei is a red giant star in the constellation Cepheus. Analysis of the radiation emitted by Mu Cephei gives the value of its peak wavelength as λ_{max} = 828.6 nm. The intensity of the radiation detected was 5.1×10^{-9} W m^{-2}.

   2.1   Explain what is meant by a black body.

(1 mark)

   2.2   Estimate the surface temperature of Mu Cephei in kelvin, stating any assumptions made.

(3 marks)

        Mu Cephei is estimated to have a radius of around 1500 solar radii. The radius of the Sun is 6.96×10^8 m.

   2.3   Calculate the power output of Mu Cephei in watts.

(3 marks)

   2.4   Calculate the distance to Mu Cephei in light years.

(2 marks)

   2.5   Suggest one reason why this value may be inaccurate.

(1 mark)

3 The properties of some stars in the constellation Canis Major are given in the table.

Name	Apparent magnitude	Absolute magnitude	Spectral class
Adhara	1.50	−4.11	B
Aludra	2.45	−7.51	B
Omicron[1]	3.89	−5.02	K
Sirius A	−1.44	1.45	A
Wezen	1.83	−6.87	F

3.1 Which of these stars appears dimmest?

(1 mark)

3.2 Define the term absolute magnitude.

(1 mark)

3.3 Which of these stars is closer than 10 parsecs to Earth? Explain your answer.

(2 marks)

3.4 Calculate the distance to Wezen in parsecs.

(2 marks)

3.5 Describe the composition and colour of Wezen, and the main absorption lines that you would expect to see in its line spectrum.

(3 marks)

4 The VLA (Very Large Array) is an array of radio telescopes in New Mexico. The array combines the power of 27 radio telescopes, each with a 25 m diameter dish. Its maximum resolving power is equivalent to that of a giant radio telescope with a diameter of 36 km.

4.1 Calculate the minimum angular resolution of a single 25 m radio telescope when observing point sources of electromagnetic radiation of wavelength 8.5×10^{-3} m.

(2 marks)

4.2 Calculate how many times smaller the minimum angular resolution of the VLA is than a single 25 m dish when observing point radio waves with a wavelength of 8.5×10^{-3} m.

(2 marks)

4.3 In 2011, the VLA detected object A, a strong radio source with a red shift of 5.95. Explain why the formula for red shift cannot be used to calculate the recessional velocity of this object.

(2 marks)

4.4 Suggest what type of object could have been observed and explain your answer.

(2 marks)

5 Menkalinan is a spectroscopic eclipsing binary system 82 light years away.
It consists of two almost identical stars of spectral class A and its orbital plane is
close to our line of sight.

Astronomers calculated the distance to Menkalinan by measuring its
angle of parallax.

5.1 Explain what is meant by the angle of parallax.

(1 mark)

5.2 Explain why parallax can't be used to measure the distance to all stars.

(1 mark)

A graph of apparent magnitude against time for Menkalinan
is shown in Figure 1.

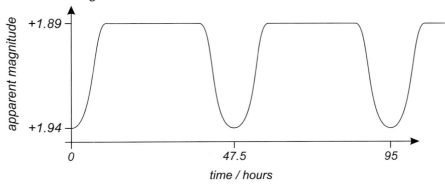

Figure 1: *Light curve for Menkalinan.*

5.3 What can you say about the relative brightness of the two stars?

(1 mark)

5.4 Write down the orbital period of Menkalinan.

(1 mark)

5.5 The spectrum from one of the stars is studied and an absorption line is seen to
fluctuate between 434.026 nm and 434.076 nm.
Explain why the wavelength of the light from the star is observed to fluctuate.

(1 mark)

5.6 Show that the orbital velocity of the star is approximately 17 km s⁻¹.

(3 marks)

5.7 The two stars in Menkalinan are classified as subgiants. Subgiants are thought
to be stars that have stopped fusing hydrogen in their cores and are contracting,
heating up, and beginning to fuse hydrogen in a shell surrounding the core.
Describe the next stage in the stars' evolution, describing any reactions in the stars
that would take place as part of this evolution.

(2 marks)

6 The diagram below shows a set of axes for a Hertzsprung-Russell (H-R) diagram.

Temperature (non-linear scale) / Spectral class

6.1 Copy and complete the diagram, marking the main sequence stars, white dwarfs, red giants and supergiants and any missing axis labels.

(3 marks)

6.2 Star X has a surface temperature of around 25 000 K and an absolute magnitude of –6.39. Plot Star X on your H-R diagram.

(1 mark)

The mass of Star X is around 30 times that of the Sun.
It will eventually become a black hole.

6.3 Explain how Star X will evolve from its red supergiant phase to become a black hole.

(3 marks)

6.4 What is meant by the Schwarzschild radius of a black hole, R_s?

(1 mark)

7 It used to be thought that the universe had always been infinite in both space and time and was static. This is now widely discredited in favour of the Big Bang theory.

7.1 Describe how the universe began according to the Big Bang theory.

(1 mark)

Hubble observed that the recessional velocity of distant objects was proportional to their distance from Earth. This observation led to Hubble's law:

$$v = Hd$$

7.2 Explain how Hubble's law supports the Big Bang theory.

(1 mark)

7.3 Assuming that $H = 65$ kms^{-1}Mpc^{-1}, estimate the age of the universe in years using Hubble's law. (1 Mpc = 3.08×10^{22} m).

(3 marks)

7.4 Red-shift data is only one piece of evidence supporting the Big Bang theory. Describe the other main pieces of evidence in support of the Big Bang theory and explain how they support the theory.

(4 marks)

1. Lenses

Learning Objectives:

- Know the properties of converging and diverging lenses.

- Understand what is meant by the principal focus and focal length of a lens.

- Know how to draw a ray diagram for an image formed by a converging or diverging lens.

  Specification Reference 3.10.1.2

There are two types of lenses that you need to know about — converging and diverging lenses. They have different shapes and different properties.

Types of lenses

Lenses change the direction of light rays by refraction — the change in direction of light as it enters a different medium. There are two main types of lens — converging and diverging. Both have different effects on light...

Converging lenses

Converging lenses are convex in shape — they bulge outwards. Converging lenses bring parallel light rays together (see Figure 1).

The **principal axis** of a lens is a straight line that passes through the centre of the lens, perpendicular to its surface on both sides. Rays parallel to the principal axis of the lens converge onto a point in front of the lens on the axis called the **principal focus**. Rays that are parallel to each other but not to the principal axis converge somewhere else on the **focal plane** — the plane perpendicular to the principal axis of the lens on which the principal focus lies. The **focal length**, f, is the distance between the lens axis and the focal plane — f is positive for a converging lens because the focal point is in front of the lens.

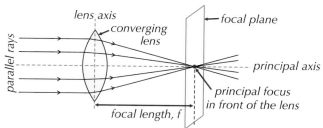

Figure 1: *Parallel rays of light are brought together by a converging lens at its principal focus.*

Diverging lenses

Diverging lenses are concave in shape — they cave inwards. They cause parallel rays of light to diverge (spread out). The principal focus of a diverging lens is at a point behind the lens. The principal focus is the point that rays from a distant object, assumed to be parallel to the principal axis, appear to have come from. The focal length, f, is the distance between the lens axis and the principal focus. f is negative for a diverging lens because it is behind the lens.

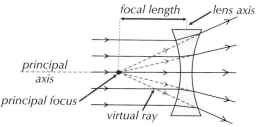

Figure 3: *Parallel rays of light are spread out by a diverging lens.*

Tip: A light ray travelling along the principal axis won't be refracted — it will just carry on in a straight line.

Tip: Refraction is covered in year 1 of A-level.

Figure 2: *A diverging lens causes parallel rays of light to diverge.*

Ray diagrams

When light rays from an object pass through a lens, an **image** is formed where the rays meet. To work out where an image will appear, you can draw a **ray diagram**. A ray diagram shows how light rays travel from the object through a lens. Two rays coming from one point on an object will meet where that part of the image is formed.

For objects sitting on the principal axis, you only need to draw two rays to work out where the image will be: one passing through the centre of the lens that doesn't get refracted at all, and one initially parallel to the principal axis that passes through the principal focus (see previous page). The principal focus for a converging lens is on the other side of the lens to the object, but the principal focus for a diverging lens is on the same side as the object (see Figures 1 and 3), so the methods of drawing the ray diagrams are slightly different.

Tip: For ray diagrams where the object is not sitting on the principal axis, see page 262.

Tip: All rays parallel to the principal axis pass through, or appear to have come from, a principal focus — see Figures 1 and 3.

Tip: Remember — the principal focus is one focal length away from the lens axis.

Tip: In reality each refracted ray will refract twice, as it enters and leaves the lens — but in ray diagrams you can just draw it as one refraction at the lens axis.

Tip: Make sure you add arrows to your rays when drawing ray diagrams.

Converging lenses

When drawing light being refracted by a converging lens, there are two things to remember:

- An incident ray parallel to the principal axis refracts through the lens and passes through the principal focus on the other side.
- An incident ray passing through the centre of the lens carries on in the same direction.

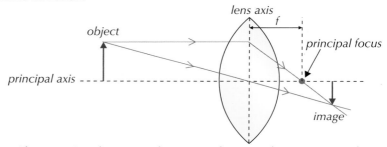

Figure 4: Ray diagram to show image formation by a converging lens.

Diverging lenses

When drawing light being refracted by a diverging lens, remember:

- An incident ray parallel to the principal axis refracts through the lens and appears to have come from the principal focus.
- An incident ray passing through the centre of the lens carries on in the same direction.

Tip: In Figure 5, the dashed orange ray coming from the image is a virtual ray — it <u>appears</u> to be coming from that point, but isn't really.

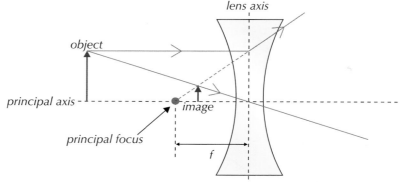

Figure 5: Ray diagram to show image formation by a diverging lens — the image is formed where the virtual ray and the central ray cross.

Real and virtual images

Lenses can produce real or virtual images. A **real image** is formed when light rays from an object are made to pass through another point in space. The light rays are actually there, and the image can be captured on a screen.

A **virtual image** is formed when light rays from an object appear to have come from another point in space. The light rays aren't really where the image appears to be, so the image can't be captured on a screen.

You can tell if an image is real or virtual by drawing a ray diagram (see previous page). If the image is formed on the other side of the lens to the object — the image is real. If the image is formed on the same side of the lens as the object — the image is virtual. Again, there are different rules for converging and diverging lenses.

Figure 6: *Plane mirrors also produce virtual images — here the image of the unlit candle appears to come from behind the mirror.*

Converging lenses

Converging lenses can form both real and virtual images, depending on where the object is.

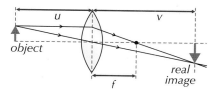 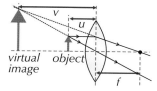

Figure 7: *A ray diagram showing a converging lens forming a real image (left) and a virtual image (right).*

In Figure 7, u = distance between object and lens axis (always positive), v = distance between image and lens axis (positive if image is real, negative if image is virtual) and f = focal length. The focal length of a converging lens is always positive and a diverging lens always has a negative focal length. The values u, v and f are related (see page 264).

The distance of the object from the lens axis, u, in relation to the focal length, f, affects the image formed — whether it is real or virtual, its orientation, its size and its position (see Figures 8 and 9).

u	Image type	Image orientation	Image size	v
beyond 2f	real	inverted	smaller than object	between f and 2f
At 2f	real	inverted	same as object	2f
between 2f and f	real	inverted	bigger than object	beyond 2f
between f and 0 (the lens axis)	virtual	right way up	bigger than object	negative

Figure 8: *Table showing the properties of an image formed by a converging lens when the object is at various positions.*

Diverging lenses

A diverging lens always produces a virtual image, so v is always negative. The image is the right way up, smaller than the object and on the same side of the lens as the object — no matter where the object is (see Figure 5 on the previous page).

Tip: An image that's formed on the same side of the lens as the object will have a negative v. Diverging lenses always have a negative focal length, f, because the rays appear to go through the focal point of the lens on the same side as the object.

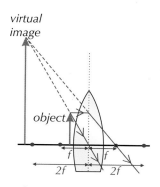

Figure 9: *An object nearer to a converging lens than the focal length will produce a virtual image, the right way up, bigger than the object and on the same side of the lens.*

Objects off the principal axis

Tip: You can do this with any part of an object — if you draw two rays coming from one part of the object, they'll converge at the same point on the image... but it's usually most useful to find the top and bottom so that you can see the image's size.

When the bottom of an object is on the principal axis, the bottom of the image is also on the axis. But if the bottom of the object isn't on the principal axis, you need to find where the bottom of the image is too. You just need to repeat the process you followed on page 260 of drawing two rays from the top of the object to find the position of the top of the image. Only this time, you draw the rays from the bottom of the object to find the bottom of the image.

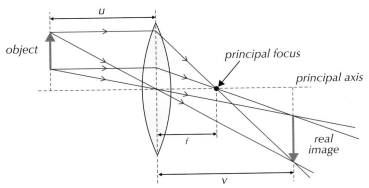

Figure 10: *A ray diagram showing where the image of an object positioned off the principal axis will be formed.*

Practice Questions — Application

Tip: Use something simple as the object in Q1 — like a vertical arrow.

Q1 An object is placed 3 m from a diverging lens on its principal axis. The lens has a focal length of −1.5 m. Draw a ray diagram to show the image formed.

Q2 Copy and complete the ray diagram below to show where the image of the object is formed. State whether the image is real or virtual.

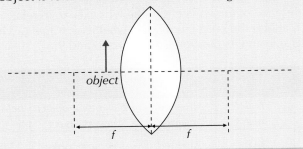

Practice Questions — Fact Recall

Q1 Describe the different properties of converging and diverging lenses.

Q2 What is the principal focus of a lens?

Q3 Describe how to use a ray diagram to find where an image is formed by a converging lens.

Q4 Describe the main properties of real and virtual images.

2. Calculations with Lenses

Apart from ray diagrams, there are some equations you can use to work out the position of an image, as well as its magnification and the lens power.

Lens equation

The focal length of a lens, f, the distance between the object and the lens axis, u, and the distance between the image and the lens axis, v, are linked by the lens equation. This equation can be derived by looking at the ray diagram in Figure 1 for an image formed by a converging lens.

The two triangles that are shaded in Figure 1 each have the same angle θ and a right angle — they are similar triangles.

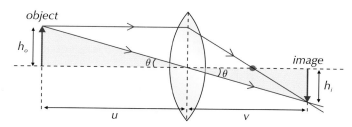

Figure 1: *A ray diagram of a real image formed by a converging lens. The shaded triangles are similar.*

Since the triangles are similar, the ratio of corresponding sides is always the same, so you can write:

$$\frac{h_o}{h_i} = \frac{u}{v} \quad \text{which rearranges to} \quad \boxed{\frac{h_o}{u} = \frac{h_i}{v}}$$

where h_o and h_i are the heights of the object and the image respectively.

There are two other triangles within the ray diagram that are similar, shown by the shaded triangles in Figure 2.

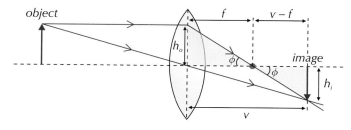

Figure 2: *A ray diagram of a real image formed by a converging lens. The shaded triangles are similar.*

Using the same method as above, you can write the equation:

$$\frac{h_o}{h_i} = \frac{f}{v-f} \quad \text{which rearranges to} \quad \frac{h_o}{f} = \frac{h_i}{v-f}$$

This rearranges to give the equation $h_i = \frac{h_o(v-f)}{f}$ which can then be substituted into the equation in the purple box above to give:

$$\frac{h_o}{u} = \frac{\left(\frac{h_o(v-f)}{f}\right)}{v} = \frac{h_o(v-f)}{fv}$$

Learning Objectives:

- Be able to use the lens equation: $\frac{1}{u} + \frac{1}{v} = \frac{1}{f}$.
- Know how to calculate linear magnification using $m = \frac{v}{u}$.
- Understand what is meant by the power of a lens and be able to calculate it, in dioptres, using: power $= \frac{1}{f}$.

Specification Reference 3.10.1.2

Tip: Similar triangles have corresponding angles that are equal and corresponding sides that are in the same proportion — i.e. they're the same shape but potentially different sizes.

Tip: You don't need to know this derivation for the lens equation — this is just to show you where the equation has come from.

The h_o on both sides cancels out, and the right-hand side can be written as two separate fractions to give:

$$\frac{1}{u} = \frac{v}{fv} - \frac{f}{fv}$$

Cancelling the fractions down and rearranging gives the lens equation:

u = distance between object and lens axis in m

$$\frac{1}{u} + \frac{1}{v} = \frac{1}{f}$$

f = focal length in m

v = distance between image and lens axis in m

Tip: You don't have to have u, v and f in metres to use this equation, but they do all need to be in the <u>same</u> distance units.

Tip: Remember... v is positive if the image is real and negative if the image is virtual. f is positive if the lens is converging and negative if the lens is diverging.

Tip: Diverging lenses always produce images between the lens and the object, so v is always a smaller distance than u (see page 260).

Tip: Don't forget that diverging lenses always produce virtual images, behind the lens — that's why v here is negative (see page 261).

⌐ **Example** ── Maths Skills ───────────────

An object is placed 0.25 m from a diverging lens with a focal length of –0.40 m. Calculate the distance from the lens to the image formed.

To find the image's distance from the lens, v, use the lens equation:

$$\frac{1}{u} + \frac{1}{v} = \frac{1}{f}$$

$f = -0.40$ m and $u = 0.25$ m, so:

$$\frac{1}{0.25} + \frac{1}{v} = \frac{1}{-0.40}$$

$$4 + \frac{1}{v} = -2.5$$

$$\frac{1}{v} = -6.5$$

$$v = \frac{1}{-6.5} = -0.15\,\text{m (to 2 s.f.)}$$

So the image is formed 0.15 m away from the lens, on the same side as the object (shown by the negative sign in the answer).

Linear magnification

A lens can be used to magnify an image — this could be an increase or a decrease in the size of an image compared to the object. The linear magnification is just the ratio of the height of the image to the height of the object. Using the two similar triangles shown in Figure 1 on the previous page, you get:

$$\frac{h_i}{h_o} = \frac{v}{u}$$

So the linear magnification produced by a lens can be calculated using the equation:

Tip: Linear magnification is just a number — it has no units. If the image is smaller than the object, the magnification will be less than 1. If the image is on the same side as the object (virtual), the magnification will be negative.

m = linear magnification

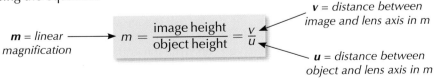

$$m = \frac{\text{image height}}{\text{object height}} = \frac{v}{u}$$

v = distance between image and lens axis in m

u = distance between object and lens axis in m

⌐ **Example** ── Maths Skills ───────────────

An object is placed 1.20 m from a lens. The lens produces an image at a distance of 1.75 m on the other side of the lens. Find the magnification.

$u = 1.20$ and $v = 1.75$, so $m = \frac{1.75}{1.20} = 1.4583... = 1.46$ (to 3 s.f.)

Power

The more powerful a lens is, the shorter its focal length. The power of a lens is the reciprocal of its focal length. So it can be calculated using:

$$P = \frac{1}{f}$$

P = lens power in dioptres, D

f = focal length in m

A more powerful (thicker) lens converges the rays more strongly and will have a shorter focal length. Diverging lenses have a negative power because they have a negative focal length.

Example — Maths Skills

A lens has a focal length of 2.5 cm. Find the power of the lens.

First, convert the focal length from cm to m: 2.5 cm = 0.025 m.

So: $P = \frac{1}{f} = \frac{1}{0.025} = 40\,D$

Practice Questions — Application

Q1 A lens is used to magnify an object of height 0.20 m. The height of the image is 0.36 m. Find the linear magnification produced.

Q2 The power of a lens is 2.4 D. Find the focal length of the lens.

Q3 An object is placed 65 cm from a lens. An image is formed 105 cm from the lens on the opposite side to the object. Find f for the lens.

Practice Questions — Fact Recall

Q1 What equation links object distance, image distance and focal length?

Q2 What is the equation that links the power of a lens and its focal length?

Tip: The cornea has a high refractive index compared to air, which causes light entering the eye to be highly refracted.

Tip: Remember, the shape of a lens affects its focusing power (see page 265).

Tip: The image formed on the retina is upside down, but it's interpreted by the brain to seem the right way up.

3. Physics of the Eye

Our eyes contain converging lenses which are used to focus light onto the retina at the back of the eye. There are, however, many different parts of the eye that all play an important role in helping us to see...

The structure of the eye

Our eyes focus light to form images that can be detected and allow us to see. There are many different parts that make up the eye which enable us to do this (see Figure 1).

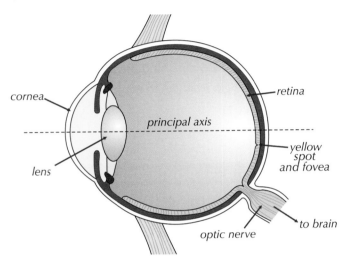

Figure 1: *The basic structure of an eye.*

The **cornea** is a transparent 'window' with a convex shape and a high refractive index. The cornea does most of the eye's focusing.

The **lens** acts as a fine focus and is controlled by muscles. When the muscles contract, tension is released and the lens takes on a fat, more spherical shape. When they relax, the lens is pulled into a thin, flatter shape. This changes the focal length of the eye.

Images are formed on the **retina**, which contains light-sensitive cells called **rods** and **cones** (see page 267). The **yellow spot** is a particularly sensitive region of the retina. In the centre of the yellow spot is the **fovea**. This is the part of the retina with the highest concentration of cones. The optic nerve carries signals from the rods and cones to the brain.

The eye as an optical refracting system

The eye can focus light from a range of distances. The **far point** is the furthest distance that the eye can focus comfortably — for normally sighted people that's infinity. When your eyes are focusing at the far point, they're unaccommodated. The **near point** is the closest distance that the eye can focus on. For young people it's about 9 cm.

You can add together the powers of the cornea, lens and other parts of the eye. That means you can think of the eye as a single converging lens of power 59 D at the far point. This gives a focal length of 1.7 cm.

When looking at nearer objects, the eye's power increases, as the lens changes shape and the focal length decreases — but the distance between the lens and the image, v, stays the same, at 1.7 cm (see Figure 2 below).

Tip: See page 265 for a recap on the power of a lens.

Tip: Don't worry about learning all these numbers, but make sure you know how the power of the lens changes depending on the position of the object, in order to focus.

Example

This ray diagram shows the eye focusing on an object. Rays from the top and bottom of the object show where the top and bottom of the image is formed. The focal length tells you the distance the object is at — you can calculate it using the lens equation (see page 264).

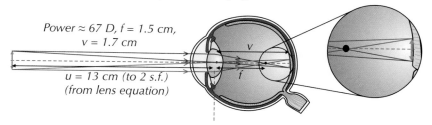

Power ≈ 67 D, f = 1.5 cm,
v = 1.7 cm

u = 13 cm (to 2 s.f.)
(from lens equation)

Figure 2: Diagram showing image formation of a nearby object by the eye.

Tip: Don't get confused by all the extra eye bits — this is just a ray diagram for a lens. Remember... if an object does not sit on the principal axis, you need to draw two sets of rays to find the top and bottom of the image — see page 262 for more on ray diagrams like this.

Rods and cones

Sensitivity of the eye

Rods and cones are cells at the back of the retina that respond to light. They're called **photoreceptors** as they detect light and convert it to an electrical signal.

Light travels through the retina to the rods and cones at the back. Rods and cones all contain chemical pigments that bleach when light falls on them. This bleaching stimulates the cell to send signals to the brain via the optic nerve. The cell is said to be active or stimulated when this happens. The cells are reset (i.e. unbleached) by enzymes using vitamin A from the blood. Rods are more sensitive to light, but don't detect colour — we rely on rods to see in darker conditions. There's only one type of rod but there are three types of cone, which are sensitive to red, green and blue light.

Tip: If the eye is focused, the image is formed on the retina — if not it will appear blurry since the image is formed somewhere else (see vision defects — pages 270-272).

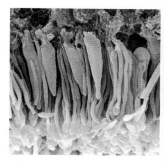

Figure 3: A coloured image of retina rods and cones taken with a scanning electron microscope. The rod cells are shown in white and the cone cells in green.

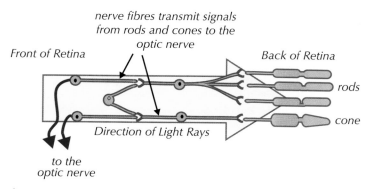

nerve fibres transmit signals from rods and cones to the optic nerve

Front of Retina

Back of Retina

rods

cone

Direction of Light Rays

to the optic nerve

Figure 4: Light travels to the rods and cones at the back of the retina, which then send electrical signals to the brain via the optic nerve.

The red, green and blue cones each absorb a range of wavelengths (see Figure 5). The eye is less responsive to blue light than to red or green, so blues often look dimmer. The brain processes signals from the three types of cone and interprets their weighted relative strengths as colour — see Figure 5.

Tip: The eye is a photodetector, meaning it detects light.

Exam Tip
If you're asked to sketch three curves to show the spectral response of the eye as a photodetector, it's the graph on the left you're being asked to draw.

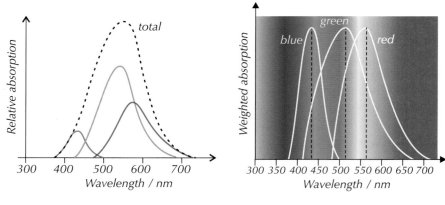

Figure 5: *Graph to show relative absorption of different wavelengths of light by the cones (left) and the weighted absorption of light against wavelength (right).*

Any colour can be produced by combining different intensities of red, green and blue light.

Example

Yellow light produces almost equal responses from the red and green cones. Yellow light can therefore be 'faked' by combining red and green light of almost equal intensity — the electrical signal from the retina will be the same and the brain interprets it as 'yellow'.

Spatial resolution

Tip: You may have heard resolution talked about in terms of image resolution, e.g. of an image taken with a digital camera. The greater the resolution, the greater the detail in the image.

In order to see something in detail, you need good **spatial resolution** — this is a measure of the ability to form separate images of objects that are close together.

Two objects can only be distinguished from each other if there's at least one rod or cone between the light from each of them. This is true as long as the rod or cone between the image doesn't share an optic nerve with any of the rods and cones detecting the images. Otherwise the brain can't resolve the two objects and it 'sees' them as one — see Figure 6.

Example

In Figure 6, each square represents a cone and each blue square represents a stimulated cone (that is detecting light).

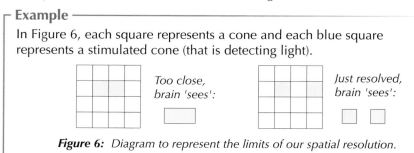

Figure 6: *Diagram to represent the limits of our spatial resolution.*

Spatial resolution is best at the yellow spot (see p.266) — the cones are very densely packed here and each cone always has its own nerve fibre, which means that the signals sent are more detailed. There are no rods in the yellow spot, though. This means that in dim light, when cones don't work well, resolution is best slightly off the direct line of sight, where the rods are more densely packed.

Away from the yellow spot, resolution is much worse. The light-sensitive cells are not as densely packed and the rods share nerve fibres, meaning that the signals from them are much less detailed — there are up to 600 rods per fibre at the edges of the retina.

Tip: The fovea, within the yellow spot, is the part of the retina which gives the very best spatial resolution. It is responsible for our central sharpest vision that we use for reading, watching TV and driving.

Practice Questions — Fact Recall

Q1 Describe how images are focused by the eye.

Q2 a) What is meant by the far point of the eye?

b) What is meant by the near point of the eye?

Q3 a) Where are rods and cones found in the eye?

b) How many types of rods and cones are there?

Q4 Sketch a graph to show the spectral response of the eye as a photodetector at different wavelengths.
Include extra curves to show the response of each separate cone type.

Q5 a) What is meant by spatial resolution?

b) What is required for an eye to detect two objects as being separate?

Tip: There's more about near points and far points on pages 266-267.

- Know what is meant by the vision defects myopia and hypermetropia and be able to draw ray diagrams to show how correction lenses work.

- Know how to calculate the power (in dioptres) of a corrective lens for myopia or hypermetropia.

- Know what is meant by the vision defect astigmatism and know the format of a corrective lens prescription for astigmatism.

    Specification Reference 3.10.1.2

4. Defects of Vision

Converging lenses and diverging lenses can both be used to help correct problems people have with their vision. Whether a converging or a diverging lens is used depends on the condition the person has...

Myopia

Short-sighted (myopic) people are unable to focus on distant objects — this happens if their far point is closer than infinity (see page 266).

   Myopia occurs when the cornea and lens are too powerful or the eyeball is too long. The focusing system is too powerful and images of distant objects are brought into focus in front of the retina. A lens of negative power is needed to correct this defect, to reduce the overall power of the system — so a diverging lens is placed in front of the eye (see Figure 1).

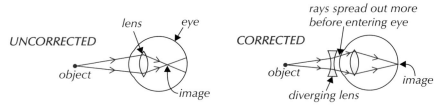

Figure 1: *Ray diagrams to show where an image is formed for a person with uncorrected myopia (left) and for corrected myopia (right).*

Tip: Light is actually refracted in stages by different parts of the eye, but for simplicity we just show the total refraction at the lens. You might be asked to assume it all refracts at the cornea instead.

   Choosing a lens to correct for short sight (myopia) depends on the far point. To correct for short sight, a diverging lens is chosen which has its principal focus at the eye's faulty far point. The principal focus is the point that rays from a distant object appear to have come from (see page 259).

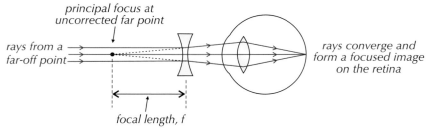

Figure 2: *A ray diagram to show how a diverging lens can be used to correct short sight — the lens has its principal focus at the eye's faulty far point.*

Tip: As well as correcting the far point, the diverging lens also makes the near point a little further away than it was. This isn't usually a problem — short-sighted people usually have a near point that is closer than normal anyway.

   The lens must have a negative focal length which is the same as the distance to the eye's far point. This means that objects at infinity, which were out of focus, now seem to be in focus at the far point.

Tip: Diverging lenses always have a negative focal length (p.261).

Tip: The power of a lens to correct for short sight is always negative.

┌─ **Example** ── **Maths Skills** ────────────────────

A short-sighted person has a far point of 5 m.
Calculate the power of lens needed to correct their vision.

Focal length, f = far point = −5 m

Power needed = $\frac{1}{f} = \frac{1}{-5}$ = −0.2 D

Hypermetropia

Long-sighted (hypermetropic) people are unable to focus on near objects. This happens if their near point is further away than normal (25 cm or more).

Hypermetropia occurs because the cornea and lens are too weak or the eyeball is too short. The focusing system is too weak and images of near objects are brought into focus behind the retina. A lens of positive power is needed to correct the defect, to increase the overall power of the system — so a converging lens is placed in front of the eye (see Figure 4).

Tip: Long-sightedness is common among young children whose lenses have grown quicker than their eyeballs.

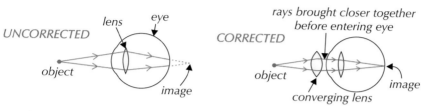

Figure 4: Ray diagrams to show where an image is formed for a person with uncorrected hypermetropia (left) and for corrected hypermetropia (right).

Figure 3: Image to show how a long-sighted person would see this scene. They would be unable to focus on the statue, but able to focus on things further away.

People with hypermetropia have a near point which is too far away. An 'acceptable' near point is 25 cm. A converging lens is used to produce a virtual image of objects that are 0.25 m away at the eye's (uncorrected) near point (see Figure 5). This means that close objects, which were out of focus, now seem to be in focus at the near point.

Exam Tip
You might see lenses drawn as a single straight line, instead of a lens shape, in the exam.

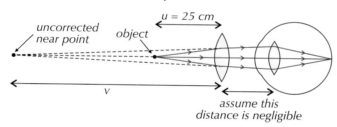

Figure 5: A ray diagram to show how a converging lens can be used to correct long sight — the lens produces a virtual image of objects 25 cm away at the patient's uncorrected near point.

You can work out the focal length, and hence the power of lens needed, using the lens equation $\frac{1}{u} + \frac{1}{v} = \frac{1}{f}$ (see p.264). The space between the eye lens and the correcting lens does affect the overall power of the system, but the effect is usually just a fraction of a dioptre if the lens is close to your eye and is unnoticeable to most patients.

Figure 6: People with hypermetropia may need to wear glasses for reading.

Tip: The distance between the eye lens and the correcting lens for myopia also affects the overall power of the system. Again, the effect is usually negligible.

Example — `Maths Skills`

A long-sighted person has a near point of 5 m.
What power of lens is needed to correct this?

$u = 0.25\,\text{m}$, $v = -5\,\text{m}$ (v is negative because the image is on the same side of the lens as the object — i.e. virtual)

$$\frac{1}{f} = \frac{1}{0.25} - \frac{1}{5} = 3.8$$

So the lens power needed = +3.8 D

Tip: The power of a lens used to correct for long sight is always positive.

Astigmatism

Astigmatism is caused by an irregularly shaped cornea or lens which has different focal lengths for different planes. For instance, when vertical lines are in focus, horizontal lines might not be.

The condition is corrected with a cylindrical lens that adds power in one plane, but not the plane perpendicular to it. The prescription for the cylindrical lens will state the power needed to correct long or short-sightedness (known as the sphere, SPH), as well as the additional power needed to correct the astigmatism (know as the cylinder, CYL) and the angle to the horizontal of the plane that does not need correcting for astigmatism (known as the axis, an angle between 0° and 180°).

Tip: "Format of a prescription" sounds a bit posh, but it just means what an optician would prescribe for you if you needed some correction for your eyesight, e.g. glasses.

┌─ **Example** ─────────────────────────────

The cylindrical lens below has an axis angle of 90°. Astigmatism is not corrected in the plane 90° to the horizontal.

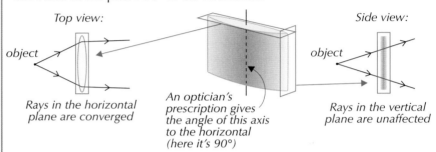

Top view:
object
Rays in the horizontal plane are converged

An optician's prescription gives the angle of this axis to the horizontal (here it's 90°)

Side view:
object
Rays in the vertical plane are unaffected

Figure 7: *A diagram showing what is meant by the axis angle (part of the format of a prescription for astigmatism).*

Practice Questions — Application

Q1 a) Copy and complete the ray diagram below to show where an image of an object 25 m away is formed in the eye of a person with a far point of 15 m.

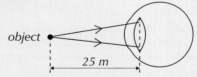

object

25 m

   b) What defect of vision does this person suffer from?

   c) Explain how an external lens could be used to correct this defect.

Q2 A woman is short-sighted and has a far point of 4.2 m.
   Calculate the power of lens needed to correct her far point.

Q3 A correcting lens for hypermetropia has a focal length of 0.27 m.
   Calculate the patient's uncorrected near point.

Tip: It might not seem clear at first how to answer Q2. First think about what equation you'll need to use to calculate the power.

Practice Questions — Fact Recall

Q1 What is meant by myopia?

Q2 a) What is meant by hypermetropia?

   b) What type of lens can be used to correct it?

Q3 a) What does it mean if someone suffers from astigmatism?

   b) What extra information would you see on the prescription of someone who suffers from astigmatism?

5. Physics of the Ear

Our ears transfer sound energy into electrical signals which the brain then interprets as sound. You need to know the roles of different parts of the ear...

Learning Objectives:

- Know the definition of intensity of sound.
- Know the simple structure of the ear.
- Be able to describe the transmission processes involved in the ear detecting sound.

Specification References 3.10.2.1 and 3.10.2.2

Intensity of sound

The **intensity** of a sound wave is defined as the amount of sound energy that passes per second per unit area (perpendicular to the direction of the wave). That's power per unit area. If the sound energy transferred per second is P, then the intensity of the sound is:

I = intensity in Wm^{-2} ⟶ $$I = \frac{P}{A}$$ ⟵ P = power in W

A = area through which the sound waves pass in m^2

The SI derived unit of intensity is Wm^{-2}, but you'll often see decibels used instead (see page 276). For any wave, intensity $\propto$ amplitude^2 — so doubling the amplitude will result in four times the intensity. Intensity is related to the loudness of sound (see page 275).

Tip: The amplitude of a wave is its maximum displacement from its rest position:

> ### Example — Maths Skills
>
> **A fire alarm sounds with a power of 4.5 W. Calculate the intensity of sound a person standing 3.0 m away would hear, assuming the sound waves spread equally in all directions.**
>
> Surface area of a sphere = $4\pi r^2$ so $A = 4\pi \times 3.0^2 = 36\pi$
>
> Intensity = $I = \frac{P}{A} = \frac{4.5}{36\pi} = 0.03978...Wm^{-2} = 0.040\,Wm^{-2}$ (to 2 s.f.)

The structure of the ear

The ear consists of three sections:

- the outer ear (pinna and auditory canal)
- the middle ear (**ossicles** and **Eustachian tube**)
- the inner ear (semicircular canals, **cochlea** and **auditory nerve**)

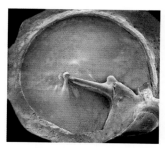

***Figure 1:** A coloured scanning electron micrograph of an eardrum, shown in red.*

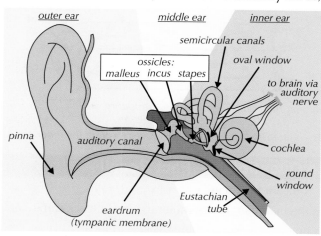

***Figure 2:** The basic structure of the ear.*

The **tympanic membrane** (eardrum) separates the outer and middle ears. Although separated, the outer and middle ears both contain air at atmospheric pressure, apart from slight pressure variations due to sound waves.

Tip: The pressure in the middle ear is maintained at atmospheric pressure by yawning and swallowing — the middle ear is opened up to the outside via the Eustachian tube (which is connected to the throat).

The oval and round windows separate the middle and inner ears. The inner ear is filled with fluid called perilymph (or endolymph in the cochlear duct). This fluid allows vibrations to pass to the basilar membrane in the cochlea. The semicircular canals are involved with maintaining balance.

The transmission process

The pinna (external ear) acts like a funnel, channelling the longitudinal sound waves into the auditory canal. The sound energy is now concentrated onto a smaller area, which increases the intensity. The sound waves consist of variations in air pressure, which force the tympanic membrane (eardrum) to vibrate.

The tympanic membrane is connected to the malleus — one of the three tiny bones (ossicles) in the middle ear. The three bones act as a lever system. The malleus passes the vibrations of the eardrum on to the incus and the stapes (which is connected to the oval window).

As well as transmitting vibrations, the ossicles have other functions, such as amplifying the force of the vibrations and reducing the energy reflected back from the inner ear. The ossicles amplify the force of the vibrations by around 50% — so the force of the vibrations is multiplied by about 1.5. The oval window has a much smaller area than the tympanic membrane. Together with the increased force produced by the ossicles, this results in greater pressure variations at the oval window — the pressure variations at the oval window are about 20 times greater than those at the eardrum. The oval window transmits vibrations to the fluid in the inner ear (see Figure 3).

Pressure waves in the fluid of the cochlea make the **basilar membrane** vibrate. Different regions of this membrane have different natural frequencies, from 20 000 Hz near the middle ear to 20 Hz at the other end. When a sound wave of a particular frequency enters the inner ear, one part of the basilar membrane resonates and so vibrates with a large amplitude. Hair cells attached to the basilar membrane trigger nerve impulses at this point of greatest vibration. These electrical impulses are sent, via the auditory nerve, to the brain, where they are interpreted as sounds.

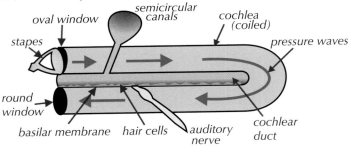

Figure 3: *The transmission of pressure waves through the inner ear (unrolled).*

Tip: Remember, $I = \frac{P}{A}$.

Tip: You might see these vibrations referred to as mechanical vibrations — it just means the same thing.

Tip: Malleus, incus and stapes are the Latin words for hammer, anvil and stirrup — this might help you to remember which one is which.

Practice Questions — Fact Recall

Q1 What is meant by the intensity of sound?

Q2 Name the main parts of the outer ear, middle ear and inner ear.

Q3 Describe the main roles of the ossicles in the ear.

Q4 How are the pressure variations at the oval window different to those at the eardrum?

Q5 Describe how sound travels through and is detected by the ear.

6. Intensity and Loudness

How loud we perceive a sound to be depends on the intensity of the sound and its frequency. Our ears can be damaged if they're exposed to an excessive amount of noise. They also naturally deteriorate as we get older.

Range of hearing

Humans can hear a limited range of frequencies. Young people can hear frequencies ranging from about 20 Hz (low pitch) up to 20 000 Hz (high pitch). As you get older, the upper limit decreases.

Our ability to discriminate between frequencies depends on how high that frequency is. For example, between 60 and 1000 Hz, you can hear frequencies 3 Hz apart as different pitches. At higher frequencies, a greater difference is needed for frequencies to be distinguished. Above 10 000 Hz, pitch can hardly be discriminated at all.

The loudness of sound you hear depends on the intensity (see p.273) and frequency of the sound waves. The weakest intensity you can hear depends on the frequency of the sound wave. The ear is most sensitive at around 3000 Hz. For any given intensity, sounds of this frequency will be loudest.

Humans can hear sounds at intensities ranging from about 10^{-12} Wm^{-2} to 100 Wm^{-2}. Sounds over a level called the threshold of feeling, equal to about 1 Wm^{-2}, can be felt.

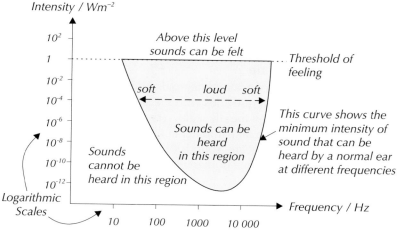

Figure 1: *A graph to show the range of human hearing.*

Perceived loudness

The perceived loudness of a sound depends on its intensity (and its frequency — see above). However, the relationship between perceived loudness and the intensity is not linear, but logarithmic:

ΔL = increase in decibels, dB $\Delta L \propto \log\left(\frac{I_2}{I_1}\right)$ I_2 = new intensity in Wm^{-2}
I_1 = original intensity in Wm^{-2}

This means that loudness, L, goes up in equal intervals if intensity, I, increases by a constant factor (provided the frequency of the sound doesn't change). E.g. if you double the intensity, double it again and so on, the loudness keeps going up in equal steps. The ear is logarithmic in its response to changes in sound intensity.

Decibels

Threshold of hearing

Since the ear's response is logarithmic (see previous page), we use a logarithmic scale called the **decibel scale** to measure it.

You can often measure loudness using a decibel meter. The decibel scale is a logarithmic scale which actually measures intensity level. The intensity level of a sound of intensity I is defined as:

intensity level in decibels, dB → $$\text{intensity level} = 10\log\left(\frac{I}{I_0}\right)$$

I = intensity in Wm^{-2}
I_0 = threshold of hearing in Wm^{-2}

I_0 is the **threshold of hearing** — the minimum intensity of sound that can be heard by a normal ear at a frequency of 1000 Hz (1 kHz). The value of I_0 is 1.0×10^{-12} Wm^{-2}. The units of the intensity level are decibels (dB). Intensity level can be given in bels — one decibel is a tenth of a bel — but decibels are usually a more convenient size.

Example — Maths Skills

A siren emits sound waves. The intensity of the sound 5.0 m from the siren is 0.94 Wm^{-2}. Calculate the intensity level at this distance in decibels.

$$\text{Intensity level} = 10\log\left(\frac{I}{I_0}\right) = 10\log\left(\frac{0.94}{1 \times 10^{-12}}\right)$$

$$= 119.73... = 120 \text{ dB (to 2 s.f.)}$$

Adjusted decibel scale

The perceived loudness of a sound depends on its frequency as well as its intensity. Two different frequencies with the same loudness will usually have different intensity levels on the dB scale. The **dBA scale** is an adjusted decibel scale which is designed to take into account the ear's response to different frequencies. On the dBA scale, sounds of the same intensity level have the same loudness for the average human ear.

Equal loudness curves

Figure 2 shows **equal loudness curves** for a normal ear.

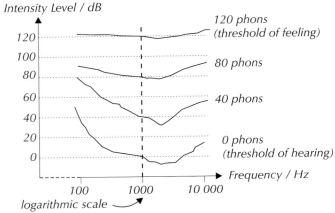

Figure 2: *Equal loudness curves.*

To generate equal loudness curves, start by generating a control frequency of 1000 Hz using a signal generator set at a particular intensity level. Then take a second signal generator and set the signal to a different frequency. Play alternating sounds from each generator, altering the intensity of (only) the second generator each time, until the sounds appear to have the same loudness. Make a note of the intensity level at this volume.

Repeat this for several different frequencies in the second generator, and plot the resulting curve on a graph. Change the intensity level of the control frequency and repeat the experiment again. If you measure intensity level in decibels, then the intensity of the control frequency gives you the loudness of the sound in phons.

Defects of hearing

As you get older, your hearing deteriorates generally, but higher frequencies are affected most. Your ears can be damaged by excessive noise. This results in general hearing loss, but frequencies around 4000 Hz are usually worst affected. People who've worked with very noisy machinery have most hearing loss at the particular frequencies of the noise causing the damage. Equal loudness curves can show hearing loss (see Figure 4).

For a person with hearing loss, higher intensity levels are needed for the same loudness, when compared to a normal ear. A peak in the curve shows damage at a particular range of frequencies.

Figure 3: *A hearing aid can be used to amplify the range of frequencies most needed by a person with a hearing problem.*

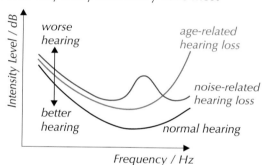

Figure 4: *Equal loudness curves at 0 phons.*

Exam Tip
You might be asked in the exam how to generate an equal loudness curve, so learn this method.

Tip: These curves show the general shapes for normal hearing and hearing defects, but the actual curves are a bit more wiggly. So don't panic if the curve you see in an exam is a little less smooth.

Practice Questions — Application

Q1 In some cities, high-frequency alarm systems have been used to deter young people from gathering in certain areas by emitting an annoying sound. Explain why the system only deters young people.

Q2 A smoke detector emits sound waves at a frequency of 3000 Hz. At a distance of 15 m, the intensity level is 40 dB. Calculate the intensity at a distance of 15 m in Wm^{-2}.

Practice Questions — Fact Recall

Q1 A student is listening to a sound. What two things does the perceived loudness of the sound depend on?

Q2 Explain why a logarithmic scale is needed to reflect the response of the human ear.

Q3 What logarithmic scale is used to measure sound intensity?

Q4 What is the dBA scale used for?

Q5 What is meant by the threshold of hearing? Give the value of the threshold of hearing for a normal ear.

Q6 Explain how equal loudness curves can be generated.

Q7 Give two reasons why your hearing might deteriorate.
Draw equal loudness curves for normal hearing and for hearing which has deteriorated due to each of the two reasons given.

Tip: Don't get the dB and dBA scales confused.

The heart's a pretty important organ, and it's beating constantly without us even noticing (most of the time). It's controlled by electrical impulses in nerve cells, which we can measure with detectors on the skin surface.

The structure of the heart

The heart is a large muscle. It acts as a double pump, with the left-hand side pumping blood from the lungs to the rest of the body and the right-hand side pumping blood from the body back to the lungs.

Traditionally, a diagram of the heart is drawn as though you're looking at it from the front, so the right-hand side of the heart is drawn on the left-hand side of the diagram and vice versa (just to confuse you).

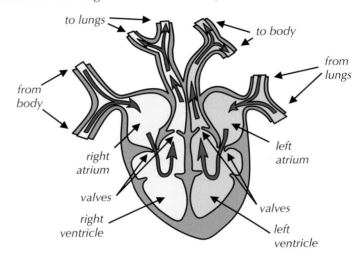

Figure 1: Simple structure of the heart.

Each side of the heart has two chambers — an **atrium** and a **ventricle** — separated by a valve. Blood enters the atria from the veins, then the atria contract, squeezing blood into the ventricles. The ventricles then contract, squeezing the blood out of the heart into the arteries. The valves ensure that the blood doesn't flow in the wrong direction when the ventricles or atria contract.

Electrical signals of the heart

When the heart is stimulated by an electrical signal it contracts and then relaxes, enabling the heart to pump blood around the body (see above).

The electrical signals that cause the heart to contract are produced by a group of cells in the heart in the sino-atrial (S-A) node in the right atrium (see Figure 2). These cells produce electrical signals that pulse about 70 times a minute.

These signals first spread through all the nerve cells of the atria, causing the atria to contract, forcing blood into the ventricles. The signals then pass to the atrioventricular (AV) node, which delays the pulse for about 0.1 seconds before passing it on to the nerve cells of the ventricles. These cells cause the ventricles to contract and blood is pumped through two valves to the lungs and body (see Figure 1). As this happens, the nerve cells of the atria relax, which is followed by the relaxation of the ventricles — this is one beat of the heart.

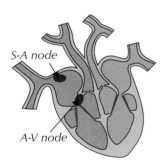

Figure 2: The positions of the sino-atrial (S-A) and atrioventricular (AV) nodes of the heart.

Tip: You might see the sino-atrial node called the pacemaker. An artificial pacemaker can be fitted in someone to regulate their heartbeat.

Electrocardiograms

ECG waveforms

The electrical signals of the heart cells can be detected as weak electrical signals at the surface of the body. A machine called an electrocardiograph detects these signals and produces an **ECG** — an **electrocardiogram**. An ECG is a plot of the potential difference between electrodes against time. They're used to find out the condition of the heart being examined.

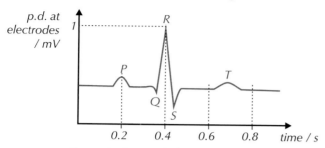

Figure 4: A typical electrocardiogram (ECG).

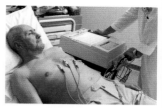

Figure 3: A patient undergoing a procedure to obtain an ECG — electrodes are attached to his chest to measure the electrical activity of his heart.

A normal ECG, covering a single heartbeat, has three separate parts: a P wave, a QRS wave and a T wave:

- The P wave corresponds to the contraction of the atria.

- The QRS wave (about 0.2 seconds later) corresponds to the contraction of the ventricles. This signal is much greater than the P wave and completely swamps the trace produced by the relaxation of the atria.

- Finally, the T wave (another 0.2 seconds later) corresponds to the relaxation of the ventricles.

Obtaining an ECG trace

When obtaining an ECG, electrodes are placed on the body and the variation in potential difference between the sites is measured. The signal is heavily attenuated (absorbed and weakened) by the body and needs to be amplified by a high impedance amplifier.

Electrodes are placed on the chest (so they're close to the heart) and on the limbs (where the arteries are close to the surface). In order to reduce the electrical resistance at the point of contact, hairs and dead skin cells are removed (e.g. using sandpaper), a conductive gel is used and the electrodes are securely attached. To reduce unwanted signals, the patient should also remain relaxed and still during the procedure, and the leads used should be shielded from any possible interference from a.c. sources in the area.

Tip: Dead skin cells can be removed from the surface of the skin using sandpaper.

Practice Question — Application

Q1 Sketch an ECG waveform for a person with a healthy heart on a graph of potential difference at the electrodes against time. Label and explain the shape of the graph in terms of what is happening to the atria and ventricles in the heart.

Practice Questions — Fact Recall

Q1 Explain how electrical signals in the heart are used to produce an ECG.

Q2 Suggest two things that could be done to help get a good trace when obtaining an ECG.

8. Ultrasound

You'll have heard of ultrasound from its wide use in prenatal scans. The main reason for its use is that it's the safest and cheapest form of non-invasive imaging we have — but, naturally, that does mean it comes with its downsides.

What is ultrasound?

Ultrasound waves are sound waves with higher frequencies than humans can hear (>20 000 Hz). For medical purposes, pulses of ultrasound waves are produced by an ultrasound scanner, usually with frequencies from 1 to 15 MHz. The scanner is placed on the surface of the patient's skin. When an ultrasound wave meets a boundary between two different materials, some of it is reflected and some of it is transmitted (undergoing refraction if the angle of incidence is not 0°). The reflected waves are detected by the ultrasound scanner and are used to generate an image — see page 282.

Acoustic impedance

The amount of reflection an ultrasound wave experiences at a boundary depends on the difference in **acoustic impedance**, Z, between the materials. The acoustic impedance of a material is defined as:

Z = acoustic impedance in $kgm^{-2}s^{-1}$ $Z = \rho c$ ρ = density in kgm^{-3}
    c = speed of sound in the medium in ms^{-1}

Example — **Maths Skills**

At 20 °C the density of air is approximately 1.2 kgm^{-3} and the speed of sound in air is 340 ms^{-1}. Find the acoustic impedance of air.

$Z = \rho c = 1.2 \times 340 = 408 = 410\ kgm^{-2}s^{-1}$ (to 2 s.f.)

Say an ultrasound wave travels through a material with an impedance Z_1. It hits the boundary between this material and another material with an impedance Z_2. The incident wave has an intensity of I_i. If the two materials have a large difference in impedance, then most of the energy is reflected (the intensity of the reflected wave I_r will be high). If the impedance of the two materials is the same then there is no reflection. The equation for the fraction of wave intensity that is reflected is:

I_r = intensity of reflected wave in Wm^{-2} Z_1 = acoustic impedance of first material in $kgm^{-2}s^{-1}$

$$\frac{I_r}{I_i} = \left(\frac{Z_2 - Z_1}{Z_2 + Z_1}\right)^2$$

I_i = intensity of incident wave in Wm^{-2} Z_2 = acoustic impedance of second material in $kgm^{-2}s^{-1}$

The fraction of wave intensity that is reflected has no units — it's just a ratio. You can also calculate the fraction of wave intensity that is transmitted by subtracting the fraction of wave intensity that is reflected from 1.

Example — **Maths Skills**

An ultrasound wave with an intensity of 7.1 × 10^{-3} Wm^{-2} hits a boundary, and a wave with an intensity of 4.9 × 10^{-4} Wm^{-2} is reflected back. What's the intensity reflection coefficient for the boundary?

$\frac{I_r}{I_i} = \frac{4.9 \times 10^{-4}}{7.1 \times 10^{-3}} = 0.06901... = 0.069$ (to 2 s.f.)

Ultrasound waves undergo **attenuation** when they travel through a material. This is when the waves are absorbed and scattered, making them harder to detect. This means the intensity of the reflected wave will always be lower the further the boundary is from the transducer (see below). Higher-frequency waves give better resolution, but the higher the frequency of the wave, the more it is attenuated. This means lower-frequency waves have to be used to image tissue deeper within the body.

The acoustic impedance also affects the attenuation — the larger the impedance, the greater the attenuation of the ultrasound moving through the material.

The piezoelectric effect

Ultrasound is produced and detected in ultrasound imaging using a transducer:

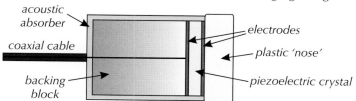

Figure 1: An ultrasound transducer.

Tip: A transducer is a device which converts one form of energy into another — here electrical energy is converted into sound.

Tip: The plastic nose is what's in contact with the patient's skin.

The transducer contains **piezoelectric crystals**, which produce a potential difference (p.d.) when they are deformed (squashed or stretched) — the rearrangement in structure displaces the centres of symmetry of their electric charges. This is called the piezoelectric effect.

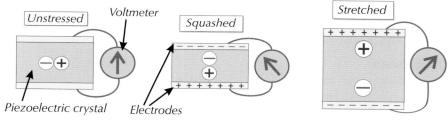

Figure 2: A p.d. is produced when piezoelectric crystals are deformed.

When you apply a p.d. across a piezoelectric crystal, the crystal deforms. If the p.d. is alternating, then the crystal vibrates at the same frequency. A piezoelectric crystal can act as a receiver of ultrasound, converting sound waves into alternating voltages, and also as a transmitter, converting alternating voltages into sound waves.

Ultrasound devices use lead zirconate titanate (PZT) crystals. The thickness of the crystal is half the wavelength of the ultrasound that it produces. Ultrasound of this frequency will make the crystal resonate and produce a large signal. The PZT crystal is heavily damped using the backing material, to produce short pulses and increase the resolution of the device.

Figure 3: An ultrasound transducer.

Tip: The vibrations of the crystal are greatest when the frequency of the potential difference equals the natural frequency of the crystals (see page 40).

Coupling media

Soft tissue has a very different acoustic impedance from air, so almost all the ultrasound energy is reflected from the surface of the body if there is air between the transducer and the body. To avoid this, you need a **coupling medium** between the transducer and the body — this displaces the air and has an impedance much closer to that of body tissue. The use of coupling media is an example of impedance matching. The coupling medium is usually an oil or gel that is smeared onto the skin.

Types of ultrasound scan

There are two types of ultrasound scan you need to know about. A-scans are mostly used for measuring distances, while B-scans are used to form images.

The A-scan

The amplitude scan (A-scan) sends a short pulse of ultrasound into the body simultaneously with an electron beam sweeping across a cathode ray oscilloscope (CRO) screen. The scanner receives reflected ultrasound pulses that appear as vertical deflections on the CRO screen.

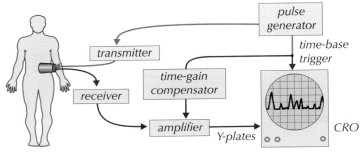

Figure 4: *The set-up of an amplitude scan.*

Weaker pulses (that have travelled further in the body and arrive later) are amplified more to avoid the loss of valuable data — this process is called time-gain compensation (TGC). The horizontal positions of the reflected pulses indicate the time the 'echo' took to return, and are used to work out distances between structures in the body (e.g. the diameter of a baby's head in the uterus, or the depth of an eyeball). A stream of pulses can produce the appearance of a steady image on the screen, although modern CROs can store a digital image after just one exposure.

The B-scan

In a brightness scan (B-scan), the electron beam sweeps down the screen rather than across. The amplitude of the reflected pulses is displayed as the brightness of the spot.

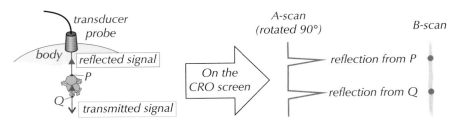

Figure 5: *The CRO output of a B-scan compared to an A-scan.*

You can use a linear array of transducers to produce a two-dimensional image, such as in the prenatal scanning of a fetus. They can also be used to check the heart is functioning properly (the image is called an echocardiogram). The B-scan allows the doctor to look at an image of the heart in real time.

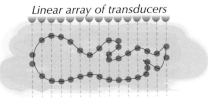

Figure 7: *A linear array of (B-scan) transducers can form a 2D image.*

Tip: A-scans can be used to monitor a baby's growth in the uterus by measuring the diameter of the head at different stages of pregnancy.

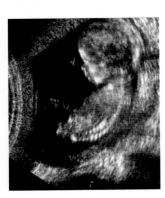

Figure 6: *An ultrasound scan of a fetus at 13 weeks.*

Tip: Don't get echocardiogram confused with electrocardiogram (ECG) (see page 279). An echocardiogram is often called an 'echo' for short.

Advantages and disadvantages

As with most medical procedures, ultrasound has advantages and disadvantages:

Advantages

- There are no known hazards or side effects — in particular, no exposure to ionising radiation, unlike X-ray imaging (see page 295).

- It's good for imaging soft tissues, since you can obtain real-time images — X-ray fluoroscopy (see page 294) can achieve this, but involves a huge dose of ionising radiation.

- Ultrasound devices are relatively cheap and portable, unlike MR scanners (see page 298), which cost millions of pounds (and X-ray machines aren't cheap either).

- The scan is a quick procedure (10-15 minutes) and the patient can move during the scan.

Disadvantages

- Ultrasound doesn't penetrate bone — so it can't be used to detect fractures or examine the brain.

- Ultrasound cannot pass through air spaces in the body (due to mismatch in impedance) — so it can't produce images from behind the lungs.

- The resolution is poor, so you can't see fine detail.

- Ultrasound cannot give information about any solid masses found.

Tip: Because the images are obtained in real time and the scan isn't harmful, the patient can watch the images as they are formed.

Tip: X-rays are covered on pages 288-295 and MR scans are covered on pages 296-298. See page 302 for a full comparison of types of medical imaging.

Tip: Remember, nearly all of the sound wave is reflected back at boundaries with air (see page 281).

Practice Question — Application

Q1 a) The density of water at 20 °C is 1000 kgm^{-3} (to 2 significant figures) and the acoustic impedance is 1.5×10^6 kgm^{-2}s^{-1}. What's the speed of sound, c, in water at this temperature?

   b) A sound wave travels from bone ($Z = 8.0 \times 10^6$ kgm^{-2}s^{-1}) to water. What will the intensity of the reflected wave be as a percentage of the intensity of the incident wave?

Practice Questions — Fact Recall

Q1 What is ultrasound?

Q2 What's the acoustic impedance of a material?

Q3 What's the piezoelectric effect?

Q4 Explain how an ultrasound wave is generated and detected.

Q5 Why is a coupling medium needed to obtain a clear ultrasound image?

Q6 a) Briefly describe what an A-scan is.

   b) Briefly describe what a B-scan is.

   c) Give an example of a use for both A-scans and B-scans.

Q7 a) Give three advantages of ultrasound scanning.

   b) Give three disadvantages of ultrasound scanning.

9. Endoscopy

Sometimes imaging methods don't quite fit the bill, and you need something that gets a clearer view. Endoscopes let you see deep inside a patient while causing minimal amounts of damage, and using no ionising radiation.

Learning Objectives:

- Know the properties of an optical fibre.
- Understand that total internal reflection occurs at the core-cladding interface.
- Know the difference between coherent and non-coherent optical fibre bundles, and know what each is used for.
- Know how a flexible endoscope works.
- Understand the use of endoscopy for internal imaging and the related advantages.

Specification Reference 3.10.4.2

Optical Fibres

Optical fibres are a bit like electric wires — but instead of carrying current they transmit light. A typical optical fibre consists of a glass core (about 5 μm to 50 μm in diameter) surrounded by a cladding, which has a slightly lower refractive index. The difference in refractive index means that light travelling along the fibre will be reflected at the core-cladding interface.

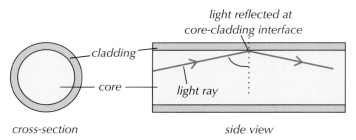

Figure 1: *A cross-section and side view of an optical fibre.*

If the light ray's angle of incidence is less than or equal to a critical angle, some light will be lost out of the fibre. But if the angle of incidence is larger than the critical angle, the light ray will be completely reflected inside the fibre. This phenomenon is called **total internal reflection** and means that the ray zigzags its way along the fibre — so long as the fibre isn't too curved. Total internal reflection can only happen when light is travelling from a more optically dense to a less optically dense medium.

Tip: You covered refractive index, angle of incidence, critical angle and total internal reflection in year 1.

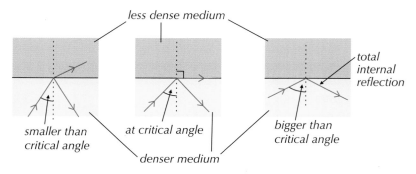

Figure 2: *Light rays hitting the core-cladding boundary at various angles.*

Figure 3: *Total internal reflection of a laser beam.*

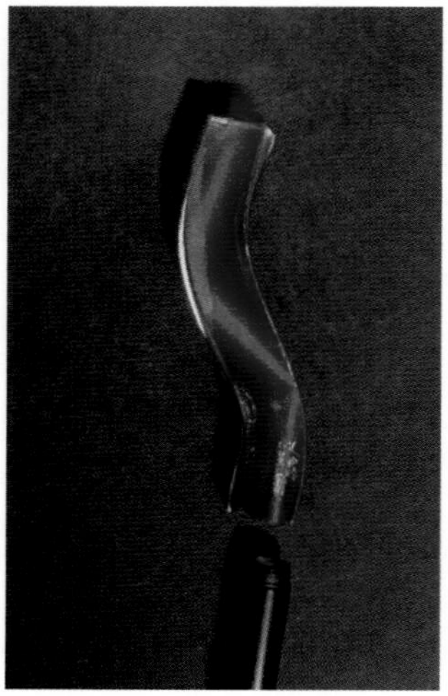

Calculating the critical angle

The **critical angle**, θ_c, depends on the refractive index of the core, n_1, and cladding, n_2, in an optical fibre.

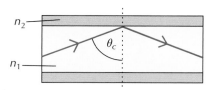

Figure 4: *The critical angle of an optical fibre.*

You can work out the critical angle using this formula:

θ_c = the critical angle in °

n_2 = the refractive index of the cladding

n_1 = the refractive index of the core

$$\sin \theta_c = \frac{n_2}{n_1}$$

Tip: The equation for the critical angle is derived from Snell's law — you covered that in year 1.
$n_1 \sin \theta_1 = n_2 \sin \theta_2$, but at the critical angle $\theta_2 = 90°$ (see Figure 2).
$\sin 90° = 1$, so
$\sin \theta_1 = \frac{n_2}{n_1}$, where $\theta_1 = \theta_c$.

Examples — Maths Skills

An optical fibre consists of a core with a refractive index of 1.5 and cladding with a refractive index of 1.4.

a) What is the critical angle of the core-cladding boundary?

You need to find the critical angle θ_c, so make that the subject by taking the inverse sin (sin⁻¹) of both sides:

$$\sin \theta_c = \frac{n_2}{n_1} \Rightarrow \theta_c = \sin^{-1}\left(\frac{n_2}{n_1}\right) = \sin^{-1}\left(\frac{1.4}{1.5}\right) = 68.960...$$
$$= 69° \text{ (to 2 s.f.)}$$

b) Would total internal reflection occur if the incident angle of light is 70°?

$70° > \theta_c$, so total internal reflection would occur.

Optical fibre bundles

An image can be transmitted along a bundle of optical fibres. This can only happen if the relative positions of fibres in a bundle are the same at each end (otherwise the image would be jumbled up) — a fibre-optic bundle in this arrangement is said to be **coherent**.

The resolution (i.e. how much detail can be seen) depends on the thickness of the fibres. The thinner the fibres, the more detail that can be resolved — but thin fibres are more expensive to make. Images can be magnified by making the diameters of the fibres get gradually larger along the length of the bundle.

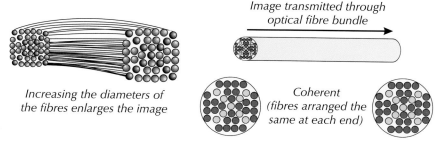

Increasing the diameters of the fibres enlarges the image

Image transmitted through optical fibre bundle

Coherent (fibres arranged the same at each end)

Figure 5: *Coherent optical fibre bundles are used to transmit images.*

Figure 6: *An optical fibre bundle, spread out.*

If the relative position of the fibres does not remain the same between each end the bundle of fibres is said to be **non-coherent**. Non-coherent bundles are much easier and cheaper to make. They can't transmit an image but they can be used to get light to hard-to-reach places — kind of like a flexible torch.

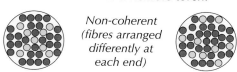

Non-coherent (fibres arranged differently at each end)

Figure 7: *Non-coherent optical fibre bundles have fibres in different relative positions at each end.*

Endoscopes

Endoscopes use optical fibres to create an image. An endoscope consists of a long tube containing two bundles of fibres — a non-coherent bundle to carry light to the area of interest and a coherent bundle to carry an image back to the eyepiece. Endoscopes are widely used by surgeons to examine inside the body.

An objective lens is placed at the distal end (furthest from the eye) of the coherent bundle to form an image, which is then transmitted by the fibres to the proximal end (closest to the eye), where it can be viewed through an eyepiece.

Endoscopes are flexible to give them more reach inside the body. The more an endoscope bends, the more likely light is to escape because bending reduces the angle of incidence — the radius of curvature needs to be kept above a certain level (usually around 20 times the fibre diameter).

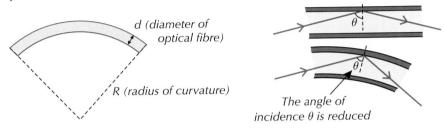

Figure 9: *Decreasing the radius of curvature decreases the angle of incidence for an incident beam of light on the core-cladding boundary.*

The endoscope tube can also contain a water channel for cleaning the objective lens, a tool aperture to perform keyhole surgery (see below) and a CO_2 channel. This allows CO_2 to be pumped into the area in front of the endoscope, making more room in the body.

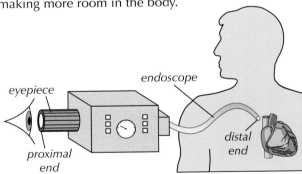

Figure 10: *An endoscope can be used to see inside the body.*

Keyhole surgery

Traditional surgery needs a large cut to be made in the body so that there's room for the surgeons to get in and perform an operation. This means that there's a large risk of infection to the exposed tissues and that permanent damage could be done to the patient's body.

New techniques in minimally invasive surgery (MIS, or keyhole surgery) mean that only a few small holes need to be cut in the body. An endoscope can be used in keyhole surgery to show the surgeon an image of the area of interest. Surgical instruments are passed through the endoscope tube, or through additional small holes in the body, so that the operation can be carried out.

Figure 8: *A medical endoscope.*

Tip: Often the image is displayed on a screen rather than through an eyepiece. This makes it easier for more than one person to see what is happening during an operation.

Common procedures include the removal of the gall bladder, investigation of the middle ear, and removal of abnormal polyps in the colon so that they can be investigated for cancer. Recovery times tend to be quicker for keyhole surgery, so the patient can usually return home on the same day — which makes it much cheaper for the hospital and nicer for the patient.

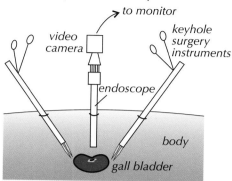

Figure 12: *Endoscopes are used in keyhole surgery.*

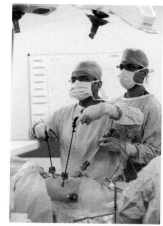

Figure 11: *Endoscopes being used in an operation to remove the gall bladder.*

Practice Questions — Application

Q1 a) Calculate the critical angle for an optical fibre if the core has $n = 1.61$ and the cladding has $n = 1.54$.

b) Say whether a beam of light with the following angles of incidence would be totally internally reflected along the fibre:

(i) 70.0°
(ii) 75.6°
(iii) 80.1°

Q2 An endoscope is being designed so that it can carry light signals that strike the core-cladding boundary at an angle greater than 65°. If the refractive index of the core is 1.70, what should the refractive index of the cladding be?

Q3 In a gastroscopy, an endoscope is sent down the patient's throat to see inside their digestive tract. Say what role each of the following would play in a gastroscopy:

a) Non-coherent optical fibre bundles.

b) Coherent optical fibre bundles.

> **Exam Tip**
> The formula for finding the critical angle is given in the data and formulae booklet. Remember, n_1 refers to the material light is travelling <u>from</u> (i.e. the core) and n_2 refers to the material light is travelling <u>to</u> (i.e. the cladding).

Practice Questions — Fact Recall

Q1 Draw and label a cross-section of an optical fibre strand.

Q2 What's the name of the process where a beam of light bounces off the core-cladding interface of an optical fibre without leaving the core?

Q3 a) What's the difference between a coherent and a non-coherent optical fibre bundle?

b) Which type of optical fibre bundle is needed to transmit images?

Q4 What's an endoscope?

Q5 Give two examples of uses for endoscopes in medicine.

Q6 Give two advantages of keyhole surgery compared to traditional methods of surgery for the patient.

Learning Objectives:

- Know the basic structure of a rotating-anode X-ray tube.
- Know how to find the maximum photon energy in an X-ray beam.
- Understand that the energy spectrum for an X-ray beam from a rotating-anode X-ray tube is a combination of a continuous spectrum and a characteristic spectrum.
- Know how to improve the sharpness and contrast of an X-ray image.
- Understand how the beam intensity and the photon energy of X-ray photons in a rotating-anode X-ray tube can be controlled.

Specification Reference 3.10.5.1

Tip: 'Evacuated' means all (or most of) the air has been removed.

10. X-ray Production

X-rays are really important in medical imaging. You need to know how to produce them first though — the next few pages will show you how it's done.

X-ray tubes

The X-rays used for diagnostic imaging are produced in an evacuated **rotating anode X-ray tube**. In the X-ray tube, electrons are emitted from a filament when it is heated by a current. The electrons are accelerated through a high potential difference (the tube voltage) towards a rotating tungsten anode.

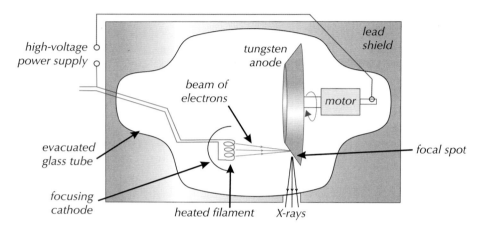

Figure 1: *A rotating-anode X-ray tube.*

Continuous spectrum

When the electrons smash into the tungsten anode, they decelerate and some of their kinetic energy is converted into electromagnetic energy, as X-ray photons. X-rays form part of the electromagnetic spectrum and have a wavelength between 0.01 and 10 nanometres. The tungsten anode emits a continuous spectrum of X-ray radiation — this is called bremsstrahlung ('braking radiation').

An X-ray photon will have maximum energy when all of the kinetic energy of an electron is converted into electromagnetic radiation. This can be calculated by multiplying the potential difference of the X-ray tube by the charge of an electron (as this gives the maximum kinetic energy of the electron). So, if a potential difference of 50 kV is used in the tube, the maximum X-ray energy will be 50 keV.

Tip: The X-ray photon is released when the outer electron moves into a lower energy level, not when the inner electron is ejected.

Characteristic spectrum

X-rays are also produced when beam electrons knock out electrons from the inner shells of the tungsten atoms. This is known as ionisation (or excitation). Electrons in the atoms' outer shells fall into the vacancies in the inner energy levels, and release energy in the form of X-ray photons.

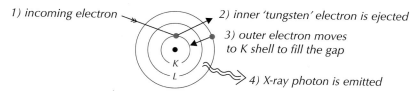

1) incoming electron

2) inner 'tungsten' electron is ejected

3) outer electron moves to K shell to fill the gap

4) X-ray photon is emitted

K
L

Figure 2: *X-rays are emitted when outer electrons move to inner energy shells to fill vacancies in tungsten atoms.*

Because there are fixed energy gaps between the electrons in an atom, photons are released at certain fixed energies, which produces a characteristic spectrum.

X-ray spectrum

So combining the continuous spectrum from bremsstrahlung and the characteristic spectrum from ionisation, you get a line spectrum superimposed on a continuous spectrum (see Figure 3).

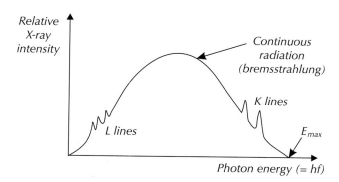

Relative X-ray intensity

Continuous radiation (bremsstrahlung)

K lines

L lines

E_{max}

Photon energy (= hf)

Figure 3: *An X-ray spectrum for a tungsten anode.*

> **Tip:** The K and L lines show which shell the electrons are filling vacancies in to cause the release of energy.

Only about 1% of the electrons' kinetic energy is converted into X-rays. The rest is converted into heat, so, to avoid overheating, the tungsten anode is rotated at about 3000 rpm. It's also mounted on copper — this conducts the heat away effectively.

> **Tip:** The focal spot also needs to be kept above a certain size to avoid overheating (see page 290).

┌─ **Example** ── **Maths Skills** ────────────────

An electron tube operates with a peak voltage of 25 kV and an electron beam current of 35 mA. Find the number of electrons arriving at the tungsten anode every second and the maximum kinetic energy of the X-ray photons released.

Current is the rate of flow of charge, i.e. $I = \dfrac{\Delta Q}{\Delta t}$.

So the number of electrons per second can be found by dividing the current by the charge on an electron:

$$\text{Electrons per second} = \frac{35 \times 10^{-3}}{1.60 \times 10^{-19}}$$
$$= 2.1875 \times 10^{17}$$
$$= 2.2 \times 10^{17} \text{ (to 2 s.f.)}$$

The maximum energy the photons can have is the maximum kinetic energy of the electrons. $E_{max} = e \times V$, so:

$$E_{max} = (1.60 \times 10^{-19}) \times (25 \times 10^3)$$
$$= 4.0 \times 10^{-15} \text{ J}$$

> **Exam Tip**
> The magnitude of the charge on an electron (1.60×10^{-19} C) is given in the exam data and formulae booklet.

> **Tip:** Remember, 1 eV (1.60×10^{-19} J) is the energy an electron would have if it was accelerated from rest through 1 V.
> So the energy in J is just voltage × (1.60×10^{-19}).

Improving image sharpness and contrast

The X-rays produced from the anode are emitted from a focal spot. The X-rays form a shadow (umbra) behind the object (see Figure 4). This results in a dark shadow between B and C, where none of the X-rays reach, and a partial shadow between A and B and between C and D. An image is therefore formed with fuzzy or unclear edges, and low sharpness. There are two ways of improving the image sharpness:

- Increasing the distance between the anode and the object, and decreasing the distance between the object and the screen.
- Reducing the width of the focal spot. This can be done by decreasing the slope of the anode. A value for θ of 17° is usually used for diagnostic X-rays, with a focal spot width of around 1 mm.

However, too small a focal spot can lead to the anode overheating. The minimum size you can feasibly use depends on the potential difference between the filament and the target, the tube current and the exposure time — these factors all affect the heating of the anode.

Tip: In diagnostic X-rays, the distance between the anode and the object is usually around 1 m. The distance between the object and the screen varies but is kept as low as possible (e.g. the screen is pressed against the patient).

Figure 5: *William David Coolidge, American physicist, holding an early version of the X-ray tube he invented.*

Tip: You met attenuation on page 279.

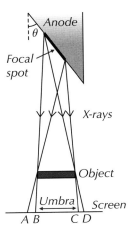

Figure 4: *Shadows cast by X-rays have a 'fuzzy' region around the edges.*

The contrast of an image is how much an image changes from black to white for different tissues and different tissue thicknesses. In order to improve the image contrast, the tube voltage can be changed (see below) — the lower the tube voltage, the lower the energy of the X-rays. The attenuation of an X-ray with lower energy is affected more by changes in tissue thickness than an X-ray with higher energy, which in turn improves the contrast of an image produced. However, the energy of the X-rays can't be so low that they can't pass through the object (because they are attenuated too much).

Varying beam intensity and photon energy

The intensity of the X-ray beam is the energy per second per unit area passing through a surface (at right angles). There are two ways to increase the intensity of the X-ray beam:

Tip: Because X-rays are harmful to the patient, the intensity needs to be carefully controlled — you need the lowest possible intensity that will give a clear image.

- Increase the tube voltage. This gives the electrons more kinetic energy. Higher energy electrons can knock out electrons from shells deeper within the tungsten atoms — giving more 'spikes' on the graphs. Individual X-ray photons also have higher maximum energies. Intensity and maximum photon energy are approximately proportional to voltage squared.

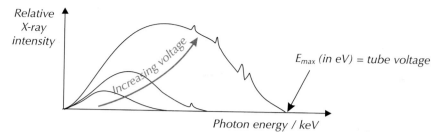

Figure 6: Increasing the tube voltage increases the intensity of the photons as well as their maximum kinetic energy.

Tip: The intensity of the X-ray beam is related to the area under the graph.

▪ Increase the current supplied to the filament. This liberates more electrons per second, which then produce more X-ray photons per second. Individual photons have the same energy as before. Intensity is approximately proportional to current.

Tip: Current is a measure of the number of electrons passing a point, not of their energy. This is why the maximum photon energy stays the same.

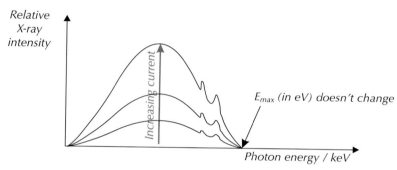

Figure 7: Increasing the tube current increases the intensity of the photons but not their maximum kinetic energy.

Practice Question — Application

Q1 In an X-ray tube, the electrons are accelerated through a potential difference of 85.0 kV towards the anode. What's the maximum energy an emitted X-ray photon could have in joules?

Practice Questions — Fact Recall

Q1 a) Draw and label a diagram of a rotating-anode X-ray tube.

b) Explain why the anode in the X-ray tube rotates.

c) Give one way of improving sharpness for a diagnostic X-ray image produced with a rotating-anode X-ray tube.

Q2 What causes the continuous spectrum of photon energies and the spikes at certain photon energies in a graph of relative X-ray intensity against photon energy for a rotating-anode X-ray tube?

Q3 a) What effect does increasing the X-ray tube voltage have on the relative intensity and maximum photon energy of emitted X-rays?

b) What effect does increasing the X-ray tube current have on the relative intensity and maximum photon energy of emitted X-rays?

11. X-ray Imaging

Now you've seen how X-rays are produced, here are some ways in which they're used in medicine. There are various types of X-ray imaging used today, from the basic X-ray tube and film set-up to advanced CT scans.

Producing X-ray images

To produce basic X-ray images, the patient is placed between an X-ray tube (p.288) and a detector plate. Traditionally this detector plate is some kind of photographic film, which is then processed to produce an image.

Nowadays in some places these are being replaced by digital sensors that require a lower exposure to produce clear images. X-rays are absorbed by certain parts of the body (e.g. bones) and pass through others (e.g. skin, muscle), so a 'shadow' is cast on the detector plate around harder tissues (see below and the next page for more on X-ray absorption). The differences in exposure across the plate can be processed to form a black and white image.

Medical X-rays are a compromise between producing really sharp, clear images (see page 290), and keeping the amount of radiation the patient is exposed to as low as possible. To do this, radiographers need to:

- Put the detection plate close to the patient and the X-ray tube far away from the patient, to improve image sharpness (see page 290).
- Make sure the patient keeps still — moving will make the image blurry.
- Put a lead collimator grid between the patient and film to stop scattered radiation 'fogging' the image and reducing contrast.
- Use **intensifying screens** (see Figure 1) next to the film surface. These contain crystals that fluoresce — they absorb X-rays and re-emit the energy as visible light, which helps to develop the photograph image quickly. Because the intensifying screen is close to the film, the photons of visible light hit the film in the correct place and little to no image quality is lost. Intensifying screens mean a shorter exposure time is needed, keeping the patient's radiation dose lower.

plastic cover
front intensifying screen
double-sided film
back intensifying screen
metal back

Figure 1: *An X-ray film cassette made up of a cover, a metal back, photographic film and two intensifying screens.*

X-ray attenuation

When X-rays pass through matter (e.g. a patient's body), they are absorbed and scattered. The intensity of the X-ray beam decreases (attenuates) exponentially with the distance from the surface, according to the material's **linear attenuation coefficient** (or linear coefficient):

I = intensity of the X-ray beam in Wm^{-2}

I_0 = initial intensity of the X-ray beam in Wm^{-2}

$$I = I_0 e^{-\mu x}$$

μ = the material's linear attenuation coefficient in m^{-1}

x = the distance from the surface in m

Example — Maths Skills

The linear attenuation coefficient of a tissue is 20 m⁻¹. How far will an X-ray travel inside the tissue before its intensity is 40% of the original intensity?

You know that the intensity is 40% of the original intensity, i.e. $\frac{I}{I_0} = 0.4$, so:

$$I = I_0 e^{-\mu x} \Rightarrow e^{-\mu x} = \frac{I}{I_o} = 0.4$$

Take the natural logarithm of both sides and rearrange to make x the subject:

$$-\mu x = \ln 0.4 \quad \Rightarrow \quad x = -\frac{\ln 0.4}{\mu} = -\frac{\ln 0.4}{20} = 0.0458... = 0.05 \, \text{m (to 1 s.f.)}$$

> **Tip:** 'ln' is the natural logarithm and you can use it to get rid of exponentials because $\ln e^x = x$. You might see 'ln' written as '$\log_e$'. There's more on this on page 402.

Half-value thickness

Half-value thickness, $x_{\frac{1}{2}}$, is the thickness of material required to reduce the intensity to half its original value. It can be derived from the equation for the intensity of the X-ray beam. Rearranging the equation gives:

$$\ln\left(\frac{I}{I_0}\right) = -\mu x$$

If the half-value thickness is the point at which the intensity is half of its original value, $\frac{I}{I_0} = \frac{1}{2}$. Substituting this and $x = x_{\frac{1}{2}}$ in gives:

$$\ln\left(\frac{1}{2}\right) = -\mu x_{\frac{1}{2}}$$

But $\ln\left(\frac{1}{2}\right) = \ln(2^{-1}) = -\ln 2$ (see page 402), so rearranging gives:

$x_{\frac{1}{2}}$ = half-value thickness in m $\longrightarrow \boxed{x_{\frac{1}{2}} = \frac{\ln 2}{\mu}} \longleftarrow$ μ = linear attenuation coefficient in m⁻¹

> **Tip:** You don't need to know the derivation for the half-value thickness.

> **Tip:** The half-value thickness and linear attenuation coefficient of a material are different for different wavelengths of radiation (i.e. different energies).

Example — Maths Skills

What's the half-value thickness for a bone with a linear attenuation coefficient of 48 m⁻¹?

$$x_{\frac{1}{2}} = \frac{\ln 2}{\mu} = \frac{\ln 2}{48} = 1.444... \times 10^{-2} = 1.4 \times 10^{-2} \, \text{(to 2 s.f.)}$$

The mass attenuation coefficient

The **mass attenuation coefficient**, μ_m, is a measure of how much radiation is absorbed per unit mass. For a material of density ρ, it's given by:

μ_m = mass attenuation coefficient in m²kg⁻¹ $\longrightarrow \boxed{\mu_m = \frac{\mu}{\rho}} \longleftarrow$ μ = linear attenuation coefficient in m⁻¹
ρ = density in kgm⁻³

> **Tip:** ρ is the Greek letter 'rho' and it's used for density.

X-ray absorption

X-rays are **attenuated** by absorption and scattering. How much energy is absorbed by a material depends on its atomic number. So tissues containing atoms with different atomic numbers (e.g. soft tissue and bone) will contrast in the X-ray image. X-rays are absorbed more by bone than soft tissue, so bones show up brightly in X-rays.

　　　If the tissues in the region of interest have similar attenuation coefficients then artificial contrast media can be used — e.g. a barium meal. Barium has a high atomic number, so it shows up clearly in X-ray images and can be followed as it moves along the patient's digestive tract. It's said to be X-ray opaque.

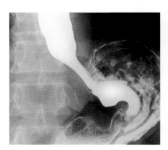

Figure 2: An X-ray image of a stomach after a barium meal has been swallowed.

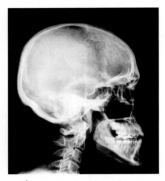

Figure 3: *An X-ray image of a human skull. X-ray photographs are negative images — the 'white' parts of the image are where the film has remained clear as it has not been exposed to radiation (which would cause it to develop and turn black).*

Flat panel detectors

Flat panel (FTP) detectors are a digital method of X-ray imaging and can be used instead of photographic detection.

X-rays are fired at the patient, who has an FTP detector behind them. The FTP detector contains a scintillator material that produces light when hit by an X-ray. The light it produces has an intensity proportional to the energy of the incident X-ray photon. The FTP detector also contains photodiode pixels which generate a voltage when light from the scintillator hits them. This voltage is proportional to the intensity of the light. For each pixel there is a thin-film transistor which reads the digital signal (from the voltage generated). From this, a digital image of inside the patient is created.

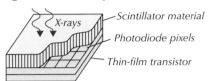

Figure 4: *Simple diagram of a flat panel (FTP) detector*

There are many advantages to using FTP detectors rather than photographic detection:

- They are more lightweight and compact, making them much more convenient as they can be moved around a hospital or positioned around an immobile patient.
- They have a higher resolution so can detect finer details.
- There is less distortion of the final image.
- The electron scan can easily be copied, stored or shared.
- They require a lower exposure to produce clear images.

Fluoroscopy

Moving images can be created using a fluorescent screen and an image intensifier. This is useful for imaging organs as they work. X-rays pass through the patient and hit the fluorescent screen, which emits light.

The light causes electrons to be emitted from the photocathode (see Figure 5). The electrons travel through the glass tube towards the fluorescent viewing screen. Electrons in the glass tube are focused onto the viewing screen and gain kinetic energy as they are accelerated by a potential difference between the photocathode and the viewing screen. Both of these things mean the image on the viewing screen is about 5000 times brighter than on the first screen.

Imaging can last several minutes, so image intensifiers are particularly important — they reduce the patient's dose of radiation by around a thousand times.

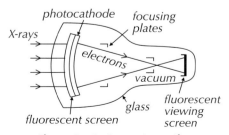

Figure 5: *An image intensifier.*

CT scanning

Computed tomography (CT) scans produce an image of a two-dimensional slice through the body. An X-ray tube produces a narrow, monochromatic (single wavelength) X-ray beam. The tube rotates around the body and the X-rays are picked up by an array of detectors. The detectors feed the signal to a computer.

The computer works out how much attenuation has been caused by each part of the body and produces a very high quality image. However, the machines are expensive and the scans involve a high radiation dose for the patient.

Advantages and disadvantages

There are both advantages and disadvantages to the use of X-rays as an imaging technique.

Advantages

- X-rays produce a good resolution and provide clear imaging of bones.
- CT scans are much quicker than MR scans (see page 298) and so the patient doesn't have to lie still for as long.
- CT scanners are cheaper than MR scanners.

Disadvantages

- X-rays are a form of ionising radiation — which can damage cells and in rare cases lead to the development of cancer.
- Investigating soft tissue with fluoroscopy requires a larger dose of radiation.
- X-rays are generally unsuitable for pregnant women.
- Patients have to lie still for the scan.

Figure 6: *A patient undergoing an upper-body CT scan.*

Tip: Take a look at page 302 for a full comparison with other medical imaging methods.

Practice Question — Application

Q1 X-rays with initial intensity 30.0 Wm^{-2} are used to image bone with $\rho = 1900$ kgm^{-3} and $\mu_m = 0.133$ m^{2}kg^{-1}.

   a) Find the linear attenuation coefficient μ of the bone.

   b) Find the intensity of the X-rays after they pass through 0.690 cm of bone.

Practice Questions — Fact Recall

Q1 Give three things a radiographer can do to make sure a clear X-ray image is formed while minimising a patient's exposure.

Q2 Explain how an intensifying screen works.

Q3 What's the half-value thickness of a material, in terms of X-rays?

Q4 Why can X-rays be used to form images of bones and other hard tissues but not skin, muscles and other soft tissues?

Q5 Explain what a barium meal is used for.

Q6 a) Briefly describe how a flat panel detector works.

   b) Give two advantages of using a flat panel detector over a photographic detector.

Q7 Describe how a CT scanner works.

Q8 a) Give two advantages of CT scans as an imaging method.

   b) Give one disadvantage of CT scans as an imaging method.

12. Magnetic Resonance Imaging

Magnetic resonance imaging is another way of looking inside the body without needing to undertake any kind of surgery. It's generally considered much safer than X-rays but, as with anything, it has its downsides.

Learning Objective:

- Understand the basic principles of a magnetic resonance (MR) scanner.

Specification Reference 3.10.4.3

Magnetic fields in MR scans

In **magnetic resonance (MR) imaging**, the patient lies in the centre of a huge superconducting magnet that produces a uniform magnetic field. The magnet needs to be cooled by liquid helium — this is partly why the scanner is so expensive.

The uniform magnetic field generated by the machine has an effect on the protons (hydrogen nuclei) in the patient's body. Protons (and neutrons) possess a quantum property called spin, which makes them behave like tiny magnets. Initially, all of the protons are orientated randomly, but in a uniform magnetic field the protons align themselves with the magnetic field lines. Parallel alignment means a proton's spin axis points in the same direction as the external magnetic field, and antiparallel alignment means the proton's spin axis points in the opposite direction to the field. As the protons spin, they precess (wobble) about the magnetic field lines. This wobble has an angular frequency called the precession frequency, which is proportional to the magnetic field strength.

> **Tip:** MR scanners can't be used on people with pacemakers or some metal implants — the strong magnetic fields would be very harmful.

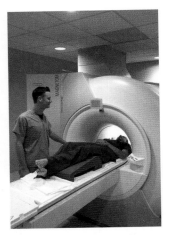

Figure 2: *A person about to undergo an MR scan is moved to the centre of the MR machine.*

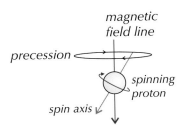

Figure 1: *Diagram of a proton which is aligned parallel to an external magnetic field. The proton precesses around the magnetic field line.*

By using smaller electromagnets called gradient field coils, smaller magnetic fields can be superimposed onto the main one, creating a gradient of magnetic field strength across the patient. This means the protons in different sections of the body will have different precession frequencies.

Producing images

Radio frequency (RF) coils are used to transmit pulses of radio waves at the same frequency as the precession frequency. Some of the protons that have their spin aligned with the magnetic field absorb the photons produced by the RF coils, which excites them and causes them to change their spin state. Changing their spin state means that they flip their alignment.

The protons in different sections of the body will have different precession frequencies (due to the gradient of the field) and will absorb RF waves with frequencies equal to their precession frequency. The RF coils can transmit pulses of different frequencies to excite protons in successive small regions of the body.

> **Tip:** It's called 'resonance' when the protons absorb the RF waves with the same frequency as their precession frequency — that's why it's called 'magnetic resonance' (MR) imaging.

In between pulses, the protons de-excite and re-emit electromagnetic energy at their precession frequency (RF waves). These RF waves are detected and the information is sent to a computer. The computer knows the positions in the body relating to each precession frequency, so it can then generate an image of a cross-section through the body or build up a 3D image. This is done by measuring various quantities of the MR signal like amplitude, frequency and phase.

Tip: A lot of the protons detected in an MR scan are found in water in the body. So MR images are particularly useful for looking at soft tissues because they contain a lot of water — for example, the brain (see Figure 4).

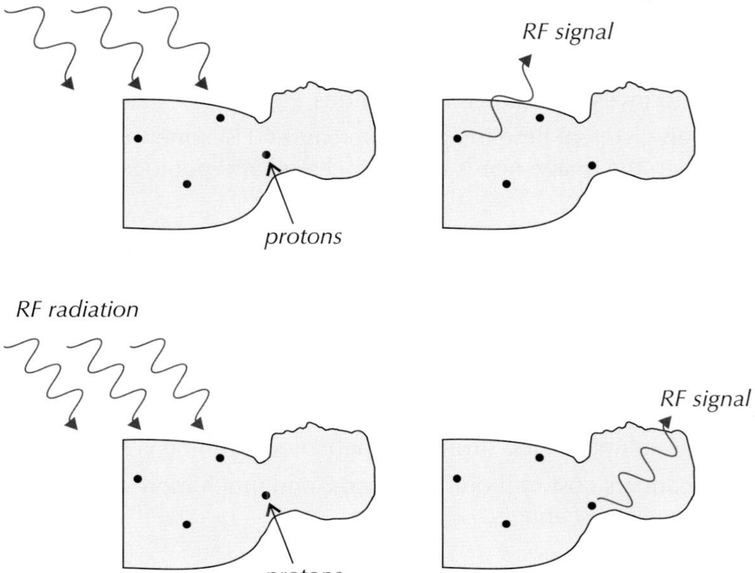

RF radiation

RF signal

protons

RF radiation

RF signal

protons

Tip: Make sure you can describe how an MR scan works — sometimes it helps to write down in bullet points the order that everything happens. That way you'll have everything straight in your head for the exam.

Figure 3: *Different RF waves excite protons in different parts of the body. (Left) RF pulses are incident on the patient, which excites some protons with a precession frequency equal to the radio frequency and causes a change of spin state. (Right) As the protons de-excite, they re-emit radio frequency radiation at their precession frequency.*

Controlling the contrast

Radio waves are applied in pulses. Each short pulse excites the hydrogen nuclei and then allows them to de-excite and emit a signal. The response of different tissue types (and therefore the contrast of the image) can be enhanced by varying the time between pulses.

Tissues consisting of large molecules such as fat are best imaged using rapidly repeated pulses. This technique is used to image the internal structure of the body. Allowing more time between pulses enhances the response of watery substances. This is used for diseased areas.

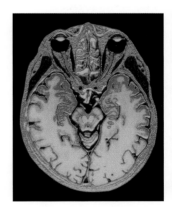

Figure 4: *An MR scan of a healthy brain.*

Advantages and disadvantages

MR scanning has advantages and disadvantages compared to CT scans and ultrasound (see page 302) that need to be taken into account when considering whether it would be appropriate for a particular type of procedure.

Advantages of MR scanning

- There are no known side effects.
- MR produces non-ionising radiation unlike the X-rays used in CT scans, so it won't damage living cells in the same way as CT scans.
- An image can be made for any slice in any orientation of the body, and multi-plane images can be made from the same scan. CT needs a new scan for each image.
- Compared to CT and ultrasound scans, MR gives higher quality images for soft tissue types (such as the brain), and better resolution between tissue types for an overall better resolution final picture.
- Contrast between different tissue types can be weighted (see previous page) to investigate different situations.
- MR can give real time images, whereas a CT scanner needs to be rotated and then the image needs to be processed and put together.

Disadvantages of MR scanning

- The imaging of bones is very poor compared to CT scans.
- Scans can be noisy and take a long time.
- Scanners are fairly narrow, so some people suffer from claustrophobia inside them.
- MR scanners usually can't be used on people with pacemakers or some metal implants — the strong magnetic fields can be very harmful.
- MR scanners cost millions of pounds, and much more than CT or ultrasound scanners.

Tip: Ionising radiation knocks electrons off atoms and molecules, which can damage body cells — this can lead to cancer developing.

Tip: Page 302 has a full comparison of different medical imaging methods.

Practice Question — Application

Q1 For each of the following, say whether an MR scan would be appropriate for diagnosis and explain your answer.
   a) A possible abscess in an otherwise healthy patient's kidney.
   b) A possible tumour deep within a patient's brain.
   c) A possible heart defect in a patient fitted with a pacemaker.
   d) A possible fracture to the radius (arm bone).
   e) A possible case of deep vein thrombosis (blood clot).

Practice Questions — Fact Recall

Q1 State the basic features of an MR scanner, explaining the role of each.

Q2 How can radiographers control the contrast of an MR image?

Q3 Which of these statements is a true disadvantage of MR imaging?
   a) The resolution is not as high as CT scanning.
   b) The radiation used is ionising.
   c) It's difficult to differentiate between different types of tissue.
   d) Scanners cost millions of pounds.

Q4 a) Give four advantages of using an MR scanner.
   b) Give four disadvantages of using an MR scanner.

13. Medical Uses of Radiation

Radiation is used in different ways for imaging — you've already seen how it is used for X-ray imaging. Although ionising radiation can be harmful to the cells in your body, it can be used to an advantage in some cases.

Medical tracers

Medical tracers are radioactive substances that are used to show tissue or organ function. Other types of imaging, e.g. X-rays (page 292), only show the structure of organs, whereas medical tracers show structure and function.

Medical tracers usually consist of a gamma-emitting radioactive isotope bound to a substance that is used by the body, e.g. glucose or water. The tracer is injected into or swallowed by the patient and then moves through the body to the region of interest. Where the tracer goes depends on the substance the isotope is bound to, i.e. it goes anywhere that the substance would normally go, and is used how that substance is normally used. The radiation emitted is recorded (e.g. by a gamma camera or PET scanner, see page 300) and an image of inside the patient produced.

You need to know about three of the main gamma-emitting radioactive isotopes used in medical tracing. Their properties are summarised in the table below:

Medical tracer	Where in the body it is used	Physical half-life	Radiation emitted	Energy of gamma radiation emitted
Technetium-99m	Range of organs	6 hours	Gamma	140 keV
Iodine-131	Thyroid	8 days	Beta and gamma	360 keV
Indium-111	Antibodies and blood cells	2.8 days	Gamma	170 or 250 keV

Technetium-99m is a very widely used isotope due to its effective half-life value (see below). It is long enough that the gamma radiation can still be detected by the time that the tracer has reached the organ being investigated, but short enough that the patient doesn't have a radioactive isotope inside them for longer than needed. Iodine-131 is used to detect and treat problems in the thyroid because iodine is naturally used by the thyroid. Indium-111 is used to label antibodies and blood cells to detect infections.

Effective half-life

All medical tracers have an **effective half-life**. Your body metabolises (uses up) the substances that medical tracers are bound to. The rate at which it manages to do this — the rate of excretion — affects how long emissions from the body can be detected for. You can think of this rate in the form of a **biological half-life**, T_B. When combined with the **physical half-life**, T_P (which only depends on the decay constant, page 175) you get an effective half-life, T_E, for the tracer:

T_E = effective half-life in s ⟶ $$\frac{1}{T_E} = \frac{1}{T_B} + \frac{1}{T_P}$$ ⟵ T_P = physical half-life in s

T_B = biological half-life in s

You can calculate the physical half life using $T_{1/2} = \dfrac{\ln 2}{\lambda}$ where the decay constant can be calculated from $A = A_0 e^{-\lambda t}$ or $N = N_0 e^{-\lambda t}$ (see page 176).

Learning Objectives:

- Know that gamma-emitting radioisotopes are used as tracers.
- Know the properties of technetium-99m, iodine-131 and indium-111, including the radiation they emit, their half-lives and the energy of their gamma radiation.
- Know that a tracer is made to travel to an organ as it labels a compound that has an affinity for that particular organ.
- Know the definitions of physical, biological and effective half-lives for a tracer and that they are related by the equation $\frac{1}{T_E} = \frac{1}{T_B} + \frac{1}{T_P}$.
- Know how a Molybdenum-Technetium generator works and why they are needed.
- Know the basic structure and working of a gamma camera, and of the photomultiplier tubes that it uses.
- Understand how PET scans work.
- Know that X-rays can be used to treat tumours, and the methods used to limit damage to healthy cells.
- Know that beta-emitting implants can be used to treat tumours.

Specification References 3.10.6.1, 3.10.6.2, 3.10.6.3, 3.10.6.4 and 3.10.6.5

The Molybdenum-Technetium generator

Technetium-99m has a physical half-life which is too short for it to be practically transported. Instead, hospitals have Molybdenum-Technetium generators delivered. Molybdenum has a much longer half-life — 66 hours, making it much better for transport.

Inside the generator, the molybdenum has been combined with aluminium oxide, which it bonds strongly with. Molybdenum then decays, producing technetium-99m which does not bond as strongly with aluminium oxide. A saline solution is placed into the generator, washing out any technetium-99m. This solution can then be injected into patients, or combined with a substance to make a specific tracer.

Gamma cameras

The gamma rays emitted by radiotracers in a patient's body are detected using a gamma camera which consist of five main parts:

- Lead shield — stops radiation from other sources entering the camera.
- Lead collimator — a piece of lead with thousands of vertical holes in it — only gamma rays parallel to the holes can pass through.
- Sodium iodide crystal — emits a flash of light (scintillates) whenever a gamma ray hits it.
- Photomultiplier tubes — each photomultiplier tube has a photocathode in it that releases an electron by the photoelectric effect when hit by a photon. Each electron is then multiplied into a cascade of electrons. So the photomultiplier tubes are used to detect the flashes of light from the crystal and turn them into pulses of electricity.
- Electronic circuit — collects the signals from the photomultiplier tubes and sends them to a computer for processing into an image.

The right radiotracer must be chosen when using a gamma camera — gamma photons with too high an energy can cause problems in the camera.

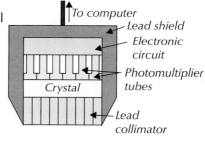

Figure 1: Diagram of a gamma camera.

Tip: You learnt about the photoelectric effect in year 1 of A-level.

Tip: Photomultiplier tubes are also used in PET scanners (see below).

Tip: The photomultiplier tubes multiply the electrons into a cascade of electrons using a bit of a complicated process — you don't need to worry about how it's done.

PET scans

Positron Emission Tomography (PET) scans involve injecting a patient with a substance used by the body, e.g. glucose, containing a positron-emitting radiotracer with a short half-life, e.g. ^{13}N, ^{15}O, ^{18}F. The patient is left for a time to allow the radiotracer to move through the body to the organs. The positrons emitted by the radioisotope collide with electrons in the organs, and annihilate. This annihilation results in high-energy gamma rays being emitted.

Detectors all around the body detect and record these gamma rays. The detectors then send the information to a computer, which builds up a map of the radioactivity in the body.

Tip: You covered electron-positron annihilation in year 1 of A-level.

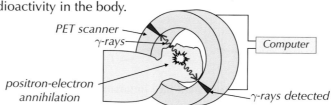

Figure 2: Diagram showing how a PET scan works.

The main advantage of PET scans is that the distribution of radioactivity matches up with metabolic activity. This is because more of the radioactive glucose (or whatever has been labelled with the medical tracer) is taken up and used by cells that are doing more work, i.e. cells with an increased metabolism. Another advantage is that brain activity can be investigated, whereas some other non-invasive methods cannot penetrate the skull. It can also give information about the malignancy of tumours and whether a tumour is spreading.

There are disadvantages too — ionising radiation is used which could damage the patient's cells. The scans can take a long time and require the patient to stay still in a narrow machine, which can be uncomfortable and claustrophobic. The machine itself is expensive and very large, so patients have to travel to their nearest hospital with a PET scanner, which could be inconvenient for them.

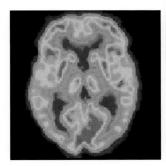

Figure 3: *Image of a 20-year-old's healthy brain using a PET scanner.*

Radiation for treatment

Ionising radiation damages cells, which is usually a bad thing. However, in some cases this property can be helpful — the ionising radiation can be directed towards tumours to destroy the tumour cells.

High-energy X-rays are fired at the tumours from outside the patient's body. This means that surrounding healthy cells are also damaged — which can lead to mutations and even a higher risk of future cancers.

To limit the radiation patients are exposed to, carefully focused beams are controlled by computers to ensure as much of the radiation as possible is hitting the tumour. Shielding is also sometimes used, and the X-ray beam may be rotated around the patient to minimise the radiation dose to healthy tissue.

Radioactive treatments can also be placed inside a patient. Implants containing beta-emitters are placed next to or inside the tumour. Beta radiation is ionising, so damages the cells in the tumour, but has a short range so the damage to healthy tissue is limited.

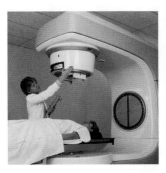

Figure 4: *Person about to undergo radiotherapy using a linear particle accelerator that emits X-rays.*

Practice Question — Application

Q1 A medical tracer has a physical half-life of 9.5 days and a biological half-life of 22 days. Explain what these terms mean and calculate the effective half-life of the tracer.

Practice Questions — Fact Recall

Q1 What is the method used to get a medical tracer to the part of the body which is being investigated?

Q2 Name a gamma-emitting isotope often used as a medical tracer and explain why it's appropriate. Include the half-life of the tracer and the energy of the gamma radiation emitted.

Q3 a) Describe how a Molybdenum-Technetium generator works.

b) Why are they used in hospitals?

Q4 Give the main components that make up a gamma camera and describe the role of each.

Q5 Describe how a PET scan can image metabolic activity.

Q6 Give two ways of treating a tumour with ionising radiation.

Q7 What are the risks of using radiation to treat tumours, and how can these risks be reduced?

14. Comparing Imaging Techniques

You need to be able to compare the main imaging techniques covered in this section. Luckily, this topic is a neat little summary of just that.

Comparison of medical imaging techniques

The main medical imaging techniques you have learnt about in this section all have different advantages and disadvantages. You need to know how they compare in terms of convenience, safety and resolution.

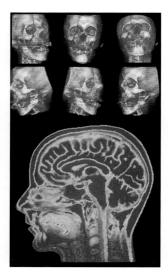

Figure 1: *CT scans (top) can show clear detail of bone damage, in this case to the skull. MR scans (bottom), on the other hand, are best suited to building a clear image of the brain and other soft tissue.*

	Ultrasound	CT	MR	PET
Safety	No known side effects	Uses damaging ionising radiation	No known side effects	Uses damaging ionising radiation
Image resolution for bones	Can't penetrate	Very good	Poor	Poor
Image resolution for tissues	Poor	Good	Very good	Very good
Convenience	Portable	Becoming more portable	Large	Large
Cost of equipment	Cheap	Expensive	Very expensive	Very expensive
Speed of scan	Up to 15 minutes	Up to 10 minutes	Up to 90 minutes	Up to 30 minutes
Patient comfort	Patient can move during scan, uses cold gel	Patient must lie still	Patient must lie still inside narrow tube, scan is very noisy	Patient must lie still inside narrow tube, can require injection

Practice Questions — Fact Recall

Q1 Give two reasons why an MR scan might be preferred to a CT scan to obtain a medical image.

Q2 Give two reasons why an MR scan might not be a feasible way of obtaining a medical image.

Q3 Give two reasons why ultrasound scanning is a preferred option for imaging an unborn fetus to each of the following:

a) CT scanning

b) MR imaging

Q4 Explain why a CT scan might be preferred to a PET scan to obtain a medical image.

Section Summary

Make sure you know...

- How converging and diverging lenses work and the definitions of principal focus and focal length.
- How to use the lens equation, and how to calculate the power of a lens and linear magnification.
- That the eye is an optical refracting system and the role of rods and cones in the eye's sensitivity.
- The spectral response of the eye as a photodetector.
- How to explain the spatial resolution of the eye in terms of rods and cones.
- What is meant by the vision defects myopia, hypermetropia and astigmatism, and how to correct them.
- How to draw ray diagrams and calculate powers for lenses used to correct defects of vision.
- What is meant by intensity of sound.
- The basic structure of the ear and the transmission process involved in detecting sound.
- What the range of human hearing is, and how it varies with frequency.
- How humans perceive loudness and why a logarithmic scale is needed to reflect this.
- How to calculate the intensity level of sound and what is meant by the threshold of hearing.
- What the dB and dBA scales are and why they're used.
- What equal loudness curves are and the effect of hearing deterioration and excessive noise on them.
- How to obtain an ECG waveform, and can explain the shape of a normal ECG waveform.
- That ultrasound waves are attenuated by materials, and reflected at tissue boundaries, and know what the acoustic impedance is and the equation for the fraction of wave intensity reflected at a boundary.
- Some examples of what piezoelectric devices are used for.
- The difference between A-scans and B-scans, and their uses.
- That optical fibres work using total internal reflection at the core-cladding interface.
- The structure and working of a flexible endoscope and its uses for internal imaging in medical physics.
- How a rotating-anode X-ray tube produces X-rays and how to improve image sharpness and contrast.
- The spectra of X-ray energies produced and how to find the maximum photon energy.
- How the beam intensity and photon energy of X-rays produced in an X-ray tube can be controlled.
- That getting an X-ray requires a balance between keeping patient dose low and image quality high.
- How intensifying screens and fluoroscopic image intensification work and why they are used.
- How flat panel (FTP) detectors work and what advantages they have over photographic detection.
- That different tissues absorb X-rays by different amounts and why a barium meal is sometimes needed.
- That X-rays undergo exponential attenuation that depends on a material's linear coefficient, μ.
- How to calculate the intensity of the X-ray beam, I, and the mass attenuation coefficient, μ_m.
- That the half-value thickness is the thickness required to reduce the intensity to half its original value.
- How a CT scanner and an MR scanner work and how they produce images.
- How gamma-emitting radioisotopes are used as medical tracers for imaging particular organs.
- The properties of technetium-99m, iodine-131 and indium-111 as medical tracers.
- The definitions of physical, biological and effective half-lives for a tracer and the equation $\frac{1}{T_E} = \frac{1}{T_B} + \frac{1}{T_P}$.
- The importance of Molybdenum-Technetium generators and how they work.
- The basic structure and working of a gamma camera, including the role of photomultiplier tubes.
- How PET scans work.
- That ionising radiation is used to treat tumours and the methods used to limit damage to healthy cells.
- The advantages and disadvantages of ultrasound, X-ray (CT), MR and PET imaging and how to compare imaging methods and their appropriateness for particular diagnostic tasks.

Exam-style Questions

Questions 1-4 are worth 1 mark each.

1 Calculate the fraction of wave intensity that is reflected when ultrasound waves are
passing from soft tissue to muscle. The acoustic impedance of soft tissue is
1.58×10^6 kgm^{-2}s^{-1} and the acoustic impedance of muscle is 1.70×10^6 kgm^{-2}s^{-1}.

    A   3.72

    B   0.269

    C   0.0366

    D   0.00134

2 Calculate the mass attenuation coefficient for a material with linear
attenuation coefficient 0.20 m^{-1} and density 1.4 kgm^{-3}.

    A   7.0 m^{2}kg^{-1}

    B   0.14 m^{2}kg^{-1}

    C   0.10 m^{2}kg^{-1}

    D   1.6 m^{2}kg^{-1}

3 Pellets containing beta-emitters can be placed next to a tumour in order to destroy
it. Which of the following is correct and an advantage of using this method to destroy
tumours?

    A   The beta radiation can be focused into a beam to target the tumour.

    B   Beta radiation is non-ionising.

    C   Beta radiation has a relatively short range.

    D   Beta radiation never damages healthy cells.

4 A ray diagram is shown below for an image formed by a converging lens. Calculate the
distance between the object and the lens to 2 s.f. (labelled *x* in the diagram).

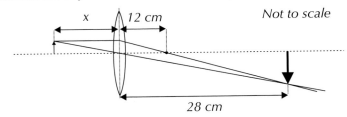

    A   21 cm

    B   0.048 cm

    C   0.025 cm

    D   8.4 cm

5 Below is a diagram of the eye.

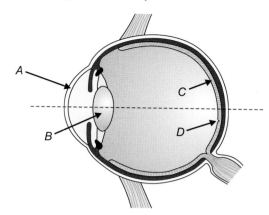

5.1 State what the letters *A-D* represent.

(2 marks)

5.2 Explain how rods and cones work. Include the role of each in your answer.

(4 marks)

5.3 Explain which part of the retina gives the best spatial resolution, and why.

(3 marks)

5.4 Explain how an A-scan can be used to measure the depth of an eyeball.

(3 marks)

6 The ears are used to detect sound waves.

6.1 Name the part of the ear which acts like a funnel for sound waves.

(1 mark)

6.2 Explain why there are greater pressure variations at the oval window of the ear than at the eardrum.

(2 marks)

6.3 State what is meant by the threshold of hearing, I_0.

(2 marks)

6.4 When a machine is in operation, the intensity level of sound it produces 5 m away is measured as 81 dB. Calculate the intensity of the sound at this point in Wm^{-2}.

(2 marks)

6.5 Explain why a dBA scale might be used instead of a dB scale when considering whether people working near the machine should be issued with ear protectors.

(2 marks)

7 Some people need corrective lenses in order to be able to see clearly.

7.1 Copy and complete the ray diagram below to show the image formed by a converging lens of the object shown.

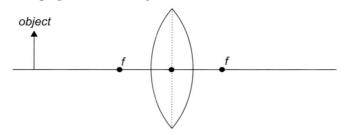

(2 marks)

7.2 A converging lens is used to produce an image with a linear magnification of 1.61 and a height of 29.3 cm. Find the height of the object.

(2 marks)

A hypermetropic person has a near point of 4.9 m.

7.3 Explain what kind of image a corrective lens for this condition will produce, and where it will produce it for objects that are 25 cm away.

(2 marks)

7.4 The person needs to be able to read a book 25 cm away from their eyes. What power of lens is needed for this to be possible?

(2 marks)

7.5 State which type of vision defect can be corrected with a cylindrical lens.

(1 mark)

8 In a rotating-anode X-ray tube, X-rays are produced in a continuous spectrum of intensity and in spikes of intensity at certain photon energies.

8.1 Explain why the anode is rotated.

(2 marks)

8.2 Explain what causes the continuous spectrum of photon energies.

(1 mark)

8.3 Explain what causes the spikes of intensity at certain photon energies.

(2 marks)

8.4 Define the half-value thickness, $x_{\frac{1}{2}}$, of a material.

(1 mark)

8.5 Calculate how far an X-ray beam will travel through a tissue with $\mu = 9.64$ m^{-1} before its intensity becomes a fifth of its initial intensity.

(3 marks)

8.6 Describe how an FTP detector works.

(3 marks)

8.7 State two alternative methods for improving the quality of an X-ray image.

(2 marks)

9 An ECG can be used to find out about the condition of the heart.

9.1 Describe where the valves in the heart are found and what their role is.

(2 marks)

The graph below shows an ECG for a healthy heart.

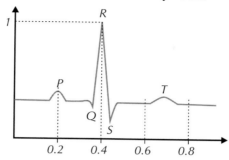

9.2 Add labels to both axes.

(2 marks)

9.3 State what is happening at the following points on the graph: P, 'QRS' and T.

(3 marks)

9.4 Explain the role of the sino-atrial node in the heart.

(2 marks)

9.5 State two methods which can ensure a good ECG waveform is obtained from a patient.

(2 marks)

10 Optical fibres are used to transmit light signals using total internal reflection.

10.1 State what is meant by the critical angle of total internal reflection.

(1 mark)

10.2 The refractive index of medium *A* is 1.25 and the refractive index of medium *B* is 1.15. Calculate the critical angle for a beam of light travelling from medium *A* to medium *B*.

(2 marks)

10.3 Explain what is meant by a coherent optical fibre bundle.

(1 mark)

10.4 Explain the roles of coherent fibre bundles and non-coherent fibre bundles in a flexible endoscope.

(2 marks)

10.5 State and explain two advantages of the use of endoscopes and keyhole surgery compared to traditional surgery methods, including safety and cost factors in your answer.

(4 marks)

11   Different medical imaging techniques are used for different diagnostic tasks.

11.1   Explain how an MR scanner uses magnetic fields and radio frequency radiation to create a cross-sectional image of the body.

(6 marks)

11.2   Give three advantages of using an MR scanner.

(3 marks)

11.3   Explain how a CT scanner works.

(4 marks)

11.4   A patient who has a pacemaker needs to have a brain scan. State a reason why it might not be safe for the patient to have an MR scan.

(1 mark)

11.5   Name two types of scan that could be used instead to scan the patient's brain.

(2 marks)

11.6   Compare the two types of scan that you gave in 11.5 in terms of convenience, resolution and patient comfort.

(5 marks)

12   Ionising radiation is often used for medical imaging.

12.1   Explain what a medical tracer is.

(1 mark)

Technetium-99m can be used as a medical tracer to produce an image of the liver.

12.2   State the physical half-life of technetium-99m.

(1 mark)

12.3   Explain how the technetium-99m is made to travel to the liver.

(1 mark)

12.4   Define the biological half-life of technetium-99m.

(1 mark)

12.5   Technetium-99m has a biological half-life of 1 day. Calculate its effective half-life. Give your answer to 2 significant figures.

(2 marks)

A hospital is using a Molybdenum-Technetium generator to produce technetium-99m.

12.6   State the reason why a hospital needs to use a Molybdenum-Technetium generator.

(2 marks)

12.7   Explain how a Molybdenum-Technetium generator works.

(3 marks)

12.8   Name the equipment that is used to detect the gamma radiation produced by the technetium-99m and briefly explain how it works.

(5 marks)

1. Inertia and Kinetic Energy

Inertia is how much an object resists a change in movement. This topic is all about the moment of inertia for rotating objects, and how you can use this to find the kinetic energy of a rotating object.

Moment of inertia

To make something start or stop moving requires a force to be applied. **Inertia** is a measure of how much an object resists a change in velocity (the larger the inertia, the larger the applied force needed to change its velocity by a given amount).

In linear systems, inertia is described by the mass of an object. The more massive it is, the harder an object is to move (or alter its motion if it's already moving). For rotating objects, this is described by the **moment of inertia**. This is a measure of how difficult it is to rotate an object, or to change its rotational speed.

The moment of inertia measures resistance to rotation, and depends on mass and its distance from the axis of rotation (the point or line around which the object is rotating).

┌─ Example ───

The flywheel in a rowing machine and the wheel of a wheelbarrow are about the same size, but have different moments of inertia.

- The flywheel in the rowing machine (see Figure 1) has a large moment of inertia — it takes a lot of effort to make it rotate.

- But the wheelbarrow wheel has a much smaller moment of inertia — it is easy to make it rotate about the central axle.

└──

Moment of inertia and mass distribution

The moment of inertia depends on how the mass is distributed. For a particle (point mass), the moment of inertia is simply:

I = moment of inertia in $kg\,m^2$

$$I = mr^2$$

m = mass of the particle in kg

r = distance from the axis of rotation in m

For an extended object, like a rod, the moment of inertia is calculated by adding up the individual moments of inertia of each point mass that makes up the object:

I = moment of inertia for an extended object

$$I = \Sigma mr^2$$

Σmr^2 = sum of mr^2 for all point masses making up the extended object

This means that the moment of inertia changes depending on the mass and how it is distributed about the axis of rotation.

For example, a hollow object will have a different moment of inertia to a solid one with the same mass, because the mass is concentrated at the edge, far from the axis of rotation.

Figure 1: *A rowing machine and a wheelbarrow.*

┌─ **Examples** ───────────────────────

The moment of inertia for a solid golf ball is $I = \frac{2}{5}mr^2$

But for a hollow tennis ball it's $I = \frac{2}{3}mr^2$

The moment of inertia for a solid wheel is $I = \frac{1}{2}mr^2$

But for a hollow ring or circular hoop it's $I = mr^2$
(The mass of a hollow ring is concentrated further from the centre, so its moment of inertia is greater, per unit mass.)

These equations are only valid for an object being rotated about its centre. An object will have a different moment of inertia if it is rotated about a point near its edge. This is because the mass will be distributed differently about the axis of rotation (i.e. the values of r will be different).

Combining moments of inertia
You can add together the individual moments of inertia of different objects to find the moment of inertia of a whole system. You'll usually be able to model individual objects as point masses.

┌─ **Example** ─ **Maths Skills** ───────────────────

a) **Calculate the moment of inertia of a 750 g bike wheel, which has a radius of 31.1 cm. A bike wheel may be modelled as a hollow cylinder for the purposes of calculating the moment of inertia.**

$I = 0.75 \times 0.311^2 = 0.0725... = 0.073\,\text{kgm}^2$ (to 2 s.f.)

b) **A 0.040 kg reflector is attached to the wheel 6.0 cm in from the outer edge. Assuming the reflector behaves like a point mass, calculate the new moment of inertia of the wheel.**

$I = \Sigma mr^2$, so $I_{new} = I_{initial} + mr^2$
(where mr^2 is the reflector's moment of inertia)
r is the distance from the centre of the wheel to the reflector, so:
$r = 31.1 - 6.0 = 25.1\,\text{cm} = 0.251\,\text{m}$
$m = 0.040\,\text{kg}$

$r = 31.1$ cm

6 cm

$I_{new} = 0.0725... + 0.040 \times 0.251^2 = 0.0750... = 0.075\,\text{kgm}^2$ (to 2 s.f.)

Rotational kinetic energy
An object's **rotational kinetic energy** depends on its moment of inertia. Just as you can find the kinetic energy of an object with linear motion, you can find the kinetic energy of a rotating object. The formula is derived like this:

- You know that kinetic energy is given by $E_k = \frac{1}{2}mv^2$.
- You also know that angular speed is given by $\omega = \frac{v}{r}$, so $v = \omega r$.
- You can substitute v for ωr to get $E_k = \frac{1}{2}m(\omega r)^2 = \frac{1}{2}m\omega^2 r^2$
- But moment of inertia $I = mr^2$, so you can write E_k as:

I = moment of inertia in kg m^2

E_k = rotational kinetic energy in J

$$E_k = \frac{1}{2}I\omega^2$$

ω = angular speed in rad s^{-1}

Example ─ Maths Skills

A dancer adds a 60.0 g mass to each end of her twirling baton. The baton rod is uniform, 70 cm long (to 2 s.f.) and has a mass of 150 g. Assume the added masses act as point masses.

Calculate the rotational kinetic energy of the baton as she spins it about its centre with an angular speed of 1.1 rad s^{-1}.

The moment of inertia for a rod of length L about its centre is $I = \frac{1}{12}mL^2$.

First, calculate the overall moment of inertia for the object:

$$I = I_{rod} + \Sigma mr^2 = \frac{1}{12}mL^2 + 2 \times \left[m \times \left(\frac{L}{2} \right)^2 \right]$$

$$= \frac{1}{12} \times 0.15 \times 0.70^2 + 2 \times [0.0600 \times 0.35^2]$$

$$= 0.0208... \text{ kg m}^2$$

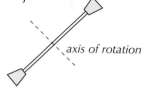

axis of rotation

Then substitute this value into the formula for kinetic energy:

$$E_k = \frac{1}{2}I\omega^2 = \frac{1}{2} \times 0.0208... \times 1.1^2 = 0.0125...$$

So the rotational kinetic energy of the baton is 0.013 J (to 2 s.f.).

Tip: You might see the kinetic energy of an object moving linearly referred to as "linear kinetic energy" or "translational kinetic energy". They both mean the same thing, i.e. $\frac{1}{2}mv^2$.

Tip: There are two masses, each a distance of $\frac{L}{2}$ from the rod's centre, so their combined moment of inertia is $\Sigma mr^2 = 2 \times \left[m \times \left(\frac{L}{2} \right)^2 \right]$

Practice Question ─ Application

Q1 A playground contains a roundabout of mass 470 kg. The roundabout has a moment of inertia of 790 kg m^2.

   a) Calculate the radius of the roundabout. Assume the roundabout is a solid wheel.

   b) A child of mass 48 kg sits 0.95 m from the centre of the roundabout. Calculate the moment of inertia of the roundabout and child combined by treating the child as a point mass.

   c) An adult pushes the roundabout while the child is sitting on it. She transfers 1350 J of energy. Assuming friction is negligible and the roundabout was originally at rest, find the angular speed of the roundabout.

Tip: The moment of inertia of a solid wheel is $I = \frac{1}{2}mr^2$.

Practice Questions ─ Fact Recall

Q1 What is meant by "moment of inertia"?

Q2 Give the equation for calculating the moment of inertia for an extended object.

Q3 What factors can affect the moment of inertia of a rotating object?

Q4 Write down the equation for the rotational kinetic energy of an object, and the corresponding linear equation. Explain the similarities between the two equations.

2. Rotational Motion

Learning Objectives:

- Understand the terms angular displacement, angular speed, angular velocity and angular acceleration.
- Be able to calculate angular velocity using $\omega = \frac{\Delta\theta}{\Delta t}$ and angular acceleration using $\alpha = \frac{\Delta\omega}{\Delta t}$
- Understand the similarities and differences between rotational motion and linear motion.
- Be able to use the equations for uniform angular acceleration:
$\omega_2 = \omega_1 + \alpha t$,
$\theta = \left(\frac{\omega_1 + \omega_2}{2}\right)t$,
$\theta = \omega_1 t + \frac{\alpha t^2}{2}$, and
$\omega_2{}^2 = \omega_1{}^2 + 2\alpha\theta$
- Be able to represent uniform and non-uniform angular acceleration on graphs.

Specification Reference 3.11.1.3

Tip: Learn the formulas for angular velocity and angular acceleration — they're not given in the exam.

Rotational motion is quite a lot like linear motion, but with more Greek letters. These equations should look quite familiar — they are similar to the equations for uniform acceleration that you met in Year 1 of A-Level physics.

Displacement, velocity and acceleration

You need to be familiar with each of these terms to do with rotational motion:

- **Angular displacement**, θ, is the angle through which a point has been rotated.
- **Angular velocity**, ω, is a vector quantity describing the angle a point rotates through per second. It's defined as:

ω = angular velocity in rad s⁻¹ $\omega = \frac{\Delta\theta}{\Delta t}$ θ = angular displacement in rad t = time in s

---
Example — Maths Skills

A blade of a fan moves through 35.0 rad in 1.67 seconds. What is the angular velocity of the fan?

$\omega = \frac{\Delta\theta}{\Delta t} = \frac{35.0}{1.67} = 20.958... = 21.0$ rad s⁻¹ (to 3 s.f.)

- **Angular speed**, ω, is just the magnitude of the angular velocity (it's a scalar quantity, so the direction of rotation doesn't matter).
- **Angular acceleration**, α, is the rate of change of angular velocity.

α = angular acceleration in rad s⁻² $\alpha = \frac{\Delta\omega}{\Delta t}$ ω = angular velocity in rad s⁻¹ t = time in s

---
Example — Maths Skills

A pilot increases the angular velocity of his helicopter's rotor blades from 15.2 rad s⁻¹ to 43.8 rad s⁻¹ in 12.1 s. Calculate the average angular acceleration of the rotor blades in this period.

$\alpha = \frac{\Delta\omega}{\Delta t} = \frac{(43.8 - 15.2)}{12.1} = 2.3636... = 2.36$ rad s⁻² (to 3 s.f.)

You may also see angular velocity related to linear velocity using $v = \omega r$ (this was derived on page 20). Angular acceleration can be related to linear acceleration using $a = \alpha r$, where r is the distance from the axis of rotation. It's derived like this:

- Angular acceleration, $\alpha = \dfrac{\text{change in angular speed}}{\text{change in time}} = \dfrac{\Delta\omega}{\Delta t}$.

- From page 20, you know that $v = \omega r$. So $\omega = \frac{v}{r}$.

- This means that $\alpha = \frac{1}{r} \times \frac{\Delta v}{\Delta t} = \frac{1}{r}a$, so $a = \alpha r$.

Figure 1: Helicopter rotor blades can rotate with a large angular speed.

Equations of motion for rotating objects

There are four main equations that you'll need to use to solve problems involving uniform (constant) angular acceleration. You need to be able to use them, but you don't have to know how they're derived — it's just shown here to help you learn them.

The derivations are pretty much the same as those for the equations for uniform linear acceleration that you met in Year 1 of A-Level physics. But instead of quantities for linear motion, they use the corresponding quantities for rotational motion given in Figure 2.

Tip: Have a look back over your Year 1 notes if you need a reminder of the equations for uniform linear acceleration.

	Linear	Rotational
Displacement	s	θ
Initial velocity	u	ω_1
Final velocity	v	ω_2
Acceleration	a	α

Figure 2: Symbols for rotational motion, and the corresponding linear quantities.

- Angular acceleration is the rate of change of angular velocity. From this you get:

$$\alpha = \frac{(\omega_2 - \omega_1)}{t} \quad \text{so} \quad \boxed{\omega_2 = \omega_1 + \alpha t} \quad \text{①}$$

where ω_1 is initial angular velocity and ω_2 is final angular velocity.

- Angular displacement = average angular velocity × time. If acceleration is constant, the average angular velocity is just the average of the initial and final angular velocities, so:

$$\boxed{\theta = \left(\frac{\omega_1 + \omega_2}{2}\right) \times t} \quad \text{②}$$

- Substitute the expression for ω_2 from equation 1 into equation 2 to give:

$$\theta = \frac{(\omega_1 + \omega_1 + \alpha t) \times t}{2} = \frac{2\omega_1 t + \alpha t^2}{2} \Rightarrow \boxed{\theta = \omega_1 t + \frac{1}{2}\alpha t^2} \quad \text{③}$$

- You can derive the fourth equation from equations 1 and 2:

Use equation 1 in the form: $\quad \alpha = \frac{(\omega_2 - \omega_1)}{t}$

Multiply both sides by θ, where: $\quad \theta = \frac{(\omega_1 + \omega_2)}{2} \times t$

This gives: $\quad \alpha\theta = \frac{(\omega_2 - \omega_1)}{t} \times \frac{(\omega_1 + \omega_2)t}{2}$

The t's on the right cancel, so: $\quad 2\alpha\theta = (\omega_2 - \omega_1)(\omega_2 + \omega_1)$

$$= \omega_2^2 - \omega_1\omega_2 + \omega_1\omega_2 - \omega_1^2$$

So: $\quad \boxed{\omega_2^2 = \omega_1^2 + 2\alpha\theta} \quad \text{④}$

So the angular equations (shown with their corresponding linear ones) are:

$$v = u + at \longrightarrow \boxed{\omega_2 = \omega_1 + \alpha t} \qquad s = ut + \frac{1}{2}at^2 \longrightarrow \boxed{\theta = \omega_1 t + \frac{1}{2}\alpha t^2}$$

$$v^2 = u^2 + 2as \longrightarrow \boxed{\omega_2^2 = \omega_1^2 + 2\alpha\theta} \qquad s = \left(\frac{u+v}{2}\right)t \longrightarrow \boxed{\theta = \left(\frac{\omega_1 + \omega_2}{2}\right)t}$$

Example — Maths Skills

A wheel initially at rest begins to spin with uniform angular acceleration. After 2.5 revolutions, it has an angular velocity of 4.9 rad s⁻¹. Calculate its angular acceleration.

First, see what variables you have to tell you which equation to use:
$\alpha = ?$, $\omega_1 = 0$, $\omega_2 = 4.9$ rad s⁻¹, $\theta = 2.5$ revolutions
So you should use $\omega_2^2 = \omega_1^2 + 2\alpha\theta$.
Next, make sure all values are in the correct units:
$\theta = 2.5$ revolutions $= 2.5 \times 2\pi = 15.7...$ radians
Rearrange the formula for α and substitute in the given values:
$$\alpha = \frac{\omega_2^2 - \omega_1^2}{2\theta} = \frac{4.9^2 - 0}{2 \times 15.7...} = 0.764...$$
$$= 0.76 \text{ rad s}^{-2} \text{ (to 2 s.f.)}$$

Tip: Angular displacement might be given in revolutions, and angular velocity in revs min⁻¹ or revs s⁻¹. So make sure you always convert to radians for displacement and rad s⁻¹ for velocity.

Angular motion-time graphs

Just like for linear motion, you can plot angular displacement and angular velocity against time, then use those graphs to find out more about the motion. Have a look back over your notes from Year 1 of A-Level physics so that you know what the gradient of and area under different motion-time graphs represent. There's more on how to calculate gradients in this book on p.8, and how to find the area under a graph on page 404.

Angular displacement-time graphs

When you plot angular displacement against time for a constant angular acceleration, you get a curve showing that displacement is proportional to t^2. This is because $\theta = \omega_1 t + \frac{1}{2}\alpha t^2$. If the object starts from rest, $\omega_1 = 0$, so $\theta = \frac{1}{2}\alpha t^2$.

Tip: The gradient of an angular displacement-time graph gives the angular velocity. So a straight line would show constant angular velocity (i.e. no angular acceleration).

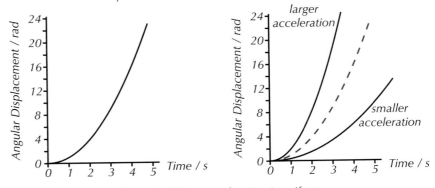

Figure 3: When acceleration is uniform, angular displacement is proportional to t^2.

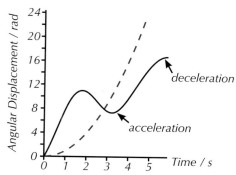

Figure 4: *When the angular acceleration is not constant, the displacement is no longer proportional to t^2.*

Tip: The gradient of a tangent to the curve gives the instantaneous angular velocity at that point.

Tip: The average angular velocity can be found from overall change in angular displacement ÷ time taken.

Angular velocity-time graphs

When you plot angular velocity against time for a constant angular acceleration, you get a straight line. This is because $\omega_2 = \omega_1 + \alpha t$ is of the form $y = mx + c$, where m (the gradient) is α and c (the y-intercept) is ω_1. For motion that starts from rest (i.e. $\omega_1 = 0$), the line will start at the origin.

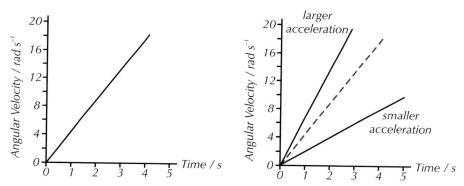

Figure 5: *When acceleration is uniform, angular velocity is directly proportional to t.*

Tip: A straight horizontal line would show constant angular velocity (i.e. there's no acceleration).

As you can see from the equation of the straight line ($\omega_2 = \omega_1 + \alpha t$), the gradient of the line gives the body's acceleration. So to find a uniform angular acceleration, you simply find the gradient of the angular velocity-time graph.

For non-uniform angular acceleration (as shown in Figure 6), the graph won't be a straight line. So to find the instantaneous acceleration at a given point, you'd find the gradient of the tangent to the curve at that point. To find the average acceleration between two points, you'd find the change in angular velocity over time between those points.

Tip: A negative gradient would show angular deceleration.

Like with linear velocity-time graphs, the area under the curve between two points gives the body's angular displacement in that time period.

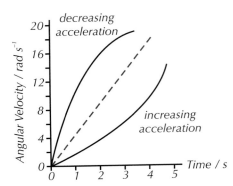

Figure 6: When the angular acceleration is not constant, the graph has a changing gradient.

Practice Questions — Application

Tip: Remember that there are 2π radians in a complete circle — see page 19.

Q1 Find the angular velocity in rad s^{-1} of the hour hand on a clock.

Q2 A spinning disc increases angular speed by 13.7 rad s^{-1} over a period of 41.2 s. Calculate the angular acceleration of the disc.

Q3 A computer stops working in a power cut. Its hard disk slows to a halt at a constant rate of 630 rad s^{-2} from spinning at full speed. In this time the disk rotates through 450 rad. At what speed was the disk spinning before the power was cut? Give your answer in revolutions per minute (rpm).

Q4 A Ferris wheel accelerates uniformly from 0.014 rad s^{-1} to 0.030 rad s^{-1} in the time it takes for the Earth to rotate by 7.0×10^{-3} rad about its axis.

a) What is the angular acceleration of the Ferris wheel?

b) Draw an angular velocity-time graph for the period in which this acceleration takes place.

c) What is the change in angular displacement of the Ferris wheel during the acceleration?

Figure 7: Large Ferris wheels have a small angular velocity to enable riders to board and disembark without stopping the wheel.

Practice Questions — Fact Recall

Q1 What is angular velocity?

Q2 Sketch an angular displacement-time graph for a spinning object that:

a) rotates at a constant angular velocity,

b) rotates at a constant angular acceleration,

c) rotates with a changing angular acceleration.

You can use the same axes for each graph.

Q3 Write down the four equations for uniform angular acceleration. State what each symbol means, and write the corresponding equation for uniform linear acceleration next to each equation.

3. Torque, Work and Power

A torque is like a moment for a body that rotates about an axis. You've seen work and power before — now it's time to apply them to rotational motion.

Torque

You should remember from Year 1 of A-Level that a **couple** is a pair of forces which cause no resultant linear motion, but which cause an object to turn.

When a force (or couple) causes an object to turn, the turning effect is known as **torque**. A torque is a bit like a moment, but it usually refers to a turning object. 'Moment' is generally used when an object is in equilibrium, and all the potential turning forces are balanced.

Like most things to do with rotating objects, torque is related to how far from the axis of rotation the force is applied. It is defined as:

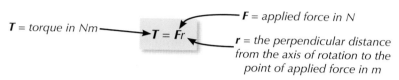

T = torque in Nm — $T = Fr$

F = applied force in N

r = the perpendicular distance from the axis of rotation to the point of applied force in m

Torque is also related to angular acceleration and moment of inertia (page 309). The formula linking them can be found as follows:

- You know that torque is given by $T = Fr$ and force is given by $F = ma$, so torque is $T = mar$.
- For a rotating object, $a = \alpha r$. So $T = m\alpha r^2$.
- You know the formula for moment of inertia is $I = mr^2$, so the equation for torque can also be written as:

T = torque in Nm — $T = I\alpha$

I = moment of inertia in kg m²

α = angular acceleration in rad s⁻²

Example — Maths Skills

Four 100 g (to 2 s.f.) masses are suspended from the axle of a wheel, as shown in the diagram. The perpendicular distance from the point of the applied weight to the centre of the axis of rotation is 0.15 m. When the masses are released, the wheel spins with an angular acceleration of 1.3 rad s⁻². Calculate the moment of inertia of the wheel. Friction is negligible.

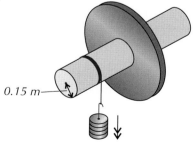

0.15 m

$T = Fr = mgr = 4 \times 0.10 \times 9.81 \times 0.15 = 0.5886$ Nm

$T = I\alpha$ so $I = T \div \alpha$

$\qquad = 0.5886 \div 1.3 = 0.452...$

$\qquad = 0.45$ kg m² (to 2 s.f.)

Learning Objectives:

- Know that torque, **T**, can be calculated using $T = Fr$ and $T = I\alpha$.
- Know that for a rotating body work is given by $W = T\theta$.
- Know that for a rotating body power is given by $P = T\omega$.
- Be aware that frictional torque must be accounted for in rotating machinery.

Specification References 3.11.1.4 and 3.11.1.6

Tip: $T = Fr$ is just like the equation for moments, $M = Fd$, only r is the distance from the axis of rotation rather than from a pivot.

Tip: The formula $a = \alpha r$ was derived on page 312.

Torque, work and power

When you rotate an object, you have to do **work** to make it move. Work in a linear system is the force multiplied by the distance. This can be rewritten for a rotating system using torque and angular displacement. The formula is derived like this:

- You know that, for linear motion, work (W) = force (F) × displacement (s).
- For a rotating system, $s = r\theta$.
- So $W = Fr\theta$.
- You know that $T = Fr$, so you can write the equation as:

$$W = T\theta$$

W = work in J T = torque in Nm θ = angular displacement in rad

Figure 1: *Exercise bikes allow the user to vary the torque so they can change the work required to turn the pedals.*

Example — Maths Skills

657 J of work is done to turn a wheel through 146 rad. What is the torque applied to the wheel to turn it through this angle?

$W = T\theta$, so $T = \dfrac{W}{\theta} = \dfrac{657}{146} = 4.50$ Nm

---

Power is the amount of work done in a given time. You know that power is given by $P = \dfrac{\Delta W}{\Delta t}$, so $P = \dfrac{\Delta (T\theta)}{\Delta t}$. But $\dfrac{\Delta \theta}{\Delta t} = \omega$, so you can write the equation as:

$$P = T\omega$$

P = power in W T = torque in Nm ω = angular velocity in rad s^{-1}

Example — Maths Skills

Louise applies a torque of 0.2 Nm to turn a doorknob 90° with an angular speed of 3.1 rad s^{-1}. Calculate the work done and the power exerted by Louise to turn the doorknob.

The doorknob is turned $90° = 90 \times \dfrac{\pi}{180} = \dfrac{\pi}{2}$ radians.

So $W = T\theta = 0.2 \times \dfrac{\pi}{2} = 0.314... = 0.3$ J (to 1 s.f.)

Power $= T\omega = 0.2 \times 3.1 = 0.62 = 0.6$ W (to 1 s.f.)

Tip: Remember — you need to convert the angle into radians for angular motion.

Frictional torque

In real-world mechanical systems, machines with rotating parts will experience an opposing **frictional torque**. Some of the power of the machine has to be used to overcome this frictional torque. So the net torque of a machine is:

$$T_{net} = T_{applied} - T_{frictional}$$

Tip: This equation isn't in your data and formulae booklet — make sure you learn it. Luckily it's quite a simple one.

You may have to use the principle of the conservation of energy to find the work done against friction for a rotating system.

A cog has a moment of inertia of 0.0040 kgm^2 and a diameter of 20.0 cm. A driving force of 0.070 N acts at the edge of the cog, in the direction of rotation at that point, causing it to accelerate whilst overcoming frictional torque. Find the power needed to overcome the frictional torque at the point that the cog has an angular velocity of 120 revs min^{-1}, if the angular acceleration at that instant is 1.25 rad s^{-2}.

Tip: The cog's diameter is 20.0 cm, so the distance, r, from the axis of rotation to the point at which the force is applied is half of this — i.e. 0.100 m.

First calculate the net torque on the cog: $T_{net} = T_{applied} - T_{frictional}$
$$= I\alpha$$
$$= 0.0040 \times 1.25$$
$$= 0.0050 \text{ Nm}$$

Then calculate the applied torque: $T_{applied} = Fr$
$$= 0.070 \times 0.100 = 0.0070 \text{ Nm}$$

Rearrange the equation for net torque to find the frictional torque:
$$T_{frictional} = T_{applied} - T_{net}$$
$$= 0.0070 - 0.0050 = 0.0020 \text{ Nm}$$

Finally, calculate the power needed to overcome friction:
$$P = T\omega = 0.0020 \times \frac{120 \times 2\pi}{60}$$
$$= 0.0251... = 0.025 \text{ W (to 2 s.f.)}$$

Tip: Remember to always convert to radians per second. 1 revolution = 2π radians.

Example — **Maths Skills**

A stationary wheel has two 0.20 kg masses suspended from it, as shown in the diagram. The masses are released. Just before they hit the ground, the masses have velocity 1.70 ms^{-1} and the wheel has 0.73 J of rotational kinetic energy, having turned through 0.90 radians. There is 0.10 Nm of frictional torque acting on the system. Calculate the height at which the masses were initially suspended above the ground.

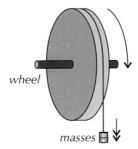

wheel

masses

Energy is always conserved, so the gravitational potential energy lost by the masses is equal to the total kinetic energy gained by the masses and the wheel, plus the work done to overcome frictional torque.

The total kinetic energy is the linear kinetic energy of the masses, plus the rotational kinetic energy of the wheel, so:
$$E_P = E_K + W \Rightarrow mgh = \tfrac{1}{2}mv^2 + E_{K rotational} + T\theta$$
$$= \tfrac{1}{2} \times 2 \times 0.20 \times 1.70^2 + 0.73 + 0.10 \times 0.90 = 1.398$$

So $h = \dfrac{1.398}{mg} = \dfrac{1.398}{2 \times 0.20 \times 9.81} = 0.356...$
$$= 0.36 \text{ m (to 2 s.f.)}$$

Practice Questions — Application

Q1 A circular saw blade accelerates from rest to 510 rad s^{-1} in 2.5 seconds. It has a moment of inertia of 5.4×10^{-4} kg m^2. Assuming friction is negligible, find the torque on the blade.

Q2 A circular cog with diameter 0.20 m is spinning in a machine. The net torque is 2.6 Nm.

a) What is the work done by the cog as it rotates through 32 radians?

b) The force applied to the cog to cause it to spin is 29 N. Find the frictional torque.

c) The cog took 0.80 s to complete the rotation of 32 radians. Find the power exerted by the machine on the cog for this rotation. Assume that the machine is rotating with a constant angular velocity.

Q3 A ship drops its anchor at sea. The anchor is attached to a chain wrapped around a wheel, and has a mass of 950 kg, as shown below. The frictional torque on the wheel is 425 Nm.

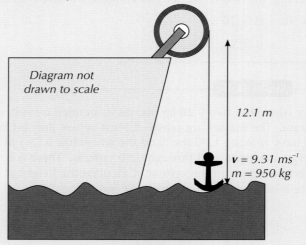

Diagram not drawn to scale

12.1 m

$v = 9.31 \ ms^{-1}$
$m = 950 \ kg$

At the moment the anchor hits the surface of the water:

- The anchor is moving vertically downwards at 9.31 ms^{-1}.
- The anchor has fallen 12.1 m.
- The rotational kinetic energy of the wheel is 61 000 J.

What angle has the wheel rotated through at this point?

Practice Questions — Fact Recall

Q1 For a rotating object, write down equations that link:

a) force, perpendicular distance from the axis of rotation and torque,

b) angular acceleration, moment of inertia and torque.

Label and give appropriate units for each quantity.

Q2 Use the equation for calculating work from torque and angular displacement to derive an expression for power in terms of torque and angular velocity.

4. Flywheels

Flywheels are wheels, usually quite large and heavy, that are fitted to machinery to smooth out torque and angular speed. They can also be used to store energy for use later on, for example in vehicles and power grids.

What are flywheels?

A **flywheel** is a heavy wheel that has a high moment of inertia (page 309) in order to resist changes to its rotational motion. This means that once it is spinning, it's hard to make it stop spinning (it has a high angular momentum).

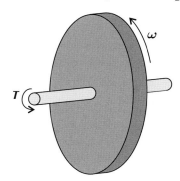

Figure 1: *A torque **T** is applied to a flywheel rotating about an axle with angular velocity **ω**.*

Flywheels are 'charged' as they are spun, turning the input torque (see page 317) into **rotational kinetic energy** (see page 310). As long as the flywheel keeps spinning at this rate, it stores the energy for later use. Rotational kinetic energy is given by:

$$E_k = \frac{1}{2}I\omega^2$$

Just enough power is continuously input to overcome frictional torque (see page 318), keeping the flywheel fully charged.

When extra energy is needed in a machine, the flywheel decelerates, transferring some of its kinetic energy to another part of the machine.

Flywheels designed to store as much energy as possible are called flywheel batteries.

Factors that affect energy storage capacity

Rotational kinetic energy is related to the moment of inertia and the square of the angular speed, which means these both affect how much energy a flywheel can store.

To increase the energy storage capacity of a flywheel, you can:

- Increase its mass — the moment of inertia, and hence kinetic energy stored, is directly proportional to the mass, so the heavier the flywheel is, the more energy it can store.

- Increase its angular speed — the energy stored increases with angular speed squared, so increasing the speed at which the flywheel spins greatly increases the amount of energy it can store.

Learning Objectives:

- Know what flywheels are and how they can be used to store energy.
- Know the factors that can affect the energy storage capacity of a flywheel.
- Understand how flywheels can be used to smooth torque and speed.
- Know how flywheels can be used in machines.
- Be able to describe how flywheels are used to store energy in vehicles and in machines used in manufacturing.

Specification Reference 3.11.1.2

Figure 2: *Traction engines often have exposed flywheels — the red arrow is pointing to the solid flywheel on this engine.*

- Use a wheel with spokes — compared to a solid wheel, a spoked wheel of the same mass stores much more energy (assuming everything else is kept constant). This is because the moment of inertia for a point mass is given by $I = mr^2$ (i.e. $I \propto r^2$). If more mass is concentrated further from the axis of rotation (i.e. r is larger), the moment of inertia will increase.

Examples — Maths Skills

A machine contains a solid, uniform flywheel of mass 3.0 kg and radius 0.45 m, giving it a moment of inertia of 0.30 kg m². It spins at 8.0 rad s⁻¹.

a) Find the rotational kinetic energy stored by the flywheel.

$$E_k = \frac{1}{2}I\omega^2 = \frac{1}{2} \times 0.30 \times 8.0^2 = 9.6 \text{ J}$$

b) The flywheel is replaced with one of mass 4.0 kg. Assuming all other variables are unchanged, find the rotational kinetic energy stored by the flywheel.

$$I = \frac{1}{2}mr^2 = \frac{1}{2} \times 4.0 \times 0.45^2 = 0.405$$
$$E_k = \frac{1}{2}I\omega^2 = \frac{1}{2} \times 0.405 \times 8.0^2 = 12.96 = 13 \text{ J (to 2 s.f.)}$$

c) The speed the new flywheel rotates at is increased to 11 rad s⁻¹. Find the rotational kinetic energy stored by the flywheel.

$$E_k = \frac{1}{2}I\omega^2 = \frac{1}{2} \times 0.405 \times 11^2 = 24.5025 = 25 \text{ J (to 2 s.f.)}$$

d) The new flywheel is replaced by a spoked flywheel with a moment of inertia of 0.60 kg m². It continues to rotate at 11 rad s⁻¹. Find the rotational kinetic energy stored by the flywheel.

$$E_k = \frac{1}{2}I\omega^2 = \frac{1}{2} \times 0.60 \times 11^2 = 36.3 = 36 \text{ J (to 2 s.f.)}$$

Limits on energy storage capacity

There is a limit to how much you can increase the factors that control the storage capacity before they become impractical — a giant, heavy wheel taking up half of your machine is unlikely to be useful. And if you increase the angular speed, the centrifugal force can increase to a point where it starts breaking the flywheel apart.

Modern flywheels are often made out of carbon fibre to allow higher angular speeds to be used without causing the flywheel to disintegrate. Although it is lighter than steel, carbon fibre is far stronger and so the wheel can be spun much faster.

Friction in flywheels

Even though a flywheel is very efficient at storing energy, it still loses some energy to air resistance and friction between the wheel and the bearings on which it spins. To combat this, modern flywheels can be:

- Lubricated to reduce friction between the bearings and the wheel.

- Levitated with superconducting magnets so there is no contact between the bearings and the wheel.

- Operated in vacuums or inside sealed cylinders to reduce the drag from air resistance.

Uses of flywheels

Smoothing torque and angular velocity

Flywheels are used in machines to balance, or 'smooth', the engine torque and the load torque. They help to keep systems running smoothly where either the power supplied or the load torque can vary.

In systems where the power supplied to the system can vary, e.g. if an engine only kicks in intermittently, flywheels are used to keep the angular velocity of any rotating components constant. The flywheel uses each spurt of power to charge up. It then delivers the energy it has stored smoothly to the rest of the system, instead of in bursts.

Flywheels are also used if the force that the system has to exert can vary, e.g. a riveting machine that needs to apply a large force intermittently. If at any time the load torque is too high (i.e. the driving force isn't big enough to power the system), then the flywheel decelerates, releasing some of its energy to top-up the system.

If the engine torque is higher than the load torque, then the flywheel will 'charge up' by accelerating so that the spare energy is stored until it is needed.

Tip: Engine torque is the torque exerted by a machine's engine. Load torque is the torque due to resistance forces that a machine must oppose to be useful.

Flywheel applications

Flywheels are used in a lot of everyday things. You need to know examples of flywheels in action in both vehicles and machines used in production processes.

┌─ **Examples** ─────────────────────────────

Some of the most common examples of flywheels are:

- Potter's wheels — Traditional potters' wheels are powered by a foot pedal, making it hard to apply a constant force to it. A flywheel is used to keep the speed of the wheel constant in order to make ceramic pots (see Figure 3).

- Regenerative braking — In regular cars, when the brakes are applied, friction causes the wheels to slow down, generating lots of heat. However, in some electric vehicles like cars and buses, when the brakes are applied, a flywheel is engaged. The flywheel then charges up with energy that would otherwise be lost. When vehicle is ready to accelerate, the flywheel uses its energy to turn the vehicle's wheels faster, before being disengaged until it's needed again.

- Power grids — When lots of electricity is used in an area, the electricity grid sometimes cannot meet that demand. Flywheels can be used to store surplus power in times of low demand and then provide extra energy while backup power stations are started up in times of high demand.

- Wind turbines — Flywheels can be used to store excess power on windy days or during off-peak times, and to give power on days without wind.

- Riveting machines — An electric motor charges up a flywheel, which then rapidly transfers a burst of power as the machine presses down on the rivet and fixes two sheets of material together. This is useful as it stops rapid changes of power going through the motor, which could cause it to stall, and means a less powerful motor can be used.

Figure 3: *Flywheels in pottery are used to maintain angular speed without the potter needing to apply a constant force to the potter's wheel.*

Advantages and disadvantages of flywheels

The properties of flywheels are quite different from other energy storage methods. Some of the advantages and disadvantages of flywheels are summarised in Figure 4.

Tip: Chemical batteries can only be discharged and recharged a finite number of times before they degrade — you may have noticed this effect in mobile phones. Flywheels don't degrade in the same way and can have a much longer lifespan.

Tip: Flywheels opposing changes in direction of motion can be an advantage in some applications — it can be used to improve the stability of vehicles.

Advantages	Disadvantages
They are very efficient.	They are much larger and heavier than other storage methods (e.g. batteries).
They last a long time without degrading.	They pose a safety risk as the wheel could break apart at high speeds. Protective casing to protect against this results in extra weight.
The recharge time is short.	Energy can be lost through friction.
They can react and discharge quickly.	If used in moving objects, they can oppose changes in direction, which can cause problems for vehicles.
They are environmentally friendly (they don't rely on chemicals to store energy).	

Figure 4: *A selection of advantages and disadvantages of flywheels.*

Practice Questions — Application

Q1 A riveting machine contains a solid flywheel of radius 5 cm. An engineer wants to replace the flywheel to improve the performance of the machine. Give and explain three factors the engineer should consider when specifying a replacement flywheel.

Q2 Explain how flywheels can be used to increase the efficiency of cars.

Practice Questions — Fact Recall

Q1 What is a flywheel?

Q2 Explain how a flywheel can be used to smooth angular velocity in a system if the engine torque or load torque are variable.

Q3 Give three examples of situations in which flywheels can be deployed, and explain why a flywheel is useful for each situation.

5. Angular Momentum

Angular momentum is a bit like linear momentum — it just involves rotational, rather than straight-line, motion. Whatever happens though, it must be conserved.

The angular momentum equation

You already know that **linear momentum** is equal to mass × velocity. You can replace mass with the moment of inertia, and linear velocity with angular velocity, and you get the formula that defines **angular momentum**:

Angular momentum has units Nms. → **angular momentum = $I\omega$**

I = moment of inertia in kgm^2

ω = angular velocity in rad s^{-1}

Learning Objectives:

- Know that angular momentum = $I\omega$.
- Know and be able to apply the law of the conservation of angular momentum.
- Be able to apply the concept of angular momentum to examples from sport.
- Understand that angular impulse is the change in angular momentum.
- Know that if the torque T is constant, $T\Delta t = \Delta(I\omega)$.

Specification Reference 3.11.1.5

Conservation of angular momentum

Like linear momentum, angular momentum is always conserved. When no external forces are applied (torque, friction, etc.), the total angular momentum of a system remains constant. This is known as the law of conservation of angular momentum, and it's useful to write it as:

$$I_{initial}\omega_{initial} = I_{final}\omega_{final}$$

This can be seen if you put two objects with different moments of inertia or angular velocities together.

Tip: Look back over your notes from Year 1 of A-Level physics if you need a reminder of linear momentum.

Example — Maths Skills

A disc has a moment of inertia I and is rotating at an angular velocity of 4 rad s^{-1}. A second identical disc that is not spinning is placed on top of the spinning disc, where it is held in place and begins to spin.
Calculate the angular velocity of the combined discs as they spin together at the same speed. Assume that frictional losses are negligible.

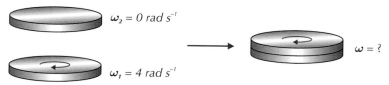

$\omega_2 = 0$ rad s^{-1}

$\omega_1 = 4$ rad s^{-1}

$\omega = ?$

Before the discs are put together, angular momentum = $I_1\omega_1 + I_2\omega_2$

Once they are put together, angular momentum = $(I_1 + I_2)\omega$

You can then equate these: $I_1\omega_1 + I_2\omega_2 = (I_1 + I_2)\omega$

So $\omega = \dfrac{I_1\omega_1 + I_2\omega_2}{(I_1 + I_2)}$

The discs are identical, so $I_1 = I_2$ and the equation becomes:

$\omega = \dfrac{I \times 4 + I \times 0}{2I} = \dfrac{4}{2} = 2$ rad s^{-1}

Figure 1: *The position of an ice skater's arms affects their moment of inertia and their angular velocity.*

Tip: Remember that 1 revolution per second is equivalent to 2π rad s^{-1}.

Another common example is an ice skater doing a spin. At the start of the spin, her arms are out away from her body. She then pulls her arms closer towards her, and begins to spin faster. This is due to the conservation of angular momentum — as she pulls in her arms, she decreases her moment of inertia, so her angular velocity must increase in order to conserve angular momentum.

Example — **Maths Skills**

An ice skater is spinning, with her arms out, at an angular velocity of 13 rad s^{-1}. With her arms out, her moment of inertia is 3.5 kgm^2. She then tucks in her arms, changing her moment of inertia to 1.2 kgm^2. Calculate her angular velocity in revolutions per second as she spins with her arms tucked in.

You can write the conservation of angular momentum as:

$$I_{initial}\omega_{initial} = I_{final}\omega_{final}$$

$$\omega_{final} = \frac{I_{initial}\omega_{initial}}{I_{final}} = \frac{3.5 \times 13}{1.2} = 37.91... \text{ rad s}^{-1}$$

$37.91... \div 2\pi = 6.03... = 6.0$ revolutions per second (to 2 s.f.)

Angular impulse

Angular impulse is the change in angular momentum. You can write the equation for angular impulse as:

Angular impulse has units Nms. → **angular impulse** $= \Delta(I\omega)$

If the torque (see page 317) on the system is constant, you can derive another equation for angular impulse as follows (assuming I is constant):

- You know that $\boldsymbol{T} = I\alpha$, and that $\alpha = \frac{\Delta\omega}{\Delta t}$.
- This means that $\boldsymbol{T} = I\frac{\Delta\omega}{\Delta t}$.
- $I\Delta\omega$ is a change in angular momentum, so:

$$\Delta(I\omega) = \boldsymbol{T}\Delta t$$

$\boldsymbol{T}$ = torque in Nm

Δt = time for which the torque is applied in s

Tip: This equation corresponds to the equation for the linear impulse (the change in linear momentum), given by $\boldsymbol{F}\Delta t = \Delta(m\boldsymbol{v})$. You met this in Year 1 of A-Level physics.

Example — **Maths Skills**

A spanner, initially at rest, has a constant torque of 0.3 Nm applied to it for 2 seconds. Calculate the angular impulse acting on the spanner and the angular velocity of the spanner at the end of the 2 seconds. The moment of inertia of the spanner is 0.2 kgm^2.

$\Delta(I\boldsymbol{\omega}) = \boldsymbol{T}\Delta t = 0.3 \times 2 = 0.6$ Nms

$I\omega_{final} - I\omega_{initial} = \boldsymbol{T}\Delta t$

$\omega_{final} = \frac{\boldsymbol{T}\Delta t + I\omega_{initial}}{I}$

$= \frac{0.3 \times 2 + 0.2 \times 0}{0.2}$

$= 3$ rad s^{-1}

Q1 a) A vinyl record of mass 0.180 kg and diameter 0.304 m is dropped onto a turntable that is rotating at $33\frac{1}{3}$ revolutions per minute. When the record lands it begins to move with the turntable. The turntable without the record has a moment of inertia of 5.12×10^{-3} kgm². Find the angular velocity of the record as it turns on the turntable. You may assume that frictional losses are negligible and that the driving force of the turntable can be ignored. The vinyl record can be modelled as a solid disc with a moment of inertia of $I = \frac{1}{2}mr^2$.

b) Power is applied to the turntable so that the angular speed returns to $33\frac{1}{3}$ revolutions per minute. Find the angular impulse of this change.

Q2 A flywheel with a moment of inertia of 0.75 kgm² accelerates from 1.0 rad s⁻¹ to 1.4 rad s⁻¹ in 25 seconds due to a constant torque. Calculate the torque acting on the flywheel in this time.

Q3 An acrobatic diver jumping from a diving board is somersaulting slowly. She wants to complete as many spins as possible before entering the water. Should the diver tuck herself into a ball or extend her arms as far as possible above her head? Explain your answer.

Tip: To answer Q3, you'll need to remember that in the formula for moment of inertia, $I \propto r^2$ (see pages 309-310).

Q1 Give the formula for angular momentum.

Q2 What is angular impulse?

Q3 Give an expression for angular impulse in situations involving a constant torque.

Q4 Give and explain an example where angular momentum is relevant to a sport.

Learning Objectives:

- Know the equation for the first law of thermodynamics — $Q = \Delta U + W$, where Q is the energy transferred to the system by heating, ΔU is the increase in the internal energy and W is the work done by the system.

- Understand applications of the first law of thermodynamics.

- Know that the equation for an ideal gas is $pV = nRT$.

Specification References 3.11.2.1 and 3.11.2.2

6. The First Law of Thermodynamics

Thermodynamics can sometimes be a bit tricky. Luckily, you've already covered some thermodynamics stuff earlier in this book. Take a look back at pages 47-71 to remind yourself of it before starting this section.

What is the first law of thermodynamics?

The first law of thermodynamics describes how energy is conserved in a system through heating, cooling and doing work. A system is a volume of space filled with gas. Systems can be either open or closed — open systems allow gas to flow in, out or through them, e.g. water vapour leaving a boiling kettle. Closed systems don't allow gas to enter or escape — the gas in a balloon can be modelled as a closed system. The balloon will lose air over a longer period of time, but in the short term, these losses are negligible.

The first law of thermodynamics can be written as:

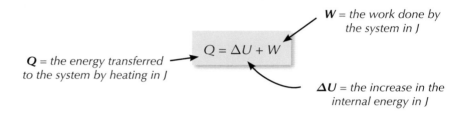

W = the work done by the system in J

$$Q = \Delta U + W$$

Q = the energy transferred to the system by heating in J

ΔU = the increase in the internal energy in J

Q can be a positive or negative value. If energy is transferred <u>to</u> the system (i.e. the system is heated), Q is positive. If energy is transferred <u>away</u> from the system (i.e. the system is cooled), Q is negative.

ΔU is the increase in **internal energy**. The internal energy of a system is the sum of the potential and kinetic energies of all the particles in the system (see page 47 for more).

W is the work done by the system (the work the gas does) — for example a gas in a cylinder expanding and moving a piston. If work is done on the gas, for example by compressing it, then the value of W is negative.

Figure 1: *The gas in a balloon can be modelled as a closed system.*

Tip: This stuff on the first law of thermodynamics is pretty important. Make sure you understand what's happening with Q, ΔU and W in all these different situations — it'll really help with the rest of this section.

Examples

- Say you supply energy to a gas by heating it. If it expands as it is heated, then the gas does work. So some of the heat energy you've supplied goes into increasing the internal energy of the gas, while some of it is transferred into work done.

- Alternatively, say you compress a gas. Work is done on the system, which means the value of W is negative. The work done on the system goes into some combination of increasing the internal energy of the gas and being transferred to the surroundings as heat energy.

- Another scenario could be heating a gas and compressing it at the same time. Then the increase in internal energy of the gas is equal to the heat energy you've put in plus the work you've done in compressing it.

A cylinder is sealed by a moveable piston. The gas in the cylinder is heated with 60 J of heat to move the piston. The internal energy of the gas increases by 5 J.

a) **Calculate the work done by the gas to move the piston.**
As heat is being input and the gas is doing work, both Q and W are positive.
$Q = \Delta U + W \Rightarrow 60 = 5 + W$
So $W = 60 - 5 = 55$ J
55 J of work is done by the gas to move the piston.

piston

gas

b) **Now the piston does 60 J of work on the gas to compress it. No heat is lost. Calculate the change in the internal energy of the gas.**
$Q = \Delta U + W = 0$
No heat is supplied or lost, so $Q = 0$, and work is done <u>on</u> the gas, so $W = -60$ J (i.e. it's negative). Therefore:
$0 = \Delta U - 60 \Rightarrow \Delta U = 60$ J
The internal energy of the gas increases by 60 J.

piston

gas

Tip: Make sure you know which letters represent which numbers — if you're getting in a flap, try drawing everything out and labelling everything. You need to make sure you <u>really</u> understand what's going on here, including the reasons for all the positive and negative numbers.

The first law and ideal gases

You need to know how to apply the first law to changes in closed systems. These are also known as **non-flow processes** because the gas doesn't go anywhere (none of the gas is allowed to flow in or out of the system). To apply the law, you have to assume that the gas in a system is an **ideal gas**. This means that the internal energy is only dependent on the temperature — as the temperature increases, the internal energy increases. You also need to assume that if any work is done, this causes a change in volume.

A change in heat energy can cause a change in volume, pressure and temperature. So you can use the ideal gas equation, which you should recognise from page 58. It's given by:

Tip: An ideal gas is theoretical — hence the "ideal". However, in a lot of situations, a gas can be approximated to (or assumed to be) an ideal gas. See pages 53-64 for more on ideal gases.

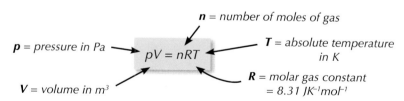

n = number of moles of gas

p = pressure in Pa $\longrightarrow$ $pV = nRT$ $\longleftarrow$ T = absolute temperature in K

V = volume in m³

R = molar gas constant = 8.31 JK⁻¹mol⁻¹

Tip: 0 K = −273 °C. See page 47.

R is constant, and for a change in a closed system, n is too (no gas is lost or added). So, $\frac{pV}{T}$ = constant. You could also write this as:

$$\frac{p_1 V_1}{T_1} = \frac{p_2 V_2}{T_2}$$

Where p_1, V_1 and T_1 are the initial pressure, volume and temperature of the gas respectively, and p_2, V_2 and T_2 are the pressure, volume and temperature after the system has experienced a change.

A closed system contains 7.50 mol of an ideal gas. Initially, the gas has a volume of 0.22 m³ and a temperature of 320 K.

a) Calculate the pressure of the gas.

Using $pV = nRT$, rearrange for p and substitute in the numbers:

$$p = \frac{nRT}{V} = \frac{7.50 \times 8.31 \times 320}{0.22}$$
$$= 9.0654... \times 10^4 \text{ Pa}$$
$$= 9.1 \times 10^4 \text{ Pa (to 2 s.f.) } (= 91 \text{ kPa})$$

b) The volume of the gas is then decreased to 0.15 m³ and the temperature increased to 372 K. Calculate the new pressure of the gas.

Using $\frac{p_1 V_1}{T_1} = \frac{p_2 V_2}{T_2}$, rearrange for p_2 and then substitute in the numbers:

$$p_2 = \frac{p_1 V_1 T_2}{T_1 V_2} = \frac{9.0654... \times 10^4 \times 0.22 \times 372}{320 \times 0.15}$$
$$= 1.54566 \times 10^5 \text{ Pa}$$
$$= 1.5 \times 10^5 \text{ Pa (to 2 s.f.) } (= 150 \text{ kPa})$$

Tip: You could also have used $pV = nRT$ again for part b).

Practice Questions — Application

Q1 A balloon containing air can be compressed. The internal energy of the air in the balloon increases by 0.300 J when the work done on the balloon is 1.98 J. Calculate the energy lost from the balloon as heat, stating any assumptions you make.

Q2 A sealed cube box contains 22 mol of an ideal gas. The length of one of the sides of the box is 0.82 m and the temperature of the gas is 308 K. Calculate the pressure of the gas.

Practice Questions — Fact Recall

Q1 Give the equation for the first law of thermodynamics, defining any symbols you use.

Q2 A gas in a sealed container is being compressed. No heat is supplied to the system while it is being compressed.

a) State whether work is done on or by the system.

b) The internal energy of the gas doesn't change during the compression. State whether the value of W and the energy transferred to the system take positive or negative values.

7. Non-flow Processes

There are a few different non-flow processes you need to know. They depend on whether temperature, heat energy, pressure or volume remain the same.

Isothermal changes

Isothermal is a fancy way of saying that temperature stays the same. The absolute temperature of an ideal gas determines its internal energy. As the temperature during an isothermal change remains constant, you can write for an isothermal process that:

$$\Delta U = 0$$

Substituting this into the first law of thermodynamics gives $Q = 0 + W$, which means that:

$$Q = W$$

This means that supplying heat energy to the system will result in an equivalent amount of work being done by the gas (as its volume increases). Basically, you can think of this as saying that heating a gas will cause it to expand, in order for its temperature (internal energy) to stay the same. In this case, both Q and W are positive.

Similarly, if work is done on the system (i.e. W is negative), then an equal amount of heat energy will be lost by the gas (so Q will also be negative). For example, if you compress a gas, it will have to give off heat in order for its temperature to stay the same.

Learning Objectives:

- Know that for an isothermal change pV = constant.
- Know that for an adiabatic change pV^γ = constant.
- Know that the work done for a change at constant pressure is $W = p\Delta V$.
- Be able to apply the first law of thermodynamics to isothermal, adiabatic, constant pressure and constant volume processes.

Specification Reference 3.11.2.2

Example —— **Maths Skills**

An ideal gas is contained in a sealed cylinder with a moveable piston. 42 J of heat energy is supplied to the gas so that it expands isothermally.

a) **What is the work done?**
The gas expands isothermally, so $\Delta U = 0$ and therefore $Q = W$.
So $W = 42$ J.

b) **Is the work done on or by the gas?**
The gas is expanding, so work is done by the gas.

Using the ideal gas equation (see page 58), you can see that a constant temperature T means that:

p = pressure in Pa ⟶ pV = constant ⟵ V = volume in m³

So you can also write:

$$p_1 V_1 = p_2 V_2$$

Tip: This should look familiar — it's Boyle's law from page 53.

Example —— **Maths Skills**

An ideal gas inside a sealed container undergoes an isothermal expansion. It has an initial volume of 1.8 m³, an initial pressure of 120 kPa and an initial temperature of 290 K. The final pressure of the gas is 98 kPa. Calculate the new volume of gas.

The ideal gas is undergoing an isothermal expansion, so you can use pV = constant and therefore $p_1 V_1 = p_2 V_2$. Rearranging this equation for V_2 and substituting the values in gives:

$$V_2 = \frac{p_1 V_1}{p_2} = \frac{120 \times 10^3 \times 1.8}{98 \times 10^3} = 2.2040... = 2.2 \text{ m}^3 \text{ (to 2 s.f.)}$$

Tip: Don't forget to convert from kPa to Pa.

Adiabatic changes

An **adiabatic** process is when no heat is transferred in or out of the system:

$$Q = 0$$

Using the first law of thermodynamics, if $Q = 0$ then $\Delta U = -W$. This means that any change in the internal energy of the system is caused by work done by/on the system. For example, if work is done by the system (it expands), then the internal energy of the system will decrease by an equivalent amount (so W is positive and ΔU is negative) — see the example below. Similarly, if work is done on the system (i.e. the gas is compressed), then the internal energy will increase by an equivalent amount. So W is negative and ΔU is positive.

Figure 1: *The pistons in a motor vehicle engine are an example of an application of adiabatic changes. There's more on this on page 342.*

Example — **Maths Skills**

An ideal gas in a sealed container undergoes an adiabatic expansion. The work done is equal to 22 J. Calculate the change in the internal energy of the system.

Work is done <u>by</u> the system (because the gas is expanding) and so W is positive: $W = 22$ J. Substitute this into $\Delta U = -W$:

$$\Delta U = -22 \text{ J}$$

This shows that work done <u>by</u> the system (and therefore a positive value for W) gives a negative value for ΔU — the internal energy decreases.

As internal energy only depends on temperature, a change in temperature will occur if work is done. The maths behind this process is pretty hard, but thankfully you just need to know that for an adiabatic change:

p = pressure in Pa
V = volume in m³
γ = adiabatic constant
$$pV^{\gamma} = \text{constant}$$

Tip: The value of the adiabatic constant, γ, depends on the type of gas in the system. For example, a monatomic gas has an adiabatic constant of $\gamma = \frac{5}{3}$.

Exam Tip
You don't need to remember any values for γ. A value will be given in the exam if you need it.

The equation $pV^{\gamma} = $ constant means you can also write:

$$p_1 V_1^{\gamma} = p_2 V_2^{\gamma}$$

Example — **Maths Skills**

A container full of helium (a monatomic gas) is sealed by a moveable piston (so gas cannot escape). The container is cylindrical, with a radius of 20 cm and a height of 20 cm. The initial pressure inside the container is 1.2×10^5 Pa. The piston moves downwards by 15 cm, adiabatically compressing the gas. Calculate the pressure inside the container after the piston has moved. (Assume that helium has an adiabatic constant of $\frac{5}{3}$.)

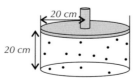
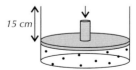

The volume of a cylinder is given by $V = \pi r^2 h$. As the gas is compressed, r stays constant so the change in volume only depends on the change in height. This means you can rearrange $p_1 V_1^{\gamma} = p_2 V_2^{\gamma}$ to give:

$$p_2 = \frac{p_1 V_1^{\gamma}}{V_2^{\gamma}} = p_1 \left(\frac{h_1}{h_2}\right)^{\gamma} = (1.2 \times 10^5) \times \left(\frac{20}{5}\right)^{\frac{5}{3}} = 1.2095... \times 10^6$$

$$= 1.2 \times 10^6 \text{ Pa (to 2 s.f.)}$$

Tip: If a diagram isn't given for these kind of questions, it always helps to do a quick sketch of what's happening. It will help you to visualise what on earth is going on.

Tip: Don't worry if you didn't spot that you could do the question this way — you'd still get all of the marks if you calculated each volume separately and then substituted them into the equation for p_2.

Changes at constant pressure

For processes where the pressure doesn't change, you can calculate the work done as a gas expands or contracts by using:

p = pressure in Pa

W = work done by the system in J $\longrightarrow$ $W = p\Delta V$ $\longleftarrow$ ΔV = change in volume in m^3

Tip: You first saw this on page 59.

You can easily see where this equation comes from:

- Work = force × distance, $W = F\Delta x$.
- Pressure = force ÷ area, so force is pressure times area, $F = pA$.
- Substituting pA for F in the work done equation gives you $W = pA\Delta x$.
- $A\Delta x$ is simply the change in volume, which gives $W = p\Delta V$.

For an expansion, ΔV and W are positive. For a compression, both are negative (remember that work is done on the system here).

From the ideal gas equation (see page 58), if p is constant then $\frac{V}{T}$ is constant, and so:

Tip: To maintain a constant pressure, a change in temperature must cause a change in volume. So, heating a gas will cause it to expand, and cooling it will cause it to decrease in volume.

$$\frac{V_1}{T_1} = \frac{V_2}{T_2}$$

where T is the absolute temperature (see page 47).

Examples — Maths Skills

A closed system contains an ideal gas that is compressed so that its volume changes from 2.5 m^3 to 1.8 m^3. The gas remains at a constant pressure of 87 kPa throughout the process. The initial temperature of the gas is 290 K.

a) Calculate the final temperature of the gas.

Rearrange $\frac{V_1}{T_1} = \frac{V_2}{T_2}$ for T_2 and substitute in the numbers:

$T_2 = \frac{V_2 \times T_1}{V_1} = \frac{1.8 \times 290}{2.5} = 208.8$ K = 210 K (to 2 s.f.)

b) Calculate work done by the system.

$W = p\Delta V = 87 \times 10^3 \times (1.8 - 2.5)$

$= -6.09 \times 10^4 = -6.1 \times 10^4$ J (to s.f.)

Tip: The value of W is negative because work is done <u>on</u> the gas.

c) The heat energy given out during this process is 1.52 × 10⁵ J. Calculate the change in internal energy.

$Q = \Delta U + W$, so $\Delta U = Q - W$

$= (-1.52 \times 10^5) - (-6.09 \times 10^4)$

$= -9.11 \times 10^4 = -9.1 \times 10^4$ J (to 2 s.f.)

Changes at constant volume

In changes where the volume of the system is kept constant, the work done is zero:

$$W = 0$$

From the first law of thermodynamics, if $W = 0$, then:

$$Q = \Delta U$$

This means that all of the heat energy transferred to the system goes directly into increasing the internal energy, U, of the system. None of it is transferred as work done.

The ideal gas equation, $pV = nRT$, can be rearranged to get $\frac{p}{T} = \frac{nR}{V}$. For a change at a constant volume, this means that:

$$\frac{p}{T} = \text{constant}$$

And therefore:

$$\frac{p_1}{T_1} = \frac{p_2}{T_2}$$

┌─ Examples ── Maths Skills ──────────────────────

A container is filled with an ideal gas. 15 000 J of heat energy is extracted from the gas so that the temperature decreases from 525 K to 462 K. The initial pressure of the gas is 132 kPa. The volume remains constant.

a) **Calculate the final pressure of the gas.**

$p_2 = \frac{p_1 T_2}{T_1} = \frac{132 \times 10^3 \times 462}{525} = 1.1616 \times 10^5 = 1.16 \times 10^5$ Pa (to 3 s.f.)

b) **Find the change in internal energy of the gas.**

The volume is constant, so $W = 0$ and $Q = \Delta U + W$ becomes $Q = \Delta U$. Heat energy is being extracted from the system, so Q is negative: $Q = -15\ 000$ J and therefore $\Delta U = -15\ 000$ J.

Practice Questions — Application

Q1 A container contains an ideal gas that undergoes an isothermal compression. The initial volume of the gas is 6.8 m³ and the final volume is 4.3 m³. The initial pressure of the gas is 112 kPa. Calculate the final pressure of the gas.

Q2 An ideal gas in a sealed container is expanded so that its volume changes from 0.60 m³ to 0.80 m³. The final pressure of the gas is 43 kPa. Assuming the process is adiabatic and the adiabatic constant of the gas is 1.28, calculate the initial pressure of the gas.

Q3 A cylinder with a moveable piston contains an ideal gas. The gas does 275 J of work so that the piston moves out. The final temperature of the gas is 281 K and the final volume of the gas is 0.050 m³. The pressure has a constant value of 88 kPa throughout this process.

a) Calculate the initial volume of the gas.

b) Calculate the initial temperature of the gas.

Q4 The pressure of an ideal gas increases from 24 kPa to 124 kPa. The initial temperature of the gas is 115 K and the volume of the gas doesn't change. Calculate the final temperature of the gas.

Practice Questions — Fact Recall

Q1 Define an isothermal change.

Q2 Define an adiabatic change.

Q3 Give the equation for the work done by a system at constant pressure.

Q4 What is the work done by a system at constant volume?

8. p-V Diagrams

p-V (pressure-volume) diagrams look different for different non-flow processes — you'll need to be able to recognise and sketch them for each process.

p-V diagrams for non-flow processes

All of the different non-flow processes you've seen on the last few pages can be represented on a p-V diagram — a graph of pressure against volume. Pressure is plotted on the vertical axis, volume is plotted on the horizontal axis, and a curve is drawn to show how the two are related as a change (non-flow process) occurs.

An arrow can be drawn on a p-V curve to indicate the direction in which the change is happening (i.e. whether each of the pressure or volume are increasing or decreasing). The area under a p-V curve represents the work done during that process. Estimating the area under a p-V graph is covered on page 404.

Isotherms

A p-V curve for an isothermal process (page 331) is called an **isotherm**. An isotherm is a smooth curve (remember pV = constant, so $p \propto \frac{1}{V}$ — they're inversely proportional). p-V diagrams for an isothermal compression and an isothermal expansion are shown in Figure 2. The arrows show the direction the change happens in — e.g. for a compression, V decreases and p increases, and for an expansion, V increases and p decreases.

The shaded area under the curve is the magnitude of the work done during the process, between V_1 and V_2. Remember, whether it's positive or negative will depend on whether the work is done <u>by</u> or <u>on</u> the system. For a compression (arrow pointing up and to the left), work is done on the system, so it's negative. For an expansion (arrow is pointing down and to the right) the system is doing work, so it's positive.

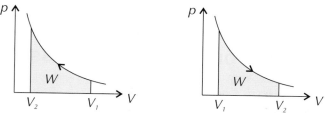

Figure 2: *p-V curves for an isothermal compression (left) and an expansion (right).*

The position of an isotherm on a p-V diagram indicates the temperature at which the process happens. The higher the temperature of the system, the further from the origin the isotherm will be.

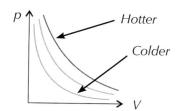

Figure 3: *p-V curves for isothermal processes at different temperatures.*

Learning Objectives:

- Be able to represent different non-flow processes on a p-V diagram.

- Know that the work done by a system can be estimated from the area under a p-V graph.

- Know that calculating the work done from a p-V graph for a change at constant pressure uses the equation $W = p\Delta V$.

- Know that in a p-V diagram of a cyclic process, the work done per cycle = the area of the loop.

Specification Reference 3.11.2.3

Figure 1: *The British chemist Robert Boyle first discovered the relationship between pressure and volume for an isothermal change.*

The *p-V* diagram below shows the isothermal expansion of a closed system. Calculate the work done by the system between pressures A and B.

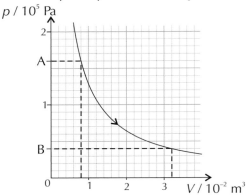

The work done is equal to the area under the graph. You can estimate the area by counting the total number of squares under the curve.

To find the work done between A and B, go across from the vertical axis to the curve and find the value of *V* at each of those points.

Then find the area under the graph between those two values of *V*.

First find out how much energy each square is worth:

The height of each square represents $(1 \times 10^5) \div 10 = 1 \times 10^4$ Pa.

The width of each square represents $(1 \times 10^{-2}) \div 5 = 0.002$ m³.

$W = p\Delta V$, so the amount of work done represented by each square is:

$1 \times 10^4 \times 0.002 = 20$ J.

Next count the number of small squares under the curve and multiply it by the work per square. The number of small squares under the line between the values of *V* that correspond to pressures A and B is around 89.

So about $89 \times 20 \approx 1780$ J of work is done between pressures A and B.

Tip: You can also estimate the area under the graph by counting the number of large squares under the graph or by using the trapezium rule.

p-V diagrams for adiabatic processes

The *p-V* curves for adiabatic processes are similar to those for isothermal processes, but they have a steeper gradient. Figure 4 shows how an isothermal and adiabatic compression between two volumes would look if they had the same initial temperature. The area under the adiabatic curve is larger than the area under the isothermal curve, so more work is done to compress gas adiabatically than isothermally.

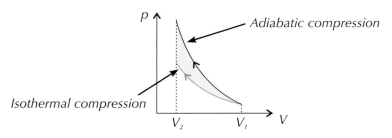

Figure 4: *p-V diagram comparing an adiabatic compression curve and an isothermal compression curve. The shaded area is the extra work needed to compress the gas adiabatically.*

Alternatively, the gas does less work if it expands adiabatically instead of isothermally. Figure 5 shows how an adiabatic and isothermal expansion would look if they had the same initial temperature.

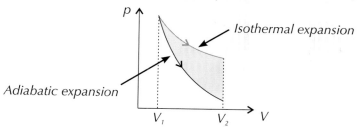

Figure 5: *p-V diagram comparing an adiabatic expansion curve and an isothermal expansion curve. The shaded area is the extra work done by the gas as it expands isothermally.*

Constant volume and constant pressure

Unsurprisingly, *p-V* diagrams for changes with a constant volume are straight vertical lines. For these processes, there is no work done as the volume doesn't change — you saw this on page 333. You can also see this from the *p-V* diagram as there is no area under the line. As a system is kept at a constant volume but heated between temperatures T_1 and T_2, its pressure will increase. If it is cooled at a constant volume, the pressure will decrease.

For a process where the pressure doesn't change, the *p-V* diagram is a horizontal straight line. The work done is the area of the rectangle under the graph — $W = p\Delta V$. You saw this equation on page 59.

Tip: An increase in pressure at a constant volume means an increase in temperature (see page 55). This is shown in Figure 6 by the initial and final temperatures: T_1 and T_2.

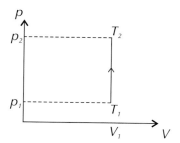

Figure 6: *p-V diagram for an increase in pressure at a constant volume.*

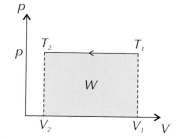

Figure 7: *p-V diagram for a decrease in volume at a constant pressure. The shaded area is the work done on the system.*

Tip: For a decrease in pressure at a constant volume, the *p-V* diagram would be the same as Figure 6, except the arrow would be pointing downwards and T_1 and T_2 would swap over, as would p_1 and p_2.

Cyclic processes

A system can undergo different processes one after another which form a cycle (loop). They start at a certain pressure and volume and return to them at the end of each cycle.

┌─ **Example** ────────────────────────

The *p-V* diagram shows:

A → B: Isothermal compression

B → C: Expansion at a constant pressure

C → D: Cooling at a constant volume

D → E: Expansion at a constant pressure

E → F: Adiabatic expansion

F → A: Compression at a constant pressure

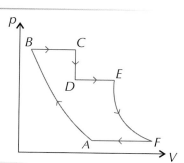

Tip: One example of a cyclic process is the working of an engine — this is covered on p.340.

To find the net work done during a cyclic process, you find the difference between the work done by the system and the work done to the system. This is equal to the area inside the loop created by the cyclic process.

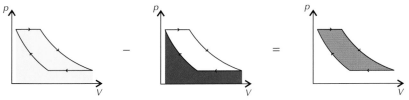

Figure 8: *The shading on the p-V diagrams shows the following: work done by the system – work done on the system = net work done by the system*

> **Work done per cycle = Area of loop**

If a question asks you to calculate the net work done per cycle of a cyclic process from a *p-V* diagram, you will need to either calculate or estimate the area of the loop. If the loop is only made up of straight lines you will be able to calculate the area of the rectangle(s). If the loop has any curved lines you will need to estimate the area by counting the number of squares in the loop (using the same method as the example on page 404) or by averaging out the area of the loop as a sum of trapeziums.

Tip: The area in the first graph shows the work done by the system as it expands. The area in the second graph shows the work done on the system as it is compressed. The area in the final graph is the difference between the two.

┌─ **Example** ── **Maths Skills** ─────────────────────────

The *p-V* diagram below shows a cyclic process for a gas. Estimate the work done per cycle of this process.

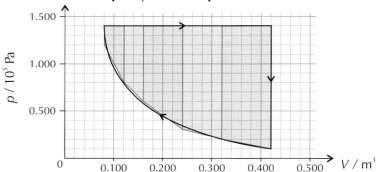

Estimating by using trapeziums:

The trapeziums that are used have been shaded on the graph above. The work done is approximately the sum of the areas of the shaded trapeziums.

Remember the area of a trapezium is given by $\frac{a+b}{2} \times h$, so you can write:

Total area = work done

$$\approx \left(\left(\frac{0.20+0.60}{2} \times 0.04\right) + \left(\frac{0.60+0.80}{2} \times 0.04\right) + \left(\frac{0.80+1.10}{2} \times 0.08\right)\right.$$
$$\left. + \left(\frac{1.10+1.20}{2} \times 0.08\right) + \left(\frac{1.20+1.30}{2} \times 0.10\right)\right) \times 10^5$$
$$= (0.016 + 0.028 + 0.076 + 0.092 + 0.125) \times 10^5 = 33\ 700\ \text{J}$$

Estimating by counting the small squares:

Value of work done for one small square $= 0.02 \times 0.10 \times 10^5 = 200$ J
Total number of small squares ≈ 168
So work done $\approx 168 \times 200 = 33\ 600$ J

Tip: A factor of 10^5 has been taken out when calculating the sum of all the trapeziums as the values of pressure on the vertical axis are $\times 10^5$.

Tip: You could also have estimated the work done by counting the large squares. One large square = 5000 J and there are approximately 7 large squares in the loop, so work done $\approx 7 \times 5000 = 35\ 000$ J.

Practice Questions — Application

Q1 The graph below shows the adiabatic expansion of a gas inside a sealed container. Estimate the work done by the system as it expands from 0.2 m³ to 0.8 m³.

Tip: Don't forget to think about whether the work done will be positive or negative.

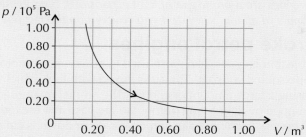

Q2 The graph below shows a cyclic process.

a) Estimate the work done by the system in one cycle.

b) Estimate the net work done by the system in one cycle.

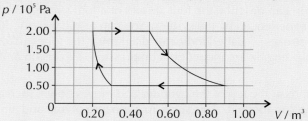

Q3 Sketch a p-V graph for a cyclic process that undergoes the following steps:

1. Cooling at a constant volume.
2. Isothermal expansion.
3. Adiabatic compression.
4. Compression at a constant pressure.

Practice Questions — Fact Recall

Q1 Define an isotherm.

Q2 A p-V diagram is used to represent a gas undergoing an isothermal expansion. If the temperature of the gas was higher, what effect would this have on the p-V diagram?

Q3 Will a gas do more work to double its volume if it expands isothermally or adiabatically?

Q4 How can you tell from a p-V diagram that no work is done by a system undergoing a change at a constant volume?

9. Four-stroke Engines and Indicator Diagrams

Four-stroke engines are a prime example of expansions, compressions, heating, cooling and all the other stuff you covered on the last few pages.

Four-stroke petrol engines

Internal combustion engines (like those used to power many motor vehicles) contain cylinders filled with air. The air is mixed with fuel, which is then burned, releasing a large amount of energy. The air-fuel mixture in these cylinders is trapped by tight-fitting pistons (so the gas can't escape), which move up and down. Each time a piston moves up or down is called a stroke. **Four-stroke engines** are engines which burn fuel once every four strokes of a piston. (Two-stroke engines burn it every two strokes, and so on...)

In an engine, the gas inside a cylinder can be thought of as a thermodynamic system. By applying everything you've learnt from the last few pages, you can describe the four strokes of a four-stroke petrol engine and the changes experienced by the system. You can also draw *p-V* diagrams to illustrate the strokes — called **indicator diagrams**. The four strokes of a piston in a four-stroke engine are induction, compression, expansion and exhaust.

Induction

The piston starts at the top of the cylinder and moves down, increasing the volume of the gas above it. This sucks in a mixture of fuel and air through an open inlet valve. The pressure of the gas in the cylinder remains constant, just below atmospheric pressure. The indicator diagram for this stroke is a straight horizontal line, showing an increase in volume at constant pressure.

Tip: A piston is a disc or short cylinder which fits closely inside a larger cylinder. The piston moves up and down as a trapped liquid or gas expands and compresses.

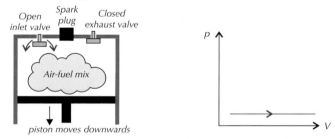

Figure 1: *Stroke 1 (induction) of a four-stroke petrol engine.*

Compression

The inlet valve is closed and the piston moves back up the cylinder. This does work on the gas, increasing the pressure. Just before the piston is at the end of this stroke, the spark plug creates a spark which ignites the air-fuel mixture. The temperature and pressure suddenly increase at an almost constant volume. The indicator diagram shows a decrease in volume and increase in pressure. Following the spark, the line is almost vertical.

Figure 2: *Nikolaus Otto was one of the engineers who helped develop the four-stroke engine.*

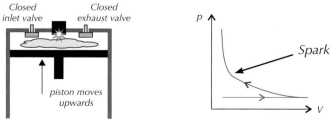

Figure 3: *Stroke 2 (compression) of a four-stroke petrol engine.*

Expansion

The hot air-fuel gas mixture expands and does work on the piston, pushing it downwards. The work done by the gas as it expands is more than the work done to compress the gas, as it is now at a higher temperature. There is a net output of work. Just before the piston is at the bottom of the stroke, the exhaust valve opens and the pressure reduces. The indicator diagram shows an increase in volume and decrease in pressure.

Tip: There is a downturn in the curve at the end of the expansion stroke. This is when the exhaust valve opens, causing the pressure to reduce whilst the volume remains constant.

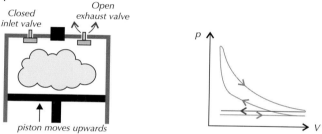

Figure 4: Stroke 3 (expansion) of a four-stroke petrol engine.

Exhaust

The piston moves up the cylinder, and the burnt gas leaves through the exhaust valve. The pressure remains almost constant, just above atmospheric pressure. The indictor diagram shows a horizontal line as volume decreases at a constant pressure.

Tip: The difference in the areas below the compression and expansion curves in the indicator diagram gives the net output of work.

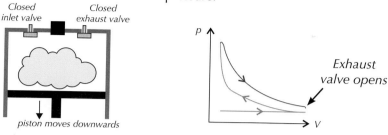

Figure 5: Stroke 4 (exhaust) of a four-stroke petrol engine.

Tip: Indicator diagrams show the processes for a single cylinder. Engines are normally made up of multiple cylinders.

Four-stroke diesel engines

Whilst four-stroke diesel engines undergo the same four strokes, they work slightly differently to four-stroke petrol engines.

In the induction stroke, only air is pulled into the cylinder, not an air-fuel mixture. Diesel engines don't have a spark plug, so in the compression stroke the air is compressed until it reaches a temperature high enough to ignite diesel fuel. Just before the end of the stroke, diesel is sprayed into the cylinder through a fuel injector and ignites. The expansion and exhaust strokes are then the same as for a petrol engine.

The indicator diagram for a diesel engine is also slightly different — there is no sharp peak at the start of the expansion stroke. The peak is much flatter, showing the point at which the diesel fuel is injected and heats up to combustion temperature — see Figure 7.

Figure 6: The cylinders containing pistons make up part of a car engine.

Exam Tip
As well as petrol and diesel engine four-stroke cycles, you could be asked about other cycles. If you are, all the information you need will be given — you'll just have to use what you know to interpret what's going on.

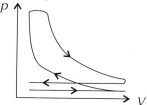

Figure 7: Indicator diagram for a four-stroke diesel engine

Theoretical indicator diagrams

Theoretical indicator diagrams assume that the cycle takes place under perfect conditions. The theoretical cycle for a four-stroke petrol engine is called the Otto cycle. For a four-stroke diesel engine, it is called the diesel cycle. Both of these theoretical models make the following assumptions:

Tip: You first saw the adiabatic constant on page 332.

1. The same gas is taken continuously around the cycle.
   The gas is pure air, with an adiabatic constant $\gamma = 1.4$.
2. Pressure and temperature changes can be instantaneous.
3. The heat source is external.
4. The engine is frictionless.

Petrol engine

The theoretical cycle for the four-stroke petrol engine is made up of four processes:

Tip: Remember, adiabatic compressions are where no heat is transferred to or from the system. See page 332.

A	The gas is compressed adiabatically.
B	Heat is supplied whilst the volume is kept constant.
C	The gas is allowed to cool adiabatically.
D	The system is cooled at a constant volume.

Tip: The theoretical and actual indicator diagrams are different because of the assumptions listed above, and because some heat is transferred to other parts of the system.

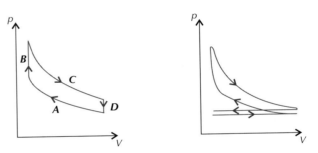

Figure 8: *A theoretical indicator diagram (left) and an actual indicator diagram (right) for a four-stroke petrol engine*

Diesel engine

The four processes in the theoretical cycle for a four-stroke diesel engine are:

A	The gas is adiabatically compressed.
B	Then heat is supplied, but this time pressure is kept constant.
C	The gas is allowed to cool adiabatically.
D	Then the system is cooled at a constant volume.

Figure 9: *Petrol and diesel engines work differently, so it's important that you don't put the wrong type of fuel in your car — this can be expensive to fix.*

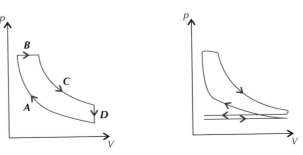

Figure 10: *A theoretical indicator diagram (left) and an actual indicator diagram (right) for a four-stroke petrol engine*

Comparing theoretical and real engines

Engineers compare indicator diagrams of real engines to theoretical models in order to see how well they are performing. The main differences between theoretical and real-life diagrams are:

Tip: The differences between a theoretical and real engine affect the engine's efficiency and power output. This is covered in more detail on the next page.

- The corners of theoretical indicator diagrams are not rounded. This is because it is assumed that the same air is used continuously. For real engines, these corners are rounded. One of the reasons for this is that the inlet and exhaust valves take time to open and close and let new air in and burnt gas out.

- In a real four-stroke petrol engine, heating doesn't take place at a constant volume (process B on the petrol engine indicator diagram in Figure 8). This is because the increase in pressure and temperature would have to be instantaneous to do this (or the piston would have to pause for a moment).

- The theoretical model doesn't include the small amount of negative work caused by the loop between the exhaust and the induction lines because it assumes the same air cycles around the system continuously. That's why the theoretical diagram is a closed loop but the actual one isn't.

- Engines have an internal heat source (the burning air-fuel mixture), not an external one. This means the temperature rise is not as large as in the theoretical model because the fuel used to heat the gas is never completely burned in the cylinder, so you can never get the maximum energy out of it. This means that theoretical engines can achieve higher pressures (and so have a higher peak).

- Energy is needed to overcome friction caused by the moving parts of a real engine, so the net work done will always be less than for a theoretical engine. This means that the area inside the loop is smaller for real four-stroke engines.

Practice Questions — Fact Recall

Q1 Describe the four strokes of a four-stroke petrol engine. Include a labelled indicator diagram in your answer.

Q2 Describe the four strokes of a four-stroke diesel engine. Include a labelled indicator diagram in your answer.

Q3 Sketch the theoretical indicator diagram for a four-stroke petrol engine, and state what each part of the cycle represents.

Q4 Sketch the theoretical indicator diagram for a four-stroke diesel engine, and state what each part of the cycle represents.

Q5 Why do real indicator diagrams for engines have rounded corners?

Q6 What assumption is made so that the theoretical indicator diagram for an engine is a continuous loop?

Q7 Give two reasons why the theoretical energy gained from an engine is higher than the actual amount of energy that is gained from an engine.

10. Engine Power and Efficiency

Now you know how a four-stroke engine works, this topic is all about the power and efficiency of an engine. You'll soon see that not all of the energy from the fuel burnt by a petrol or diesel engine is transferred into useful work.

Engine Power

Indicated power

You saw on page 338 that the area of a loop for a cyclic process gives the work done. For engines, the small amount of negative work (see page 343) is negligible, so the net work done by an engine cylinder for one cycle is the area of the loop on the indicator diagram.

The **indicated power** of an engine cylinder is the net work done by the cylinder in one second (i.e. work done in one cycle × number of cycles per second). If an engine has more than one cylinder, you need to multiply the cylinder's indicated power by the number of cylinders to get the engine's indicated power.

> Indicated power of engine =
> (area of *p-V* loop) × (no. of cycles per second) × (no. of cylinders)

You can think of indicated power as being the maximum theoretical power generated by the gases in the engine cylinders.

Output (brake) power

A piston moving up and down in an engine is connected to a crankshaft by a rigid rod — see Figure 1. This converts the up and down motion of a piston into a rotational motion. The greater the pressure on the piston (and hence the larger the force generating the up and down motion), then the greater the rotational motion (torque) generated.

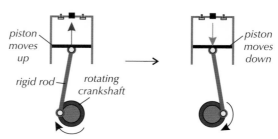

Figure 1: *Each piston in an engine is connected to the crankshaft by a stiff rod.*

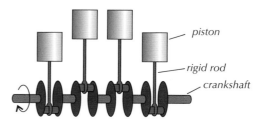

Figure 3: *Pistons are attached to the same crankshaft in a line.*

Figure 2: *A crankshaft can be seen at the top of this diesel engine.*

The **output (or brake) power** of an engine is the useful power output at the crankshaft. It can be calculated from:

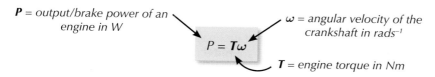

P = output/brake power of an engine in W

ω = angular velocity of the crankshaft in rad s^{-1}

$$P = T\omega$$

T = engine torque in Nm

Tip: You've already seen this equation on page 318.

Example — Maths Skills

The crankshaft of an engine rotates with frequency 34 cycles per second and torque 256 Nm. Calculate the output power of the engine.

First calculate the angular velocity of the crankshaft and then substitute into the equation for power:

$\omega = 2\pi f = 2 \times \pi \times 34 = 213.62...$ rad s^{-1}

$P = T\omega = 256 \times 213.62...$

$\qquad = 54\ 688.8...$

$\qquad = 55\ 000$ W (to 2 s.f.) (= 55 kW)

Tip: Take a look back at pages 312-318 for a reminder of the rotational motion equations.

Friction power

Friction occurs between many different moving parts of an engine, for example between the piston and the cylinder, at the bearings and when the valves are opened or closed. Work needs to be done to overcome friction in the engine and the power needed to do this is called the **friction power**. This means that the brake power of the engine is less than the indicated power that was calculated for the engine.

Tip: Remember — power is just the amount of energy transferred per second (1 W = 1 Js^{-1}).

friction power = indicated power – brake power

Example — Maths Skills

The power of an engine with 6 cylinders is being investigated. When the engine is operating at 31 cycles per second, the area of a *p-V* graph loop for one of the engine's cylinders is found to be 82 J. The manufacturer of the engine claims a brake power of 12 600 W.
Calculate the friction power of the engine.

First calculate the indicated power:

Indicated power =

$\qquad$ = (area of *p-V* loop) × (no. of cycles per second) × (no. of cylinders)

$\qquad$ = 82 × 31 × 6

$\qquad$ = 15 252 W

Then substitute into the equation for friction power:

Friction power = indicated power – brake power

$\qquad$ = 15 252 – 12 600 = 2652

$\qquad$ = 2700 W (to 2 s.f.) (= 2.7 kW)

Tip: The answer is rounded to 2 s.f. in this example. Although it might seem that the lowest number of significant figures in the question is one (6 cylinders), this is an exact value. So it's not been rounded and doesn't affect the number of significant figures in your answer.

Section 7 — Option C: Engineering Physics   345

Engine efficiency

All efficiencies are just a measure of how much of the input power is transferred usefully. An engine's **input power** is the amount of heat energy per unit time it could potentially gain from burning fuel. The calorific value of the fuel tells you how much energy the fuel has stored in it per unit volume, so the input power is the rate of fuel supplied multiplied by its calorific value.

Tip: You might be given the calorific value in terms of energy per unit mass. If this happens, you'll need the flow rate to be in terms of mass per second rather than volume per second.

$$\text{input power} = \text{calorific value} \times \text{fuel flow rate}$$

There are three kinds of engine efficiency you need to know. The **mechanical efficiency** of an engine is affected by the amount of energy lost through moving parts (for example, through friction). It's a ratio of the power generated by the cylinders to the power output at the crankshaft.

$$\text{mechanical efficiency} = \frac{\text{brake power}}{\text{indicated power}}$$

Thermal efficiency describes how well heat energy is transferred into work.

$$\text{thermal efficiency} = \frac{\text{indicated power}}{\text{input power}}$$

The equation for the **overall efficiency** is:

$$\text{overall efficiency} = \frac{\text{brake power}}{\text{input power}}$$

Example — **Maths Skills**

An engine with an overall efficiency of 42% has an input power of 123 kW. The indicator diagram shows the engine has an indicated power of 92 kW. Calculate the mechanical efficiency of the engine.

Overall efficiency $= \dfrac{\text{brake power}}{\text{input power}}$

Rearranging this gives:

Brake power = overall efficiency × input power
$= 0.42 \times 123\ 000$
$= 51\ 660\ \text{J}$

Tip: Unless a question specifies, you can give an efficiency either as a decimal or as a percentage.

Mechanical efficiency $= \dfrac{\text{brake power}}{\text{indicated power}}$
$= \dfrac{51\ 660}{92\ 000}$
$= 0.561...$
$= 56\%$ (to 2 s.f.)

Practice Questions — Application

Q1 The theoretical indicator diagram for a single cylinder petrol engine is shown below. The engine operates at 24 cycles per second. The torque of the crankshaft is 260 Nm.

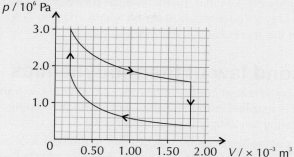

a) Estimate the indicated power of the engine.

b) Calculate the output power of the engine.

c) Using your answers from a) and b), calculate the following:
   (i) the friction power of the engine,
   (ii) the mechanical efficiency of the engine.

Q2 a) An engine uses fuel with a calorific value of 43.2 MJkg^{-1}. The engine burns the fuel at a rate of 0.018 kg s^{-1}. Calculate the input power of the engine.

b) Explain what would happen to the input power if each cylinder within the engine took twice as long to complete one cycle.

Tip: Don't forget to convert MJkg^{-1} into Jkg^{-1}.

Practice Questions — Fact Recall

Q1 Give the equation for calculating indicated power.

Q2 What is the function of a crankshaft?

Q3 What is the equation for the output power of an engine? Define all the terms you use.

Q4 What equations are used for calculating the friction power and input power of an engine?

Q5 Give the equations for mechanical, thermal and overall efficiency.

Figure 2: *This candle ornament is a simple example of a heat engine. The candles are the heat source and the colder surroundings are part of the heat sink.*

Tip: T_H must be larger than T_C in order for work to be done in an engine.

11. The Second Law of Thermodynamics

The next two pages show what the theoretical efficiency of a heat engine is. You'll also see why an engine can never be 100% efficient — for this you need to know about the second law of thermodynamics.

The second law of thermodynamics

Heat engines convert heat energy into work. No engine can transfer all the heat energy it is supplied with into useful work though — some heat always ends up increasing the temperature of the engine.

If the engine temperature reaches that of the heat source, then no heat flows and no work is done. This means that the working of a heat engine can't be described using only the first law of thermodynamics (see page 328) — Q and W would both be zero. So engines also have to obey the second law of thermodynamics, which can be expressed as:

> Heat engines must operate between a **heat source** and a **heat sink**.

A heat sink is a region which absorbs heat from the engine. A diagram of a heat engine operating between a heat source and a heat sink is shown in Figure 1.

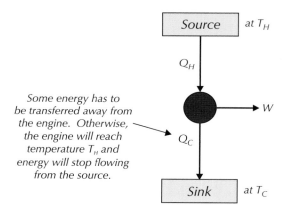

Some energy has to be transferred away from the engine. Otherwise, the engine will reach temperature T_H and energy will stop flowing from the source.

Figure 1: *Diagram to show the energy transfers in an engine — from a heat source to a heat sink. (H stands for hot and C stands for cold.)*

The heat energy transferred to the engine from the heat source is Q_H. Some of this energy is converted into useful work, W, but some of the energy (Q_C) must be transferred to the heat sink, which has a lower temperature (T_C) than the heat source. This means engines can never be 100% efficient.

If the working of an engine could just be described by the first law of thermodynamics, theoretically all of the heat energy supplied to a heat engine could be transferred into useful work.

The second law and engine efficiency

You can think of the efficiency of an engine as a measure of the amount of useful output energy from the original input energy. So the energy efficiency of the engine is just the ratio of work done by the engine to the energy supplied by the heat source. The work done by the engine is the difference in the energy given by the heat source and the energy transferred to the heat sink, so you can calculate the **efficiency** of a heat engine using the following equation:

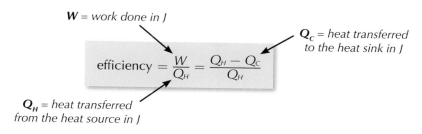

W = work done in J

Q_C = heat transferred to the heat sink in J

$$\text{efficiency} = \frac{W}{Q_H} = \frac{Q_H - Q_C}{Q_H}$$

Q_H = heat transferred from the heat source in J

By assuming perfect conditions, you can also calculate the maximum theoretical efficiency:

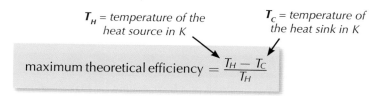

T_H = temperature of the heat source in K

T_C = temperature of the heat sink in K

$$\text{maximum theoretical efficiency} = \frac{T_H - T_C}{T_H}$$

In practice, a real heat engine's efficiency is lower than its theoretical maximum. This is because energy is lost through the engine's components due to friction and the energy that is needed to move them. Also, the fuel that is used doesn't burn entirely, and so the value of Q_H is actually lower in practice.

Example — **Maths Skills**

A heat engine is operating between a heat source at 422 K and a heat sink at 308 K. Calculate the maximum theoretical efficiency of the engine.

Maximum theoretical efficiency $= \dfrac{T_H - T_C}{T_H} = \dfrac{422 - 308}{422} = 0.270142...$
$= 27.0\ \%$ (to 3 s.f.)

Maximising efficiency

To maximise the efficiency of an engine, as much as possible of the input heat energy must be transferred usefully. However, engines are very inefficient — there is usually a lot of waste heat, which is transferred to the surrounding area and lost.

Combined heat and power (CHP) plants try to limit energy waste by using this waste heat for other purposes, for example heating houses and businesses nearby. An example of this is the Markinch Biomass CHP plant which was recently built in Fife, Scotland. It generates electricity which it supplies to a local papermaker and the National Grid. The excess heat is then used to create steam to dry paper in the paper mill. Another example is the Shotton CHP station — see Figure 3.

Figure 3: *Shotton CHP station in Flintshire, Wales. The plant was designed to capture waste heat and use it to provide steam to a turbine and generator.*

Practice Question — Application

Q1 The diagram shows a heat engine operating between a heat source and a heat sink.

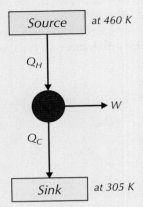

a) Calculate the maximum theoretical efficiency of the engine.

b) If the engine actually has an efficiency of 26% and draws 790 J from the heat source, calculate the energy lost to the heat sink.

c) Calculate the useful work done by the engine.

Practice Questions — Fact Recall

Q1 Why can't the working of a heat engine be described by the first law of thermodynamics alone?

Q2 What does the second law of thermodynamics say about heat engines? Use a diagram to help illustrate your answer.

Q3 Give the equations for efficiency (in terms of thermal energy transferred) and maximum theoretical efficiency (in terms of source and sink temperatures) of a heat engine.

Q4 Give two reasons why a real heat engine's efficiency is lower than the maximum theoretical efficiency of an engine.

Q5 What is a combined heat and power plant and why are they used?

12. Reversed Heat Engines

You've already seen that heat engines transfer heat from a hot area to a cold area. Reversed heat engines are, you guessed it, the reverse of a heat engine — the heat is transferred from a cold area to a hot area.

What are reversed heat engines?

Reversed heat engines operate between hot and cold reservoirs like other engines. However, the big difference is the direction of energy transfer — heat energy is taken from the cold reservoir and transferred to the hot reservoir. For reversed heat engines, these reservoirs are called spaces (instead of sources and sinks).

Heat naturally flows from hotter to colder spaces, so to transfer heat from a colder space to a hotter space, work (W) must be done.

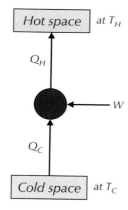

Figure 1: *Diagram to show the energy transfers in a reversed heat engine — from a cold space to a hot space.*

Compare the direction of the arrows for work (W) in the diagrams for a heat engine (page 348) and a reversed heat engine. A reversed heat engine diagram shows that there is an input of work. A heat engine diagram shows that there is an output of work.

It might seem strange that heat energy can be taken from a cold space — but remember that as long as the cold space isn't at absolute zero, it will have some heat energy (even if it's just a small amount).

Heat pumps and **refrigerators** are both reversed heat engines, but they have different functions.

Refrigerators

In this case, the cold space is inside the refrigerator, whilst the hot space is the refrigerator's surroundings (the room the refrigerator is in). A refrigerator aims to extract as much heat energy from the cold space as possible for each joule of work done. The work is done to transfer heat energy away from the cold interior, via pipes on the back of the appliance. Refrigerators keep enclosed spaces cool, so they can be used to keep perishable food fresh for longer.

Learning Objectives:
- Understand the principles and uses of refrigerators and heat pumps.
- Be able to draw a diagram showing energy transfers for a reversed heat engine.
- Understand the term coefficient of performance.
- Know the equations for the coefficient of performance for a refrigerator,
$$COP_{ref} = \frac{Q_C}{W} = \frac{Q_C}{Q_H - Q_C},$$
and a heat pump,
$$COP_{hp} = \frac{Q_H}{W} = \frac{Q_H}{Q_H - Q_C}.$$
- Know the equations for the maximum theoretical coefficient of performance for a refrigerator,
$$COP_{ref} = \frac{T_C}{T_H - T_C},$$
and a heat pump,
$$COP_{hp} = \frac{T_H}{T_H - T_C}.$$

Specification Reference 3.11.2.6

Heat pumps

Here, the cold space is usually the outdoors and the hot space is the inside of a house. A heat pump aims to pump as much heat as possible into the hot space per joule of work done. They are used to heat rooms and water in homes.

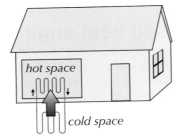

Figure 3: Diagram showing how a heat pump is used to heat a house.

Coefficient of performance (COP)

Reversed heat engines are judged on how well they can transfer heat based on the amount of work done on them. The **coefficient of performance** is a measure of how well this work is converted into heat transfer. For example, a heat pump with a coefficient of performance of 4 transfers 4 J of heat energy for every 1 J of work done. It's similar to efficiency (see page 346), but it's not called that as the coefficient of performance can be above 1. A COP greater than 1 means that the amount of energy transferred is greater than the amount of work needed to transfer it. The higher the COP, the lower the running cost.

Tip: The efficiency of an engine can never be above 100%.

Coefficient of performance for refrigerators

As it's the heat removed from the cold space that's important for a refrigerator, its coefficient of performance (COP) is:

Q_C = heat transferred from the cold space in J

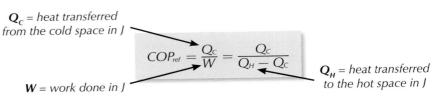

$$COP_{ref} = \frac{Q_C}{W} = \frac{Q_C}{Q_H - Q_C}$$

W = work done in J

Q_H = heat transferred to the hot space in J

If it is running at the maximum theoretical efficiency, the coefficient of performance becomes:

Tip: Don't forget to check the units — questions often have the temperatures in °C rather than K.

T_H = temperature of the hot space in K

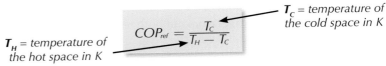

$$COP_{ref} = \frac{T_C}{T_H - T_C}$$

T_C = temperature of the cold space in K

Coefficient of performance for heat pumps

As it's the heat transferred to the hot space that's important for a heat pump, its coefficient of performance (COP) is:

Q_H = heat transferred to the hot space in J

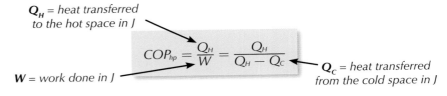

$$COP_{hp} = \frac{Q_H}{W} = \frac{Q_H}{Q_H - Q_C}$$

W = work done in J

Q_C = heat transferred from the cold space in J

The maximum theoretical coefficient of performance is:

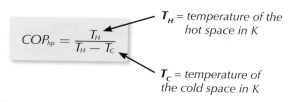

$$COP_{hp} = \frac{T_H}{T_H - T_c}$$

T_H = temperature of the hot space in K

T_c = temperature of the cold space in K

Example — Maths Skills

A house installs a heat pump to keep its rooms at 23 °C by pumping heat in from the outside. In theory, how much does the coefficient of performance change if the outside temperature rises from 2 °C to 10°C?

If outside temperature is 2 °C, the theoretical coefficient of performance is:

$$COP_{hp} = \frac{T_H}{T_H - T_c} = \frac{296}{296 - 275} = \frac{296}{21} = 14.09...$$

If outside temperature is 10 °C, the theoretical coefficient of performance is:

$$COP_{hp} = \frac{T_H}{T_H - T_c} = \frac{296}{296 - 283} = \frac{296}{13} = 22.76...$$

So the coefficient of performance increases by:

$$22.76... - 14.09... = 8.67... = 8.7 \text{ (to 2 s.f.)}$$

Practice Questions — Application

Q1 A company makes refrigerators. The company claims that when one of their refrigerators is placed in a room at 22.0 °C, the food in the refrigerator can be kept at 4.0 °C.

    a) Calculate the maximum theoretical COP for this refrigerator using the company's claim.

    b) Tests done on the refrigerator find that the heat transferred from the inside of the refrigerator in one hour is 4.9×10^5 J and the heat transferred to the surroundings in one hour is 6.2×10^5 J. Calculate the actual COP of the refrigerator.

Q2 A factory that makes clay pots uses an oven to bake the clay. The oven used by the factory is a heat pump that takes heat energy from outside the building.

    a) The heat pump is working at its maximum theoretical COP of 1.40. The work done by the heat pump in one hour is 6250 J. Calculate the heat energy transferred to the hot space in one hour.

    b) The temperature outside the building is 280 K. Find the temperature inside the oven.

Practice Questions — Fact Recall

Q1 What is a reversed heat engine?

Q2 Draw a diagram to show the energy transfers in a reversed heat engine.

Q3 Describe the difference between refrigerators and heat pumps.

Section Summary

Make sure you know...

- That moment of inertia is the resistance to rotation or changes in rotation, and how to calculate it.
- That an object's radius and distribution of mass affect its moment of inertia.
- How to find the rotational kinetic energy using $E_k = \frac{1}{2}I\omega^2$.
- What is meant by angular displacement, angular speed, angular velocity and angular acceleration.
- How to find angular velocity using $\omega = \frac{\Delta\theta}{\Delta t}$, and angular acceleration using $\alpha = \frac{\Delta\omega}{\Delta t}$.
- How to use the equations for uniform angular acceleration.
- How to represent uniform and non-uniform angular acceleration on angular displacement-time graphs and angular velocity-time graphs.
- That torque can be calculated using $T = Fr$ or $T = I\alpha$.
- That, for a rotating body, work done is given by $W = T\theta$ and power is given by $P = T\omega$.
- How and why frictional torque must be accounted for in rotating machines.
- What a flywheel is, and the factors that affect the storage capacity of a flywheel.
- How flywheels are used to store energy in vehicles and in machines used in manufacturing.
- That angular momentum is given by $I\omega$, and that it must always be conserved.
- That angular impulse is change in angular momentum, $\Delta(I\omega)$, and the formula $\Delta(I\omega) = T\Delta t$.
- The equation for the first law of thermodynamics — $Q = \Delta U + W$.
- What a non-flow process is.
- That the equation $pV = nRT$ can be used for an ideal gas undergoing a non-flow process.
- What isothermal and adiabatic changes are.
- That pV = constant for a system undergoing an isothermal change.
- That pV^γ = constant for a system undergoing an adiabatic change.
- That $\Delta U = 0$ for an isothermal change in an ideal gas, $Q = 0$ for an adiabatic change, $W = p\Delta V$ for a change at constant pressure and $W = 0$ for a change at constant volume.
- How to recognise and sketch p-V diagrams for different types of non-flow processes.
- That the area under a p-V graph is equal to the work done by or on a system.
- That a closed loop on a p-V graph represents a cyclic process, where the area enclosed by the loop is equal to the work done per cycle.
- The four strokes of four-stroke petrol and diesel engines, and what happens in each stroke.
- How to sketch and recognise the real indicator diagrams for four-stroke petrol and diesel engines.
- How theoretical indicator diagrams compare to real indicator diagrams for four-stroke engines.
- How to calculate the indicated power, output power, friction power and input power of an engine.
- How to calculate the mechanical, thermal and overall efficiency of an engine.
- That p-V diagrams can be used to measure and predict the power and efficiency of an engine.
- Why the working of a heat engine can't be described by the first law of thermodynamics alone.
- That the second law of thermodynamics says that a heat engine must operate between a heat source and a heat sink, and the diagram used to illustrate this.
- How to calculate the efficiency and maximum theoretical efficiency of a heat engine.
- Why a practical engine has an efficiency lower than the theoretical efficiency of the engine.
- How W and Q_H can be maximised by the use of combined heat and power (CHP) schemes.
- That a reversed heat engine pumps heat energy from a cold space to a hot space.
- How to calculate the coefficient of performance and maximum theoretical coefficient of performance for a refrigerator and a heat pump.

1 A bubble of air is initially at the bottom of a large tank of water. As it rises towards the surface of the water, it expands adiabatically.

At the bottom of the tank, the air bubble has a volume of 125 mm³, a temperature of 380 K and a pressure of 1820 kPa. When the bubble is just below the surface of the water, it has a volume of 986 mm³.

The adiabatic constant of air is 1.4.

1.1 Define the phrase 'expands adiabatically'.

(1 mark)

1.2 Calculate the pressure of the air bubble when it is just below the surface of the water.

(2 marks)

1.3 Calculate the temperature of the air bubble when it is just below the surface of the water.

(2 marks)

The *p-V* diagram for this process is shown below.

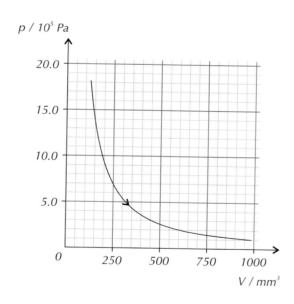

1.4 Estimate the change in internal energy of the air from the bottom of the tank to the surface of the water.

(3 marks)

1.5 Sketch a *p-V* curve showing how the process would have appeared if the bubble of air had expanded isothermally as it rose to the surface.

(4 marks)

2 The diagram shows a water wheel with 7 buckets. The water wheel has radius $r = 3.15$ m, and when no buckets are attached its moment of inertia is 1980 kg m². Each bucket has mass $m_b = 4.20$ kg when empty.
As the wheel rotates, each bucket is filled with $m_w = 12.7$ kg of water. When a bucket reaches the bottom of the wheel, it is upturned and the water drops out.
At any moment, exactly three buckets are always full of water.

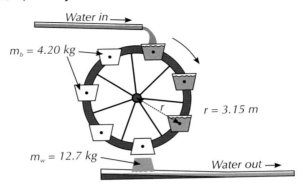

2.1 Calculate the moment of inertia for the wheel as depicted in the diagram.

(2 marks)

2.2 The wheel rotates at 4.95 revolutions per minute. Find its rotational kinetic energy.

(2 marks)

2.3 The wheel speeds up when a constant torque of 3.00 Nm is applied for 95 seconds. Calculate the number of rotations made by the wheel during this time. Frictional torque is negligible.

(3 marks)

2.4 Find the angular impulse acting on the wheel during this 95 second period.

(1 mark)

3 A balloon filled with 1620 cm³ of helium is heated by increasing the temperature of the surroundings. As the helium inside the balloon gets hotter, the balloon expands at a constant pressure of 101 kPa to 1930 cm³.

3.1 35.0 J of heat is supplied to the helium. Calculate its change in internal energy.

(2 marks)

3.2 The final temperature of the helium is 272 K. Calculate its initial temperature.

(2 marks)

3.3 The balloon is then compressed isothermally. The helium gives out 16 J of heat energy. Calculate the work done as the gas is compressed and state whether the work is done **on the gas** or **by the gas**.

(2 marks)

3.4 On a p-V graph, sketch two isotherms for the balloon being compressed between V_1 and V_2 at two different temperatures. Label your graph to show which curve represents the process happening at a higher temperature and which curve represents the process happening at a lower temperature.

(3 marks)

4   A revolving restaurant at the top of a tower completes exactly two full rotations every hour. It has a moment of inertia of 2.70×10^6 kg m^2.

4.1   Find the angular momentum of the revolving restaurant.

(2 marks)

4.2   When the restaurant is cleaned, all the furniture is pushed against the outside wall of the restaurant and the moment of inertia increases to 3.00×10^6 kg m^2. Calculate the resulting change in angular speed of the restaurant, assuming that the restaurant is a freely rotating system.

(2 marks)

4.3   The computer system rotating the restaurant detects the change in angular velocity and automatically accelerates the restaurant back to its initial angular speed over a 15.0 second period.
Calculate the net torque required to carry out this acceleration.

(2 marks)

4.4   The frictional torque experienced during this period of acceleration is 22.9 Nm. Calculate the torque applied by the system's motor to achieve the acceleration.

(1 mark)

4.5   Find the work done by, and average power output of, the motor as it causes this change in angular speed.

(3 marks)

5   An engineer is investigating different types of engine. The theoretical indicator diagram for a cylinder from a small motor is shown below.

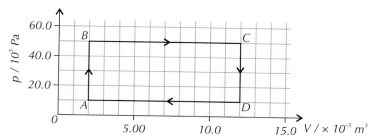

5.1   Calculate the work done by the cylinder in one cycle.

(1 mark)

The engineer modifies the cylinder so that between points B and D, the system undergoes an adiabatic expansion. The time taken to complete one cycle stays the same.

5.2   Sketch the new theoretical indicator diagram for this engine.

(2 marks)

5.3   Explain whether the modification to the engine will increase or decrease the output power of the engine. (You can assume there are no frictional forces.)

(2 marks)

6   A lot of cars use four-stroke petrol engines.

6.1   Explain how a four-stroke petrol engine works.

(6 marks)

6.2   Sketch a theoretical indicator diagram for a four-stroke petrol engine,
labelling all the different steps.

(3 marks)

Many cars use four-stroke diesel engines instead. A car contains a four-stroke
diesel engine with four cylinders that uses 3.2 kg of diesel for every 70.0 km
travelled at a constant speed of 110 kmh^{-1}. The diesel used in the car has a calorific
value of 45.4 MJkg^{-1}. The engine has a thermal efficiency of 42.8% and an output
power of 24 kW.

6.3   Calculate the fuel flow rate of the engine when it is travelling at 110 kmh^{-1}.

(2 marks)

6.4   Calculate the input power, indicated power, mechanical efficiency and
overall efficiency of the engine when it is travelling at 110 kmh^{-1}.

(4 marks)

The theoretical indicator diagram for the engine is shown below.

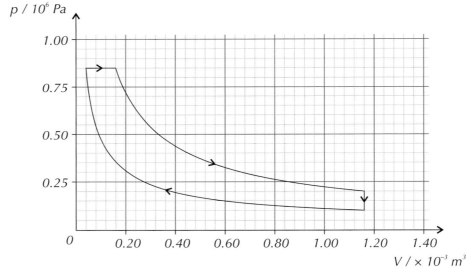

6.5   Calculate how many cycles the crankshaft rotates through in one second.

(3 marks)

There is a flywheel attached to the end of the crankshaft.

6.6   State what a flywheel is and explain why having one attached to the end of the
crankshaft helps to ensure a smooth rotational motion.

(3 marks)

6.7   Calculate the torque of the crankshaft.

(2 marks)

6.8   Calculate the work done by the crankshaft after the car has travelled
at 110 kmh^{-1} for 2 minutes.

(2 marks)

7   A company makes heat engines and reversed heat engines. The company claims that when working between 350 K and 422 K, their heat engine has efficiency 0.5.

7.1   Are the company's claims true? Explain your answer.

(2 marks)

A student investigates the operation of the heat engine and finds that it extracts 29 J of heat from a heat source at 422 K, and 25 J of waste energy is output to a heat sink at 350 K.

7.2   Calculate the work done by the engine.

(1 mark)

7.3   Calculate the efficiency of the engine.

(1 mark)

7.4   Explain why no engine can be 100% efficient. Include the definition of the second law of thermodynamics in your answer.

(2 marks)

The student finds that when operating between 565 K and 480 K, the reversed heat engine can extract 16 J of heat from the cold space and pump 38 J of heat to the hot space.

7.5   Draw a diagram to show the energy transfers in this reversed heat engine.

(3 marks)

7.6   Calculate the coefficient of performance of the reversed heat engine when it is acting as a refrigerator and when it is acting as a heat pump.

(2 marks)

7.7   If the reversed heat engine is used as a heat pump, what is the maximum theoretical coefficient of performance?

(1 mark)

The student sets up the system shown below using a heat engine and reversed heat engine provided by a different company. The output of work from the heat engine is used as the input of work for the reversed heat engine. Using the new heat engine, the heat energy extracted from the heat source at 422 K is 32 J. The heat energy extracted from the cold space at 480 K is 16 J. The new heat engine operates at maximum efficiency.

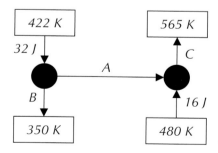

7.8   Calculate the values of A, B and C.

(4 marks)

Learning Objectives:

- Understand what a cathode ray is and how they are produced in a discharge tube.

- Understand the principle of thermionic emission of electrons.

- Know that the work done accelerating an electron through a p.d. is equal to the kinetic energy it has gained when it reaches the anode.

- Understand the equation $\frac{1}{2}mv^2 = eV$, where e is the charge on an electron.

Specification References 3.12.1.1 and 3.12.1.2

Tip: The cathode rays don't actually glow. They ionise atoms in the glass when they collide with them at high speeds, causing electrons to excite and de-excite — which releases photons.

1. Discovering Electrons

This section is a bit different to the ones that you've seen so far. It covers a few topics that you've already seen, but here in much more historical detail. You'll see how three important ideas in physics developed through a series of important discoveries and experiments. The first one is electrons...

Discovery of cathode rays

During the 1800s, scientists thought that the atom was the smallest particle that existed. In the late 1800s a series of experiments changed their minds.

The phrase '**cathode ray**' was first used in 1876, to describe what causes the glow that appears on the wall of a discharge tube like the one in Figure 1, when a high potential difference is applied across the terminals. The **cathode** is connected to the negative terminal of the battery and becomes negatively charged as electrons flow from the battery to the cathode. The **anode** is connected to the positive battery terminal and becomes positively charged.

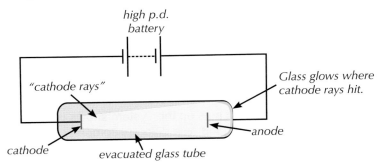

Figure 1: A cathode ray discharge tube.

The rays seemed to come from the cathode (hence their name) and there was a lot of argument about what the rays were made of. J. J. Thomson ended the debate in 1897, when he demonstrated (see pages 363-365) that cathode rays:

a)   have energy, momentum and mass,

b)   have a negative charge,

c)   have the same properties, no matter what gas is in the tube and what the cathode is made of,

d)   have a specific charge much bigger than that of hydrogen ions. So they either have a tiny mass, or a much higher charge — Thomson assumed they had the same size charge as hydrogen ions and a tiny mass.

Thomson concluded that all atoms contain these 'cathode ray particles', or **electrons** as they were soon known — cathode rays are just beams of electrons. He had discovered the first subatomic particle.

Thermionic emission

When you heat a metal, its free electrons gain kinetic energy. Give them sufficient energy and they'll break free from the surface of the metal — this is called **thermionic emission**. Once they've been emitted, the electrons can be accelerated by an electric field in an **electron gun** — see Figure 2.

Tip: Try breaking the word down to remember what it means — think of it as 'therm' (to do with heat) + 'ionic' (to do with charge) + 'emission' (giving off) — so it's 'giving off charged particles when you heat something'.

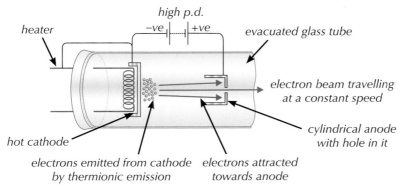

Figure 2: An electron gun.

A heating coil heats the metal cathode. The electrons that are emitted are accelerated towards the cylindrical anode by the electric field set up by the high voltage (see next page).

Electrons are tiny compared to the smallest atom, so they are easily stopped or deflected by atoms. So the glass tube in an electron gun has to be evacuated of air so that the electrons can travel freely in the electric field.

Some electrons pass through a little hole in the anode, making a narrow electron beam. The electrons in the beam move at a constant velocity because there's no field beyond the anode — i.e., there's no force.

Electron guns are combined with fluorescent screens in cathode ray tubes (CRTs). The electron beam is directed at the screen, causing it to emit light and produce a picture on the screen. CRTs are used in old-fashioned TV screens and computer monitors.

Tip: If you increase the potential difference between the anode and cathode, the electrons will move faster (they'll have more kinetic energy — see page 364). If you increase the current through the heater, more electrons will be emitted per second by the cathode.

Work done and electronvolts

You've met the equation for the work done to move a charge through a potential difference on page 93. It's just:

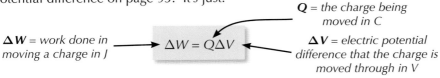

$$\Delta W = Q\Delta V$$

ΔW = work done in moving a charge in J

Q = the charge being moved in C

ΔV = electric potential difference that the charge is moved through in V

Here, Q is the charge of a single electron, e, and ΔV is just the potential difference between the cathode and anode, V. So you get that the work done in accelerating an electron through a p.d. V is $\Delta W = eV$.

The kinetic energy that the electron will have as it leaves the anode (through the hole) is equal to the work done in accelerating it through the potential difference between the cathode and anode, and is given by $\frac{1}{2}mv^2$:

Tip: The magnitude of the charge on an electron, $e = 1.60 \times 10^{-19}$ C.

Tip: This equation was used by Thomson to work out the specific charge of an electron, $\frac{e}{m_e}$ (see next topic).

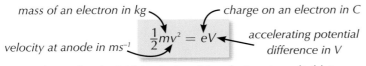

mass of an electron in kg

charge on an electron in C

$$\frac{1}{2}mv^2 = eV$$

velocity at anode in ms⁻¹

accelerating potential difference in V

(assuming the initial velocity of the electron is negligible)

Tip: The mass and charge of an electron are given as constants in the data and formulae booklet in the exam.

From this formula you can define a new unit of energy called the **electronvolt** (eV):

Tip: 1 eV = 1.60 × 10⁻¹⁹ J

> 1 electronvolt is the kinetic energy carried by an electron after it has been accelerated from rest through a potential difference of 1 volt.

Tip: The unit MeV is the mega-electronvolt (equal to 1.60 × 10⁻¹³ J) and GeV is the giga-electronvolt (1.60 × 10⁻¹⁰ J).

So, the energy gained or work done in eV on an electron accelerated by a potential difference is:

> energy gained by electron (in eV) = accelerating voltage (in V)

Tip: When the first CRTs were made, we didn't know the value for the charge on an electron — that was worked out much later on.

Practice Question — Application

Q1 An electron gun uses thermionic emission to produce electrons which are accelerated by a high potential difference.

  a) With reference to a suitable equation, explain what would happen to the beam of electrons if you increased the potential difference between the anode and cathode.

  b) Calculate the kinetic energy, in J, of an electron that has been accelerated from rest through a potential difference of 12.0 kV.

  c) Calculate the force needed to accelerate the electron in part b) if the distance over which it is accelerated is 0.250 m.

Tip: To answer Q1 c), remember the equation *Work done = Force × displacement* from the Mechanics section in year 1 of A-level physics.

Practice Questions — Fact Recall

Q1 What did scientists in the late 1800s mean by cathode rays?

Q2 Describe what is meant by thermionic emission.

Q3 Sketch a labelled diagram of an electron gun.

Q4 Why does the glass tube in an electron gun have to be evacuated?

Q5 Explain where the formula $\frac{1}{2}mv^2 = eV$ comes from.

Tip: The magnitude of the charge on an electron is 1.60 × 10⁻¹⁹ C and is given in the data and formulae booklet in the exam.

2. Specific Charge of an Electron

Learning Objectives:
- Know one method of determining the specific charge, $\frac{e}{m_e}$, of an electron.
- Understand the significance of J. J. Thomson's determination of $\frac{e}{m_e}$ and its comparison with the specific charge of the hydrogen ion.

    Specification Reference 3.12.1.3

So scientists thought that subatomic particles existed, but they had yet to prove it. It was the measurement of the ratio of charge to mass for an electron that convinced them that they were probably much smaller than the atom...

Finding the specific charge of the electron

The **specific charge** or charge-to-mass ratio of a charged particle is just its charge per unit mass measured in $C\,kg^{-1}$. For an electron, the specific charge is denoted $\frac{e}{m_e}$. In 1897, Thomson measured $\frac{e}{m_e}$ and showed that subatomic particles exist. There are a few different ways of measuring it, and you need to know one method inside out for your exam — it doesn't matter which one though (see the Tip at the top of the next page).

Experiment to find the specific charge of an electron ($\frac{e}{m_e}$):

Electrons are charged particles, so they can be deflected by an electric or a magnetic field. This method uses a magnetic field in a piece of apparatus called a fine-beam tube (see Figure 1).

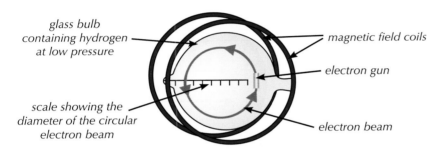

glass bulb containing hydrogen at low pressure

magnetic field coils

electron gun

scale showing the diameter of the circular electron beam

electron beam

Figure 1: *A fine-beam tube.*

A beam of electrons from an electron gun (see page 361) is passed through low-pressure hydrogen gas. The electrons in the beam collide with the hydrogen atoms along its path and transfer some of their energy, causing electrons in the atoms to move into higher energy levels. This is known as excitation. As the electrons in these excited hydrogen atoms fall back to the ground state, they emit light. The electron beam is seen as a glowing trace through the gas.

Two circular magnetic field coils either side of the tube generate a uniform magnetic field inside the tube. The electron beam is fired at right angles to the magnetic field, so the beam curves round in a circle.

The circular motion of the electron beam means that the magnetic force on the electron (see p.129) is acting as a centripetal force (see p.23). Magnetic force is given by $F = Bev$ and centripetal force is given by $F = \frac{mv^2}{r}$. So equating these gives:

$$\frac{mv^2}{r} = Bev$$

Tip: You studied specific charge way back in year 1 of A-Level physics — just remember you need to have the mass in kg and the charge in C and it should be easy.

Tip: It's important that the hydrogen gas is at a low pressure so that there are as few hydrogen atoms as possible to inhibit the electron beam. You want just enough hydrogen atoms for the beam to be visible, but not enough to stop the beam.

Tip: Check out Section 5 — Magnetic Fields (p.122) if you're having trouble with the experiment.

Tip: These formulas are found on:
p.129 for magnetic force
p.24 for centripetal force

Tip: Your teacher might have shown you a different method for calculating $\frac{e}{m_e}$. That's fine... just choose the one you understand the best and learn it.

Tip: Remember, $\frac{1}{2}mv^2 = eV$ comes from $\Delta W = Q\Delta V$ (see page 361).

Figure 2: J. J. Thomson won the 1906 Nobel Prize in Physics for discovering the electron. He balanced a beam of electrons between a magnetic and an electric field to find the specific charge of an electron (it was a little bit like Millikan's experiment in the next topic).

Tip: You don't need to know how to measure these quantities using the fine-beam tube — just make sure you can use them to calculate $\frac{e}{m_e}$.

Tip: You can check that this value is right in the data and formulae booklet.

So the radius of the circular path of the electrons is given by:

m_e = mass of an electron in kg

v = velocity of the electron in ms^{-1}

r = radius of the circle in m

$$r = \frac{m_e v}{Be}$$

e = magnitude of the charge on an electron in C

B = magnetic field strength in T

You can rearrange the equation for the kinetic energy of the electron to find the velocity, in terms of the accelerating potential of the electron gun.

$$\frac{1}{2}m_e v^2 = eV \Rightarrow v = \sqrt{\frac{2eV}{m_e}}$$

If you substitute this into the equation for the radius of the path above (and tidy it all up a bit) you get:

$$r = \frac{m_e v}{Be} \Rightarrow r = \frac{m_e\sqrt{\frac{2eV}{m_e}}}{Be} \qquad \text{substituting } v = \sqrt{\frac{2eV}{m_e}}$$

$$\Rightarrow r^2 = \frac{m_e^2\left(\frac{2eV}{m_e}\right)}{B^2 e^2} \qquad \text{squaring both sides}$$

$$\Rightarrow r^2 = \frac{2m_e V}{B^2 e} \qquad \text{cancelling } m_e \text{ and } e$$

$$\Rightarrow \frac{e}{m_e} = \frac{2V}{B^2 r^2} \qquad \text{making } \frac{e}{m_e} \text{ the subject}$$

So the specific charge of an electron can be found using the equation:

V = accelerating potential

$\frac{e}{m_e}$ = specific charge of an electron

$$\frac{e}{m_e} = \frac{2V}{B^2 r^2}$$

r = radius of the circle

B = magnetic field strength

You can measure all the quantities on the right-hand side of the equation using the fine-beam tube, leaving you with the specific charge, $\frac{e}{m_e}$.

Example — Maths Skills

A beam of electrons is accelerated through a potential difference of 200.0 kV by an electron gun in a fine-beam tube. The magnetic field strength in the tube is 0.0754 T and the radius of the circular beam is measured to be 2.00 cm. Calculate the specific charge of an electron.

Use the equation for specific charge: $\frac{e}{m_e} = \frac{2V}{B^2 r^2}$

$$= \frac{2(200\times10^3)}{(0.0754)^2(2.00\times10^{-2})^2}$$

$$= 1.7589...\times10^{11}$$

$$= 1.76\times10^{11} \text{ C kg}^{-1} \text{ (to 3 s.f.)}$$

The significance of Thomson's findings

The largest specific charge that had ever been measured before was the specific charge of an H^+ ion. Using his own method, Thomson found in 1897 that $\frac{e}{m_e}$ is much greater than the specific charge of the H^+ ion, meaning that it either has a much greater charge, or is much lighter. He assumed that electrons had the same charge and that they were very light.

It turns out that $\frac{e}{m_e}$ (1.76×10^{11} C kg^{-1}) is about 1800 times greater than the specific charge of a hydrogen ion or proton (9.58×10^7 C kg^{-1}). And the mass of a proton is about 1800 times greater than the mass of an electron — Thomson was right, electrons and protons do have the same size charge.

Practice Question — Application

Q1 An experiment to determine the specific charge of an electron uses the apparatus below.

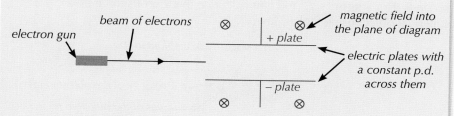

A beam of electrons is passed between two deflecting plates with a potential difference applied across them. A magnetic field is applied perpendicular to the electric field and the electron beam using an electromagnet. The whole apparatus is contained within a vacuum.

a) Before the electromagnet is turned on, what would you expect to happen to the beam of electrons as it travels between the plates?

b) The electromagnet is turned on and provides a magnetic field such that the electron beam travels in a straight line (i.e. it isn't deflected). Show that the magnetic field strength, B, is related to the potential difference applied across the plates, V, by the equation: $B = \frac{V}{vd}$, where v is the velocity of the electron beam, and d is the distance between the plates.

c) The kinetic energy of an electron upon entering the field is 4.20 keV and these measurements were taken: $V = 5.00 \times 10^3$ V, $d = 50.0$ mm, $B = 2.60$ mT.
Calculate the specific charge of an electron.

Tip: For Q1 b), you need to think about the two forces that are acting on the electron. Remember, in an electric field, $Fd = Q\Delta v$ (see page 94).

Tip: For Q1 c) use the equation $\frac{1}{2}mv^2 = eV$ and remember that 1 eV = e J.

Practice Questions — Fact Recall

Q1 Describe an experimental set-up that could be used to determine the specific charge of an electron, including the quantities that you would need to measure.

Q2 Derive an equation for the radius of a beam of electrons travelling at a velocity v in a magnetic field of strength B.

Q3 Explain the significance of the discovery that the specific charge of an electron is much greater than that of the hydrogen ion.

Tip: You only need to be able to describe one experiment for finding the specific charge of an electron — it doesn't matter which one.

3. Millikan's Oil-Drop Experiment

Learning Objectives:

- Know that Stokes' law is $F = 6\pi\eta rv$.
- Be able to use Stokes' law for the viscous force on an oil droplet to calculate the droplet radius.
- Understand the principle of Millikan's experiment to determine the charge of an electron.
- Be able to explain the motion of a falling oil droplet with and without an electric field, and use its terminal speed to determine the mass and charge of the droplet.
- Know that the condition for holding a charged oil droplet of charge Q stationary between two oppositely charged plates is $\frac{QV}{d} = mg$.
- Know the significance of Millikan's findings in relation to the quantisation of electric charge.

Specification Reference 3.12.1.4

So in 1897, J. J. Thomson had discovered that electrons exist and after measuring the specific charge he thought that they must be very, very light. But evidence didn't come until 1909, when Robert Millikan estimated the charge of an electron, meaning the mass could also be estimated.

Stokes' law

Before you start thinking about Millikan's experiment, you need a bit of extra theory.

When you drop an object into a fluid, like air, it experiences a **viscous drag force**. This force acts in the opposite direction to the velocity of the object, and is due to the **viscosity** of the fluid. Viscosity, in simple terms, is just how thick the fluid is. Water is more viscous than air, honey is more viscous than water. Viscosity is measured in pascal seconds (Pas = $kg\,m^{-1}\,s^{-1}$). You can calculate this viscous force on a spherical object using Stokes' law:

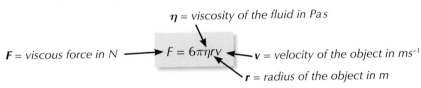

η = viscosity of the fluid in Pa s

F = viscous force in N ⟶ $F = 6\pi\eta rv$ ⟵ v = velocity of the object in ms^{-1}

r = radius of the object in m

Example — **Maths Skills**

Find the viscous drag force acting on a spherical drop of oil with a radius of 0.25 cm falling through air with a viscosity of 1.8×10^{-5} Pas at 4.5 m s⁻¹.

Radius = 0.25 cm = 2.5×10^{-3} m. Substitute into $F = 6\pi\eta rv$:

$F = 6 \times \pi \times (1.8 \times 10^{-5}) \times (2.5 \times 10^{-3}) \times 4.5 = 3.8170... \times 10^{-6}$
$= 3.8 \times 10^{-6}$ N (to 2 s.f.).

Millikan's experimental set-up

This is the apparatus that Millikan used to calculate the charge on an electron:

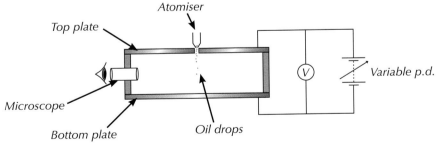

Figure 1: The basic apparatus for Millikan's oil-drop experiment.

The atomiser created a fine mist of oil drops that were charged by friction as they left the atomiser (positively if they lost electrons, negatively if they gained electrons). Some of the drops fell through a hole in the top plate and could be viewed through the microscope.

When he was ready, Millikan could apply a potential difference between the two plates, producing a field that exerted a force on the charged drops. By adjusting the p.d., he could vary the strength of the field.

Tip: To give you a feel for the size of the apparatus, Millikan's plates were circular, with a diameter of about the width of this page. They were separated by about 1.5 cm.

Tip: The eyepiece carried a scale to measure distances (and so velocities) accurately.

Forces on an oil drop with no electric field between the plates

With the electric field turned off, the forces acting on each oil drop are:

- the weight of the drop, $F = mg$ — acting downwards
- the viscous force from the air, $F = 6\pi\eta rv$ (Stokes' law — see previous page) — acting upwards

   The drop will reach **terminal velocity** (i.e. it will stop accelerating) when these two forces are equal. So:

$$mg = 6\pi\eta rv$$

   Since the mass of the drop is the volume of the drop multiplied by the density, ρ, of the oil, and the drop is spherical, this can be rewritten as:

$$\frac{4}{3}\pi r^3 \rho g = 6\pi\eta rv$$

So the radius of the oil drop (or any sphere moving through a fluid) is given by:

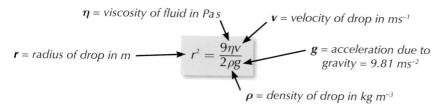

$\boldsymbol{\eta}$ = viscosity of fluid in Pa s
v = velocity of drop in ms^{-1}
r = radius of drop in m
$$r^2 = \frac{9\eta v}{2\rho g}$$
g = acceleration due to gravity = 9.81 ms^{-2}
$\boldsymbol{\rho}$ = density of drop in $kg\, m^{-3}$

   Millikan measured η and ρ in separate experiments, and v could be measured using the microscope, so he could now calculate r — ready to be used when he switched on the electric field (see below).

Example — Maths Skills

A drop of water, with radius 4.7×10^{-7} m, is falling vertically through air at a steady speed of 2.7×10^{-5} ms^{-1}. The viscosity of air is 1.8×10^{-5} Pa s. Calculate the density of the water drop.

$r^2 = \dfrac{9\eta v}{2\rho g}$ rearranged gives $\rho = \dfrac{9\eta v}{2r^2 g}$

$$= \frac{9 \times (1.8 \times 10^{-5}) \times (2.7 \times 10^{-5})}{2 \times (4.7 \times 10^{-7})^2 \times 9.81}$$

$$= 1009.21.... = 1000\ kg\, m^{-3}\ \text{(to 2 s.f.)}$$

Forces on an oil drop with an electric field between the plates

Millikan's next step was to apply a p.d. across the plates, creating an electric field. The field introduced a third major factor — an electric force on the drop.

   He adjusted the applied p.d. until the drop was stationary. Since the viscous force is proportional to the velocity of the object, once the drop stopped moving, the viscous force disappeared.

Now the only two forces acting on the oil drop were:

- the weight of the drop — acting downwards
- the force due to the uniform electric field — acting upwards

---

Tip: Millikan had to take account of things like upthrust as well, but you don't have to worry about that — keep it simple.

Tip: $g = 9.81$ ms^{-2}, the acceleration due to gravity.

Tip: You learnt about terminal velocity in year 1 of A-Level physics — it's just the velocity of an object when the driving forces equal the frictional forces.

Tip: η is the Greek letter 'eta'.

Exam Tip
You won't be given this one in the data and formulae booklet, so you need to remember how to derive it.

Tip: The applied p.d. had to be adjusted so that the force due to the electric field was acting upwards. This means that if the droplet was negatively charged, the top plate would need to be positively charged, and vice versa.

Tip: If you're struggling to understand the electric fields in this experiment, go to pages 87-97.

The weight of the drop is just $F = mg$ as before, and the electric force is given by:

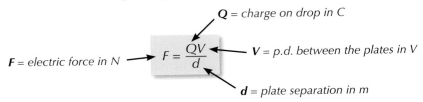

F = electric force in N $F = \dfrac{QV}{d}$ Q = charge on drop in C

V = p.d. between the plates in V

d = plate separation in m

Since the drop is stationary, this electric force must be equal to the weight, so:

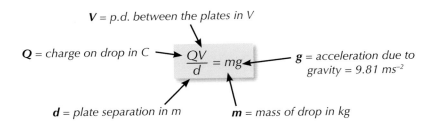

V = p.d. between the plates in V

Q = charge on drop in C $\dfrac{QV}{d} = mg$ g = acceleration due to gravity = 9.81 ms^{-2}

d = plate separation in m m = mass of drop in kg

Tip: You'll get this equation in the data and formulae booklet, but you need to remember and understand when and how to use it.

Which gives the following equation:

$$\frac{QV}{d} = mg = \frac{4}{3}\pi r^3 \rho g$$

Tip: All that's happened here is that mass has been substituted for volume × density (ρ). The drop is spherical, so volume = $\frac{4}{3} \times \pi \times$ radius^3.

The first part of the experiment gave a value for r, so the only unknown in this equation is Q. This meant that Millikan could find the charge on the drop. He repeated the experiment for hundreds of drops. The charge on any drop was always a whole number multiple of -1.60×10^{-19} C.

Example — Maths Skills

An oil drop of mass 1.63×10^{-14} kg is held stationary in the space between two charged plates. The plates are 3.00 cm apart and have a 5.00×10^3 V potential difference across them. Find the charge on the oil drop in terms of e, the charge on an electron.

$\dfrac{QV}{d} = mg$, so $Q = \dfrac{mgd}{V}$

$= \dfrac{(1.63 \times 10^{-14}) \times 9.81 \times (3.00 \times 10^{-2})}{5.00 \times 10^3}$

$= 9.59... \times 10^{-19}$ C

Divide this by the charge on an electron to find the charge in terms of e.

$Q = \dfrac{9.59... \times 10^{-19}}{1.60 \times 10^{-19}} e$

$= 5.996...e$

$= 6.00e$

Tip: Remember, the magnitude of the charge on an electron, $e = 1.60 \times 10^{-19}$ C.

Quantisation of electric charge

The results of the oil-drop experiment were really significant.
Millikan concluded that charge can never exist in smaller quantities than 1.60×10^{-19} C. He assumed that this was the charge carried by an electron. Later experiments confirmed that both these things are true.

> Charge is "quantised". It exists in "packets" of size 1.60×10^{-19} C
> — the **fundamental unit of charge**.
> This is the size of the charge carried by one electron.

This discovery meant that the mass of an electron could be calculated exactly, proving that it was the lightest particle ever discovered (at the time).

Figure 2: *Robert Andrews Millikan won the 1923 Nobel Prize for Physics for determining the charge on an electron and for his work on the photoelectric effect (see page 378-381).*

Practice Question — Application

Q1 A scientist repeated Millikan's oil-drop experiment and observed an oil drop's motion between two horizontal plates. The plates were 4.0 mm apart and were oppositely charged by applying a p.d. across them. The density of the oil used was 980 kg m^{-3}.
The viscosity of air was 1.8×10^{-5} Pa s.

a) The scientist observed that an oil drop fell at a steady speed of 2.5×10^{-5} m s^{-1} when no p.d. was applied across the plates. Calculate the radius of the oil drop.

b) The scientist then applied a p.d. across the plates and adjusted it until an oil drop identical to the one in part a) was stationary. The applied p.d. was 98 V. Find the charge on the oil droplet.

c) The scientist kept everything in the experiment the same except the plate separation, which he doubled. He then adjusted the applied p.d. again, so that another oil droplet with an identical radius was stationary. The applied p.d. was half the applied p.d. in part b). Find the charge on the oil droplet.

Practice Questions — Fact Recall

Q1 Write down Stokes' law, defining any variables.

Q2 Sketch the basic set-up for Millikan's oil-drop experiment.

Q3 What is the name and formula for the force that causes an oil drop's velocity to change when there is an electric field between the plates in Millikan's oil-drop experiment?

Q4 Name all the forces acting on a charged oil drop in Millikan's experiment when:

a) the electric field is turned off

b) the electric field is turned on and the drop is in motion

c) the electric field is turned on and the drop is stationary

Q5 Derive an equation linking charge, applied p.d., plate separation, density, gravitational field strength and oil-drop radius for an oil drop held stationary in an electric field.

Q6 Describe the result that led Millikan to believe that the size of the charge on an electron was 1.60×10^{-19} C.

Q7 Explain the significance of Millikan's discovery of the quantisation of charge in determining properties of the electron.

4. Light — Newton vs Huygens

Learning Objectives:

- Understand Newton's corpuscular theory of light.
- Understand Huygens' wave theory of light.
- Be able to compare Newton and Huygens' theories of light and explain why Newton's was preferred.
- Understand the significance of Young's double-slit experiment.
- Be able to explain the production of fringes by Young's double-slit experiment.
- Know that Huygens' theory of light was eventually widely accepted after a delay.
- Understand the nature of electromagnetic waves.

Specification References 3.12.2.1, 3.12.2.2 and 3.12.2.3.

The next four topics are all about how physicists developed the theory of light. The big question is — what is light... a wave or a particle? Scientists have disagreed over this question time and time again. The answer they've settled on for now — it's a bit of both...

Newton's corpuscular theory

In 1671, Newton published his *New Theory about Light and Colors*. In it he suggested that light was made up of a stream of tiny particles that he called '**corpuscles**'.

One of his major arguments for light being a particle was that light was known to travel in straight lines, yet waves were known to bend in the shadow of an obstacle (**diffraction**). Experiments weren't accurate enough then to detect the diffraction of light. Light was known to reflect and refract, but that was it. His theory was based on the principles of his laws of motion — that all particles, including his 'corpuscles', will 'naturally' travel in straight lines.

Newton believed that reflection was due to a force that pushed the particles away from the surface — just like a ball bouncing back off a wall. He thought refraction occurred because the corpuscles travelled faster in a denser medium like glass.

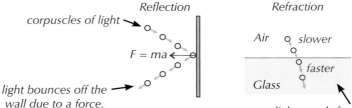

Figure 1: Reflection and refraction of light explained by Newton's corpuscular theory.

Huygens' wave theory

The idea that light might be a wave had existed for some time before it was formalised by Huygens in 1678 — just 7 years after Newton first publicly stated his belief that light was a particle. At the time, nobody took much notice of him because his theory was different from Newton's.

Huygens developed a general model of the propagation of waves in what is now known as **Huygens' principle**:

> Every point on a wavefront may be considered to be a point source of secondary **wavelets** that spread out in the forward direction at the speed of the wave. The new wavefront is the surface that is tangential to all of these secondary wavelets.

Figure 2: English physicist Sir Isaac Newton, who considered light to be made up of a stream of particles ('corpuscles').

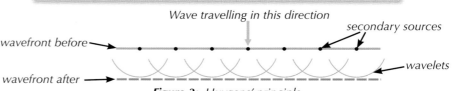

Figure 3: Huygens' principle.

In other words... if you consider a snapshot in time of a wavefront, each point along it is considered as a point wave source that then spreads out in front in a hemispherical 'wavelet' to form the new wavefront in front. The wavelets spread out at the speed of the wave. See Figure 3 for a diagram of Huygens' principle.

By applying his theory to light, Huygens found that he could explain reflection and refraction easily. He predicted that light should slow down when it entered a denser medium, rather than speed up. He also predicted that light should diffract around tiny objects and that two coherent light sources should interfere with each other (see below).

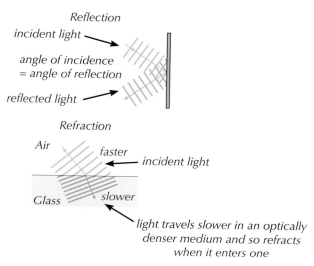

Figure 5: *Reflection and refraction of light explained by Huygens' wave theory.*

Figure 4: *Dutch physicist Christiaan Huygens disagreed with Newton on the nature of light.*

Tip: If you can't remember refraction, reflection, diffraction and interference, it'll really help to go back and look at your year 1 A-Level physics notes before tackling this stuff.

Tip: A key difference between Newton's theory and Huygens' theory is that Huygens predicted that light would slow down in an optically denser medium, whereas Newton predicted that light would speed up.

Up until the end of the 18th century, most scientists sided with Newton. He'd been right about many things before, so it was generally assumed that he must be right about light being corpuscular. The debate raged for over 100 years until Thomas Young carried out experiments on the interference of light in Cambridge in around the year 1800...

Why Newton's theory was preferred

Newton's corpuscular theory was much more popular at the time because imagining light as a stream of particles explained reflection and refraction in a way that more intuitively fitted in with the existing understanding of physics. It couldn't explain **diffraction**, but the equipment of the time wasn't capable of demonstrating diffraction in light.

Scientists thought **double refraction** (a polarisation effect, where shining light through certain crystals makes two images instead of one) couldn't be explained by thinking of light as a wave. Newton's corpuscular theory explained it in terms of the corpuscles having 'sides'.

Additionally, there was no experimental evidence to support Huygens' theory that light was a wave until Young's interference experiments more than 100 years later.

Over time, Newton's reputation grew as his ideas on maths, gravity, forces and motion revolutionised physics. By the time of Thomas Young a century later, he was a figure scientists didn't want to disagree with.

Young's double-slit experiment

Diffraction and interference are both uniquely wave properties. If it could be shown that light showed interference patterns, that would help decide once and for all between corpuscular theory and wave theory.

In order to get clear interference patterns, you need two coherent sources. The problem with showing light interfering was getting two coherent light sources, as light is emitted from most sources in random bursts.

Young solved this problem by using only one point source of light (a light source passed through a narrow slit). In front of this was a slide with two narrow slits in it — a double slit. Light spreading out by diffraction from the slits was equivalent to two coherent point sources.

Light from the double slit was projected onto a screen and bright and dark 'fringes' were formed where light from the two slits overlapped (see Figure 6).

Tip: If you need a recap on Young's double-slit experiment, it was covered in full detail in year 1 of A-Level physics.

Tip: You may remember from year 1 physics that coherent means that they have the same wavelength and frequency and a fixed phase difference between them.

Figure 7: *Young used light from a window and a piece of card with a slit to create a coherent source. Nowadays this experiment is reproduced with laser light because it is monochromatic (all the same wavelength) and coherent.*

Tip: Path difference and phase difference are different things. The two waves below are in phase — they have 0 phase difference, but their path difference is λ.

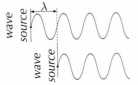

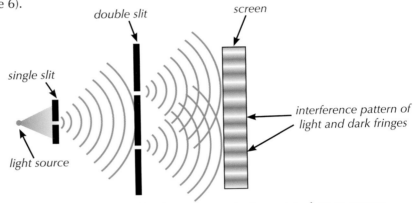

Figure 6: *Young's double-slit system producing an interference pattern.*

The dark and bright fringes are produced by **constructive** and **destructive interference**. When the waves from the two slits are in phase (meaning that they are both at the same point in their wave cycle) the waves will reinforce each other. This is called constructive interference and produces a bright fringe on the screen. Bright fringes are formed when the path difference between the waves at the screen is zero, or any multiple of the wavelength.

When the waves from the two slits are exactly out of phase (meaning their phase difference is 180°, where the full wavelength is represented by 360°) the waves will cancel each other out. This is called destructive interference and produces a dark fringe on the screen.

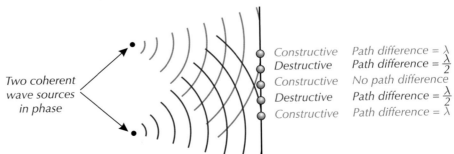

Figure 8: *Two coherent wave sources creating an interference pattern of constructive and destructive interference, depending on the path difference.*

Young's experiment was proof that light could both diffract (through the narrow slits) and interfere (to form the interference pattern on the screen). Newton's corpuscular theory predicted that there would be only two fringes, corresponding to the two slits that the corpuscles could pass through. Young's experiment showed that this clearly wasn't happening, and Huygens' theory could explain everything.

Tip: For this section, you only need to know about Young's double-slit experiment and how it was evidence for light being a wave. You don't need to do any of the nasty calculations that you had to do in year 1 of A-Level physics.

Why Huygens' theory was not accepted

Despite the evidence from the double-slit experiment, Huygens' and Young's ideas weren't widely accepted. Newton's work had revolutionised physics, and by this point he was an established historical figure who other scientists didn't want to contradict.

There were also problems with Huygens' wave theory. It used longitudinal waves, but light was known to be able to be polarised — a property of transverse waves only. Also, Huygens' theory failed to explain both double refraction and why sharp shadows were formed by light.

It took more than a decade before Young (at about the same time as the French scientist Fresnel) realised that transverse waves could explain the behaviour of light. Following this, other scientists soon started agreeing with Huygens that light was a wave.

Tip: A polarised wave is just a wave in which all the vibrations are in one plane. Have a look at your year 1 physics notes if you're still not sure.

Practice Question — Application

Q1 In the 18th century, most scientists believed that light was made up of small particles. An experiment in which light was shone through a pinhole onto a screen is shown below. According to the corpuscular theory, the light beam reaching the screen should have got thinner as the pinhole width was decreased, but instead it was observed to get bigger.

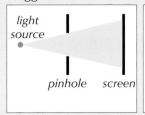

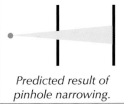

Predicted result of pinhole narrowing.

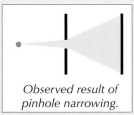

Observed result of pinhole narrowing.

a) Explain why the width of the beam was predicted to decrease.
b) Explain why the width of the beam actually increased.

Practice Questions — Fact Recall

Q1 Briefly describe Newton's corpuscular theory of light.
Q2 Explain briefly the concept of Huygens' principle.
Q3 Give a reason why Newton's theory was preferred to Huygens' theory.
Q4 What difficulty did scientists face when trying to observe the interference and diffraction of light?
Q5 Explain Young's double-slit experiment and how he got around the problem described in Q4.
Q6 Explain the significance Young's double-slit experiment.
Q7 What discovery finally convinced many scientists of Huygens' wave theory of light?

5. Electromagnetic Waves

Learning Objectives:

- Understand the nature of electromagnetic waves.
- Know Maxwell's formula for the speed of electromagnetic waves in a vacuum:
  $c = \dfrac{1}{\sqrt{\mu_0 \varepsilon_0}}$ where
  μ_0 is the permeability of free space and ε_0 is the permittivity of free space.
- Appreciate that ε_0 relates to the electric field strength due to a charged object in free space.
- Appreciate that μ_0 relates to the magnetic flux density due to a current-carrying wire in free space.
- Know how Fizeau measured the speed of light and its implications for Maxwell's equations.
- Know that Hertz's discovery of radio waves, and the measurement of their speed, was further evidence of light as an electromagnetic wave.

Specification Reference 3.12.2.3

As the popularity of Huygens' wave theory grew, scientists began to research the wave nature of light. Several developments in the 19th century led them to believe that light was in fact a transverse, electromagnetic wave...

Light as an electromagnetic wave

In the second half of the 19th century, James Clerk Maxwell was trying to unite the ideas of magnetism and electricity. He created a mathematical model of magnetic and electric fields. This model said that a change to these fields would create an **electromagnetic** (EM) **wave**, radiating out from the source of the disturbance.

We now know that electromagnetic waves are transverse waves, made up of oscillating electric and magnetic fields that are perpendicular to each other and the direction of travel.

Maxwell's prediction came before any experimental evidence for the existence of EM waves. He predicted that there would be a spectrum of EM waves, travelling at the same speed with different frequencies. Maxwell's model showed theoretically that all electromagnetic waves should travel at the same speed in a vacuum, c. He calculated the speed of an electromagnetic wave in a vacuum using:

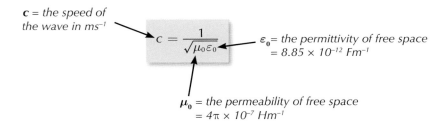

c = the speed of the wave in ms^{-1}

$$c = \frac{1}{\sqrt{\mu_0 \varepsilon_0}}$$

ε_0 = the permittivity of free space = 8.85×10^{-12} Fm^{-1}

μ_0 = the permeability of free space = $4\pi \times 10^{-7}$ Hm^{-1}

μ_0 relates to the magnetic flux density due to a current-carrying wire in free space, while ε_0 relates to the electric field strength due to a charged object in free space. The unit H in the value for μ_0 stands for 'henry' — it's the standard unit for inductance. The values of both μ_0 and ε_0 will be given in the data and formulae booklet you'll get in your exams.

┌─ **Example** ── Maths Skills ─────────────

The speed of an electromagnetic wave in a vacuum is:

$$\frac{1}{\sqrt{\mu_0 \varepsilon_0}} = \frac{1}{\sqrt{(4\pi \times 10^{-7}) \times (8.85 \times 10^{-12})}}$$
$$= 2.9986... \times 10^8 \ ms^{-1}$$
$$= 3.00 \times 10^8 \ ms^{-1} \ \text{(to 3 s.f.)}.$$

Figure 1: *James Clerk Maxwell was a Scottish physicist who made great contributions to the theory of electromagnetism.*

Fizeau's measurement of the speed of light

In the mid 1800s, a French physicist called Hippolyte Fizeau measured the speed of light by passing a beam of light through the gap between two cog teeth to a reflector about 9 km away. The cog was rotated at exactly the right speed so that the reflected beam was blocked by the next cog tooth.

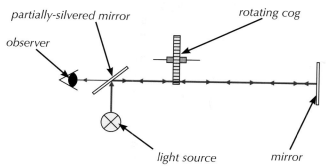

Figure 2: *Fizeau's experiment to determine the speed of light.*

Figure 3: *French physicist Hippolyte Fizeau.*

Using the frequency of rotation and the number of gaps, Fizeau was able to calculate the time taken for the light to travel to the reflector and back. Because Fizeau knew both the time taken and the distance travelled, he could use these to calculate an estimate for the speed of light.

When Maxwell calculated his value of *c*, he found that it was very close to the value measured by Fizeau years earlier. So this provided strong evidence that light, as well as ultraviolet and infrared radiation beyond the visible spectrum, is an electromagnetic wave.

> **Tip:** A partially-silvered mirror splits a beam of light into two by reflecting some of the beam and allowing the rest to pass through as if the mirror were transparent.

Radio waves

In the late 1880s, Heinrich Hertz produced and detected radio waves using electric sparks. He used an induction coil and a capacitor to produce a high voltage, and showed that radio waves were produced when high voltage sparks jumped across a gap of air.

He detected these radio waves using a loop of wire with a gap in it in which sparks were induced by the radio waves (see Figure 4). The fact that a potential difference was induced in the loop showed that the waves had a magnetic component (as a changing magnetic field is needed to induce a potential difference — see pages 139-142).

> **Tip:** An induction coil is like a transformer — whenever the current is interrupted a high voltage is induced in the secondary coil.

> **Tip:** Induction is covered earlier in the book, on page 133.

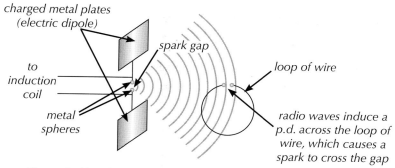

Figure 4: *Hertz's experimental set-up used to discover radio waves.*

You can show that radio waves have an electric component by replacing the wire loop with a second dipole parallel to the first. The radio waves will create an alternating current in the second dipole.

Figure 5: *Heinrich Hertz discovered radio waves using a set-up similar to the one in Figure 4.*

Tip: A progressive wave is a wave that carries energy from one place to another.

Tip: You've met stationary waves before in year 1 of A-Level physics, so hopefully this will all sound familiar. If not, have a look back at your year 1 notes.

Tip: You've seen the equation for the speed of a wave in year 1 of A-Level physics:
c = wave speed in ms⁻¹
λ = wavelength in m
$$c = f\lambda$$
f = frequency in Hz

In later experiments, Hertz used a flat metal sheet to create stationary radio waves, showing that they could both be reflected and show interference. He also went on to show that radio waves can be refracted, diffracted and polarised.

As well as showing that radio waves show all the properties of electromagnetic waves, he measured their velocity using stationary radio waves.

Stationary radio waves

A **stationary wave** is the superposition of two progressive waves with the same **frequency** (or wavelength) and amplitude, moving in opposite directions. They can be created by reflecting a progressive wave back on itself. They can be demonstrated by oscillating a piece of string — see Figure 6.

The wave generated by the oscillator is reflected back and forth. For most frequencies the resultant pattern is a jumble. However, if the oscillator happens to produce an exact number of waves in the time it takes for a wave to get to the end and back again, then the original and reflected waves reinforce each other.

The frequencies at which this happens are called **resonant frequencies** and it causes a stationary wave where the overall pattern doesn't move along — it just vibrates up and down, so the string forms oscillating 'loops'. The points on the wave with zero amplitude are known as nodes, and the bits of the wave with maximum amplitude are called antinodes.

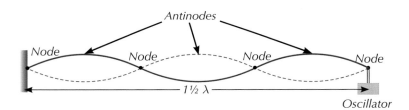

Figure 6: *A stationary wave with four nodes and three antinodes.*

Using the set-up in Figure 7, Hertz produced stationary radio waves at a fixed resonant frequency. He moved the radio wave detector between the transmitter and the reflecting sheet and measured the distance between nodes. Since the distance between nodes is half a wavelength, he could work out the wavelength of the waves and use $c = f\lambda$ to calculate the wave speed.

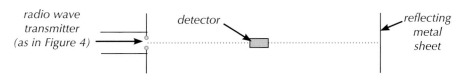

Figure 7: *Experiment to measure the velocity of radio waves.*

He measured the speed of radio waves to be the same in a vacuum as the rest of the electromagnetic spectrum. This helped to confirm that radio waves, like light, are electromagnetic waves.

Practice Question — Application

Q1 An experimental set-up to determine the wave speed of microwave radiation is shown in Figure 8 below.

Tip: This experiment sets up stationary microwaves in a very similar way to Hertz's experiment with radio waves on p.375.

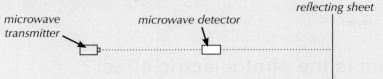

reflecting sheet

microwave transmitter

microwave detector

Figure 8: *An experiment to determine the speed of microwave radiation.*

The frequency of the microwaves is adjusted until the detector can detect nodes and antinodes as it is moved between the transmitter and reflector. The frequency is 1.5×10^{10} Hz.

a) The distance between two adjacent nodes is measured to be 0.01 m. Calculate the wavelength of the microwaves.

b) Calculate the speed of the microwaves.

Practice Questions — Fact Recall

Q1 What is an electromagnetic wave?

Q2 Write down the equation derived by Maxwell for the speed of an electromagnetic wave.

Q3 Name the constants μ_0 and ε_0, explain to what each refers, and give the value of both constants.

Q4 Describe Fizeau's experiment, and explain how it provided evidence that light is an electromagnetic wave.

Q5 Describe how Hertz discovered radio waves.

Q6 What properties of radio waves did Heinrich Hertz discover?

6. The Photoelectric Effect

In 1705, light was a particle. In 1805, light was a wave. You should remember from year 1 of A-Level physics that observations of photoelectricity presented a few problems that wave theory just couldn't explain. By 1905, Einstein's photon theory of light had been published — and light was a particle again. I wish they'd make their minds up...

What is the photoelectric effect?

If you shine electromagnetic radiation of a high enough frequency onto the surface of a metal, it will instantly emit electrons (see Figure 1). For most metals, this frequency falls in the ultraviolet range.

Because of the way atoms are bonded together in metals, metals contain 'free electrons' that are able to move about the metal. The free electrons on or near the surface of the metal can sometimes absorb energy from the radiation.

If an electron absorbs enough energy, the bonds holding it to the metal break and the electron is released. This is called the **photoelectric effect** and the electrons emitted are called photoelectrons.

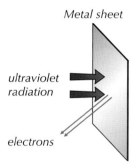

Figure 1: *The photoelectric effect.*

You don't need to know the details of any experiments on this — you just need to learn the main conclusions:

Conclusion 1   For a given metal, no photoelectrons are emitted if the radiation has a frequency below a certain value — called the **threshold frequency**.

Conclusion 2   The photoelectrons are emitted with a variety of kinetic energies ranging from zero to some maximum value. This value of maximum kinetic energy increases with the frequency of the radiation, but is unaffected by the intensity of the radiation.

Conclusion 3   The number of photoelectrons emitted per second is directly proportional to the intensity of the radiation.

The photoelectric effect and wave theory

You can't explain all the observations and conclusions of the photoelectric effect experiment if EM radiation only acts as a wave...

Tip: Remember — photoelectrons are just the electrons released from a metal's surface.

Tip: The intensity of radiation is the amount of energy per second hitting a unit area of the metal.

Tip: Don't get the photoelectric effect confused with thermionic emission (page 361). The difference is that the photoelectric effect is the result of incident electromagnetic radiation on a metal, whereas thermionic emission is due to heating.

Threshold frequency

Wave theory says that for a particular frequency of EM wave, the energy carried should be proportional to the intensity of the beam. The energy carried by the EM wave would also be spread evenly over the wavefront.

This means that if EM radiation were shone on a metal, each free electron on the surface of the metal would gain a bit of energy from each incoming wave. Gradually, each electron would gain enough energy to leave the metal. If the EM waves had a lower frequency (i.e. were carrying less energy) it would take longer for the electrons to gain enough energy, but it would happen eventually. However, electrons are never emitted unless the waves are above a threshold frequency — so wave theory can't explain the threshold frequency.

Tip: The key thing about the photoelectric effect is that it shows that light <u>can't just act as a wave</u>. Certain observations of the photoelectric effect can't be explained by classical wave theory.

Kinetic energy of photoelectrons

The higher the intensity of the waves, the more energy they should transfer to each electron — the kinetic energy should increase with intensity.

Wave theory can't explain the fact that the kinetic energy depends only on the frequency in the photoelectric effect.

Exam Tip
You might have to explain how changing the intensity and frequency of the light affects the photoelectrons emitted — so make sure you learn it.

The photon model of light

Max Planck's wave packets

Max Planck was the first to suggest that EM waves can only be released in discrete packets, or quanta. The energy, E, carried by one of these wave packets is:

Tip: The singular of quanta is 'quantum'.

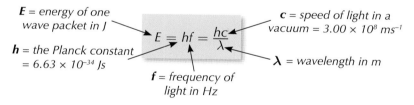

E = energy of one wave packet in J

h = the Planck constant = 6.63×10^{-34} Js

f = frequency of light in Hz

c = speed of light in a vacuum = 3.00×10^8 ms^{-1}

$$E = hf = \frac{hc}{\lambda}$$

λ = wavelength in m

Tip: Remember, $c = f\lambda$, so $f = \frac{c}{\lambda}$.

Example —— Maths Skills

Calculate the wavelength of a wave packet with an energy of 4.12×10^{-19} J.

Rearrange $E = \frac{hc}{\lambda}$ into $\lambda = \frac{hc}{E}$ and substitute in the values for E, h and c to calculate the wavelength.

$$\lambda = \frac{hc}{E} = \frac{(6.63 \times 10^{-34}) \times (3.00 \times 10^8)}{4.12 \times 10^{-19}} = 4.827... \times 10^{-7}$$
$$= 4.83 \times 10^{-7} \text{ m (to 3 s.f.)}$$

Exam Tip
Remember that for calculations like this you'll be given Planck's constant, h, and the speed of light, c — hurrah.

Einstein's photons

Einstein went further by suggesting that EM waves (and the energy they carry) can only exist in discrete packets. He called these wave packets **photons**.

He saw these photons of light as having a one-on-one, particle-like interaction with an electron in a metal surface. Each photon would transfer all its energy to one specific electron. The photon model could be used to explain the photoelectric effect.

Figure 2: Albert Einstein, the physicist who explained the photoelectric effect using photons.

Figure 3: Solar cells use the photoelectric effect to convert light energy into electricity.

Explaining the photoelectric effect

The photon model of light can explain the observations and conclusions for the photoelectric effect that the wave model of light can't...

Work function and threshold frequency

When electromagnetic radiation hits a metal, the metal's surface is bombarded by photons. If one of these photons collides with a free electron, the electron will gain energy.

Before an electron can leave the surface of the metal, it needs enough energy to break the bonds holding it there. This energy is called the **work function** energy — its value depends on the metal. The symbol for work function is ϕ, the Greek letter 'phi'.

If the energy gained from the photon is greater than the work function energy, the electron can be emitted. If it isn't, the metal will heat up, but no electrons will be emitted.

The **threshold frequency** is the minimum frequency a photon can have and still cause a photoelectron to be emitted. The energy of a photon at the threshold frequency is equal to the work function.

Maximum kinetic energy

The energy transferred from electromagnetic radiation to an electron is the energy it absorbs from one photon, hf. The kinetic energy the emitted photoelectron will be carrying when it leaves the metal is hf minus any other energy losses.

The minimum amount of energy an electron being emitted can lose is the work function energy, so the maximum kinetic energy of a photoelectron, E_k, is given by the equation $E_k = hf - \phi$.

Electrons from deeper down in the metal lose more energy than electrons on the surface, so the emitted photoelectrons have a range of energies even if all the incident photons have the same energy. Photoelectrons have a maximum kinetic energy — electrons have this energy when they are on the surface of the metal and the only energy lost is in escaping from the material (i.e. the work function).

The kinetic energy of the photoelectrons is independent of intensity, as they only absorb one photon at a time. However, increasing the intensity increases the number of photons hitting the metal, so increases the number of photoelectrons emitted.

The significance of Einstein's work

Einstein's work was hugely significant. He'd demonstrated that light is a stream of particles called photons, and that photons are the smallest possible unit of electromagnetic radiation — a quantum.

As well as winning him the Nobel Prize in 1921, Einstein's photon model opened up a whole new branch of physics called quantum theory.

The ultraviolet catastrophe

Photoelectricity wasn't the only phenomenon that scientists were struggling to explain by treating light as a wave. Black body radiation also can't be explained without thinking about particles.

Black body radiation

All objects that are hotter than **absolute zero** (see page 47) emit electromagnetic radiation due to their temperature. At room temperature, this radiation lies mostly in the infrared part of the spectrum (which we can't see) — but heat something up enough and it'll start to glow and emit visible light. This is because the wavelengths of radiation emitted depend on the temperature of the object.

A **black body** is an object with a pure black surface that emits radiation strongly and in a well-defined way. It absorbs all light incident on it — we call them black bodies because they don't reflect any light.

> A black body is a body that absorbs all electromagnetic radiation of all wavelengths and can emit all wavelengths of electromagnetic radiation.

Because they emit all wavelengths of electromagnetic radiation, they emit a continuous spectrum of electromagnetic radiation. We call it **black body radiation**. A graph of radiation power output against wavelength for a black body varies with temperature, but they all have the same general shape (as shown in Figure 4). They're known as black body curves.

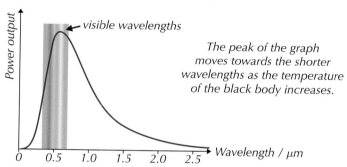

Figure 4: Black body radiation curve for a black body at 5000 K.

The peak of the graph moves towards the shorter wavelengths as the temperature of the black body increases.

Tip: The Sun can be assumed to be a black body. It absorbs almost all of the light incident on it, and emits the most intense radiation in the yellow part of the visible region of the electromagnetic spectrum.

Wave theory and the ultraviolet catastrophe

Wave theory could explain the slope of black body radiation curves like the one in Figure 4 at long wavelengths (low frequencies). Classical wave theory suggested that the power radiated was proportional to λ^{-4} — but this meant that the power output was predicted to head towards infinity in the ultraviolet region.

This was the ultraviolet catastrophe — wave theory, then widely accepted, had predicted something that was impossible, and nobody could work out how to adapt the theory to explain it.

It wasn't until Einstein built on Planck's interpretation of radiation in terms of quanta and came up with the photon model of light that physics was able to explain black body curves — even though Planck wasn't actually trying to solve the ultraviolet catastrophe at the time. Using the assumption that the object absorbed and radiated discrete packets of energy, he derived a formula that correctly matched the observed black body curve.

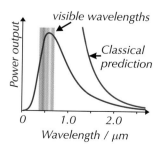

Figure 5: The classical wave theory prediction (blue line) tends to infinity in the ultraviolet region because it assumed that power $\propto \frac{1}{\lambda^4}$.

Q1 Light of wavelength 550 nm is shone on a piece of metal which has a work function of 2.1 eV. Photoelectrons are emitted from the metal.

　　a) Find the energy of a photon of the incident light.

　　b) Explain why the emitted photoelectrons have a range of kinetic energies.

Tip: Remember, the magnitude of the charge on an electron is $e = 1.60 \times 10^{-19}$ C and the speed of light in a vacuum is $c = 3.00 \times 10^8$ m s^{-1}. They are both given in the data and formulae booklet in the exam.

Practice Questions — Fact Recall

Q1 What were the three main conclusions drawn from detailed experimentation on the photoelectric effect?

Q2 Describe Einstein's photon theory of light.

Q3 What did Max Planck suggest about light?

Q4 When low-frequency light is shone on a piece of metal, no photoelectrons are emitted at all. Explain how wave theory fails to explain this result, and how photon theory explains it.

Q5 What is the work function of a metal?

Q6 What does wave theory predict should happen to the kinetic energy of photoelectrons as the intensity of the light is increased? What actually happens? Explain why photoelectrons have a maximum kinetic energy according to Einstein's photon theory.

Q7 Describe what is meant by the 'ultraviolet catastrophe'.

7. Wave-Particle Duality

*Is it a wave? Is it a particle? You're probably a bit tired of this debate...
that's understandable. Don't worry — scientists have basically agreed
to disagree... there's evidence for both arguments, so in the end they just
decided it could be either a wave or a particle... and so can everything else.
Hopefully some of this will be familiar from year 1 of A-Level physics...*

Is light a particle or a wave?

As you saw on page 372-373, diffraction and interference of light can only be explained using waves. If the light was acting as a particle, the light particles in the beam would either not get through the gap (if they were too big), or just pass straight through and the beam would be unchanged.

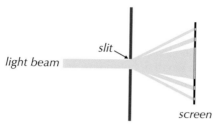

Figure 1: *Diffraction of light waves as they pass through a narrow slit.*

But the results of photoelectric effect experiments (see p.378) can only be explained by thinking of light as a series of particle-like photons. If a photon of light is a discrete bundle of energy, then it can interact with an electron in a one-to-one way. All the energy in the photon is given to one electron.

The photoelectric effect and diffraction show that light behaves as both a particle and a wave — this is known as **wave-particle duality**.

Wave-particle duality theory

Louis de Broglie made a bold suggestion in his PhD thesis. He said if 'wave-like' light showed particle properties (photons), 'particles' like electrons should be expected to show wave-like properties.

The de Broglie equation relates a wave property (wavelength, λ) to a moving particle property (momentum, $p = mv$).

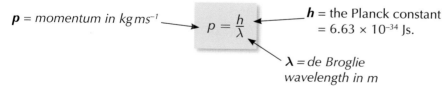

p = *momentum in* $kg\,ms^{-1}$ $p = \frac{h}{\lambda}$ **h** = the Planck constant = 6.63×10^{-34} Js.

λ = *de Broglie wavelength in m*

It comes from assuming that a photon of energy $E = hf$ has a mass, m, given by $mc^2 = hf$. Then momentum is given by $mc = \frac{hf}{c}$ and rearranging $c = f\lambda$ gives $\frac{f}{c} = \frac{1}{\lambda}$. So $p = mc = \frac{h}{\lambda}$.

The **de Broglie wave** of a particle can be interpreted as a 'probability wave'. Many physicists at the time weren't very impressed — his ideas were just speculation. But later experiments confirmed the wave nature of electrons and other particles.

Electron diffraction

Diffraction patterns are observed when accelerated electrons in a vacuum tube interact with the spaces in a graphite crystal. As they pass through the spaces, they diffract — just like waves passing through a narrow slit — and produce a pattern of rings. This provides evidence that electrons have wave properties.

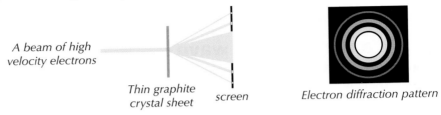

A beam of high velocity electrons

Thin graphite crystal sheet

screen

Electron diffraction pattern

Figure 2: *Electron diffraction — an experiment that shows electrons have wave properties.*

According to wave theory, the spread of the lines in the diffraction pattern increases if the wavelength of the wave is greater. In electron diffraction experiments, a small accelerating voltage, i.e. slow electrons, gives widely spaced rings. Increase the electron speed and the diffraction pattern circles squash together towards the middle, i.e. the amount of diffraction decreases. This fits in with the de Broglie equation — if the velocity is higher, the wavelength is shorter and the spread of lines is reduced.

In general, λ for electrons accelerated in a vacuum tube is about the same size as electromagnetic waves in the X-ray part of the spectrum. The de Broglie wavelength of an electron is related to the accelerating voltage by:

Tip: You can also calculate λ from $\frac{1}{2}mv^2 = eV$. Rearrange this to get $v = \sqrt{\frac{2eV}{m}}$, then substitute this into $\lambda = \frac{h}{p} = \frac{h}{mv}$ to get $\lambda = \frac{h}{m\sqrt{\frac{2eV}{m}}} = \frac{h}{\sqrt{2meV}}$.

h = the Planck constant in Js.

λ = de Broglie wavelength $\longrightarrow$ $\lambda = \dfrac{h}{\sqrt{2meV}}$ $\longleftarrow$ V = accelerating voltage in V

m = mass of an electron in kg

e = magnitude of the charge on an electron in C

Tip: $e = 1.60 \times 10^{-19}$ C, m (of an electron) $= 9.11 \times 10^{-31}$ kg and $h = 6.63 \times 10^{-34}$ Js.

Example — Maths Skills

An electron of mass 9.11×10^{-31} kg is fired from an electron gun and has a de Broglie wavelength of 1.07×10^{-10} m. What anode voltage is needed to produce this wavelength?

Rearranging $\lambda = \dfrac{h}{\sqrt{2meV}}$ gives $V = \dfrac{h^2}{2me\lambda^2}$.

So $V = \dfrac{(6.63 \times 10^{-34})^2}{2(9.11 \times 10^{-31})(1.60 \times 10^{-19})(1.07 \times 10^{-10})^2}$

$= 131.70... \text{ V} = 132 \text{ V}$ (to 3 s.f.)

Tip: For more on electron guns, pop back to page 361.

You only get diffraction if a particle interacts with an object of about the same size as its de Broglie wavelength.

Example

A tennis ball with a mass of 0.058 kg and speed 100 ms^{-1} has a de Broglie wavelength of 10^{-34} m. That's 10^{19} times smaller than the nucleus of an atom. There's nothing that small for it to interact with, and so it only acts as a particle.

Tip: Electrons can be used to investigate the spacing between atoms in a crystal — an electron beam will diffract when the de Broglie wavelength of the electrons is roughly the same size as the spaces between the atoms (see pages 386-387).

Example — Maths Skills

An electron of mass 9.11×10^{-31} kg is fired from an electron gun at 5.4×10^6 ms^{-1}. Roughly what size of object will give the most noticeable electron diffraction?

An electron will diffract when the size of the object is roughly the same size as its de Broglie wavelength, so you need to find λ.

Momentum of electron $= mv = 9.11 \times 10^{-31} \times 5.4 \times 10^6$
$ = 4.9194 \times 10^{-24}$ kg ms^{-1}

Substitute this into de Broglie's equation:

$\lambda = \dfrac{h}{mv} = \dfrac{6.63 \times 10^{-34}}{4.9194 \times 10^{-24}} = 1.3 \times 10^{-10}$ m (to 2 s.f.)

So, only crystals with atom layer spacing around this size are likely to cause noticeable diffraction of this electron.

Tip: Be careful when you're dealing with numbers in standard form — make sure you punch everything into your calculator correctly. Always check that the magnitude of your answer seems about right.

Example — Maths Skills

Electrons with a wavelength of 2.13×10^{-10} m are diffracted as they pass between atoms in a crystal lattice. Estimate the velocity of the electrons.

Rearranging the de Broglie equation, $\lambda = \dfrac{h}{mv}$, gives $v = \dfrac{h}{m\lambda}$.

Substitute in $\lambda = 2.13 \times 10^{-10}$ m, $h = 6.63 \times 10^{-34}$ Js and $m = 9.11 \times 10^{-31}$ kg to get:

$v = \dfrac{6.63 \times 10^{-34}}{9.11 \times 10^{-31} \times 2.13 \times 10^{-10}} = 3.4167... \times 10^6$

$ = 3.42 \times 10^6$ ms^{-1} (to 3 s.f.)

Practice Questions — Application

Q1 a) Electrons fired from an electron gun have a de Broglie wavelength of 9×10^{-11} m. How will increasing the velocity of the electrons affect their de Broglie wavelength?

   b) Describe the effect increasing the velocity will have on the diffraction pattern of the electrons.

Q2 Calculate the de Broglie wavelength of an electron accelerated from rest through a potential difference of 4200 V.

Q3 An electron has a de Broglie wavelength of 1.71×10^{-10} m.

   a) Calculate the momentum of the electron.

   b) Calculate the kinetic energy of the electron.

Q4 An alpha particle has a mass of 6.64×10^{-27} kg and travels at a velocity of 75.0 ms^{-1}. Calculate the speed of an electron that has the same de Broglie wavelength as this alpha particle.

Tip: For these questions you will need to know that the mass of an electron is 9.11×10^{-31} kg and the magnitude of the charge on an electron is 1.60×10^{-19} C.

Practice Questions — Fact Recall

Q1 Describe what is meant by wave-particle duality.

Q2 Name two effects that show electromagnetic waves have both wave and particle properties.

Q3 Name the phenomenon that shows that electrons have wave-like properties.

8. Electron Microscopes

Electron microscopes take advantage of the wave-particle duality of electrons to produce high-resolution images of very small things — like atoms.

Basic operation

Electron microscopes work by firing electrons at a sample and seeing how they interact with it (see below).

Electron microscopes get much better resolution than normal optical microscopes because they can interact with (i.e. be diffracted by) much smaller objects. This is because their de Broglie wavelength is so small.

The de Broglie wavelength depends on the anode potential used to accelerate the electrons. To resolve detail around the size of an atom, the electron wavelength needs to be similar to the diameter of an atom — around 0.1 nm. Using $\lambda = \frac{h}{\sqrt{2meV}}$ from page 384, this means an anode voltage of about 150 V.

Transmission electron microscopes (TEM)

A **transmission electron microscope** (**TEM**) works a bit like a slide projector, but uses electrons instead of light. The electrons are accelerated towards the sample under test using a positive electric potential (e.g. an **electron gun**). The beam is focused onto the sample using magnetic fields. Any interactions of the electrons with the sample are transformed into an image which is projected onto a screen. A very thin specimen must be used for a TEM to work.

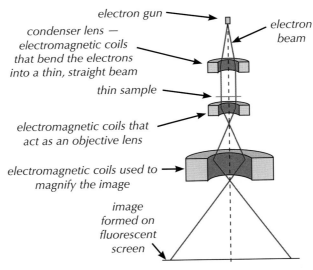

electron gun
electron beam
condenser lens — electromagnetic coils that bend the electrons into a thin, straight beam
thin sample
electromagnetic coils that act as an objective lens
electromagnetic coils used to magnify the image
image formed on fluorescent screen

Figure 1: *The set-up of a transmission electron microscope (TEM), showing an electron beam passing from an electron gun through a sample to form a magnified image on a screen.*

The magnetic field of the condenser lens focuses the electrons into a thin, straight beam which passes through the thin sample. The structure of the sample causes some of the electrons to diffract.

The magnetic fields of the coils that act as objective and magnifier lenses deflect the electrons so that they eventually form a magnified image on a fluorescent screen.

Tip: Electromagnetic coils are used as 'magnetic lenses' which focus the beam of electrons.

Tip: The tube that the electrons travel in must be evacuated so that they don't interact with anything other than the sample.

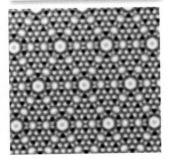

Figure 2: *A TEM image of the atomic surface of silicon.*

Tip: The electrons are produced by thermionic emission in the electron gun (page 361).

The smaller the de Broglie wavelength of the electrons, the better the resolution. A higher anode potential means that the electrons travel at a higher velocity, and so their de Broglie wavelength is smaller. So increasing the anode potential, or the speed of the electrons, gives a better resolution.

Tip: Smaller de Broglie wavelengths mean the electrons can interact with smaller things, so you can see more detail.

Scanning tunnelling microscope (STM)

A scanning tunnelling microscope (STM) is a different kind of microscope that uses principles of quantum mechanics. A very fine probe is positioned very close (around 1 nm away) to the surface of a sample. A high voltage is applied between the probe and the surface, making the probe negatively charged in relation to the sample.

Electrons "tunnel" from the probe to the surface, resulting in a weak electric current. The smaller the distance between the probe and the surface, the greater the current. By scanning the probe over the surface and measuring the current, you produce a 3D image of the surface of the sample.

There are two ways you can image a sample surface:

- Fix the distance — keep the probe height the same and measure the changes in current.
- Fix the current — keep the current the same by adjusting the probe height and measure the changes in probe height.

The position of the probe is controlled by three piezoelectric transducers made from materials which experience a tiny change in length when a potential difference is applied to them. This tiny change in length allows the probe to be moved by tiny distances when scanning the sample surface and altering the distance of the probe from the sample.

Tip: If the probe is too high, there will be no detectable current. If it is too close, it may hit the sample surface and be damaged.

Tip: If the height of the probe is fixed and the probe reaches a raised bit of the surface, the gap will decrease and the current will increase. If the current is fixed and the probe reaches a raised bit, the height of the probe will be increased in order to keep the current constant.

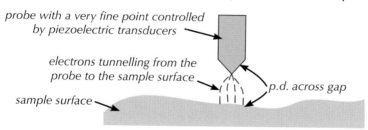

probe with a very fine point controlled by piezoelectric transducers

electrons tunnelling from the probe to the sample surface

p.d. across gap

sample surface

Figure 3: *A scanning tunnelling microscope (STM) scanning a surface.*

Tip: The likelihood of an electron tunnelling across the gap is small, but finite. Fortunately, you don't need to know anything else about tunnelling — it's a complicated business.

Practice Questions — Fact Recall

Q1 What anode potential is needed to produce electrons with de Broglie wavelengths similar to the size of an atom (around 0.10 nm)?

Q2 Describe how a transmission electron microscope (TEM) produces an image of a very thin sample.

Q3 Describe how a scanning tunnelling microscope (STM) produces a 3D image of the surface of a sample.

Q4 Describe how the position of an STM probe is made to move by tiny distances.

Figure 4: *An STM image of gold atoms (red, yellow and brown) on a graphite substrate (green).*

- Understand the
  principle of the
  Michelson-Morley
  interferometer.

- Be able to outline the
  Michelson-Morley
  experiment that was
  designed to detect
  absolute motion.

- Understand the
  significance of
  the failure of the
  Michelson-Morley
  experiment to detect
  absolute motion.

- Appreciate the
  invariance of the
  speed of light.
  **Specification
  Reference 3.12.3.1**

9. Michelson-Morley Experiment

*The final fascinating development in this section covers Einstein's theory of
special relativity. Maxwell's theory of EM waves (p.374) led scientists to
believe all EM waves were vibrations in an invisible 'ether'. The Michelson-
Morley experiment tried to measure the speed of the Earth through the ether,
with surprising results...*

Absolute motion

During the 19th century, most physicists believed in the idea of **absolute
motion**. They thought everything, including light, moved relative to a fixed
background — something called the ether.

    Michelson and Morley tried to measure the absolute speed of the
Earth through the ether using a piece of apparatus called an interferometer.
They expected the motion of the Earth to affect the speed of light they
measured in certain directions. According to Newton, the speed of light
measured in a lab moving parallel to the light would be $(c + v)$ or
$(c - v)$, where v is the speed of the lab.

    By comparing variations in the speed of light parallel and
perpendicular to the motion of the Earth, Michelson and Morley hoped to find
v, the absolute speed of the Earth.

The Michelson-Morley interferometer

The interferometer was basically two mirrors and a semi-silvered glass block
which acts as a partial reflector. When you shine light at a partial reflector,
some of the light is transmitted and the rest is reflected, making two separate
beams. The mirrors were at right angles to each other, and an equal distance
from the partial reflector.

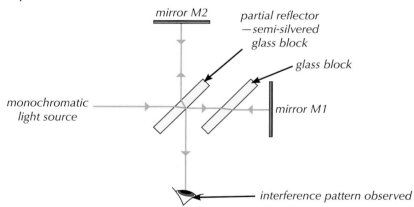

Figure 1: *The set-up of the Michelson-Morley interferometer*

    Monochromatic light is sent towards the partial reflector. Some
light is reflected, and some transmitted, so the light is split into two beams
travelling at right angles to each other. The other glass block is needed so that
both light beams travel through the same amount of glass and air.

    The beams are reflected at mirrors M1 and M2. When the reflected
beams meet back at the beam-splitter, they have a phase difference that
depends on the difference in their path lengths. They form an interference
pattern (see page 372). This interference pattern is recorded by the observer.

    Then the whole interferometer is rotated through 90° and the
experiment repeated.

Tip: This experiment is
a simplified version of
the one actually used —
Michelson and Morley
used more mirrors to
increase the path length
of the light.

Tip: Remember,
monochromatic light
is light that is all of the
same wavelength and
frequency — it's used
here because different
wavelengths of light are
refracted by different
amounts.

Tip: Both beams must
travel through the same
amount of glass and
air, so that they are
both slowed down by
the glass for the same
amount of time. This
means that, in the
absence of ether, both
beams take the same
amount of time to get to
the observer.

Expected outcome

According to Newton's laws, light moving parallel to the motion of the Earth should take longer to travel to the mirror and back than light travelling at right angles to the Earth's motion. So rotating the apparatus should have changed the travel time for the two beams. This would cause a tiny shift in the interference pattern.

Observed outcome

They repeated the experiment over and over again — at different times of day and at different points in the year. Taking into account the expected range of the experimental error, they could detect absolutely no shift in the interference pattern. The time taken by each beam to travel to each mirror was unaffected by rotating the apparatus.

So, Newton's laws didn't work in this situation. Most scientists were really puzzled by this "null result". Eventually, the following conclusions were drawn:

> - It's impossible to detect absolute motion — the ether doesn't exist.
> - The speed of light has the same value for all observers (it is invariant).

Tip: Einstein used this result to come up with his theory of special relativity — see next topic.

Practice Questions — Fact Recall

Q1 Describe the idea of absolute motion.

Q2 How was it hoped that the Michelson-Morley experiment would provide evidence for absolute motion?

Q3 In the Michelson-Morley experimental set-up, explain what the following objects were used for:
   a) semi-silvered glass block
   b) plane mirrors
   c) unsilvered glass block

Q4 What results were expected from the Michelson-Morley experiment?

Q5 The Michelson-Morley experiment failed to detect absolute motion. Explain why this is significant and what conclusions can be drawn from it.

10. Special Relativity

The invariance of the speed of light, as demonstrated by the Michelson-Morley experiment, is one of the cornerstones of special relativity. The other key concept is that of an inertial frame of reference.

Frames of reference

A **frame of reference** is just a space or system of coordinates that we decide to use to describe the position of an object — you can think of a frame of reference as a set of coordinates.

> An **inertial frame** of reference is one in which Newton's first law is obeyed. (Newton's first law says that objects won't accelerate unless they're acted on by an external force.)

Imagine sitting in a carriage of a train waiting at a station. You put a marble on the table. The marble doesn't move, since there aren't any horizontal forces acting on it. Newton's first law applies, so it's an inertial frame. You'll get the same result if the carriage moves at a steady speed (as long as the track is smooth, straight and level) — another inertial frame.

As the train accelerates out of the station, the marble moves without any force being applied. Newton's 1st law doesn't apply. The accelerating carriage isn't an inertial frame.

accelerating frame

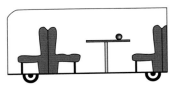 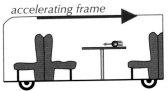

Figure 1: *A train carriage that is stationary or moving at a constant speed (left), is an inertial frame. An accelerating train carriage (right) is not an inertial frame.*

Rotating or accelerating reference frames aren't inertial. In most cases, though, you can think of the Earth as an inertial frame — it's near enough.

Special relativity

Einstein's theory of special relativity only works in inertial frames and is based on two postulates (assumptions):

- Physical laws have the same form in all inertial frames.
- The speed of light in free space is invariant.

The first postulate says that if we do any physics experiment in any inertial frame we'll always get the same result. That means it's impossible to use the result of any experiment to distinguish between a stationary reference frame and one moving at a constant velocity.

The second postulate says that the speed of light (in a vacuum) always has the same value. It isn't affected by the movement of the person measuring it or by the movement of the light source.

Learning Objectives:

- Understand the concept of an inertial frame of reference.
- Know the two postulates of Einstein's theory of special relativity.
- Understand what proper time is and that time dilation is a consequence of special relativity.
- Be able to use the equation for time dilation: $t = \dfrac{t_0}{\sqrt{1 - \frac{v^2}{c^2}}}$
- Be able to explain how muon decay gives evidence of time dilation.
- Understand length contraction as a consequence of special relativity and be able to use the equation: $l = l_0 \sqrt{1 - \dfrac{v^2}{c^2}}$.
- Understand the equivalence of mass and energy and be able to use the formulas $E = mc^2$ and $E = \dfrac{m_0 c^2}{\sqrt{1 - \frac{v^2}{c^2}}}$.
- Draw and understand graphs of variation of mass and kinetic energy with speed.
- Understand how Bertozzi's experiment provided direct evidence for the relativistic variation of kinetic energy with speed.

Specification References 3.12.3.2, 3.12.3.3, 3.12.3.4 and 3.12.3.5

Tip: Special relativity only works in inertial frames.

Time dilation

Einstein's theory of special relativity has a few consequences — one of which is that 'a moving clock runs slower than a stationary clock'. Hopefully this will soon make sense...

The **proper time** between two events, t_0, is the time interval between the events as measured by an observer that is stationary in relation to the events. Special relativity says that an observer of the two events that is moving at a constant velocity in relation to the events will measure a longer time interval, t, between the two events. This is known as **time dilation**, and means that a clock that is moving in relation to an observer will appear to run slow.

Imagine again a train carriage moving at a constant velocity. Set up an experiment in which an observer on the train reflects a beam of light vertically between two mirrors so that it bounces from one mirror to the other and back again (see Figure 2).

To the stationary observer:

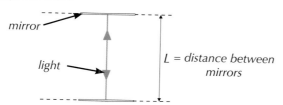

mirror

light

L = distance between mirrors

Figure 2: *View of an observer on the train.*

The stationary observer relative to this event is the person on the train. The stationary observer sees that the light has travelled $2L$.

Time = distance ÷ speed, so the proper time taken for this event is: $t_0 = \dfrac{2L}{c}$

To the moving observer:

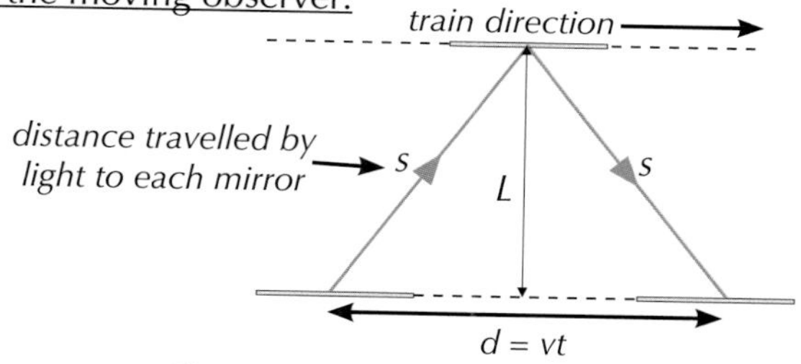

train direction →

distance travelled by light to each mirror → s s

L

$d = vt$

Figure 3: *View of an observer on the platform.*

The moving observer in relation to this event is the observer on the platform. They are moving in relation to the event (and the train) at velocity v.

The light has travelled $2s = 2\sqrt{L^2 + \left(\dfrac{vt}{2}\right)^2}$ (using Pythagoras' theorem) and time = distance ÷ speed, so the time taken for this event, t, is:

$$t = 2s \div c = \frac{2\sqrt{L^2 + \left(\frac{vt}{2}\right)^2}}{c}$$

Multiplying by c, squaring both sides and solving for t gives: $t = \dfrac{2L}{c\sqrt{\left(1 - \frac{v^2}{c^2}\right)}}$.

Using $t_0 = \dfrac{2L}{c}$ (see above) you can substitute in for L to get:

time in s → $t = \dfrac{t_0}{\sqrt{1 - \frac{v^2}{c^2}}}$ ← *proper time in s*

velocity of observer in $m\,s^{-1}$

the speed of light in free space in $m\,s^{-1}$

Tip: Anti-aging tip: If you were to go travelling in space at close to the speed of light for several years, you'd come back and everyone on Earth would have aged more than you.

Tip: The mirrors are stationary relative to the train carriage.

Tip: $d = vt$ is the distance the mirrors have moved in the time measured by the moving observer for the light to bounce back to the first mirror.

Tip: The train must be moving at a velocity $-v$, i.e. the same speed but in the opposite direction.

Tip: Don't get confused here — the person on the platform may seem stationary to you, but they are moving in relation to the event.

Tip: To get this equation you also need to factorise the square root, but you won't need to do any of this in the exam.

This is the formula for time dilation. The $\dfrac{1}{\sqrt{1 - \frac{v^2}{c^2}}}$ bit is known as the Lorentz factor.

Example — **Maths Skills**

Anne is on a high-speed train travelling at $0.90c$. She switches on a torch for exactly 2 seconds. Claire is standing on the platform and sees the same event, but records a longer time. It appears to Claire that Anne's clock is running slow.

In this experiment, Anne is the stationary observer, so she measures the proper time, t_0. Claire is moving at $0.90c$ relative to the event, and so measures a time t given by:

$$t = \frac{t_0}{\sqrt{1 - \frac{v^2}{c^2}}} = \frac{2}{\sqrt{1 - \frac{(0.90c)^2}{c^2}}} = \frac{2}{\sqrt{1 - 0.90^2}} = 4.588... = 4.6\text{ s (to 2 s.f.)}$$

To the external observer (e.g. Claire) moving clocks appears to run slowly. Likewise, if Anne observed Claire, it would appear that Claire's clock was running slowly.

Tip: This is a really fast train... the moving observer has to be going really fast for time dilation to have much of an effect.

Tip: It's really important that you get the "stationary observer" right.

Muon decay

Muons are particles created in the upper atmosphere that move towards the ground at speeds close to c. In the laboratory (at rest) they have a **half-life** of less than 2 µs. From this half-life, you would expect most muons to decay between the top of the atmosphere and the Earth's surface, but that doesn't happen.

Scientists can measure the decay of muons (see Figure 4). First the average speed, v, of the muons is determined (it's about $0.99c$). Two detectors are used to measure the muon count rate at high altitude (MR1) and at ground level (MR2). The two count rates are then compared. By knowing the half-life of the muons, you can predict the change in count rate between the two detectors.

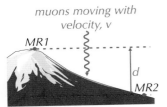

muons moving with velocity, v

MR1

d

MR2

Figure 4: *An experiment to measure the decay of muons as they travel from the upper atmosphere towards the ground at velocity v.*

Example

Here are some typical results from the muon decay experiment in Figure 4:

- Count at MR1 = 500 per minute
- Count at MR2 = 325 per minute
- Distance between detectors (d) = 2000 m
- Time as determined by a stationary observer = $d \div v = 6.73$ µs
- Half-life of muons at rest = 1.53 µs

We can do some calculations using the data above. In the reference frame of the observer the muons seemed to have travelled for 4.4 half-lives between the two detectors. You would expect the count rate at the second detector to be only about 25 counts per minute.

However, in a muon's reference frame, travelling at $0.99c$, the time taken for the journey is just $t_0 = 0.95$ µs. From the point of view of the muons, the time elapsed is less than their half-life. From the point of view of the observer, it appears that the half-life of the muons has been extended.

Length contraction

Another consequence of Einstein's theory of special relativity is that an object moving in the same direction as its length looks shorter to an external observer. It's known as **length contraction** and is very similar to time dilation.

A stationary observer measures the length of an object as l_0. An observer moving at a constant velocity, v, will measure a shorter length, l. l is given by the equation:

l_0 = length as measured by the stationary observer in m

l = length as measured by the moving observer in m

v = velocity of moving observer relative to object in ms^{-1}

c = speed of light in free space in ms^{-1}

$$l = l_0\sqrt{1 - \frac{v^2}{c^2}}$$

This equation comes from the equation for time dilation on page 391. It is the formula for length contraction.

This equation comes from the equation for time dilation on page 391.

Example — Maths Skills

Anne (still in the train moving at 0.90c from the previous page) measures the length of her carriage as 3.0 m. Claire, on the platform, measures the length of the carriage as it moves past her.

Claire measures a length:

$$l = l_0\sqrt{1 - \frac{v^2}{c^2}} = 3.0\sqrt{1 - \frac{(0.90c)^2}{c^2}} = 3.0\sqrt{1 - 0.90^2} = 1.3076...$$
$$= 1.3 \text{ m (to 2 s.f.)}$$

Tip: The length contraction equation is in the data and formulae booklet in the exam.

Tip: Remember — a stationary observer would be travelling at the same velocity as the object, so they're stationary relative to the object.

Tip: Length contraction affects all lengths measured by a stationary and a moving observer — including distances travelled.

Tip: It's only the length in the direction of travel that appears to shrink. The height of the carriage will appear the same to both observers.

Relativistic mass and energy

The last consequence of special relativity you need to know about is that the faster an object moves, the more massive it gets. An object with rest mass m_0 moving at a velocity v has a **relativistic mass**, m, given by the equation:

m_0 = rest mass of the object in kg

m = mass of a moving object in kg

v = velocity of moving object in ms^{-1}

c = speed of light in free space in ms^{-1}

$$m = \frac{m_0}{\sqrt{1 - \frac{v^2}{c^2}}}$$

As the relative speed of an object approaches c, the mass approaches infinity (see Figure 5). So, in practice, no massive object can move at a speed greater than or equal to the speed of light.

Tip: So increasing an object's kinetic energy increases its mass — but it's only noticeable near the speed of light.

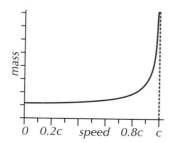

Figure 5: A graph of the variation of mass with speed approaching c.

Einstein extended his idea of relativistic mass to write down the most famous equation in physics:

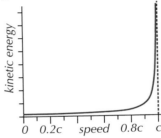

m = mass in kg

E = energy in J ⟶ $E = mc^2$ ⟵ c = speed of light in free space in ms^{-1}

This equation says that mass is a form of energy and can be converted into other forms. Or, alternatively, any energy you supply to an object increases its mass — it's just that the increase is usually too small to measure. The total energy of a relativistic object is given by the equation:

$$E = \frac{m_0 c^2}{\sqrt{1 - \dfrac{v^2}{c^2}}}$$

Just like mass, the kinetic energy of an object tends to infinity as its speed tends to infinity — see Figure 6.

kinetic energy (y-axis), *speed* (x-axis) with markings 0, 0.2c, 0.8c, c

Figure 6: *A graph of the variation of kinetic energy with speed approaching c.*

Example — **Maths Skills**

A beam of protons is accelerated to a speed of 1.69×10^8 m s^{-1}. Calculate the relativistic mass and the total energy of one of the protons. Rest mass of a proton = 1.67×10^{-27} kg

$$m = \frac{m_0}{\sqrt{1 - \dfrac{v^2}{c^2}}} = \frac{1.67 \times 10^{-27}}{\sqrt{1 - \dfrac{(1.69 \times 10^8)^2}{(3.00 \times 10^8)^2}}}$$

$$= 2.021... \times 10^{-27} = 2.02 \times 10^{-27} \text{ kg (to 3 s.f.)}$$

Total energy = $E = mc^2 = (2.021... \times 10^{-27})(3 \times 10^8)^2$
$$= 1.819... \times 10^{-10} = 1.82 \times 10^{-10} \text{ J (to 3 s.f.)}$$

Figure 7: *A very long particle accelerator is required to get particles moving as close to the speed of light as possible. This one in California is 2 miles long. Particle accelerators can also be circular, such as the Large Hadron Collider at CERN.*

Bertozzi's demonstration of mass increasing with speed

In the 1960s William Bertozzi used linear particle accelerators to accelerate pulses of electrons to a range of energies from 0.5 MeV to 15 MeV. The particles were smashed into an aluminium disc 8.4 metres away (see Figure 8).

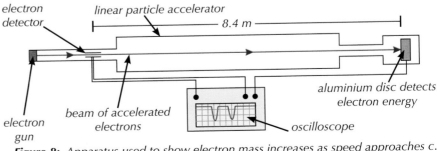

electron detector — *linear particle accelerator* — 8.4 m — *aluminium disc detects electron energy* — *electron gun* — *beam of accelerated electrons* — *oscilloscope*

Figure 8: *Apparatus used to show electron mass increases as speed approaches c.*

The time taken by electrons of each energy to reach the aluminium disc was measured so that their speeds could be calculated. As the energy of the electrons was increased, the speed of the electrons didn't increase as you would expect from $E = \frac{1}{2}mv^2$, but instead tailed off towards a maximum value approaching 3×10^8 ms^{-1} (c). This showed that as the energy increased, the mass increased (as the velocity didn't increase past c).

To check that the electrons had the energy he thought they had, Bertozzi used the heat generated by the collisions at each energy to calculate the kinetic energy of the particles immediately before impact. Bertozzi found that plotting a graph of kinetic energy against speed gave a curve that closely matched that predicted by Einstein's formula. This was the first direct evidence for special relativity.

Practice Questions — Application

Q1 A ball with a diameter of 6.7×10^{-2} m travels at $0.98c$ past a stationary observer. What diameter would the ball appear to have to the observer if measured in line with the direction of travel?

Tip: Q1 hint — the observer isn't stationary relative to the ball...

Q2 A beam of electrons with a constant speed of 2.8×10^8 ms^{-1} is fired past two detectors a fixed distance d apart. A stationary observer (relative to the detectors) measured the time taken for the beam to travel between the two detectors as 77 ns.

a) Calculate the distance d in the frame of reference of:
   (i) the detectors   (ii) the electrons

b) The rest mass of an electron is 9.11×10^{-31} kg. Calculate the relativistic mass of the electrons in the beam.

c) Calculate the total energy of each electron.

Q3 Unstable particles with a half-life of 11.0 ns are accelerated to $0.95c$ and fired in a thin beam past two detectors spaced 20.0 m apart.

Approximate the ratio: $\dfrac{\text{intensity of the beam at second detector}}{\text{intensity of the beam at first detector}}$.

Assume that, once decayed, the particles no longer contribute to the intensity of the beam.

Tip: If you're struggling with Q3, start off working out the time taken for the particles to travel from one detector to the other.

Practice Questions — Fact Recall

Q1 a) What is meant by a frame of reference?
   b) What is meant by an inertial frame of reference?

Q2 Write down the two postulates of special relativity.

Q3 Describe what is meant by time dilation. Write down a formula for it.

Q4 Explain how muon decay provides evidence for time dilation.

Q5 Write down the formula that you would use to calculate the observed length of a fast-moving object.

Q6 What is meant by relativistic mass?

Q7 Write down two formulas for the equivalence of mass and energy.

Q8 Sketch graphs showing how mass and kinetic energy vary with speed as speed approaches c.

Q9 Describe Bertozzi's experiment and explain how it provided evidence for special relativity.

Section Summary

Make sure you know...

- What a cathode ray is and how cathode rays are produced in a discharge tube.
- How electrons are released from the surface of a metal by thermionic emission.
- That the work done in accelerating an electron (from rest) through a potential difference is equal to the kinetic energy it has at the anode, and be able to use the formula $\frac{1}{2}mv^2 = eV$.
- Any one method of determining the specific charge of an electron $\frac{e}{m_e}$, and be able to calculate the specific charge of an electron, given experimental values.
- How to use the equations for centripetal force $F = \frac{mv^2}{r}$ and magnetic force $F = Bev$.
- Why the measurement of the specific charge of an electron was significant.
- How to use Stokes' law for the viscous force on a spherical object $F = 6\pi\eta rv$ to calculate the radius of an oil drop.
- Millikan's experimental set-up for measuring the charge of an oil drop.
- How a charged oil drop moves without an electric field, including its terminal velocity.
- How a charged oil drop moves within an electric field, including how it can be held stationary.
- That a stationary oil drop in an electric field must obey the equation $\frac{QV}{d} = mg$.
- What is meant by the quantisation of charge and why it was significant.
- Newton's corpuscular theory and Huygens' wave theory of light, including the differences between them, and why Newton's theory was preferred.
- The principle of Young's double-slit experiment, and its significance at the time.
- How bright and dark fringes are produced in Young's double-slit experiment.
- That Huygens' theory of light was eventually widely accepted.
- The nature of EM waves, and be able to use the formula for the vacuum speed of an EM wave $c = \frac{1}{\sqrt{\mu_0 \varepsilon_0}}$.
- How Fizeau measured the speed of light, and his measurements' implications for Maxwell's equations.
- How Hertz discovered radio waves and showed that they had all the properties of EM waves.
- How classical wave theory failed to explain observations of photoelectricity.
- Planck's interpretation of EM waves in terms of quanta, and Einstein's photon theory of light.
- What black-body radiation and the ultraviolet catastrophe are.
- The wave-particle duality theory including the evidence for particles acting as waves and vice versa.
- How to use the de Broglie equation $p = \frac{h}{\lambda}$, and that it links a wave property to a particle property.
- That an electron's de Broglie wavelength is linked to the accelerating voltage by $\lambda = \frac{h}{\sqrt{2meV}}$.
- The effect of changing electron speed on electron diffraction patterns.
- The structure and operation of TEM and STM microscopes, and how to increase their resolution.
- The structure of the interferometer used to detect absolute motion, and how the Michelson-Morley experiment intended to detect absolute motion, but failed due to the invariance of the speed of light.
- What an inertial frame of reference is.
- The concepts of proper time and time dilation, and be able to use the equation $t = \frac{t_0}{\sqrt{1 - \frac{v^2}{c^2}}}$.
- How muon decay provides evidence for time dilation.
- The concept of length contraction, and be able to use the equation $l = l_0\sqrt{1 - \frac{v^2}{c^2}}$.
- The concept of relativistic mass, and the equivalence of mass and energy by $E = mc^2$ and $E = \frac{m_0 c^2}{\sqrt{1 - \frac{v^2}{c^2}}}$.
- How to draw and interpret graphs of variation of mass and kinetic energy with speed.
- How Bertozzi's experiment provided direct evidence for the variation of kinetic energy with speed.

Exam-style Questions

1   A transmission electron microscope uses thermionic emission to release electrons, which are then accelerated through a potential difference to a speed v, to form a beam. The beam is focused onto a thin sample of silicon. The diameter of a silicon atom is 2.20×10^{-10} m.

1.1   Describe what is meant by thermionic emission.

(2 marks)

1.2   Explain why a transmission electron microscope tube must be evacuated of air.

(1 mark)

1.3   Explain why electrons must behave as a wave in order for a transmission electron microscope to work.

(2 marks)

1.4   Show that the anode potential of the electron gun is related to the de Broglie wavelength of an electron in the beam by the equation:

$$V = \frac{1.51 \times 10^{-18}}{\lambda^2}.$$

(2 marks)

1.5   Suggest a suitable anode potential to view the silicon atoms in the sample.

(1 mark)

1.6   Explain why increasing the speed of the electrons would improve the resolution of the microscope.

(2 marks)

2.1   Write down the two postulates of Einstein's theory of special relativity.

(2 marks)

2.2   Einstein's theory of special relativity only works in inertial reference frames. What is meant by an inertial reference frame?

(1 mark)

2.3   In an experiment, a beam of electrons is accelerated to a velocity of 2.9×10^8 ms^{-1}. The rest mass of an electron is 9.11×10^{-31} kg.

Show that the relativistic mass of an electron at this velocity is approximately four times greater than its rest mass.

(1 mark)

2.4   An electron travels 10 km. A stationary observer measures the time taken for the electron to travel this distance. In the electron's reference frame, the time taken to travel this distance is different to the time measured by the observer. Describe how this time would compare to the time measured by the observer. Name this effect.

(2 marks)

3   In the 17th and 18th centuries, there was a great debate over the nature of light. The two main theories were Newton's corpuscular theory of light and Huygens' wave theory of light.

In the early 19th century, Young's double-slit experiment produced two coherent light sources by passing light from one source through a double-slit system, as shown below.

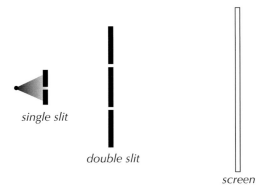

3.1   Explain what pattern you would expect to be formed on the screen according to Newton's corpuscular theory of light, giving a reason for your answer.

(2 marks)

3.2   Describe the pattern that will actually be seen on the screen and explain how it supports Huygens' wave theory.

(3 marks)

3.3.   Give one reason why Huygens' theory was not widely accepted, despite the results of Young's double-slit experiment.

(1 mark)

3.4   Light actually exhibits both wave and particle properties. State one example of a phenomenon that can only be explained by the particle theory of light.

(1 mark)

4   The diagram below shows a cathode ray discharge tube.

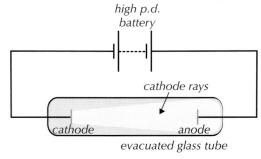

4.1   Describe how cathode rays are produced in a discharge tube and outline the key discoveries of their properties that led scientists to predict the existence of subatomic particles.

(6 marks)

5 The apparatus in the diagram below is used to determine the specific charge of an electron. The beam of electrons is accelerated to a velocity, v, of 8.44×10^6 ms^{-1} by the electron gun. The magnetic field strength, B, is 1.20×10^{-3} T and the radius, r, of the beam is determined to be 4.00 cm.

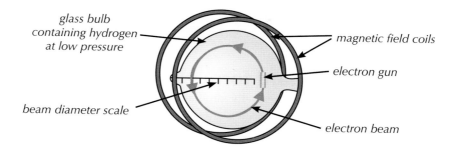

5.1 Show that the specific charge of an electron in the beam is given by:

$$\frac{e}{m_e} = \frac{v}{Br}$$

(2 marks)

5.2 Calculate the specific charge of an electron in the beam from the measurements of this experiment. Give your answer to an appropriate number of significant figures.

(2 marks)

5.3 J. J. Thomson calculated the value of the specific charge of an electron in 1897. Explain the significance of his findings.

(2 marks)

6 Heinrich Hertz used a flat metal sheet and a radio wave transmitter with a fixed frequency, f, to create stationary radio waves. He measured the distance between the antinodes of the stationary wave using a radio wave detector.

6.1 Describe how you could calculate the wave speed of a radio wave from the results of this experiment.

(2 marks)

6.2 The experiment is carried out with a radio wave frequency of 4.00×10^7 Hz. The distance between two antinodes is 3.75 m. Calculate the wave speed.

(1 mark)

6.3 Calculate the speed of an electromagnetic wave in a vacuum using only the constants:

permittivity of free space, $\varepsilon_0 = 8.85 \times 10^{-12}$ F m^{-1}
permeability of free space, $\mu_0 = 4\pi \times 10^{-7}$ H m^{-1}

(1 mark)

6.4 Explain what the results of Hertz's stationary waves experiment provided evidence for.

(1 mark)

1. Calculations

At least 40% of your exam questions will involve maths skills of some sort, and you should be familiar with all of the skills from year 1 of A-level. This section covers the extra stuff you need to know on exponentials and logarithms, as well as briefly recapping some of the maths skills you'll have used before.

Tip: If you don't write a number in standard form, it's known as decimal form — e.g. 0.00012 or 34 500.

Standard form

You should be pretty familiar with using standard form by now, but don't forget that standard form must always look like this:

This number must always be between 1 and 10.

$$A \times 10^n$$

'n' is the number of places the decimal point moves.

Make sure you know what the standard form button looks like on the calculator that you'll use in the exam. It'll probably say either 'Exp', 'EE' or '× 10^x' — see Figure 1.

Figure 1: *The 'Exp' or '×10^x' button is used to input standard form on calculators.*

Tip: Système International d'Unités is French for International System of Units.

Units

Every time you do a calculation, you need to give the correct units of the quantity you've calculated. You might need to convert quantities into the right units before using a formula. You should remember the SI (Système International) **base units**. The only one that you probably won't be familiar with from year 1 is the unit for 'quantity of matter', the mole (mol).

Lot of other units are derived from the base units. Here are some of the SI derived units that you'll come across quite often in year 2 of A-level physics:

Tip: There are alternative units for some of these quantities which you might be asked to use (e.g. MeV for energy) but these are the proper 'SI derived units'.

Quantity	SI derived unit	Symbol	Written in SI base units
Energy, work, heat	joule	J	kgm^2s^{-2}
Resistance	ohm	Ω	$kgm^2s^{-3}A^{-2}$
Potential difference, e.m.f.	volt	V	$kgm^2s^{-3}A^{-1}$
Charge	coulomb	C	As
Force, weight	newton	N	$kgms^{-2}$
Power	watt	W	kgm^2s^{-3}
Pressure, stress	pascal	Pa	$kgm^{-1}s^{-2}$
Frequency	hertz	Hz	s^{-1}
Capacitance	farad	F	$m^{-2}kg^{-1}s^4A^2$
Magnetic flux density	tesla	T	$kgs^{-2}A^{-1}$
Magnetic flux	weber	Wb	$m^2kgs^{-2}A^{-1}$

Figure 2: *Quantities and their SI derived units.*

Prefixes

Prefixes are scaling factors that let you write very big or small numbers without having to put everything in standard form (see previous page). You should know these from year 1, but just in case, here's a quick reminder:

Prefix	Multiple of unit
femto (f)	1×10^{-15}
pico (p)	1×10^{-12}
nano (n)	1×10^{-9}
micro (μ)	1×10^{-6}
milli (m)	$0.001 \ (1 \times 10^{-3})$
centi (c)	$0.01 \ (1 \times 10^{-2})$
kilo (k)	$1000 \ (1 \times 10^{3})$
mega (M)	1×10^{6}
giga (G)	1×10^{9}
tera (T)	1×10^{12}

Figure 4: *Common SI prefixes.*

If you need to convert between prefixes for area or volume, be very careful. To convert from m² to cm² you would need to multiply the quantity by $(1 \times 10^{2})^2$, not just 1×10^{2}.

It's the same for converting between volumes, for example from nm³ to mm³, you would multiply by $(1 \times 10^{-6})^3$ and not just by 1×10^{-6}.

Example — Maths Skills

Convert 0.083 m³ into cm³.

$1 \text{ cm} = 1 \times 10^{-2}$ m, so the scaling factor between m and cm is $1 \div (1 \times 10^{-2}) = 1 \times 10^{2}$.
So to convert from m³ to cm³, multiply by $(1 \times 10^{2})^3$.
$0.083 \times (1 \times 10^{2})^3 = 83\ 000$ cm³

Prefixes can tell you the order of magnitude of a quantity. This gives you a rough idea of the quantity's size, which is useful if you're using it to estimate another value. For example, a length of 1 m is 3 orders of magnitude greater than a length of 1 mm.

Orders of magnitude and logarithms

Scales that use orders of magnitude are known as logarithmic. They allow data that ranges over many orders of magnitude to be displayed and interpreted easily.

Example — Maths Skills

The intensity of sound is usually measured using the logarithmic decibel (dB) scale. This is because, on a linear scale of Wm⁻², the sound intensity at the front of a loud concert is 1 000 000 000 times the sound intensity of a whisper. So we use the decibel scale instead which is defined by:

$$\text{intensity in dB} = 10 \log\left(\frac{I}{I_0}\right)$$

— intensity in Wm⁻²

— threshold of hearing (minimum intensity of sound that can be heard by the human ear at 1000 Hz)

With this scale, a loud concert is 110 dB and a whisper is 20 dB.

Tip: The SI base unit kilogram, kg, is the only one that already has a prefix. Most of the time you'll need to convert into kg to do a calculation, not g.

Figure 3: *Capacitors with their capacitance values printed in μF. The farad is a huge unit, so it's almost always used with a scaling prefix.*

Tip: When converting, you should check your answer is sensible. A centimetre is smaller than a metre, so there are more of them in a given value. So if you're converting from cm to m (or cm³ to m³) the number should get smaller.

Tip: The order of magnitude of a number is the power that 10 would be raised to if the number were written in standard form. E.g. 12 340 can be written as 1.234×10^{4}, so its order of magnitude is 4.

Figure 5: *There are buttons on your calculator for log and ln. If you press shift or second function first, you'll get the $10^{\blacksquare}$ or $e^{\blacksquare}$ functions, the inverse functions of log and ln. There's also a button for the constant e.*

Tip: ln and e are inverse functions, so $\ln e^x = e^{\ln x} = x$.

A logarithm base 10 of a number is defined as the power to which ten must be raised to to get that number:

$$\log_{10} x = y \quad \text{means} \quad 10^y = x$$

You can have bases other than 10. The other common base in physics is the constant 'e'. e is equal to 2.71828... — it's stored in your calculator (see Figure 5). $\log_e$ is known as the natural logarithm, or 'ln'.

$$\ln x = y \quad \text{means} \quad e^y = x$$

There are a few log rules that work for both log and ln that you need to know:

$$\log AB = \log A + \log B \qquad \log\frac{A}{B} = \log A - \log B \qquad \log x^n = n\log x$$

Examples — **Maths Skills**

Show that $\ln y = \ln(ke^{-ax})$ forms a straight line on a $\ln y$ against x graph.

$$\begin{aligned}\ln y = \ln(ke^{-ax}) &= \ln k + \ln e^{-ax} \\ &= \ln k - a\ln e^x \\ &= \ln k - ax\end{aligned}$$

$\ln y = -ax + \ln k$ is a straight line with gradient $-a$ and intercept $\ln k$.

Significant figures

You should remember that you always give your answer to the lowest number of significant figures (s.f.) used in the calculation. Unfortunately it's not that simple when you're working with logarithms and raising to powers.

> If you're taking a logarithm, you should give your answer to the same number of decimal places as there are significant figures in the number you're taking a logarithm of.

> If you're raising to a power, you should give your answer to the same number of significant figures as there are decimal places in the power.

Exam Tip
In your exam, you might only get a mark if your answer is correct <u>and</u> is given to an appropriate number of significant figures. They probably won't tell you when they'll be taking significant figures into account, so make sure you always round your answers to the correct number of significant figures.

Exam Tip
Don't forget to show your working, there may be marks available for intermediate steps in a calculation question — so even if you get the answer wrong, you might get some credit if you've shown your working.

Examples — **Maths Skills**

The number of undecayed nuclei in a sample of a radioactive isotope is given by $N = N_0 e^{-\lambda t}$, where N_0 is the original number of undecayed nuclei, t is the time elapsed and λ is the decay constant.

Calculate N after exactly 2 minutes in a sample of an isotope with decay constant 3.40×10^{-2} s^{-1}, that originally had 4.280×10^{24} undecayed nuclei.

$$\begin{aligned}N = N_0 e^{-\lambda t} &= 4.280 \times 10^{24} \times e^{-(3.40\times10^{-2})\times(2\times60)} \\ &= 4.280 \times 10^{24} \times e^{-4.08} \\ &= 7.236... \times 10^{22}\end{aligned}$$

4.08 has 2 decimal places, so the $e^{-\lambda t}$ bit can be given to 2 s.f. and N_0 is given to 4 s.f.
The least number of s.f. is 2, so the answer is $N = 7.2 \times 10^{22}$ (to 2 s.f.)

2. Algebra

Physics involves a lot of rearranging formulas and substituting values into equations. Easy stuff, but it's also easy to make simple mistakes.

Algebra symbols

Here's a quick recap of some of the symbols that you will come across:

Symbol(s)	Meaning
$=$, $\approx$	equal to, roughly equal to
$<$, $\leq$, $<<$	less than, less than or equal to, much less than
$>$, $\geq$, $>>$	greater than, greater than or equal to, much greater than
$\propto$	proportional to
Δ	change in (a quantity)
Σ	sum of

Tip: An example of using $\propto$ can be found in the topic on the gas laws (pages 53-56).

Tip: Δ is the Greek capital letter 'delta'. An example of using Δ can be found on page 23.

Rearranging equations

When rearranging equations, remember the golden rule — whatever you do to one side of the equation, you must do to the other side of the equation.

┌─ **Example** — **Maths Skills** ──────

The equation for the charge on a discharging capacitor is $Q = Q_0 e^{-\frac{t}{RC}}$. Rearrange the equation to make t the subject.

$Q = Q_0 e^{-\frac{t}{RC}}$

⟩ Divide by Q_0

$\dfrac{Q}{Q_0} = e^{-\frac{t}{RC}}$

⟩ Take ln of both sides

$\ln\left(\dfrac{Q}{Q_0}\right) = -\dfrac{t}{RC}$

⟩ Multiply by $-RC$

$t = -RC\ln\left(\dfrac{Q}{Q_0}\right)$

└──────

Figure 1: *It can be easy to make a mistake rearranging equations when you're stressed in an exam. It's a good idea to double check rearrangements, especially if it's a tricky one where you've had to combine and rearrange equations.*

Substituting into equations

Make sure you avoid the common mistakes by putting values in the right units and putting numbers in standard form before you substitute.

┌─ **Example** — **Maths Skills** ──────

Two protons collide and annihilate to produce two gamma rays. The rest energy of one proton is equal to the energy of one of the gamma rays. Using the equation $E = \dfrac{hc}{\lambda}$, find the wavelength of one of the gamma rays.

h = the Planck constant = 6.63×10^{-34} Js, c = speed of light = 3.00×10^8 ms^{-1} and E = rest energy of a proton = 938.257 MeV

h and c are in the right units but E is not. E must first be converted from MeV to eV and then from eV to J:

$938.257 \times (1 \times 10^6) \times 1.60 \times 10^{-19} = 1.501... \times 10^{-10}$ J

To find the wavelength of the gamma ray, first rearrange the equation to make λ the subject and then substitute in the correct values:

$E = \dfrac{hc}{\lambda} \Rightarrow \lambda = \dfrac{hc}{E} = \dfrac{6.63 \times 10^{-34} \times 3.00 \times 10^8}{1.501... \times 10^{-10}} = 1.32 \times 10^{-15}$ m (to 3 s.f.)

└──────

Tip: You covered annihilation in year 1.

Tip: The values of h, c and E are given in your data and formulae booklet.

Tip: Converting all the values into the correct units <u>before</u> putting them into the equation stops you making silly mistakes.

3. Graphs

You can get a lot of information from a graph — you'll need to know what the area under a graph and the gradient represent, and be able to sketch and recognise simple graphs, given an equation.

Area under a graph

Many quantities in A-level physics can be found from the area between a curve or line and the horizontal axis of a graph. Here are a few examples from year 2:

- Gravitational potential, $\Delta V = -g\Delta r$ is the area under a *g-r* graph (see p.79).
- Electric potential difference, $\Delta V = E\Delta r$ is the area under a *E-r* graph (see p.93).
- Energy (work done) is given by $W = Q\Delta V$, so the energy stored on a capacitor is the area under a graph of *Q* against *V* (p.105).

To find an area under a graph, you'll either need to work it out exactly or estimate the area — it depends on the graph's shape. You'll need to estimate the area if the graph is not made up of straight lines.

Tip: You should try to remember all the examples you've seen in your course, but if you're ever stuck trying to remember what the area under the graph represents in an exam, sometimes you can multiply the quantities on the *x* and *y* axes to work it out. (But be careful, this isn't always the case.)

Tip: Remember, you can end up with 'negative' areas (under the horizontal axis) if the quantity on the vertical axis is a vector. Just subtract their magnitude from the positive areas to work out the quantity represented by the total area.

Tip: The charge is given in mC, but you need it in C to be able to calculate the energy. So the change in charge is 0.7×10^{-3} C.

Tip: There's more on gravitational potential on page 79.

Example — Maths Skills

A capacitor is charged and the charge-p.d. graph is shown below. Use the graph to find the energy stored by the capacitor when the p.d. across it reaches 12 V.

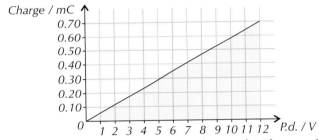

Energy stored (work done) = $Q\Delta V$ = area under graph. The area is a triangle, so the energy stored = $\frac{1}{2}$ × base × height = $\frac{1}{2} \times 12 \times (0.70 \times 10^{-3})$
$$= 4.2 \times 10^{-3} \text{ J}$$

Example — Maths Skills

The graph shows how the gravitational field strength of a mass increases with distance from its centre. Find the change in gravitational potential as an object moves from 1.0×10^5 m to 4.0×10^5 m away from the mass.

Change in gravitational potential
$\Delta V = g\Delta r$, so it's the area under the graph.
For a curved graph, you can estimate the area by counting the number of squares under the graph, which is approximately 8.
The area of one square
= 0.5 Nkg^{-1} × 0.5 × 10^5 m = 2.5 × 10^4 Jkg^{-1}.
So the change in gravitational
potential ≈ 2.5 × 10^4 × 8
$$= 2 \times 10^5 \text{ Jkg}^{-1}$$

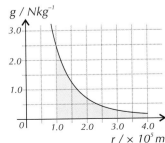

Rates of change

A graph is a plot of how one variable changes with another. The rate of change of the variable on the vertical axis with respect to the variable on the horizontal axis at any point is given by the gradient of the graph.

> Rate of change of y with $x = \dfrac{\Delta y}{\Delta x}$ = gradient of a y-x graph.

Often, the gradient represents a useful rate of change that you want to work out. For example — for an electric field created by a point charge, the electric field strength, E, is given by the gradient of the graph of absolute electric potential, V, against distance from the charge, r. That's because $E = \dfrac{\Delta V}{\Delta r}$ (see page 92).

If the graph is curved, you can find the instantaneous rate of change (at a point), by drawing a **tangent** to the curve at the point where you want to know the gradient and finding the gradient of the tangent.

Tip: You can also find the average rate of change by calculating the total change in y over the total change in x.

Example — Maths Skills

The graph below shows how the gravitational potential of the Moon's gravitational field varies with the distance from its surface.

Find the magnitude of the gravitational field strength of the Moon at a distance of 1500 km from its surface.

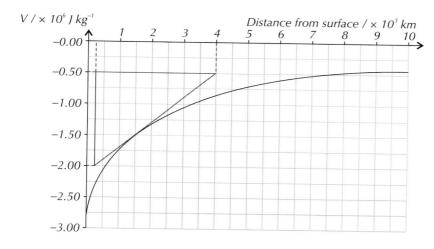

Tip: When drawing a tangent, it helps to make it long — it will be easier to draw, and the tangent line will be more likely to intersect some grid lines, making the gradient easier to calculate.

A tangent to the curve at 1.5×10^3 km is drawn on the graph. Its gradient is:

$$\frac{\Delta y}{\Delta x} = \frac{(-0.5 \times 10^6) - (-2.00 \times 10^6)}{(4 \times 10^3 \times 10^3) - (0.25 \times 10^3 \times 10^3)}$$

$$= \frac{1.50 \times 10^6}{3.75 \times 10^6}$$

$$= 0.4$$

So the magnitude of the gravitational field strength at 1500 km = 0.4 N kg^{-1}

Figure 1: *Make sure you use a really sharp pencil and a ruler whenever you're drawing graphs and tangents.*

Rate of change of a gradient

The gradient is already a 'rate of change of something', so the rate of change of a gradient is the 'rate of change of the rate of change'. Sometimes these represent useful quantities too.

A common example of this is on a displacement-time graph. The rate of change of its gradient is equal to the acceleration.

┌─ **Example** ─ **Maths Skills** ──────────────────────────

The graphs below show an object in simple harmonic motion (see pages 26-29). The rate of change of the displacement-time graph gives the velocity, and the rate of change of velocity gives the acceleration.

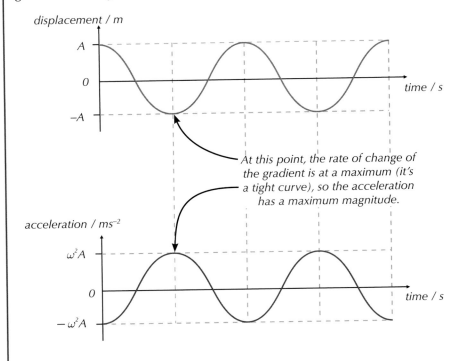

At this point, the rate of change of the gradient is at a maximum (it's a tight curve), so the acceleration has a maximum magnitude.

Tip: The maximum rate of change of gradient is where the curve is most tightly curved.

Tip: A is the maximum value of the displacement of an object oscillating with simple harmonic motion. ω is the angular frequency of the object. For more, see page 30.

Tip: Make sure you're familiar with your spreadsheet program before you start trying to model anything. You'll need to know how to reference and do calculations involving cells.

Modelling rates of change

You can use equations involving rates of change to model quantities with an iterative spreadsheet. If you know the rate of change of something, you can plot a graph for how it changes over time.

This is particularly useful for modelling relationships that don't have nice simple graphs, for example, radioactive decay $-\frac{\Delta N}{\Delta t} = \lambda N$.

Radioactive decay is an iterative process (the number of nuclei that decay in one time period controls the number that are available to decay in the next). This means that you can use a spreadsheet to model the decrease in the number of unstable nuclei in a sample of an isotope if you know the decay constant, λ, and the number of undecayed nuclei in the initial sample, N_0.

- Set up a spreadsheet with column headings for total time t, ΔN and N (number of undecayed nuclei left in the sample) as well as a single data input cell for each of Δt and λ.

- Decide on a Δt that you want to use, e.g. 0.1 s — this is the time interval between the values of N that the spreadsheet will calculate. The most sensible time interval will depend on your decay constant.

- You can then enter formulas into the spreadsheet to calculate the number of undecayed nuclei left after each time interval. You'll need to use $\Delta N = -\lambda \times N \times \Delta t$, rearranged from the radioactive decay equation above.

Tip: Data input cells are the cells in which you write the variables that aren't changing, e.g. decay constant and time interval. Make sure the references to them in your formulas are fixed when you autofill the rows (iterations) later.

Data input cells		
	Δt in s	1000
	λ in s^{-1}	1×10^{-4}

t in s	ΔN (from equation)	N
$t_0 = 0$		$N_0 =$ initial number of undecayed nuclei in sample
$t_1 = t_0 + \Delta t$	$(\Delta N)_1 = -\lambda \times N_0 \times \Delta t$	$N_1 = N_0 + (\Delta N)_1$
$t_2 = t_1 + \Delta t$	$(\Delta N)_2 = -\lambda \times N_1 \times \Delta t$	$N_2 = N_1 + (\Delta N)_2$

Figure 2: An example of the formulas that can be used to create an iterative spreadsheet giving the number of undecayed nuclei over time for a decaying radioactive isotope.

Tip: If you write the formulas properly, the spreadsheet can automatically fill them in for as many rows (iterations) as you want.

- You can then plot a graph of N against t — see Figure 3.

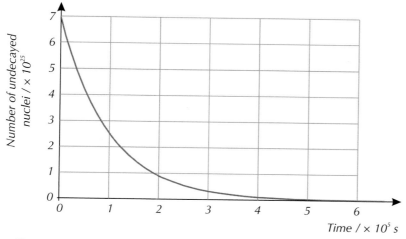

Figure 3: A graph of N against t for an isotope with a decay constant of 1×10^{-4} s^{-1} that originally contained 7×10^{25} undecayed nuclei. Plotted from an iterative spreadsheet using a time interval of 1000 s.

You may have to fiddle with your value for Δt to get a graph with a nice shape, and make sure that you iterate enough times. The graph in Figure 3 was plotted from around 650 iterations.

Sketching graphs

There are some graph shapes that crop up in Physics all the time. The following graphs are examples of the types of graphs you need to know how to recognise and sketch. k is constant in all cases.

Tip: There are loads of examples of $y = kx$ and $y = \frac{k}{x}$ in physics. For example, $Q = CV$ for a capacitor with fixed capacitance on page 103.

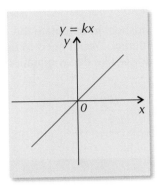

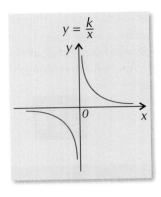

Tip: Make sure you can recognise a graph of e^{kx} too (see page 409).

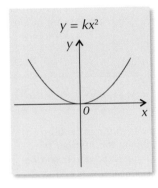

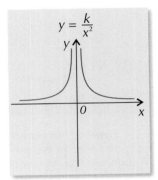

Tip: The x-t, v-t and a-t graphs for simple harmonic motion on page 27 are sin and cos graphs, e.g.:
$x = A\cos(\omega t)$.
The velocity and acceleration graphs can both be calculated from the displacement graph.

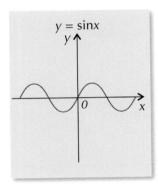

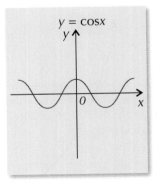

Tip: The graph for kinetic energy of an object in simple harmonic motion against time on p.28 has the same shape as the $y = \cos^2 x$ graph.

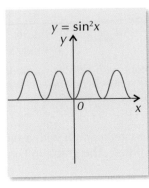

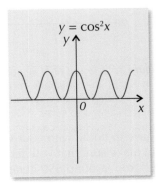

4. Exponential and Log Graphs

There are quite a few exponential relationships in year 2 of A-level, and you need to know what they'll look like as a graph. You also need to know how you can use log-graphs to plot a seemingly complicated relationship as a nice straight line. Read on...

Exponential graphs

A fair few of the relationships you need to know about in year 2 of A-level Physics are exponential — where the rate of change of a quantity is proportional to the amount of the quantity left. Here are a few that crop up in the A-level course (if they don't ring a bell, go have a quick read about them)...

- Charge on a capacitor (p.112) — the decay of charge on a discharging capacitor is proportional to the amount of charge left on the capacitor:

$$Q = Q_0 e^{\frac{-t}{CR}}$$

  There are also exponential relationships for *I* and *V*, and for charging capacitors (see page 110).

- Radioactive decay (p.176) — the rate of decay of a radioactive sample is proportional to the number of undecayed nuclei in the sample:

$$N = N_0 e^{-\lambda t}$$

  The activity of a radioactive sample behaves in the same way.

- X-ray attenuation (p.292) — the reduction of intensity of X-rays over distance is proportional to their initial intensity:

$$I = I_0 e^{-\mu x}$$

Because the rate of change is proportional to the amount, exponential growth gets faster and faster as the amount gets bigger. Exponential decay is just opposite of exponential growth — it gets slower and slower as the amount gets smaller. Because of this, exponential graphs have a characteristic shape that you need to be able to recognise and sketch:

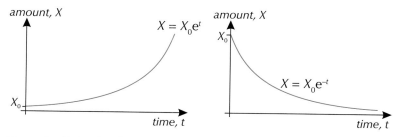

Figure 2: (Left) The characteristic graph of exponential growth. The rate of growth gets faster and faster. (Right) The characteristic graph of exponential decay. The rate of decay gets slower and slower, never reaching zero.

> **Tip:** Don't forget, e is a constant equal to 2.718... — it'll be stored in your calculator. Look for a button that says $e^{\blacksquare}$ or e^x (see page 402).

> **Tip:** These equations are all examples of exponential decay — see Figure 2.

Figure 1: The growth of bacteria is a classic example of something that can be modelled with exponential growth. The more bacterium there are, the faster the rate of growth becomes.

> **Tip:** You should recognise the decay graph from the spreadsheet modelling graph on p.407 — both show an exponential decay.

Log-linear graphs

You can plot exponential relationships using the natural log, ln. For the relationship $y = ke^{-ax}$, if you take the natural log of both sides of the equation you get:

$$\ln y = \ln (ke^{-ax}) = \ln k + \ln (e^{-ax}) \quad \text{so} \quad \ln y = \ln k - ax$$

Then all you need to do is plot $(\ln y)$ against x. You get a straight-line graph with $(\ln k)$ as the vertical intercept, and $-a$ as the gradient.

Tip: Remember the log rules from p.402 — you'll need them here.

Tip: Graphs with the natural logarithm of a function on one of the axes are called log-linear graphs. There are examples of how they can be used on pages 112-113 and 176-177.

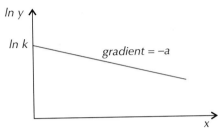

Figure 3: The function $y = ke^{-ax}$ plotted on a log-linear graph of $\ln y$ against x.

Log-log graphs

You can use logs to plot a straight-line graph for other relationships too. Say the relationship between two variables x and y is:

$$y = kx^n$$

Take the log (base 10) of both sides to get:

$$\log y = \log k + n \log x$$

So $\log k$ will be the y-intercept and n will be the gradient of the graph.

Tip: This relationship is called a power law relationship.

Examples — Maths Skills

A physicist carries out an experiment to determine the nuclear radius, R (in m) of various elements with various nucleon numbers, A. She plots a line of best fit for her results on a graph of $\log R$ against $\log A$. Part of the graph is shown. Using the equation $R = R_0 A^{1/3}$, find the value of the constant R_0 from the graph.

First take logs of both sides:

$\log R = \log (R_0 A^{1/3}) = \log R_0 + \log A^{1/3}$
$\qquad\qquad = \log R_0 + \frac{1}{3} \log A$

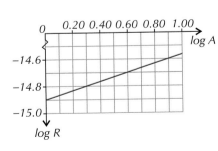

Comparing this to the equation of a straight line (in the form $y = mx + c$), you can see that the gradient of the graph is $\frac{1}{3}$ and the vertical intercept is $\log R_0$.

Tip: Look back at pages 164-165 for more on nuclear radius.

So, reading from the graph, the vertical intercept is about -14.9.

$\log R_0 = -14.9$, so $R_0 = 10^{-14.9} = 1.258... \times 10^{-15} = 1.3$ fm (to 2 s.f.)

Tip: If $a = \log_{10} b$, then $b = 10^a$.

5. Geometry and Trigonometry

You'll often find that you need to deal with different 2D and 3D shapes in Physics. You'll need to resolve forces, which could mean using all sorts of angle rules, as well as Pythagoras and trigonometry. You should have used most of this in year 1, but here's a quick recap.

Geometry basics

Angle rules

These angle rules should be familiar from year 1 — make sure you know them.

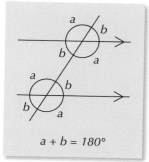

$a + b = 180°$

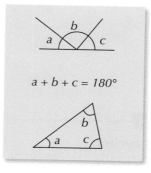

$a + b + c = 180°$

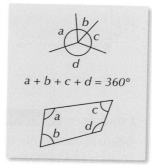

$a + b + c + d = 360°$

Tip: Remember, the arrows on the lines in the diagram mean that they're parallel.

Angles can be measured in degrees or radians:

- To convert from degrees to radians, multiply by $\frac{\pi}{180}$.

- To convert from radians to degrees, multiply by $\frac{180}{\pi}$.

Tip: You'll need to use radians when working with circular motion. Don't forget to put your calculator into either degrees or radians.

Circumference and arc length

You'll need to know the distance round a circle, or part of it, for calculations involving circular motion:

Circumference, $C = 2\pi r$

Arc length, $l = r\theta$, θ in radians

Tip: You'll be given these in the data and formulae booklet, just remember that θ is in radians.

Areas of shapes

These might seem simple, but make sure you know them off by heart — you might need to calculate the area under a graph, or split a shape into the shapes below in order to find its area.

Triangle

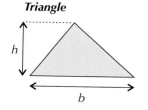

Circle

Rectangle

Trapezium

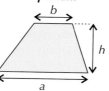

$A = \frac{1}{2} \times b \times h$

$A = \pi \times r^2$

$A = h \times w$

$A = \frac{a+b}{2} \times h$

Tip: The only area given in your data and formulae booklet in the exam is the area of a circle.

Surface areas

If you need to work out the surface area of a 3D shape, you just need to add up the areas of all the 2D faces of the shape. The exception to this is a sphere, where the surface area is given by $4\pi r^2$ — this will be given to you in the data and formulae booklet.

Volumes of shapes

Make sure you remember how to calculate the volumes of a cuboid, a sphere and a cylinder:

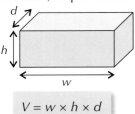 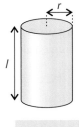

$$V = w \times h \times d \qquad V = \tfrac{4}{3}\pi r^3 \qquad V = \pi r^2 l$$

Example — Maths Skills

Estimate the density of a nucleus.

Protons and neutrons have nearly the same mass, $m_{nucleon} \approx 1.7 \times 10^{-27}$ kg. A nucleus with a nucleon number A has mass $A \times m_{nucleon}$.

Assuming nuclei are spherical, the volume of a nucleus is given by $V = \tfrac{4}{3}\pi R^3$. The radius of a nucleus, R, is given by $R = R_0 A^{1/3}$

Density = mass ÷ volume, so:

$$\rho = \frac{mass}{volume} = \frac{A \times m_{nucleon}}{\tfrac{4}{3}\pi R^3} = \frac{A \times m_{nucleon}}{\tfrac{4}{3}\pi (R_0 A^{1/3})^3} = \frac{A \times m_{nucleon}}{\tfrac{4}{3}\pi R_0^3 A} = \frac{3 m_{nucleon}}{4\pi R_0^3}$$

$$= \frac{3 \times 1.7 \times 10^{-27}}{4\pi \times (1.4 \times 10^{-15})^3} = 1.47... \times 10^{17} = 1.5 \times 10^{17} \text{ kg m}^{-3} \text{ (to 2 s.f.)}$$

Trigonometry basics

You can use Pythagoras' theorem for all right-angled triangles — the square of the hypotenuse is equal to the sum of the squares of the two smaller sides.

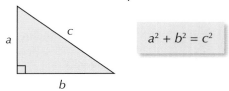

$$a^2 + b^2 = c^2$$

To work out a length or angle within a right-angled triangle, remember SOH CAH TOA:

$$\sin\theta = \frac{opposite}{hypotenuse} \qquad \cos\theta = \frac{adjacent}{hypotenuse} \qquad \tan\theta = \frac{opposite}{adjacent}$$

For really small angles, you can make the following assumptions for the values of sin, cos and tan, where θ is in radians:

$$\sin\theta \approx \theta \qquad \tan\theta \approx \theta \qquad \cos\theta \approx 1$$

This is really useful in Physics as the angles are often small (e.g. the equation for the time period of a pendulum assumes that the angle is small — p.37).

Exam Structure and Technique

Passing exams isn't all about revision — it really helps if you know how the exam is structured and have got your exam technique nailed so that you pick up every mark you can.

A-level exam structure

For AQA A-level Physics you'll sit three exam papers. Papers 1 and 2 are each worth 34% of the total marks and paper 3 is worth 32% of the total marks.

Paper 1

This paper will be 2 hours long and have 85 marks up for grabs. The questions can be based on anything you learnt in year 1 of your A-level course, as well as the Further Mechanics section from year 2 (pages 19-46 in this book). Make sure you look at how many marks each question part is worth before answering it — generally the more marks there are, the more work you'll need to put in to get them all. The paper consists of long and short answer questions worth a total of 60 marks, and 25 multiple choice questions, each worth one mark.

Paper 2

This paper will also be 2 hours long and have 85 marks available. The questions are based on all of the core topics from year 2 of your A-level course, that's sections 2-6 in this book.

Like paper 1, there are long and short answer questions worth a total of 60 marks, and 25 multiple choice questions worth a total of 25 marks.

Paper 3

This paper is 2 hours long as well, but there are 80 marks on offer here. The paper is split into 2 sections:

- **Section A (45 marks)**
  Questions are based on practical skills and data analysis and can be on anything from the entire A-level course (except for Section 7). You will likely be given an experiment that someone has done and then need to answer a series of short and long answer questions about the experiment.
  These questions could include analysis of the method and interpretation of the results, as well as calculations and explanations using the theory behind the experiment.

- **Section B (35 marks)**
  This section is made up of short and long answer questions on the optional topic you've studied.

Command words

It sounds obvious, but it's really important you read each question carefully, and give an answer that fits. Look for command words in the question — they'll give you an idea of the kind of answer you should write.

Commonly used command words for written questions are state, describe, discuss and explain.

Exam Tip
Make sure you have a good read through this exam structure. It might not seem important now but you don't want to get any nasty surprises at the start of an exam.

Exam Tip
If you don't know the answer to a multiple choice question, it's always best to make the best guess you can rather than give no answer at all. You might just guess the right answer...

Exam Tip
Although paper 2 is mainly about year 2 of your A-level course, the examiners are assuming you remember everything from year 1 — so don't forget everything after you've done paper 1.

Exam Tip
Section A could test any of your practical skills — from safety and experiment design, to identifying errors, analysing results and drawing conclusions.

Exam Tip
Part of your assessment in A-level physics is the Practical Endorsement, which isn't part of any of the written papers. There's more on this on pages 17-18.

- State — give a definition, example or fact.

- Describe — don't waste time explaining <u>why</u> a process happens — that's not what the question is after. It just wants to know <u>what</u> happens.

- Discuss — you'll need to include more detail. Depending on the question you could need to cover what happens, what the effects are, and perhaps include a brief explanation of why it happens.

- Explain — give reasons for why something happens, not just a description.

Extended responses

For some questions, you'll need to write an 'extended response'. These questions are designed to test how well you can put together a well structured and logical line of reasoning. They'll often require you to give a long answer in full written English, e.g. to explain, analyse or discuss something. To get top marks, you need to make sure that:

- your scribble, sorry, writing is legible,

- your spelling, punctuation and grammar are accurate,

- your writing style is appropriate,

- you answer the question and all the information you give is relevant to the question you've been asked,

- you organise your answer clearly and coherently,

- you use specialist scientific vocabulary where it's appropriate.

These questions could also involve an extended calculation, or a combination of a calculation and a full written answer, e.g. you could be asked to calculate a value and say how it supports a conclusion.

When doing extended calculations, make sure your working is laid out logically and it's clear how you've reached your answer. That includes making sure any estimates and assumptions you've made in your working are clearly stated, e.g. assuming air resistance is negligible.

There's usually a lot to think about with this type of question, and it can be easy to write down a lot of great and relevant physics but forget to answer all parts of the question. It's always a good idea to double check you've done everything a question has asked you to do before moving on.

Strange questions

You may get some weird questions that seem to have nothing to do with anything you've learnt. DON'T PANIC. Every question will be something you can answer using physics you know, it just may be in a new context.

Check the question for any keywords that you recognise. For example, if a question talks about acceleration, think about the rules and equations you know, and whether any of them apply to the situation in the question. Sometimes you might have to pull together ideas from different parts of physics — read the question and try to think about what physics is being used. That way you can list any equations or facts you know to do with that topic and try to use them to answer the question.

Time management

This is one of the most important exam skills to have. How long you spend on each question is really important in an exam — it could make all the difference to your grade.

Everyone has their own method of getting through an exam. Some people find it easier to go through the paper question by question and some people like to do the questions they find easiest first. The most important thing is to find out the way that suits you best before the exam — and that means doing all the practice exams you can before the big day.

Check out the exam timings given by AQA that can be found on page 413 and on the front of your exam paper. These timings give you a bit over 1 minute per mark — try to stick to this to give yourself the best chance of picking up as many marks as possible.

Some questions will require lots of work for only a few marks but other questions will be much quicker. Don't spend ages struggling with questions that are only worth a couple of marks — move on. You can come back to them later when you've bagged loads of other marks elsewhere.

Exam Tip
Make sure you read the rest of the information given on the front of the exam paper before you start. It'll help make sure you're well prepared.

Examples

The questions below are both worth the same number of marks but require different amounts of work.

1.1 Define the term 'isotope'.

(2 marks)

2.1 Draw a labelled diagram of a circuit that would be suitable for a student to measure the resistance of component A.

(2 marks)

Question 1.1 only requires you to write down a definition — if you can remember it this shouldn't take you too long.

Question 2.1 requires you to draw a diagram including a number of components — this may take you a lot longer than writing down a definition, especially if you have to add quite a few components and work out whether they should be in parallel or series.

So, if you're running out of time it makes sense to do questions like 1.1 first and come back to 2.1 if you've got time at the end.

Exam Tip
Don't forget to go back and do any questions that you left the first time round — you don't want to miss out on marks because you forgot to do the question.

Answers

Section 1 — Further Mechanics

1. Circular Motion
Page 22 — Application Questions
Q1 Period in seconds $= 28 \times 24 \times 3600 = 2.419... \times 10^6$ s

Angular speed $= \omega = \dfrac{2\pi}{T} = \dfrac{2\pi}{2.419... \times 10^6}$

$= 2.5972... \times 10^{-6}$ rad s^{-1} $= \mathbf{2.6 \times 10^{-6}\ rad\,s^{-1}}$ **(to 2 s.f.)**

$\omega = \dfrac{v}{r} \Rightarrow v = r\omega = (384\,000 \times 1000) \times 2.5972... \times 10^{-6}$

$= 997.331...$

$= \mathbf{1000\ ms^{-1}}$ **(to 2 s.f.)**

Q2 Frequency in rev s^{-1} $= f = \dfrac{\text{no. of revolutions}}{\text{time taken}} = \dfrac{4}{3600}$

$\omega = 2\pi f = 2 \times \pi \times \dfrac{4}{3600} = \dfrac{\pi}{450}$ rad s^{-1}

$\omega = \dfrac{\theta}{t} \Rightarrow \theta = \omega t = \dfrac{\pi}{450} \times 60 = \dfrac{2}{15}\pi = 0.41887...$
$= \mathbf{0.419\ rad}$ **(to 3 s.f.)**

$\omega = \dfrac{v}{r} \Rightarrow v = r\omega = \dfrac{125}{2} \times \dfrac{\pi}{450} = 0.43633...$ ms^{-1}
$= \mathbf{0.436\ ms^{-1}}$ **(to 3 s.f)**

Q3 Frequency in rev s^{-1} $= f = 460 \div 60$, 2.0 cm from centre:
Angular speed $= \omega = 2\pi f = \dfrac{46}{3}\pi = 48.171...$
$= \mathbf{48\ s^{-1}}$ **(to 2 s.f.)**

Linear speed $= v = r\omega = 0.020 \times \dfrac{46}{3}\pi$
$= 0.9634... $ ms^{-1} $= \mathbf{0.96\ ms^{-1}}$ **(to 2 s.f.)**

4.0 cm from centre: Angular speed $= \mathbf{48\ s^{-1}}$ **(to 2 s.f.)**
Angular speed is the same at any point on a solid rotating object.
Linear speed $= 0.040 \times \dfrac{46}{3}\pi = 1.926... = \mathbf{1.9\ ms^{-1}}$ **(to 2 s.f.)**

Q4 Kinetic energy $= \dfrac{1}{2}mv^2$

$\omega = \dfrac{v}{r} \Rightarrow v = \omega r$

$\omega = 2\pi f \Rightarrow v = 2\pi f r$

$f = \dfrac{1}{T}$ and r is the length of the string l, so $v = \dfrac{2\pi l}{T}$

So kinetic energy $= \dfrac{1}{2}m\left(\dfrac{2\pi l}{T}\right)^2$
$= \mathbf{\dfrac{2m\pi^2 l^2}{T^2}}$

Page 22 — Fact Recall Questions
Q1 Angle in radians = angle in degrees $\times \dfrac{2\pi}{360}$

Q2 The angle that an object rotates through per second.

Q3 $\omega = \dfrac{v}{r}$

Q4 The period is the time taken for a complete revolution. The frequency is the number of complete revolutions per second.

Q5 $\omega = 2\pi f$

2. Centripetal Force and Acceleration
Page 25 — Application Questions
Q1 Frequency in rev s^{-1} $= f = 15 \div 60.0 = 0.25$ rev s^{-1}
Angular speed $\omega = 2\pi f = 2 \times \pi \times 0.25 = 0.5\pi$ rad s^{-1}
$F = m\omega^2 r = 60.0 \times (0.5\pi)^2 \times \dfrac{8.5}{2} = 629.1...$
$= \mathbf{630\ N}$ **(to 2 s.f.)**

This might seem like a lot, but it's about the same as the force experienced by the rider due to gravity.

Q2 **C**

$a = \omega^2 r$ so rearranging gives: $\omega = \sqrt{\dfrac{a}{r}}$

$\omega = \dfrac{2\pi}{T}$ so rearranging gives: $T = \dfrac{2\pi}{\omega}$

So $T = \dfrac{2\pi}{\sqrt{\frac{a}{r}}} \Rightarrow T = 2\pi\sqrt{\dfrac{r}{a}}$

So time to complete 3 orbits $= 3 \times T = 6\pi\sqrt{\dfrac{r}{a}}$

Q3 a) $a = \dfrac{v^2}{r} = 31.1^2 \div 56.8$
$= 17.0283... = \mathbf{17.0\ ms^{-2}}$ **(to 3 s.f.)**

b) The formula for centripetal force is $F = \dfrac{mv^2}{r}$, so if the linear speed is halved (i.e. v becomes $\frac{v}{2}$), then the centripetal force is quartered. So the new centripetal force is $\dfrac{F}{4}$.

Q4 If the biker's speed is the minimum possible speed for him to not fall, the centripetal acceleration towards the centre (i.e. down) at the top of the cylinder will be 9.81 ms^{-2}, due to gravity. The motorcycle will fall if its circular motion has a centripetal acceleration smaller than this.

So $a = 9.81$ ms$^{-2} = \dfrac{v^2}{r}$
Rearranging this: $v = \sqrt{ar} = \sqrt{9.81 \times 5.0}$
$= 7.003...$ ms^{-1} $= \mathbf{7.0\ ms^{-1}}$ **(to 2 s.f.)**

This is only about 16 mph — not very fast at all.

Page 25 — Fact Recall Questions
Q1 If an object is moving in a circle, centripetal acceleration is the acceleration of the object directed towards the centre of the circle. Centripetal force is the force towards the centre of the circle responsible for the centripetal acceleration.

Q2 $a = \omega^2 r$
a = centripetal acceleration in ms^{-2}
ω = angular speed (in rad s^{-1})
r = radius of circular motion in m

Q3 $F = \dfrac{mv^2}{r}$
F = centripetal force in N
m = mass of object in kg
v = magnitude of linear velocity in ms^{-1}
r = radius of circular motion in m

3. Simple Harmonic Motion
Page 29 — Application Questions
Q1 No energy is lost in this system, so all the potential energy is converted to kinetic energy. The potential energy at the maximum displacement is the maximum potential energy.
$E_{P\,(max)} = 28$ J $= E_{K\,(max)}$

$E_{K\,(max)} = \dfrac{1}{2}mv_{max}^2$ so rearranging gives $v_{max} = \sqrt{\dfrac{2 \times E_{K\,(max)}}{m}}$

So $v_{max} = \sqrt{(2 \times 28) \div 50}$
$= 1.058...$
$= \mathbf{1.1\ ms^{-1}}$ **(to 2 s.f.)**

Q2 a) The girl's maximum kinetic energy is equal to her maximum gravitational potential energy, i.e.
$\dfrac{1}{2}mv_{max}^2 = mgh_{max} = 35 \times 9.81 \times 0.40 = 137.34$ J
So $\dfrac{1}{2}mv_{max}^2 = 137.34$
$\Rightarrow v_{max} = \sqrt{\dfrac{2 \times 137.34}{m}} = \sqrt{\dfrac{2 \times 137.34}{35}}$
$= 2.8014... = \mathbf{2.8\ ms^{-1}}$ **(to 2 s.f.)**

b)

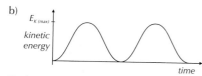

The kinetic energy starts at O as the girl starts from rest. It then varies sinusoidally (like a sine wave) between O and E K (max) as potential energy is converted to kinetic energy and back. The kinetic energy is at a maximum when the swing is at its lowest point.

Q3 a) Maximum displacement = A = 0.60 m
Maximum velocity = $\omega A = 2\pi fA$ = 0.90 ms^{-1}
Rearrange $v_{max} = 2\pi fA$ to give $f = \frac{v_{max}}{2\pi A}$
So f = 0.90 ÷ (2 × π × 0.60) = 0.2387...
= **0.24 Hz (to 2 s.f.)**

b) $f = \frac{1}{T}$ so rearranging gives $T = \frac{1}{f}$
So T = 1 ÷ 0.2387... = 4.188... = **4.2 s (to 2 s.f.)**

c) Maximum acceleration = $\omega^2 A$
= $(2\pi f)^2 A$ = (2 × π × 0.2387...)^2 × 0.60 = **1.35 ms^{-2}**

Page 29 — Fact Recall Questions

Q1 The oscillation of an object in which the object's acceleration is directly proportional to its displacement from its equilibrium position, and the acceleration is always directed towards the equilibrium.

Q2 The frequency of oscillation is the number of complete cycles per second. The period of oscillation is the time taken for a complete cycle.

Q3 The velocity is $\frac{\pi}{2}$ radians ahead of the displacement.

Q4 a) Equilibrium
b) Maximum displacement
c) Maximum displacement
d) Equilibrium

Q5 E.g.

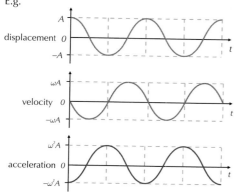

Q6 At its maximum displacement, the object's kinetic energy, E_K, is zero (so it has zero velocity). All of its energy is potential energy, E_P. As the object moves towards the equilibrium position, the restoring force does work on the object and transfers some E_P to E_K. At the equilibrium position, the object's E_P is said to be zero and its E_K is maximum — so its velocity is maximum. As the object moves away from the equilibrium, all that E_K is transferred back to E_P again.

4. Calculations with SHM

Page 32 — Application Questions

Q1 a) Acceleration at $x = A$ given by
$a = -\omega^2 A = -1.5^2 \times 1.6 = $ **–3.6 ms^{-2}**

The question asks for the acceleration, not the magnitude of acceleration, which is why you can't use $a_{max} = \omega^2 A$. The displacement given in the question is a positive value, so the acceleration is negative.

b) Angular frequency = $\omega = 2\pi f$, rearranging gives $f = \frac{\omega}{2\pi}$
= 1.5 ÷ (2 × π) = 0.2387... Hz
Time to complete 1 oscillation:
$T = \frac{1}{f} = \frac{1}{0.2387...} = $ 4.188... s
Time to complete 15 oscillations
= 15 × 4.188... = 62.83... s = **63 s (to 2 s.f.)**

Q2 a) Frequency = $f = \frac{1}{T} = \frac{1}{0.75} = $ 1.333... s^{-1}
Maximum speed is given by: $v_{max} = \omega A = 2\pi fA$
Substituting this in and rearranging for amplitude gives:
$A = \frac{v_{max}}{2\pi f}$
$= \frac{0.85}{2 \times \pi \times 1.3333...} = 0.101...$ m = **0.10 m (to 2 s.f.)**

b) Velocity at x = 0.080 m:
$v = \pm \omega \sqrt{A^2 - x^2}$
$= \pm 2\pi f \sqrt{A^2 - x^2}$
$= \pm 2 \times \pi \times 1.333... \times \sqrt{0.101...^2 - 0.080^2}$
$= \pm 0..5228...$ ms^{-1}
$= $ **± 0.52 ms^{-1} (to 2 s.f.)**

The answer has a ± sign at the front because you don't know the direction of the velocity.

Q3 Period = T = time to complete exactly 5 oscillations ÷ 5
$= \frac{15.5}{5} = $ 3.1 s
Frequency = $f = \frac{1}{T} = \frac{1}{3.1} = $ 0.3225... s^{-1}
Amplitude = A = 0.45 m
Displacement at time t: $x = A\cos(\omega t) = A\cos(2\pi ft)$
At time t = 10.0, x = 0.45 × cos(2π × 0.3225... × 10.0)
= 0.0681... m = **0.068 m (to 2 s.f.)**

Don't forget to put your calculator into radians for these calculations.

Q4 Pendulum passes through equilibrium twice every period, so if it's set to 120 ticks per minute:
Period = T = 120 ÷ 2 ÷ 60 = 1 s
Frequency = $f = \frac{1}{T} = \frac{1}{1} = $ 1 s^{-1}
Amplitude A = 6.2 cm = 0.062 m
Magnitude of max acceleration = $a_{max} = \omega^2 A = (2\pi f)^2 A$
So, $a_{max} = (2\pi \times 1)^2 \times 0.062$
= 2.447... ms^{-2} = **2.4 ms^{-2} (to 2 s.f.)**

Page 32 — Fact Recall Questions

Q1 Displacement = $x = A\cos(\omega t)$

Q2 Acceleration = $a = -\omega^2 x$
Velocity = $v = \pm \omega \sqrt{A^2 - x^2}$

Q3 **B**

Q4 Max speed = $v_{max} = \omega A$

Q5

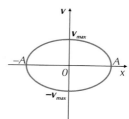

Q6

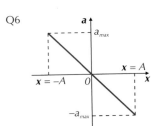

5. The Mass-Spring System as a Simple Harmonic Oscillator

Page 35 — Application Questions

Q1 Restoring force due to a spring = $F = -kx$
Extension in m = $2.5 \div 100 = 0.025$ m
Treat each force independently, and add them together:
Force due to first spring = $F_1 = -(25 \times 0.025) = -0.625$ N
Force due to second spring = $F_2 = -(45 \times 0.025) = -1.125$ N
Total force = $F_1 + F_2 = -1.75$ N
So the size of the force = **1.8 N (to 2 s.f.)**

Q2 Extension in m = $x = 45 \div 1000 = 0.045$ m
Force given by: $F = -kx$, so rearranging this for spring
constant: $k = -\dfrac{F}{x} = \dfrac{18}{0.045} = 400$ Nm^{-1}
The force is in the opposite direction to the displacement,
so the minus sign disappears.
Period = $T = 2\pi\sqrt{\dfrac{m}{k}} = 2\pi \times \sqrt{\dfrac{1}{400}} = 0.3141...$ s
= **0.31 s (to 2 s.f.)**

Page 35 — Fact Recall Question

Q1 a) Period of oscillation of a mass on a spring
= $T = 2\pi\sqrt{\dfrac{m}{k}}$

b) Put a known mass into a trolley and attach a horizontal
spring to one side of the trolley, where the other end of
the spring is attached to a wall. Have a position sensor
on the other side of the trolley to the spring — this
position sensor measures the displacement of the mass.
Set the mass oscillating and use the data logger and the
computer to find the period of oscillation. Repeat the
experiment, but vary the mass and spring constant and
observe how the period changes.

6. The Simple Pendulum and Other Types of SHO

Page 39 — Application Questions

Q1 $T = 2\pi\sqrt{\dfrac{l}{g}} = 2\pi\sqrt{\dfrac{2.50}{9.81}}$
$= 3.1718...$ s = **3.17 s (to 3 s.f.)**

Q2 a) K.E. $= \frac{1}{2}mv^2$, $\rho = \dfrac{m}{V}$, so $m = V\rho$
Substituting this in gives:
K.E. $= \frac{1}{2}V\rho v^2 = \frac{1}{2} \times 0.0120 \times 1000 \times 1.20^2$
$= $ **8.64 J**

b) (i) The angular frequency doesn't change as it is only
dependent on L and g.
(ii) If the volume of water in the tube increases, the
length, L, of the water column must have increased
(assuming the cross-sectional area stayed the same).
An increase in L means a decrease in angular
frequency.
(iii) The angular frequency doesn't change as it is only
dependent on L and g, not ρ or m.

Page 39 — Fact Recall Questions

Q1 a) Period of oscillation of a simple pendulum is $T = 2\pi\sqrt{\dfrac{l}{g}}$

b) Set up a simple pendulum made of a bob at the end of
a stiff rod. Attach it to an angle sensor and computer, so
the angle of the pendulum is recorded continuously as it
oscillates and the period is calculated. Vary the length
and observe how the period changes.

Q2 Graph of period squared
against pendulum length:

Graph of period
squared against mass:

Graph of period
squared against amplitude:

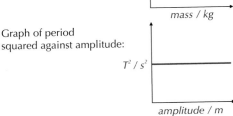

As T is independent of m and A, T^2 is too.

7. Free and Forced Vibrations

Page 43 — Fact Recall Questions

Q1 A free vibration involves no transfer of energy between
the oscillating object and its surroundings. The object will
continue to oscillate at its natural frequency and with the
same amplitude forever. A forced vibration occurs if there's
a periodic external driving force acting on the object.

Q2 a) When the driving frequency is much less than the
oscillator's natural frequency, the driver displacement
and oscillator displacement are in phase (the oscillator
can easily 'keep up' with the driver).
b) When the driving frequency is the same as the natural
frequency (resonance), then the driver displacement and
oscillator displacement are 90° out of phase.
c) When the driving frequency is much greater than
the natural frequency, then the driver and oscillator
displacement will be completely out of phase.

Q3 Resonance occurs when the driving frequency approaches
the natural frequency of an object and the object begins to
oscillate with a rapidly increasing amplitude.

Q4 E.g. any three from: a radio's electric circuit resonating
when it's tuned to the same frequency as a radio station / a
glass resonating when driven by a sound wave at its natural
frequency / a column of air in an organ pipe resonating
when driven by the motion of air at its base / a swing in a
playground resonating when it's pushed by someone at its
natural frequency.

Q5 A damping force is a force that acts on an oscillator and
causes it to lose energy to its surroundings, reducing the
amplitude of its oscillations.

Q6 Light damping — damping such that an oscillating system takes a long time to stop, and the amplitude of the system reduces by only a small amount each period.

Heavy damping — damping such that the system takes less time to stop oscillating than a lightly damped system, and the amplitude gets much smaller each period.

Critical damping — damping such that the amplitude of an oscillating system is reduced (and so the system returns to equilibrium) in the shortest possible time.

Overdamping — extremely heavy damping such that an oscillating system takes longer to return to equilibrium than a critically damped system.

Q7

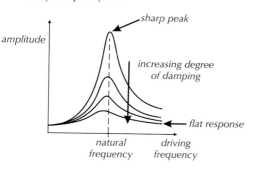

Exam-style Questions – Pages 45-46

1   D (1 mark)
Equation for maximum acceleration: $a_{max} = \omega^2 A$
$\omega = 2\pi f$, which when substituted in gives $a_{max} = (2\pi f)^2 A$
Therefore, $a_{max} \propto f^2$, so doubling the frequency will quadruple the maximum acceleration.

2   C (1 mark)
A critically damped system takes the shortest possible time to return to equilibrium. An overdamped system takes longer to return to equilibrium.

3   C (1 mark)
The correct graph needs to show 1 complete cycle (i.e. 2 'peaks') in 3 seconds and zero potential energy at 0, 1.5 s and 3 s.

4.1   Period given by: $T = 2\pi\sqrt{\dfrac{m}{k}}$
$$= 2 \times \pi \times \sqrt{\dfrac{0.60}{18}}$$
$$= 1.147... \text{ s}$$
Time taken for 5 oscillations $= 5 \times 1.147... = 5.735... $ s
$$= \textbf{5.7 s (to 2 s.f.)}$$
(2 marks for correct answer, otherwise 1 mark for calculating the time period.)

4.2   Velocity is given by: $v = \pm\omega\sqrt{A^2 - x^2}$
$$= \pm\dfrac{2\pi}{T}\sqrt{A^2 - x^2}$$
$$= \pm\dfrac{2\times\pi}{1.147...} \times \sqrt{0.20^2 - 0.15^2}$$
$$= \pm 0.7245... \text{ ms}^{-1}$$
$$= \pm \textbf{0.72 ms}^{-1} \textbf{ (to 2 s.f.)}$$
Don't forget to convert 20 cm and 15 cm into metres.
(3 marks for correct answer, otherwise 1 mark for substituting the equation for ω in terms of T into the equation for velocity.)

4.3   Maximum kinetic energy is: $E_{K\,(max)} = \dfrac{1}{2}mv_{max}^{\;2}$
Maximum velocity is: $v_{max} = \omega A$
$$= \dfrac{2\pi}{T}A = \dfrac{2\times\pi}{1.147...} \times 0.20$$
$$= 1.095... \text{ ms}^{-1}$$
Substitute this into the equation for maximum kinetic energy:
$$E_{K\,(max)} = \dfrac{1}{2} \times 0.60 \times (1.095...)^2 = \textbf{0.36 J}$$
(3 marks for correct answer, otherwise 1 mark for calculating maximum velocity, and 1 mark for substituting this into the formula for maximum kinetic energy.)

4.4   Period given by $T = 2\pi\sqrt{\dfrac{m}{k}}$
Rearrange this for k:
$$k = \dfrac{4\pi^2 m}{T^2} = (4 \times \pi^2 \times (0.6 + 1)) \div (1.147...)^2$$
$$= \textbf{48 Nm}^{-1}$$
(2 marks for correct answer, otherwise 1 mark for rearranging the period of oscillation equation for the spring constant.)

5.1   Period given by $T = 2\pi\sqrt{\dfrac{m}{k}}$
Rearrange this for k:
$$k = \dfrac{4\pi^2 m}{T^2} = (4 \times \pi^2 \times 1.8) \div 3.2^2 = 6.939... \text{ Nm}^{-1}$$
$$= \textbf{6.9 Nm}^{-1} \textbf{ (to 2 s.f.)}$$
(2 marks for correct answer, otherwise 1 mark for rearranging the period of oscillation equation for the spring constant.)

5.2   Force given by Hooke's law: $F = -kx$
Maximum force will be at maximum displacement (i.e. the amplitude).
So $F = -6.939... \times 0.22 = -1.5267... $ N $= -1.5$ N (to 2 s.f.)
So the magnitude of the force $= \textbf{1.5 N (to 2 s.f.)}$
(2 marks for correct answer, otherwise 1 mark for using the maximum displacement to find the maximum force.)

5.3   Velocity is given by: $v = \pm\omega\sqrt{A^2 - x^2}$
$$= \pm\dfrac{2\pi}{T}\sqrt{A^2 - x^2}$$
$$= \pm\dfrac{2\times\pi}{3.2} \times \sqrt{0.22^2 - 0.12^2}$$
$$= \pm 0.3620... \text{ ms}^{-1}$$
Kinetic energy is: $E_K = \frac{1}{2}mv^2 = 0.5 \times 1.8 \times (\pm 0.3620...^2)$
$$= 0.1179... \text{ J}$$
$$= \textbf{0.12 J (to 2 s.f.)}$$
(4 marks for correct answer, otherwise 1 mark for calculating angular frequency, and 1 mark for calculating velocity.)

5.4   Max acceleration $= \omega^2 A = \left(\dfrac{2\pi}{T}\right)^2 A = \left(\dfrac{2\times\pi}{3.2}\right)^2 \times 0.22$
$$= 0.8481... \text{ ms}^{-2} = \textbf{0.85 ms}^{-2} \textbf{ (to 2 s.f.)}$$

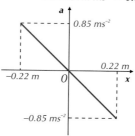

(3 marks – 1 mark for correct shape, 1 mark for calculating maximum acceleration, and 1 mark for indicating that the maximum acceleration occurs at the minimum displacement (and vice versa).)

6.1   The amplitude of the pendulum's oscillation will increase rapidly around its natural frequency *(1 mark)*. This is called resonance *(1 mark)*.

6.2 E.g. friction with the water acts to dampen the oscillation *(1 mark)*. The resonance will occur at a frequency slightly less than the pendulum's natural frequency *(1 mark)*. The pendulum's response will be flatter (i.e. the resonance peak will be less sharp) *(1 mark)*.

Section 2 — Thermal Physics

1. Thermal Energy Transfer
Page 52 — Application Questions
Q1 $Q = mc\Delta\theta$, so $Q = 0.45 \times 244 \times 3.0 = 329.4$
     = **330 J (to 2 s.f.)**

Q2 a) E.g. the thermometers may not be exactly calibrated, which would cause a systematic error. Place both thermometers in the same beaker of water and check that they agree to avoid this.
    E.g. the input temperature of the water might vary, leading to random errors during the experiment. Fill a container with water and allow it to come to room temperature, then use this water for the experiment.

   b) For the first run, $Q_1 = m_1 c\Delta\theta + H$.
      For the second run, $Q_2 = m_2 c\Delta\theta + H$.
      Combining and rearranging these gives:
      $$c = \frac{Q_2 - Q_1}{(m_2 - m_1)\Delta\theta} = \frac{7515 - 5511}{(0.450 - 0.330) \times 10.0}$$
      $$= \textbf{1670 J kg}^{-1}\textbf{K}^{-1} \text{ (or J kg}^{-1}\,^{\circ}\text{C}^{-1})$$

   Remember that if the change in temperature is the same for both runs, the heat losses to the surroundings (H) will cancel out when you combine the equations. Remember that the time period must be the same for each run too. Look back at page 50 for more on this.

Q3 a) Energy lost from 25.0 °C to 0 °C
      $= mc\Delta\theta = 0.1000 \times 4180 \times 25.0 = 10\ 450$ J
      Energy lost in freezing $= ml = 0.1000 \times 334\ 000$
      $= 33\ 400$ J
      Energy lost from 0 °C to −5.00 °C
      $= mc\Delta\theta = 0.1000 \times 2110 \times 5.00 = 1055$ J
      Total energy lost $= 10\ 450 + 33\ 400 + 1055$
      $= 44\ 905 = \textbf{44\ 900 J (to 3 s.f.)}$

   b) Energy transfer happens faster when the difference in temperature is larger. If the temperature of the freezer is brought closer to the temperature of the water, it will take longer to freeze.

Page 52 — Fact Recall Questions
Q1 a) Absolute zero.
    b) 0 K, −273 °C.
       You could also say −273.15 °C.
Q2 The internal energy of a body is the sum of the randomly distributed kinetic and potential energies of all its particles.
Q3 a)

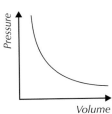

   b) E.g. The average molecule speed increases / the distribution of molecule speeds becomes more spread out.

Q4 Energy is transferred between gas molecules when they collide — one molecule speeds up and the other slows down.

Q5

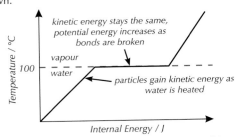

Q6 The specific latent heat of fusion is the quantity of thermal energy needed to be transferred to melt or freeze 1 kg of a substance.

2. The Three Gas Laws
Page 56 — Application Question
Q1 a) The volume has halved. As p and V are inversely proportional, the pressure will have doubled, i.e. $p = 2.8 \times 10^5$ Pa.
    b) Start temperature in K = 27 + 273 = 300 K.
       End temperature in K = −173 + 273 = 100 K.
       The temperature is divided by 3. As V and T are directly proportional, the volume will also be divided by 3, i.e. $V = 10.0$ cm^3.

Page 56 — Fact Recall Questions
Q1 It must have a fixed mass.
Q2 a) Boyle's law states that at a constant temperature the pressure p and volume V of a(n ideal) gas are inversely proportional.
    b)

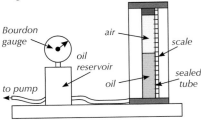

    c) Set up an experiment so that oil traps a pocket of air in a sealed tube with fixed dimensions (as shown in the diagram below).

Use a tyre pump to increase the pressure on the oil and a Bourdon gauge to measure the pressure. As the pressure increases, more oil will be pushed into the tube, the oil level will rise, and the air will compress. Measure the volume occupied by air in the tube as it reduces at regular pressure intervals by multiplying the length of the part of the tube containing air by $\pi \times$ radius of tube squared. Then plot a graph of pressure against 1/volume, you should get a straight line.

Q3 a) Charles' law states that at constant pressure, the volume V of an ideal gas is directly proportional to its absolute temperature T.

b)

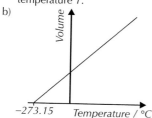

Q4 a) The pressure law states that at constant volume, the pressure p of an ideal gas is directly proportional to its absolute temperature T.

b)

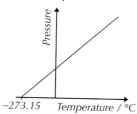

3. The Ideal Gas Equation

Page 60 — Application Questions

Q1 Molar mass = 12.0 + 16.0 + 16.0 = **44.0 g**

Q2 $pV = nRT \Rightarrow V = \dfrac{nRT}{p}$

$$= \frac{23 \times 8.31 \times (25 + 273)}{2.4 \times 10^5} = 0.23731...$$

$$= \textbf{0.24 m}^3 \textbf{ (to 2 s.f.)}$$

Q3 $pV = NkT \Rightarrow p = \dfrac{NkT}{V}$

$$= \frac{(8.21 \times 10^{24}) \times (1.38 \times 10^{-23}) \times 500}{4.05}$$

$$= 1.3987... \times 10^4$$

$$= \textbf{1.40} \times \textbf{10}^4 \textbf{ Pa (to 3 s.f.)}$$

Q4 Energy transferred = work done = $p\Delta V$

$$= 1.15 \times 10^5 \times \frac{1}{2} \times 2.0$$

$$= \textbf{1.2} \times \textbf{10}^5 \textbf{ J (to 2 s.f.)}$$

The volume of the gas doubles, and the final volume is 2.0 m³, so the change in volume is $\frac{1}{2} \times 2.0 = 1.0$ m³.

Q5 $pV = NkT \Rightarrow T = \dfrac{pV}{Nk}$

$$= \frac{(1.29 \times 10^5) \times 0.539}{(1.44 \times 10^{25}) \times (1.38 \times 10^{-23})}$$

$$= 349.894... = \textbf{350 K (to 3 s.f.)}$$

Q6 a) 1 mole of gas would have a mass of 44 g and 0.88 kg = 880 g. Moles of gas = $\dfrac{880}{44}$ = **20**.

b) $pV = nRT \Rightarrow T = \dfrac{pV}{nR}$

$$= \frac{(2.3 \times 10^5) \times 0.39}{20 \times 8.31} = 539.71...$$

$$= \textbf{540 K (to 2 s.f.)}$$

Page 60 — Fact Recall Questions

Q1 Relative molecular mass is the sum of the relative atomic masses of all the atoms that make up a single molecule.

Q2 The Avogadro constant, N_A, is the number of atoms in exactly 12 g of the carbon isotope $^{12}_{6}C$. Its value is 6.02×10^{23} mol⁻¹.

Q3 The molar mass of a substance is the mass (usually in grams) that one mole of that substance would have.

Q4 $pV = nRT$

p = pressure (in Pa)
V = volume (in m³)
n = number of moles of gas
R = molar gas constant (= 8.31 J mol⁻¹ K⁻¹)
T = temperature (in K)

Q5 Boltzmann's constant, k, is effectively the gas constant for one molecule of gas. It is equivalent to $\dfrac{R}{N_A}$, and its value is 1.38×10^{-23} JK⁻¹.

Q6 $pV = NkT$

p = pressure (in Pa)
V = volume (in m³)
N = number of molecules of gas
k = Boltzmann constant (= 1.38×10^{-23} JK⁻¹)
T = temperature (in K)

Q7 work done = $p\Delta V$

p = pressure (in Pa)
ΔV = change in volume (in m³)

4. Kinetic Theory and the Pressure of an Ideal Gas

Page 64 — Application Questions

Q1 a) The momentum of the molecule in the direction normal to the wall is equal to $m\boldsymbol{u}$. The collision is elastic, so it will rebound with velocity $-\boldsymbol{u}$ and momentum $-m\boldsymbol{u}$. So its change in momentum is $m\boldsymbol{u} - (-m\boldsymbol{u}) = 2m\boldsymbol{u}$.

b) Force is the rate of change of momentum, i.e. $2m\boldsymbol{u} \times$ number of collisions per second. The time between collisions is $\dfrac{2l}{\boldsymbol{u}}$, and so the number of collisions per second is $\dfrac{\boldsymbol{u}}{2l}$. So force = $2m\boldsymbol{u} \times \dfrac{\boldsymbol{u}}{2l} = \dfrac{m\boldsymbol{u}^2}{l}$.

c) The force exerted by N molecules with velocity $\boldsymbol{u}_1$, $\boldsymbol{u}_2$... etc. will be equal to $\dfrac{m(\boldsymbol{u}_1^2 + \boldsymbol{u}_2^2 + etc.)}{l}$, or $\dfrac{mN\overline{\boldsymbol{u}^2}}{l}$.

The pressure is given by $\dfrac{force}{area}$, so $p = \dfrac{\left(\dfrac{mN\overline{\boldsymbol{u}^2}}{l}\right)}{l^2} = \dfrac{mN\overline{\boldsymbol{u}^2}}{V}$

The volume V is equal to l^3.

d) Molecules can move in three dimensions, so their mean square speed is given by $(c_{rms})^2 = \overline{\boldsymbol{u}^2} + \overline{\boldsymbol{v}^2} + \overline{\boldsymbol{w}^2}$ where $\boldsymbol{u}$, $\boldsymbol{v}$ and $\boldsymbol{w}$ are the components of their velocity in the three directions normal to the walls of the container. As the molecules move randomly, $\overline{\boldsymbol{u}^2} = \overline{\boldsymbol{v}^2} = \overline{\boldsymbol{w}^2}$ and so $(c_{rms})^2 = 3\overline{\boldsymbol{u}^2}$, or $\overline{\boldsymbol{u}^2} = \frac{1}{3}(c_{rms})^2$.

Substituting this into the equation from part c) gives $pV = \frac{1}{3}Nm(c_{rms})^2$.

Q2 $pV = \frac{1}{3}Nm(c_{rms})^2$

$$\Rightarrow p = \frac{1}{3}\frac{Nm(c_{rms})^2}{V}$$

$$= \frac{\frac{1}{3}(5 \times 6.02 \times 10^{23}) \times (5.31 \times 10^{-26}) \times (8.11 \times 10^6)}{1.44}$$

$$= 3.0005... \times 10^5 = \textbf{3.00} \times \textbf{10}^5 \textbf{ Pa (to 3 s.f.)}$$

Q3 The temperature of a gas is related to the kinetic energy of the gas molecules — as the temperature increases, the average speed of the molecules increases. This means the rate of change of momentum of the molecules colliding with the walls of the container increases, and so the force on the walls of the container increases. If the volume of the container is fixed, this will result in an increased pressure inside the container because there will be more collisions between the molecules and the walls of the container in a given amount of time and on average, a collision will result in a larger change in momentum, and so exert a larger force on the walls of the container.

If the pressure inside the container remains constant, the volume of the container will increase to compensate for the temperature change because if the volume is larger, there will be a longer time between molecule-wall collisions, and so the rate of change of momentum and therefore the force on the walls of the container will be reduced. As the volume increases, the surface area of the walls increases. Pressure is defined as the force per unit area, and so increasing the area stops the pressure from increasing.

Page 64 — Fact Recall Questions
Q1 c_{rms} is the root mean square speed (of the molecules in a gas). I.e. the square root of the mean of the squared speeds of the molecules.
Q2 The average change of momentum in a collision is increased and the average time between collisions is reduced. These both mean that the rate of change of momentum is greater, so the force on the wall is increased.
Q3 The volume of the gas or the pressure of the gas.
Q4 Any four of e.g.: All molecules are identical / the gas contains a large number of molecules / the molecules have negligible volume compared with the volume of the container / the molecules continually move about randomly / Newtonian mechanics apply / collisions between molecules or at the walls of the container are perfectly elastic / molecules move in a straight line between collisions / the forces that act during collisions last for much less time that the time between collisions.

5. Kinetic Energy of Gas Molecules
Page 66 — Application Questions
Q1 $\frac{1}{2}m(c_{rms})^2 = \frac{3}{2}kT$
$$= \frac{3}{2} \times (1.38 \times 10^{-23}) \times 112 = 2.3184 \times 10^{-21}$$
$$= \mathbf{2.32 \times 10^{-21}\,J} \text{ (to 3 s.f.)}$$

Q2 Change in energy $= nN_A \times \frac{3}{2}kT$
$$= 2.44 \times (6.02 \times 10^{23}) \times \frac{3}{2} \times (1.38 \times 10^{-23}) \times (290 - 250)$$
$$= 1216.2... = \mathbf{1200\,J} \text{ (to 2 s.f.)}$$

Page 66 — Fact Recall Questions
Q1 For an ideal gas, you can assume that all of the internal energy is in the form of kinetic energy.
Q2 Average kinetic energy $= \frac{1}{2}m(c_{rms})^2 = \frac{3}{2}kT \left(= \frac{3RT}{2N_A} = \frac{3nRT}{2\ N}\right)$
Q3 Total kinetic energy $= \frac{3}{2}kTnN_A$ or $\frac{3}{2}RTn$

6. Development of Theories
Page 68 — Fact Recall Questions
Q1 Theoretical concepts, such as kinetic theory, are based on assumptions and derivations from knowledge — they explain why something happens. Empirical concepts, such as the gas laws, are based on observations — they can predict what will happen but do not explain why.
Q2 E.g. Ancient Greek and Roman philosophers including Democritus had ideas about gases 2000 years ago, some of which were quite close to what we now know to be true. Robert Boyle discovered the relationship between pressure and volume at a constant temperature in 1662 — this became known as Boyle's law. This was followed by Charles' law in 1787 when Jacques Charles discovered that the volume of a gas is proportional to temperature at a constant pressure. The pressure law was discovered by Guillaume Amontons in 1699, who noticed that at a constant volume, temperature is proportional to pressure. It was then re-discovered much later by Joseph Louis Gay-Lussac in 1809.
In the 18th century a physicist called Daniel Bernoulli explained Boyle's Law by assuming that gases were made up of tiny particles — the beginnings of kinetic theory. But it took another couple of hundred years before kinetic theory became widely accepted, helped by Robert Brown's discovery of Brownian motion in 1827 and Einstein's explanation of it.
Q3 The zigzag, random motion of any particles suspended in a fluid is known as Brownian motion. It provides evidence for the kinetic particle theory of the different states of matter, as the random motion is believed to be a result of collisions with fast, randomly-moving particles in the fluid.

Exam-style Questions — Pages 70-71
1   **C** *(1 mark)*
2   **A** *(1 mark)*
3.1   Increasing the temperature increases the average speed of the gas molecules *(1 mark)*. This means they collide with the walls of the container more often and on average there's a larger change in momentum during a collision *(1 mark)*. This means a greater force is exerted on the walls of the container in the same amount of time, and so the pressure is increased *(1 mark)*.

3.2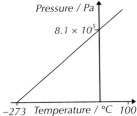

(2 marks available — 1 mark for straight-line graph and 1 mark for crossing through (−273, 0) and (0, 8.1 × 10⁵).)

3.3 $pV = NkT \Rightarrow N = \frac{pV}{kT} = \frac{(8.1 \times 10^5) \times 0.51}{(1.38 \times 10^{-23}) \times 273}$
$$= 1.096... \times 10^{26} = \mathbf{1.1 \times 10^{26}} \text{ (to 2 s.f.)}$$
(2 marks for correct answer, 1 mark for correct working if answer incorrect.)

3.4   Total mass of gas $= (1.096... \times 10^{26}) \times (2.7 \times 10^{-26})$
$$= 2.96... \text{ kg}$$
$Q = mc\Delta\theta = 2.96... \times (2.2 \times 10^3) \times 150$
$$= 9.7699... \times 10^5 = \mathbf{9.8 \times 10^5\,J} \text{ (to 2 s.f.)}$$
(2 marks for correct answer, 1 mark for finding the mass of gas if answer incorrect.)

4.1 $Q = mc\Delta\theta = (92 \times 10^{-3}) \times 2110 \times 25$
$$= 4.853 \times 10^3 \text{ J} = \textbf{4.9} \times \textbf{10}^3 \textbf{ J (to 2 s.f.) (1 mark)}$$

4.2 $Q = ml = (92 \times 10^{-3}) \times (3.3 \times 10^5)$
$$= 3.036 \times 10^4 \text{ J} = \textbf{3.0} \times \textbf{10}^4 \textbf{ J (to 2 s.f.) (1 mark)}$$

4.3 Total energy needed $= (4.853 \times 10^3) + (3.036 \times 10^4)$
$$= 3.5213 \times 10^4 \text{ J}$$
$$\text{Time taken} = \frac{\text{total energy needed}}{\text{rate of energy supplied}}$$
$$= \frac{3.5213 \times 10^4}{50.0} = 704.26$$
$$= \textbf{700 s (to 2 s.f.)}$$
(2 marks for correct answer, 1 mark for correct working if answer wrong.)

4.4 It would slow down the melting of the ice *(1 mark)*. Heat is transferred from hotter substances to colder substances — as the ice is colder than 25 °C, insulating the beaker will stop heat transfer into the ice from the surrounding air *(1 mark)*.

5.1 Absolute zero is the lowest possible temperature any substance can have, and is equal to 0 K or –273 °C *(1 mark)*.

5.2 The Avogadro constant is the number of atoms in exactly 12 g of the carbon isotope $^{12}_{6}\text{C}$ *(1 mark)*.
Allow "the number of particles in exactly 1 mole of a substance".

5.3 Molar mass is the mass that 1 mole of a substance would have *(1 mark)*.

5.4 $N = n \times N_A = 54.0 \times (6.02 \times 10^{23}) = 3.2508 \times 10^{25}$
$$= \textbf{3.25} \times \textbf{10}^{25} \textbf{ (to 3 s.f.) (1 mark)}$$

5.5 $pV = nRT \Rightarrow T = \dfrac{pV}{nR}$
$$= \frac{1.00 \times 10^5 \times 4.18}{54.0 \times 8.31}$$
$$= 931.49... = \textbf{931 K (to 3 s.f.)}$$
(2 marks for correct answer, 1 mark for correct working if answer incorrect).

5.6 Average kinetic energy $= \frac{1}{2}m(c_{rms})^2$
$$\frac{1}{2}m(c_{rms})^2 = \frac{3}{2}kT$$
$$= \frac{3}{2} \times (1.38 \times 10^{-23}) \times 931.49...$$
$$= 1.928... \times 10^{-20} \text{ J}$$
Total kinetic energy $= (1.928... \times 10^{-20})$
$$\times (3.2508 \times 10^{25})$$
$$= 6.2681... \times 10^5$$
$$= \textbf{6.27} \times \textbf{10}^5 \textbf{ J (to 3 s.f.)}$$
(3 marks for correct answer, otherwise 1 mark for using correct formula for the average kinetic energy and 1 mark for multiplying this answer by the number of molecules if answer incorrect.)
If you got the wrong answers for parts 5.4 or 5.5 but your calculations here were correct, you get all the marks for part 5.6.

Section 3 — Gravitational and Electric Fields

1. Gravitational Fields
Page 74 — Application Questions
Q1 $F = \dfrac{Gm_1m_2}{r^2}$
$$= \frac{(6.67 \times 10^{-11}) \times (2.15 \times 10^{30}) \times (2.91 \times 10^{30})}{(1 \times 10^{11})^2}$$
$$= 4.17308... \times 10^{28}$$
$$\approx \textbf{4.2} \times \textbf{10}^{28} \textbf{ N}$$

Q2 $\dfrac{2.5}{0.5} = 5$, so the force will be $5^2 = 25$ times larger:
$$25 \times 25 = \textbf{625 N}$$

Q3 a) 6 370 000 m + 10 000 m = 6 380 000 m = **6380 km**
b) The upwards force must balance the downwards force due to gravity:
$$F = \frac{Gm_1m_2}{r^2} = \frac{(6.67 \times 10^{-11}) \times (2500) \times (5.98 \times 10^{24})}{(6380 \times 10^3)^2}$$
$$= 24 497.7... \text{ N} = \textbf{24 500 N (to 3 s.f.)}$$

Page 74 — Fact Recall Questions
Q1 Any object with a mass will feel an attractive force due to gravity.

Q2 A gravitational field is a region in which an object will experience an attractive force (due to gravity).

Q3 a)

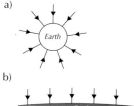

b)

Q4 a) $F = \dfrac{Gm_1m_2}{r^2}$
b) G is the gravitational constant, $6.67 \times 10^{-11} \text{ Nm}^2\text{kg}^{-2}$.
c) The masses must be point masses, or uniform spheres (where all the mass acts as if it's at the centre, i.e. acts as a point mass).

2. Gravitational Field Strength
Page 77 — Application Questions
Q1 $g = \dfrac{F}{m} = \dfrac{581}{105} = 5.5333... = \textbf{5.53 Nkg}^{-1} \textbf{ (to 3 s.f.)}$

Q2 $g = \dfrac{F}{m}$, so $F = gm$. The magnitude of the value of g is lower on the Moon, which means an object with the same mass will experience less force due to gravity (and so be easier to lift).

Q3 $g = \dfrac{GM}{r^2} = \dfrac{6.67 \times 10^{-11} \times 6.42 \times 10^{23}}{(3390 \times 10^3)^2}$
$$= 3.7261... = \textbf{3.73 Nkg}^{-1} \textbf{ (to 3 s.f.)}$$

Q4 a) $F = \dfrac{Gm_1m_2}{r^2}$
Rearrange for r:
$$r = \sqrt{\frac{Gm_1m_2}{F}} = \sqrt{\frac{6.67 \times 10^{-11} \times 7.34 \times 10^{22} \times 65}{105}}$$
$$= 1.7408... \times 10^6 = \textbf{1.7} \times \textbf{10}^6 \textbf{ m (to 2 s.f.)}$$
b) $g = \dfrac{GM}{r^2} = \dfrac{6.67 \times 10^{-11} \times 7.34 \times 10^{22}}{((1.7408... \times 10^6) + (640 \times 10^3))^2}$
$$= 0.8636... = \textbf{0.86 Nkg}^{-1} \textbf{ (to 2 s.f.)}$$

Q5 g due to the larger mass:
$$g = \frac{GM}{r^2} = \frac{3Gm}{9x^2}$$
g due to the smaller mass:
$$g = \frac{GM}{r^2} = \frac{Gm}{(3x + x)^2} = \frac{Gm}{16x^2}$$
Total g:
$$g = \frac{3Gm}{9x^2} + \frac{Gm}{16x^2} = \frac{19Gm}{48x^2} \textbf{ (or 2.64... } \times \textbf{10}^{-11} \frac{m}{x^2}\textbf{)}$$
The gravitational field for both objects is in the same direction, so it doesn't matter which way you add them up — in this case, left is positive.

Page 77 — Fact Recall Questions

Q1 g is the gravitational field strength (or the force per unit mass due to gravity), measured in Nkg^{-1}.

Q2 M is the mass of the object creating the gravitational field.

Q3 The correct graph is c).

3. Gravitational Potential

Page 81 — Application Questions

Q1 a) No change (G is a constant).

b) V will double (twice as negative) as it's related to $\frac{1}{r}$ ($V = -\frac{GM}{r}$).

c) g will be four times bigger as it's related to $\frac{1}{r^2}$ ($g = \frac{GM}{r^2}$).

d) No change (mass is the same everywhere).

Q2 $\Delta W = m\Delta V = 1.72 \times 531$
$= 913.32$
$= \textbf{913 J (to 3 s.f.)}$

Q3 Number of squares underneath the graph between 2×10^6 m and 5×10^6 m is approximately 4 squares. Therefore change in gravitational potential $\approx \textbf{4} \times \textbf{10}^6 \textbf{ Jkg}^{-1}$.
Don't forget, each square on the graph is worth 1×10^6 m across, so the answer must be multiplied by 1×10^6 too.

Page 81 — Fact Recall Questions

Q1 Gravitational potential is the gravitational potential energy that a unit mass would have at a point. It's measured in Jkg^{-1}.

Q2 The negative energy sign means that the gravitational potential is negative and increases as you move further away from the source of the gravitational field. At infinity, the gravitational potential is zero. You can think of this negative potential as being caused by you having to do work against the gravitational field to move an object out of it.

Q3

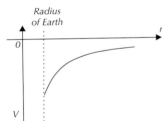

Q4 The gradient tells you the value of $-g$ at that point.

Q5 ΔW is the work done.

Q6 Equipotentials are lines (in 2D) and surfaces (in 3D) that join together all of the points with the same gravitational potential, V.

4. Orbits

Page 86 — Application Questions

Q1 $T^2 \propto r^3$

So $\dfrac{T^2}{r^3} = \dfrac{T_{new}^2}{\left(\frac{r}{2}\right)^3}$

$\dfrac{T^2}{r^3} = \dfrac{8T_{new}^2}{r^3}$

$T^2 = 8T_{new}^2$

$T_{new}^2 = \dfrac{T^2}{8}$

$T_{new} = \dfrac{T}{\sqrt{8}} = \dfrac{T}{2\sqrt{2}}$

Q2 $T^2 \propto r^3$

So $\dfrac{T_{close}^2}{r_{close}^3} = \dfrac{T_{far}^2}{r_{far}^3}$

$\dfrac{42^2}{(3.95 \times 10^5)^3} = \dfrac{T_{far}^2}{(4.84 \times 10^5)^3}$

$T_{far}^2 = \dfrac{42^2 \times (4.84 \times 10^5)^3}{(3.95 \times 10^5)^3} = 3245.214...$

$T_{far} = 56.966... = \textbf{57 hours (to 2 s.f.)}$

Q3 a) $E_K = \frac{1}{2}mv^2$

Rearrange for v:

$v = \sqrt{\dfrac{2E_K}{m}} = \sqrt{\dfrac{2 \times (2.4 \times 10^{29})}{1.31 \times 10^{22}}} = 6.053... \times 10^3$

$= \textbf{6.1} \times \textbf{10}^3 \textbf{ ms}^{-1} \textbf{ (to 2 s.f.) (= 6.1 kms}^{-1}\textbf{)}$

b) Total energy at the point when Pluto is closest to the Sun:

$E_{total} = E_K + E_P = (2.4 \times 10^{29}) + (-3.9 \times 10^{29})$
$= -1.5 \times 10^{29}$ J

The total energy is always the same at any point during the orbit, so use the same equation and rearrange for E_K:

$E_K = E_{total} - E_P = (-1.5 \times 10^{29}) - (-3.2 \times 10^{29}) = \textbf{1.7} \times \textbf{10}^{29}$ **J**

Q4 Using the equation for escape velocity:

$v = \sqrt{\dfrac{2GM}{r}} = \sqrt{\dfrac{2 \times (6.67 \times 10^{-11}) \times (7.34 \times 10^{22})}{(1.74 \times 10^6)}}$

$= 2.3722... \times 10^3$
$= \textbf{2.37} \times \textbf{10}^3 \textbf{ ms}^{-1} \textbf{ (to 3 s.f.) (= 2.37 kms}^{-1}\textbf{)}$

Page 86 — Fact Recall Questions

Q1 A satellite is a small mass that orbits a larger mass.

Q2 The orbital speed of a satellite is inversely proportional to the square root of the radius of its orbit ($v \propto \frac{1}{\sqrt{r}}$), so as the radius increases the speed decreases.

Q3 The orbital period of a satellite is proportional to the square root of the radius cubed ($T \propto \sqrt{r^3}$), so as the radius increases the orbital period increases.

Q4 It is best to use a logarithmic scale as it's a lot easier to compare data for different planets that have widely different orbital radii and orbital periods.

Q5 The total energy of a satellite is always the same. Because its gravitational potential energy is higher when its height is larger, its kinetic energy (and so orbital speed) must be lower, and vice versa.

Q6 The escape velocity is the minimum speed required for an unpowered object to leave the gravitational field of another object.

Q7 A geostationary satellite is a satellite that orbits directly over the equator and is always above the same point on Earth. It has an orbital period of 1 day. Geostationary satellites are always above the same point of the Earth, so receivers don't need to be repositioned to keep up with them.

Q8 A low orbiting satellite is a satellite which orbits between 180 and 2000 km above the Earth's surface and at a higher angular speed than Earth. They can be used for imaging (such as mapping and spying) and monitoring the weather.

5. Electric Fields

Page 91 — Application Questions

Q1 $F = \dfrac{1}{4\pi\varepsilon_0} \dfrac{Q_1 Q_2}{r^2}$

$= \dfrac{1}{4\pi \times (8.85 \times 10^{-12})} \dfrac{(-1.60 \times 10^{-19})^2}{(5.22 \times 10^{-13})^2}$

$= 8.44784... \times 10^{-4}$
$= \textbf{8.45} \times \textbf{10}^{-4} \textbf{ N (to 3 s.f.)}$

Q2

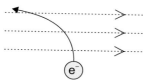

Q3 Start by rearranging the formula for the magnitude of $\boldsymbol{E}$ to make r the subject:

$\boldsymbol{E} = \dfrac{1}{4\pi\varepsilon_0}\dfrac{Q}{r^2} \Rightarrow r^2 = \dfrac{1}{4\pi\varepsilon_0}\dfrac{Q}{\boldsymbol{E}}$

$\Rightarrow r = \sqrt{\dfrac{1}{4\pi\varepsilon_0}\dfrac{Q}{\boldsymbol{E}}}$

Then put the numbers in:

$r = \sqrt{\dfrac{1}{4\pi \times (8.85 \times 10^{-12})}\dfrac{4.15 \times 10^{-6}}{15\,000}}$

$= 1.57725...$

$= \textbf{1.6 m (to 2 s.f.)}$

Q4 $\boldsymbol{E} = \dfrac{F}{Q} = \dfrac{0.080}{5.0 \times 10^{-5}} = \textbf{1600 NC}^{-1}$

Q5 If the particle isn't moving, the upwards force from the electric field must balance the particle's weight, i.e.

$mg = \boldsymbol{E}Q = \dfrac{VQ}{d}$

This comes from weight = mg, $\boldsymbol{E} = \frac{F}{Q}$ and $\boldsymbol{E} = \frac{V}{d}$.

Rearrange this to make d the subject, then put in the numbers:

$mg = \dfrac{VQ}{d} \Rightarrow d = \dfrac{VQ}{mg}$

$= \dfrac{(5.00 \times 10^{-9}) \times 2(1.60 \times 10^{-19})}{(6.64 \times 10^{-27}) \times 9.81}$

$= 0.02456...\,\text{m} = \textbf{2.46 cm (to 3 s.f.)}$

Q6 $F = \dfrac{1}{4\pi\varepsilon_0}\dfrac{Q_1Q_2}{r^2}$

Rearrange for r:

$r = \sqrt{\dfrac{1}{4\pi\varepsilon_0}\dfrac{Q_1Q_2}{F}}$

Distance between <u>centre</u> of the sphere and the point charge = $r + 3r = 4r$, charge of sphere = Q and charge of point charge = q. Substitute these in:

$4r = \sqrt{\dfrac{1}{4\pi\varepsilon_0}\dfrac{Qq}{F}}$

The change in the sphere's radius doesn't affect the electrostatic force (as a sphere is treated as a point charge acting at the centre of the sphere). So you can write the new distance between the centre of the sphere and the point charge as:

new distance = $\sqrt{\dfrac{1}{4\pi\varepsilon_0}\dfrac{9Qq}{F}} = 3 \times \sqrt{\dfrac{1}{4\pi\varepsilon_0}\dfrac{Qq}{F}} = 3 \times 4r = 12r$

The distance between the <u>surface</u> of the sphere and the point charge = $12r$ – radius of sphere = $12r - 2r = \textbf{10r}$.

Page 91 — Fact Recall Questions

Q1 $\boldsymbol{F} = \dfrac{1}{4\pi\varepsilon_0}\dfrac{Q_1Q_2}{r^2}$

Q2 $\boldsymbol{E}$ is a measure of the force per unit charge (in an electric field).

Q3

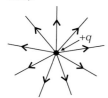

6. Electric Potential
Page 95 — Application Questions

Q1 a) $V = \dfrac{1}{4\pi\varepsilon_0}\dfrac{Q}{r} = \dfrac{1}{4\pi\varepsilon_0}\dfrac{12.6 \times 10^{-6}}{5.19 \times 10^{-2}}$

$= 2.18298... \times 10^6$

$= \textbf{2.18} \times \textbf{10}^6\textbf{ V (to 3 s.f.)}$

b) First find the new potential at the centre of the smaller sphere:

$V = \dfrac{1}{4\pi\varepsilon_0}\dfrac{Q}{r} = \dfrac{1}{4\pi\varepsilon_0}\dfrac{12.6 \times 10^{-6}}{((5.19 \times 10^{-2}) + (12.9 \times 10^{-2}))}$

$= 6.26... \times 10^5 \text{ V}$

Then find the work done:

$\Delta W = Q\Delta V$

$= (0.152 \times 10^{-6}) \times [(2.18 \times 10^6) - (6.26... \times 10^5)]$

$= 0.236163...$

$= \textbf{0.237 J (to 3 s.f.)}$

Q2 a) The change in potential between 3.0 m and 6.0 m is equal to the area under the graph between $r = 3.0$ m and $r = 6.0$ m.

Value of the area of one square = $1 \times 0.5 = 0.5$ V
Total number of squares under the graph ≈ 15.
So change in potential $\approx 15 \times 0.5 \approx \textbf{7.5 V}$.

b) Use the graph to find the value of E at some distance r. For example, $E = 5.0$ NC^{-1} at $r = 3.0$ m.

Rearrange the equation $E = \dfrac{1}{4\pi\varepsilon_0}\dfrac{Q}{r^2}$ to get Q as:

$Q = 4\pi\varepsilon_0 r^2 E$, then substitute in the values for E and r:

$Q = 4 \times \pi \times 8.85 \times 10^{-12} \times 3.0^2 \times 5.0$

$= 5.0045... \times 10^{-9}$

$= \textbf{5.0} \times \textbf{10}^{-9}\textbf{ C (to 2 s.f.)}$

Q3 $\Delta W = Q\Delta V \Rightarrow \Delta V = \dfrac{\Delta W}{Q}$

$= \dfrac{5.14 \times 10^{-6}}{83.1 \times 10^{-9}}$

$= 61.8531...$

$= \textbf{61.9 V (to 3 s.f.)}$

Page 95 — Fact Recall Questions

Q1 Absolute electric potential is the electric potential energy that a unit positive charge (+1 C) would have at a particular point in an electric field.

Q2 a) 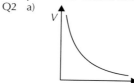   b)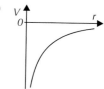

Q3 The electric potential difference between two points is the energy needed to move a unit charge between those points.

Q4 ΔW = work done in moving a charge.
Q = the charge being moved.
ΔV = electric potential difference through which the charge is moved.

Q5 a) Spherical surfaces.
b) Flat planes.

7. Comparing Electric and Gravitational Fields

Page 97 — Application Question

Q1 a) Gravitational:

$$F = \frac{Gm_1 m_2}{r^2} = \frac{6.67 \times 10^{-11} \times (9.11 \times 10^{-31})^2}{(6.00 \times 10^{-10})^2}$$
$$= 1.5376... \times 10^{-52} \text{ N} = \mathbf{1.54 \times 10^{-52} \text{ N} \text{ (to 3 s.f.)}}$$

Electrostatic:

$$F = \frac{1}{4\pi\varepsilon_0} \frac{Q_1 Q_2}{r^2} = \frac{1}{4 \times \pi \times 8.85 \times 10^{-12}} \times \frac{(1.60 \times 10^{-19})^2}{(6.00 \times 10^{-10})^2}$$
$$= 6.3941... \times 10^{-10} = \mathbf{6.39 \times 10^{-10} \text{ N} \text{ (to 3 s.f.)}}$$

b) $F = \dfrac{Gm_1 m_2}{r^2}$

As particles are assumed to be identical, $m_1 = m_2$, so:

$$F = \frac{G(m_1)^2}{r^2}$$

Rearrange for m_1:

$$m_1 = \sqrt{\frac{Fr^2}{G}} = \sqrt{\frac{6.3941... \times 10^{-10} \times (6.00 \times 10^{-10})^2}{6.67 \times 10^{-11}}}$$
$$= 1.8577... \times 10^{-9} \text{ kg} = \mathbf{1.86 \times 10^{-9} \text{ kg} \text{ (to 3 s.f.)}}$$

Page 97 — Fact Recall Questions

Q1 Any four from:
1. Gravitational field strength g is the force a unit mass would experience in a gravitational field. Electric field strength E is the force a unit positive charge would experience in an electric field.
2. Newton's law and Coulomb's law are the same but with G switched for $\frac{1}{4\pi\varepsilon_0}$ and m (or M) switched for Q.
3. Field lines for a radial gravitational field and a radial electric field have the same shape.
4. Gravitational potential V and absolute electric potential V give the energy a unit mass or charge would have at a point.
5. The equipotentials for a uniform spherical mass and a point charge both form spherical surfaces.
6. The formulas for the work done to move a unit mass through a gravitational potential or to move a unit charge through an electric potential are the same but with m switched for Q.

Q2 E.g. Gravitational forces are always attractive, but electric forces can be attractive or repulsive.

Exam-style Questions — Pages 99-101

1 **C (1 mark)**.
Electric forces can be attractive or repulsive.

2 **A (1 mark)**.
There will be a uniform electric field pointing from the 100 V plate to the 0 V plate. The alpha particle is positively charged, so there will be an electrical force on it in this direction.

3 **B (1 mark)**.
Gravitational field strength decreases as you move away from the mass by the inverse square law.

4 **B (1 mark)**.

5.1 A force field is a region in which a body experiences a non-contact force. **(1 mark)**

5.2

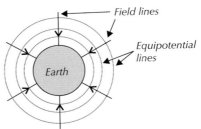

(1 mark for field lines pointing to the centre of Earth, 1 mark for equipotential lines as circles round Earth.)

5.3 No — g depends on the distance from the centre of the Earth, so the height difference in a classroom will have a negligible effect. **(2 marks — 1 mark for correct answer and 1 mark for correct reasoning.)**

5.4 $g = \dfrac{GM}{r^2} \Rightarrow r = \sqrt{\dfrac{(6.67 \times 10^{-11}) \times (5.98 \times 10^{24})}{6.31}}$
$= 7.950... \times 10^6 \text{ m}$
$= 7950.58... \text{ km}$

So altitude $= 7950.58... - 6370 = 1580.58...$
$= \mathbf{1580 \text{ km} \text{ (to 3 s.f.)}}$

(3 marks for correct answer, otherwise 1 mark for correct rearrangement of formula to make r the subject and 1 mark for attempting to find the altitude by subtracting the radius of the Earth.)

5.5 First find the distance r:
$V = -\dfrac{GM}{r} \Rightarrow r = -\dfrac{GM}{V}$
$= -\dfrac{(6.67 \times 10^{-11}) \times (5.98 \times 10^{24})}{-20.6 \times 10^6}$
$= 1.936... \times 10^7 \text{ m}$

Then find g:
$g = \dfrac{GM}{r^2} = \dfrac{(6.67 \times 10^{-11}) \times (5.98 \times 10^{24})}{(1.936... \times 10^7)^2}$
$= 1.0639... \text{ Nkg}^{-1} = \mathbf{1.06 \text{ Nkg}^{-1} \text{ (to 3 s.f.)}}$

(3 marks for correct answer, otherwise 1 mark for rearranging to make r the subject and 1 mark for correct r.)

5.6 $F = mg = (2.53 \times 10^3) \times 1.0639... = 2691.707...$
$= \mathbf{2690 \text{ N} \text{ (to 3 s.f.)}}$ *(1 mark)*

5.7 $v = \sqrt{\dfrac{2GM}{r}} = \sqrt{\dfrac{2 \times (6.67 \times 10^{-11}) \times (5.98 \times 10^{24})}{1.936... \times 10^7}}$
$= 6.4187... \times 10^3$
$= \mathbf{6.42 \times 10^3 \text{ ms}^{-1} \text{ (to 3 s.f.)}}$

(2 marks for correct answer, otherwise 1 mark for use of correct formula)
The value of r in this equation is the distance from the centre of the mass M (in this case Earth), so use the distance you found in 5.5.

5.8 A geostationary orbit **(1 mark)**.

6.1 A uniform electric field is produced, pointing from the plate with the more positive potential to the plate with the less positive potential. **(1 mark for uniform electric field, 1 mark for direction.)**

6.2 A radial electric field is produced, pointing away from the particle. **(1 mark for radial electric field, 1 mark for direction.)**

6.3 $\Delta V = \dfrac{1}{4\pi\varepsilon_0} \dfrac{Q}{r} = \dfrac{1}{4\pi\varepsilon_0} \dfrac{[79.0 \times (1.60 \times 10^{-19})]}{2.08 \times 10^{-12}}$
$= 5.46425... \times 10^4 = \mathbf{5.46 \times 10^4 \text{ V} \text{ (to 3 s.f.)}}$
(2 marks for correct answer, otherwise 1 mark for using correct value for Q (1.264×10^{-17}).)

6.4 $\Delta W = Q\Delta V = 2 \times (1.60 \times 10^{-19}) \times (5.46... \times 10^4)$
$$= 1.74... \times 10^{-14} \text{ J}$$

Work done in bringing the particle to a stop = loss in kinetic energy.

$$\Delta W = E_k = \frac{1}{2}mv^2 \Rightarrow v = \sqrt{\frac{2E_k}{m}} = \sqrt{\frac{2 \times 1.74... \times 10^{-14}}{6.64 \times 10^{-27}}}$$
$$= 2.29493... \times 10^6$$
$$= \mathbf{2.29 \times 10^6 \ ms^{-1}} \text{ (to 3 s.f.)}$$

(4 marks for correct answer, otherwise 1 mark for finding the work done, 1 mark for relating this to the kinetic energy lost and 1 mark for rearranging to find v.)

7.1 $g = \frac{F}{m} \Rightarrow F = mg = (203 \times 10^{-3}) \times 9.81$
$$= 1.99143 = \mathbf{1.99 \ N} \text{ (to 3 s.f.)} \text{ (1 mark)}$$

7.2 $F = \frac{1}{4\pi\varepsilon_0}\frac{Q_1 Q_2}{r^2} \Rightarrow r = \sqrt{\frac{1}{4\pi\varepsilon_0}\frac{Q_1 Q_2}{F}}$
$$= \sqrt{\frac{1}{4\pi\varepsilon_0}\frac{(-92.5 \times 10^{-6})(-34.7 \times 10^{-6})}{1.99...}}$$
$$= 3.8069... = \mathbf{3.81 \ m} \text{ (to 3 s.f.)}$$

(4 marks for correct answer, otherwise 1 mark for use of correct formula, 1 mark for attempt to make r the subject and 1 mark for correct rearrangement.)

If you didn't get the correct answer for 7.1 but your calculations were all correct here, you get full marks as long as you used the same number as in 7.1 for **F**.

7.3 $\Delta W = Q\Delta V \Rightarrow \Delta V = \frac{\Delta W}{Q} = \frac{18.5 \times 10^{-3}}{92.5 \times 10^{-6}} = \mathbf{200 \ V}$

(2 marks for correct answer, otherwise 1 mark for correct rearrangement to make ΔV the subject.)

7.4 $E = \frac{V}{d} = \frac{200}{10.4 \times 10^{-2}} = 1923.07...$
$$= \mathbf{1920 \ Vm^{-1}} \text{ (to 3 s.f.)} \text{ (1 mark)}$$

Section 4 — Capacitors

1. Capacitors
Page 104 — Application Questions
Q1 $C = \frac{Q}{V}$ rearranged gives $Q = CV = 0.10 \times 230 = \mathbf{23 \ C}$

Q2 E.g. it would have to be very large to provide enough power, which might hinder the portability of the media player / it could only power the device for a short time, so it would need charging very often / the voltage would decrease as the capacitor discharged, so it would be difficult to produce a constant output.

Page 104 — Fact Recall Questions
Q1 The capacitance of an object is the amount of charge it is able to store per unit potential difference (p.d.) across it.
Q2 E.g. use a variable resistor — constantly alter the resistance in the circuit to keep the current the same.

2. Energy Stored by Capacitors
Page 106 — Application Questions
Q1 $E = \frac{1}{2}CV^2 = \frac{1}{2} \times 40 \times 10^{-3} \times 230^2 = \mathbf{1060 \ J}$ (to 3 s.f.)
Q2 $Q = It = 0.025 \times 45 = 1.125 \times 10^{-3}$ C
$$E = \frac{1}{2}\frac{Q^2}{C} = \frac{1}{2} \times \frac{(1.125 \times 10^{-3})^2}{5 \times 10^{-6}} = \mathbf{0.13 \ J} \text{ (to 2 s.f.)}$$

Page 106 — Fact Recall Questions
Q1 Like charges are forced together, against their electrostatic repulsion, onto the plates. Some of the energy taken to force them together is stored as electrical potential energy.
Q2 Find the area under the Q-V graph.
Q3 $E = \frac{1}{2}QV$, $E = \frac{1}{2}CV^2$, $E = \frac{1}{2}\frac{Q^2}{C}$

3. Dielectrics
Page 108 — Application Questions
Q1 $\varepsilon_r = \frac{\varepsilon_1}{\varepsilon_0} = \frac{1.99 \times 10^{-11}}{8.85 \times 10^{-12}} = 2.248... = \mathbf{2.25}$ (to 3 s.f.)
Q2 $C = \frac{A\varepsilon_0\varepsilon_r}{d} = \frac{4.5 \times 10^{-3} \times 6.0 \times 10^{-3} \times 8.85 \times 10^{-12} \times 86}{0.5 \times 10^{-3}}$
$$= 4.109... \times 10^{-11} \text{ F} = \mathbf{41 \ pF} \text{ (to 2 s.f.)}$$
Q3 $C = \frac{A\varepsilon_0\varepsilon_r}{d}$ and A, ε_0 and d are constant.

So $C_{initial} = \text{const.} \times \varepsilon_{initial}$ and $C_{final} = \text{const.} \times \varepsilon_{final}$
So $\frac{C_{initial}}{\varepsilon_{initial}} = \frac{C_{final}}{\varepsilon_{final}}$
$$C_{final} = \frac{C_{initial} \times \varepsilon_{final}}{\varepsilon_{initial}} = \frac{16.6 \times 10^{-9} \times 2.55}{3.00}$$
$$= 1.411 \times 10^{-8} = \mathbf{14.1 \ nF} \text{ (to 3 s.f.)}$$

Page 108 — Fact Recall Questions
Q1 A dielectric material is an electric insulator.
Q2 Permittivity describes how difficult it is to generate an electric field of a given size in a medium.
Q3 The dielectric constant is also known as relative permittivity.
Q4 It represents the permittivity of a material compared to the permittivity of free space. (It is the ratio of the size of an electric field generated in a vacuum to the size of the field generated in the medium, under the same conditions.)
Q5 The polar molecules are randomly orientated.
Q6 When a charge is applied, the negative ends of the molecules are attracted to the positively charged plate, and vice versa. The molecules rotate and align themselves anti-parallel to the electric field generated by the plates.
Q7 $C = \frac{A\varepsilon_0\varepsilon_r}{d}$

4. Charging and Discharging
Page 113 — Application Questions
Q1 $V = V_0 e^{-\frac{t}{RC}} = 15.0 \times e^{-\frac{1.5 \times 10^{-3}}{(2.6 \times 10^3) \times (0.60 \times 10^{-6})}} = 5.7345...$
$$= \mathbf{5.7 \ V} \text{ (to 2 s.f.)}$$
Don't forget to convert all the values you've been given into SI units.
Q2 When the capacitor is fully charged,
the charge is $Q = CV = 15 \times 10^{-9} \times 230 = 3.45 \times 10^{-6}$ C
Then after 0.01 seconds,
$$Q = Q_0 e^{-\frac{t}{RC}} = 3.45 \times 10^{-6} \times e^{-\frac{0.01}{(50 \times 10^3) \times (15 \times 10^{-9})}}$$
$$= 5.587... \times 10^{-12} = \mathbf{6 \times 10^{-12} \ C} \text{ (to 1 s.f.)}$$
Q3 a) $\ln(V)$

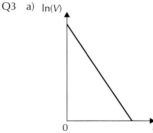

b) $\ln(V_0)$, where V_0 is the initial potential difference across the capacitor.

Page 113 — Fact Recall Questions

Q1 Electrons flow from the negative terminal of the power source to the capacitor plate connected to it. This is due to them being repelled by the negative terminal, and attracted by the positive capacitor plate (which is positively charged due to electrons flowing from it to the positive terminal of the power source). The electrical insulator between the plates lets no charge pass across, so the charge builds up on the plate.

Q2 a) 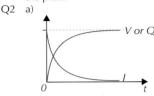   b)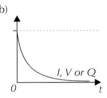

Q3 $Q = Q_0(1 - e^{-\frac{t}{RC}})$

Q4 E.g. remove the power source and reconnect the circuit.

Q5 $V = V_0 e^{-\frac{t}{RC}}$

Q6 E.g. Discharge a fully charged capacitor of initial charge Q_0 through a fixed resistor. Use a data logger connected to a voltage sensor and ammeter to measure the current through the resistor and the potential difference across the capacitor against time. Once the capacitor has fully discharged, connect the data logger to a computer to analyse the results.

5. Time Constant and Time to Halve
Page 117 — Application Questions

Q1 $\tau = RC = 4.2 \times 10^{-6}$ s
$T_{\frac{1}{2}} = 0.69RC = 0.69 \times 4.2 \times 10^{-6}$
$= 2.898 \times 10^{-6} = $ **2.9 μs (to 2 s.f.)**

Q2 a) The capacitor losing 63% of its charge is the same as the capacitor discharging to 37% of its original charge, which takes a time equal to the time constant, $\tau = RC$.
$\tau = RC = 400 \times 10^3 \times 15 \times 10^{-6} = $ **6 s**

b) $C = \frac{\tau}{R} = \frac{60}{400 \times 10^3} = $ **1.5 × 10⁻⁴ F**

c) The capacitor would have to be a lot bigger than the original one. The resistance of the resistor could be increased instead.

Q3 $\frac{Q}{Q_0} = 0.3$

$\frac{Q}{Q_0} = e^{-\frac{t}{RC}}$ so $0.3 = e^{-\frac{20.0}{RC}} \Rightarrow \ln 0.3 = \ln(e^{-\frac{20.0}{RC}})$

$\Rightarrow \ln 0.3 = -\frac{20.0}{RC} \Rightarrow RC = -\frac{20.0}{\ln 0.3} = $ **16.6 s (to 3 s.f.)**

Q4 a) $\ln(V_0)$, where V_0 is the initial potential difference across the capacitor.

b) Vertical axis intercept = $\ln V_0 = 2.708$
Rearranging to find V_0:
$V_0 = e^{2.708} = 14.99... = $ **15.0 V (to 3 s.f.)**

c) The gradient of the graph is:
$-\frac{2.708 - 1.001}{2.00 \times 10^{-6} - 0} = -853\,500$
The gradient $= -\frac{1}{RC}$ so $RC = -1 \div -853\,500$
$= 1.171... \times 10^{-6}$
$= $ **1.17 μs (to 3 s.f.)**

d) $T_{1/2} = 0.69RC = 0.69\tau = 0.69 \times 1.171... \times 10^{-6}$
$= 8.0843... \times 10^{-7}$ s
$= $ **0.81 μs (to 2 s.f.)**

Page 117 — Fact Recall Questions

Q1 The resistance of the circuit and the capacitance of the capacitor.

Q2 The time taken for the capacitor to discharge to $\frac{1}{e}$ (about 37%) of its original charge. It is given by $\tau = RC$.

Q3 The time to halve is the time taken for the charge, current or potential difference to fall to half of its initial value for a discharging capacitor. $T_{\frac{1}{2}} = 0.69RC$, or $T_{\frac{1}{2}} = \ln(2)RC$.

Q4 Use the graph to find the time at which the current has fallen to 50% of its original value.

Exam-style Questions — pages 119-121

1 **A (1 mark)**.
The charge increases quickly and then levels off as it charges, and decreases quickly and then levels off as it discharges.

2 **C (1 mark)**.
$Q = Q_0 e^{-\frac{t}{RC}}$ so $\frac{Q}{Q_0} = e^{-\frac{t}{RC}}$ where $\frac{Q}{Q_0}$ is 0.65.
So $\ln(0.65) = -\frac{1}{100 \times 10^3 \times C}$
so $C = -\frac{1}{100 \times 10^3 \times \ln(0.65)} = $ 2.3 × 10⁻⁵ F *(to 2 s.f.)*

3 **C (1 mark)**.
$Q = CV = 5 \times 10^{-3} \times 24 = 0.12\,C$
and $Q = It$, so $I = \frac{Q}{t} = \frac{0.12}{30} = 4 \times 10^{-3}$ A = 4 mA

4 **B (1 mark)**.
$E = \frac{1}{2}CV^2 = \frac{1}{2} \times 20 \times 10^{-6} \times 24^2 = 5.76 \times 10^{-3}$ J = 5.76 mJ
(Remember, the charge stored by a capacitor is half the energy supplied by the power source.)

5.1 $C = \frac{A\varepsilon_0 \varepsilon_r}{d}$
$= \frac{25 \times 10^{-2} \times 25 \times 10^{-2} \times 8.85 \times 10^{-12} \times 3.3}{1.0 \times 10^{-3}}$ *(1 mark)*
$= 1.825... \times 10^{-9} = $ **1.8 nF (to 2 s.f.)** *(1 mark)*

5.2 The molecules in the dielectric behave as polar molecules — one end is negatively charged and one end is positively charged *(1 mark)*. When charge is stored on the capacitor, an electric field is generated. The negative ends of the polar molecules are attracted to the positive plate, and vice versa. The molecules rotate and align themselves anti-parallel with the field *(1 mark)*. Each molecule has its own electric field, which now opposes the electric field generated between the plates *(1 mark)*. This causes a reduction in the overall electric field between the plates, meaning that less potential difference is needed to charge the capacitor to a given charge Q. This leads to an increase in capacitance as $C = Q \div V$ *(1 mark)*.

5.3 E.g. she could increase the surface area of the capacitor plates / she could reduce the thickness of the paper separating the plates. *(1 mark for either correct answer)*

6.1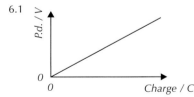

(1 mark for straight line through the origin.)

6.2 1 / capacitance *(1 mark)*.

6.3 Energy stored by the capacitor *(1 mark)*.
Notice that the area enclosed by the line and each axis is exactly the same.

6.4 $V = V_0 e^{-\frac{t}{RC}} = 12 \times e^{-\frac{0.030}{29 \times 10^3 \times 1 \times 10^{-6}}}$ *(1 mark)*
$= 4.26... = $ **4.3 V (to 2 s.f)** *(1 mark)*

6.5

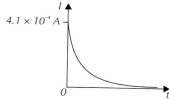

4.1×10^{-4} A

(1 mark for correct shape.)

6.6 $T_{1/2} = 0.69RC = 0.69 \times 29 \times 10^3 \times 1.0 \times 10^{-6}$
    $= 0.02001 = \textbf{0.020 s (to 2 s.f.)}$ ***(1 mark)***

7.1 The energy stored by the capacitor is half of the energy
    taken from the battery, so the capacitor stores 3.75 J of
    energy ***(1 mark)***.
    Alternatively, you could use the formula for energy stored,
    $E = \frac{1}{2}CV^2$.

7.2 $\tau = RC = 2.0 \times 10^3 \times 3.0 \times 10^{-3} = \textbf{6 s}$ ***(1 mark)***

7.3 After 14 s, the charge on the capacitor is:
    $Q = Q_0(1 - e^{-\frac{t}{RC}}) = CV(1 - e^{-\frac{t}{RC}})$ ***(1 mark)***
    $Q = 3.0 \times 10^{-3} \times 50.0(1 - e^{-\frac{14}{6}})$
    $= 0.135... = \textbf{0.14 C (to 2 s.f.)}$ ***(1 mark)***

7.4

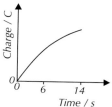

*(1 mark for shape — charge should increase to around
63% at 6 s and 90% at 14 s)*

Section 5 — Magnetic Fields

1. Magnetic Flux Density

Page 125 — Application Question
Q1 a) The force will act upwards.
    b) $F = BIl = (9.21 \times 10^{-3}) \times 1.44 \times (2.51 \times 10^{-2})$
        $= 3.328... \times 10^{-4}$
        $= \textbf{3.33} \times \textbf{10}^{-4}$ **N (to 3 s.f.)**
    c) Change the dc supply for an ac supply.
        *When the direction of the current changes, the direction of
        the force will also change. So if the current changes direction
        rapidly (as in an ac supply) the wire will be pushed rapidly in
        opposite directions and vibrate.*

Page 125 – Fact Recall Questions
Q1 a) You can use your left hand.
    b) The first (index) finger represents the direction of the
        magnetic field, the second (middle) finger represents the
        direction of the current, and the thumb represents the
        direction of the force (or motion).
Q2 The current-carrying wire must be perpendicular to the
    magnetic field.
Q3 B is the magnetic flux density, measured in teslas (T).

2. Investigating Force on a Current-Carrying Wire
Page 128 — Application Question
Q1 a)

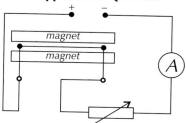

    b) A variable resistor should be used so that the current
        can be kept constant in order to determine a value for B
        while changing I.
    c) Only that length of wire will experience a vertical force,
        which will contribute to the reading on the top pan
        balance.
    d) Convert the mass readings into force using $F = mg$, then
        plot your data on a graph of force F against length I.
        Because $F = BIl$, the gradient of the line of best fit is
        equal to BI. You know the current, I, so just divide
        the gradient by I (in amps) to get the magnetic flux
        density, B.

3. Forces on Charged Particles
Page 132 — Application Questions
Q1 Only charged particles experience a force from magnetic
    fields (neutrons have no charge).
Q2 The force will act upwards (using the left-hand rule).
    *Because the particle is negatively charged, you need to point
    your second finger in the opposite direction to its actual
    motion.*
Q3 $F = BQv$
    $= (640 \times 10^{-3}) \times (3.2 \times 10^{-19}) \times (5.5 \times 10^3)$
    $= 1.126... \times 10^{-15} = \textbf{1.1} \times \textbf{10}^{-15}$ **N (to 2 s.f.)**
Q4 a) $F = \frac{mv^2}{r}$, $F = BQv$
        $\Rightarrow \frac{mv^2}{r} = BQv$
        $\Rightarrow B = \frac{mv^2}{Qvr}$
        Cancelling the 'v's:
        $B = \frac{mv}{Qr}$
    b) $B = \frac{mv}{Qr} = \frac{(1.67 \times 10^{-27}) \times (1.99 \times 10^7)}{(1.60 \times 10^{-19}) \times 5.49}$
        $= 0.03783... = \textbf{0.0378 T (to 3 s.f.)}$

Page 132 — Fact Recall Questions
Q1 It will follow a circular path.
Q2 Charged particles are produced and fired into a semicircular
    electrode, where a uniform magnetic field is applied
    perpendicular to the particle's motion. The particle
    follows a circular path and leaves the electrode, where it's
    accelerated by a potential difference before entering another
    semicircular electrode. Here it follows a slightly larger
    circular path due to its increased speed, before leaving the
    electrode. This process repeats, with the direction of the
    potential difference switching each time, until the particle
    finally exits the cyclotron at a very high speed.

4. Electromagnetic Induction
Page 136 — Application Question
Q1 a) Flux linkage is at its greatest when the area is normal to the magnetic field. If the angle of the field with the normal to the area is increased, the flux linkage decreases.

This is because the number of field lines passing through the area decreases.

b) $N\Phi = BAN \Rightarrow B = \dfrac{N\Phi}{AN}$

$$= \dfrac{1.3 \times 10^{-6}}{(25 \times 10^{-4}) \times 10}$$

$$= 5.2 \times 10^{-5}\ \text{T} = \textbf{52}\ \mu\textbf{T}$$

Don't forget to convert the area to m².

c) E.g. he could increase the area of the coil / increase the number of turns in the coil.

Increasing the area or the number of turns of the coil will increase the flux linkage of the coil, which will reduce the percentage error in the measurements.

Page 136 — Fact Recall Questions
Q1 An e.m.f. (and a current if the conductor is part of a circuit).
Q2 $\Phi = BA$
Q3 Flux linkage is measured in webers (Wb) (or weber turns).
Q4 $(N\Phi) = BAN\cos\theta$

$N\Phi$ is the flux linkage in Wb (or weber-turns).
B is the magnetic flux density in T.
A is the area of the coil inside the magnetic field in m².
N is the number of turns on a coil.
θ is the angle between the normal to the plane of the coil and the magnetic field in °.

5. Investigating Flux Linkage
Page 138 — Application Question
Q1 The amplitude of the induced e.m.f. will drop from a maximum value to zero as the search coil rotates from the 90° position to 0°. This is because when the search coil is at the 90° position, its area is perpendicular to the magnetic field of the large circular coil, so a maximum e.m.f. will be induced by the alternating current. When it's at 0°, its area is parallel to the magnetic field so no e.m.f. will be induced.

6. Faraday's Law and Lenz's Law
Page 142 — Application Questions
Q1 The plane's wings act as a conductor, and the Earth has a magnetic field that the plane cuts through, so electromagnetic induction occurs.

Q2 a) $\varepsilon_{max} = BAN\omega$

$$= 1.5 \times 0.24 \times 50 \times 4.0\pi$$

$$= \textbf{72}\pi\ \textbf{V}\ (= 230\ \text{V to 2 s.f.})$$

Remember the maximum induced e.m.f. is when $\sin\omega t = 1$ (or -1).

b)

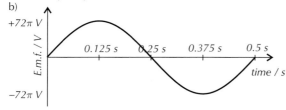

The coil does 1 full rotation in 0.5 s — 4π rad s⁻¹ is equal to 720° per second.

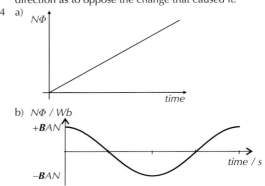

Doubling the speed will double both the frequency and the amplitude of the induced e.m.f.

Page 142 — Fact Recall Questions
Q1 Faraday's law states that the induced e.m.f. is directly proportional to the rate of change of flux linkage (magnitude of induced e.m.f., $\varepsilon = N\dfrac{\Delta\Phi}{\Delta t}$).

Q2 Rate of change of flux linkage $= N\dfrac{\Delta\Phi}{\Delta t}$.

Q3 Lenz's law states that the induced e.m.f. is always in such a direction as to oppose the change that caused it.

Q4 a)

b)

7. Alternating Current
Page 146 — Application Questions
Q1 a) $I_{rms} = \dfrac{I_0}{\sqrt{2}} = \dfrac{8.2}{\sqrt{2}} = 5.798... = \textbf{5.8 A (to 2 s.f.)}$

b) $V_{rms} = \dfrac{V_0}{\sqrt{2}} = \dfrac{9.0}{\sqrt{2}} = 6.363... = \textbf{6.4 V (to 2 s.f.)}$

Q2 a) $V_{rms} = \dfrac{V_0}{\sqrt{2}}$, so $V_0 = \sqrt{2}V_{rms}$

$V_0 = \sqrt{2} \times 120 = 169.70... = \textbf{170 V (to 2 s.f.)}$

b) $V_{peak\text{-}to\text{-}peak} = 2 \times V_0 = 2 \times 169.70... = 339.411...$
$= \textbf{340 (to 2 s.f.)}$

c) average power $= I_{rms} \times V_{rms} = 20.0 \times 120 = \textbf{2400 W}$

Q3 a) 20.0 ms cm⁻¹ = 0.0200 s cm⁻¹
Wave spans 8 cm, so:
Time period = 0.0200 × 8 = 0.16 s
Frequency = 1 ÷ T = 1 ÷ 0.16 = 6.25 = **6 Hz (to 1 s.f.)**

b) From the graph, peak voltage = 3 × 0.50 V = 1.5 V
$V_{rms} = \dfrac{V_0}{\sqrt{2}} = \dfrac{1.5}{\sqrt{2}} = 1.060... = \textbf{1 V (to 1 s.f.)}$

Page 146 — Fact Recall Questions
Q1 An electron beam moving across the screen.
Q2 There would be a straight vertical line on the voltage axis.
Q3 The voltage of an ac supply will be below the peak voltage most of the time. That means it won't have as high a power output as a dc supply with the same peak voltage. To compare them properly, you need to average the ac voltage somehow. A normal average won't work, because the positive and negative bits cancel out.
Q4 $I_{rms} = \dfrac{I_0}{\sqrt{2}}$

8. Transformers

Page 151 — Application Questions

Q1 a) $\frac{N_s}{N_p} = \frac{V_s}{V_p}$, so $V_s = V_p \frac{N_s}{N_p} = 190 \times \frac{420}{250} = 319.2$
$$= \mathbf{320\ V\ (to\ 2\ s.f.)}$$

b) $I_p V_p = I_s V_s$, so $I_p = \frac{I_s V_s}{V_p} = \frac{13 \times 75}{120} = 8.125$
$$= \mathbf{8.1\ A\ (to\ 2\ s.f.)}$$

Q2 efficiency $= \frac{I_s V_s}{I_p V_p} = \frac{P_s}{P_p} = \frac{62}{65} = 0.9538...$
$$= \mathbf{0.95\ (or\ 95\%)\ (to\ 2\ s.f.)}$$
Remember $P = IV$.

Q3 a) Any two from: e.g. resistance in the coils causes them to heat up; energy is lost when the core magnetises and demagnetises; induced eddy currents in the core cause it to heat up.

b) (i) A step-down transformer.

(ii) Efficiency $= \frac{I_s V_s}{I_p V_p} \Rightarrow I_p = \frac{[I_s V_s]}{\text{Efficiency} \times V_p}$
$$= \frac{120}{0.85 \times 230} = 0.6138...$$
$$= \mathbf{0.61\ A\ (to\ 2\ s.f.)}$$

Q4 $P = I^2 R$, so $R = \frac{P}{I^2} = \frac{12.3 \times 10^6}{1250^2} = 7.872\ \Omega$
Length = total resistance ÷ resistance per km
$$= \frac{7.872}{0.0985} = 79.91... = \mathbf{79.9\ km\ (to\ 3\ s.f.)}$$

Page 151 — Fact Recall Questions

Q1 When an alternating current flows through the primary coil, the core magnetises and demagnetises quickly. This causes a rapid change in flux through the secondary coil, which induces an e.m.f.. The more turns there are in the secondary coil compared to the primary coil, the bigger the induced e.m.f. is compared to the input e.m.f.
Remember, e.m.f. is just another way of saying voltage.

Q2 Eddy currents are looping currents induced by the changing magnetic flux in the core. They create a magnetic field that acts against the field that induced them, reducing the field strength, and dissipate energy by generating heat.

Q3 E.g. using thick copper wires to reduce resistance; using a magnetically soft core that magnetises and demagnetises quickly; laminating the core to prevent eddy currents.

Q4 Power = current × voltage, so increasing the voltage reduces the current for a given amount of power. Energy loss due to heating is proportional to I^2R, so a lower current means less energy loss through heating of the wires.

Exam-style Questions — Pages 153-156

1 **C** *(1 mark)*
This is Faraday's law.

2 **B** *(1 mark)*
Remember e.m.f. = B/v, so it's proportional to the velocity.

3 **B** *(1 mark)*
Increasing the frequency means the rate of change of flux linkage is larger (same flux change, shorter time), so it increases the maximum induced e.m.f.

4 **A** *(1 mark)*
The resistance needs to be kept as low as possible to reduce heating in the wires. Thinner wires have a higher resistance.

5.1 $F = BIl \Rightarrow l = \frac{F}{BI} = \frac{4.5 \times 10^{-2}}{1.7 \times 190 \times 10^{-3}}$
$$= 0.139... = \mathbf{14\ cm\ (to\ 2\ s.f.)}$$
(2 marks for correct answer, 1 mark for correct working if answer incorrect.)

5.2 E.g. by running an alternating current through the wire. By rotating the magnetic field (or using an alternating magnetic field). *(1 mark for each method.)*

5.3 The current flows anticlockwise *(1 mark)*.

5.4 $N\Phi = BAN \cos\theta$
$$= 1.7 \times (10.0 \times 10^{-4}) \times 1 \times \cos 41 = 0.00128...$$
$$= \mathbf{1.3 \times 10^{-3}\ Wb\ (to\ 2\ s.f.)\ (or\ weber\ turns)}$$
(2 marks for correct answer, 1 mark for correct calculation and conversion of units from cm^2 to m^2 if answer incorrect.)
Remember 1 cm^2 = 1 × 10^{-4} m^2.

5.5 $\varepsilon_{max} = BAN\omega$
$$= 1.7 \times (10.0 \times 10^{-4}) \times 1 \times 10.0\pi$$
$$= \mathbf{0.017\pi\ V\ (= 0.053\ V\ to\ 2\ s.f.)\ (1\ mark)}$$
The e.m.f. is at a maximum when $\sin \omega t = 1$ (or −1).

5.6
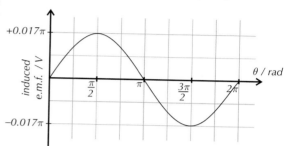

(2 marks available — 1 mark for correct sinusoidal shape and 1 mark for maximum and minimum induced e.m.f. marked correctly.)

6.1 The particle will follow a circular path *(1 mark)*.

6.2 $F = BQv$
$$= 0.93 \times (1.60 \times 10^{-19}) \times (8.1 \times 10^7) = 1.20528 \times 10^{-11}$$
$$= \mathbf{1.2 \times 10^{-11}\ N\ (to\ 2\ s.f.)}$$
The force would act downwards.
(1 mark for correct force and 1 mark for direction.)
Find the direction of the force with Fleming's left-hand rule — because electrons have a negative charge, point your second finger in the opposite direction to its velocity. The magnitude of the charge on an electron is given in the exam data and formulae booklet.

6.3 $F_{+2e} = -2 \times F_{-e} = -2 \times 1.20528 \times 10^{-11} = 2.41056 \times 10^{-11}$
The force will be $\mathbf{2.4 \times 10^{-11}\ N\ (to\ 2\ s.f.)}$ upwards.
(1 mark for correct force and 1 mark for direction.)
If you didn't get the correct answer for 6.2, you'll get the marks for 6.3 as long as the force is double the force for the electron and in the opposite direction.

7.1 The e.m.f. will be greatest when $\theta = 90°$ or $270°$ *(1 mark)*.
This is when the coil is parallel to the magnetic field.

7.2
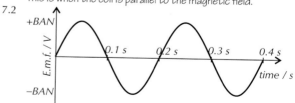

(1 mark for sinusoidal shape, 1 mark for correctly labelled axes and 1 mark for two complete wave cycles.)
Time period = 1/frequency, so one wave cycle takes 1/5 = 0.2 seconds to complete. You're not told the starting position of the coil, so it doesn't matter what part of the wave cycle the graph starts at.

8.1 An alternating current is required *(1 mark)*.

8.2 The lower the resistance, the less the wires heat up, so the more efficient the transformer. **(1 mark)**.

8.3 efficiency = $\dfrac{I_s V_s}{I_p V_p}$

$\Rightarrow$ power out = efficiency × power in
$$= 0.91 \times (1.2 \times 10^3) = 1092$$
$$= \textbf{1.1 kW (to 2 s.f.) (1 mark)}$$

8.4 $E = Pt = (1.2 \times 10^3 - 1092) \times (3.0 \times 60 \times 60)$
$$= 1\ 166\ 400\ \text{J} = \textbf{1.2 MJ (to 2 s.f.) (1 mark)}$$

Allow 1 mark here if the working is correct but the answer carried through from 8.3 is incorrect.

9.1 Downwards **(1 mark)**.

9.2 E.g. Keeping the wire length constant, the magnetic flux density can be found by varying the current using the variable dc power supply and noting the mass shown on the top pan balance for a range of currents **(1 mark)**. After converting the masses to weights using $W = mg$ **(1 mark)**, a graph of force against current can be plotted **(1 mark)**. Drawing a line of best fit and finding the gradient gives you BI, so dividing the gradient by I gives the magnetic flux density B **(1 mark)**.

9.3 $F = BIl$, so $l = \dfrac{F}{BI} = \dfrac{0.65 \times 10^{-3} \times 9.81}{25 \times 10^{-3} \times 1.6} = 0.1594...$
$$= \textbf{0.16 m (to 2 s.f.) (2 marks)}$$

Don't forget to multiply the mass reading in kg by $g = 9.81\ \text{N kg}^{-1}$ to get the force.

9.4 The top pan balance will show a negative value for mass **(1 mark)** because the force on the cradle will act upwards **(1 mark)**.

Section 6 — Nuclear Physics

1. Rutherford Scattering
Page 159 — Fact Recall Questions
Q1 A stream of alpha particles from a radioactive source were fired at very thin gold foil. A circular fluorescent screen was placed around the foil to detect the angle at which the alpha particles were deflected by the gold foil.

Q2 The nucleus must be positive to repel the positively charged alpha particles. The nucleus must also be tiny and most of the atom must be empty space as very few alpha particles are deflected by an angle larger than 90° and most of the other alpha particles just pass straight through. Most of the atom's mass must be in the nucleus as the fast alpha particles (with a high momentum) are deflected by the nucleus — so the alpha particles must be striking something more massive than themselves.

2. Measuring Nuclear Radius
Page 162 — Application Questions
Q1 $r = \dfrac{Q_{zinc} q_{alpha}}{4\pi\varepsilon_0 E_{elec}} = \dfrac{30 \times 2 \times (1.60 \times 10^{-19})^2}{4\pi \times (8.85 \times 10^{-12}) \times (1.0 \times 10^{-13})}$
$$= 1.381... \times 10^{-13}\,\text{m}$$
Therefore nuclear radius $\approx$ **1.4 × 10⁻¹³ m (to 2 s.f.)**

Q2 Initial $E_k = (4.0 \times 10^6) \times (1.60 \times 10^{-19}) = 6.4 \times 10^{-13}$ J
$$E_{elec} = \dfrac{Q_{lead} q_{alpha}}{4\pi\varepsilon_0 r} = 6.4 \times 10^{-13}\,\text{J}$$
$$r = \dfrac{(+82e)(+2e)}{4\pi\varepsilon_0 (6.4 \times 10^{-13})} = \dfrac{82 \times 2 \times (1.60 \times 10^{-19})^2}{4\pi \times (8.85 \times 10^{-12}) \times (6.4 \times 10^{-13})}$$
$$= 5.898... \times 10^{-14} = \textbf{5.9 × 10⁻¹⁴ m (to 2 s.f.)}$$

Q3 $E = 50$ MeV $= (5.0 \times 10^7) \times (1.60 \times 10^{-19}) = 8.0 \times 10^{-12}$ J
$$\lambda \simeq \dfrac{hc}{E} = \dfrac{(6.63 \times 10^{-34}) \times (3.00 \times 10^8)}{(8.0 \times 10^{-12})}$$
$$= 2.486... \times 10^{-14} = \textbf{2.5 × 10⁻¹⁴ m (to 2 s.f.)}$$

Q4 $E = 200$ MeV $= (2.0 \times 10^8) \times (1.60 \times 10^{-19}) = 3.2 \times 10^{-11}$ J
$$\lambda \simeq \dfrac{hc}{E} = \dfrac{(6.63 \times 10^{-34}) \times (3.00 \times 10^8)}{(3.2 \times 10^{-11})}$$
$$= 6.215... \times 10^{-15}\,\text{m}$$
$$R \simeq \dfrac{1.22\lambda}{2\sin\theta} = \dfrac{1.22 \times (6.215... \times 10^{-15})}{2\sin 33°} = 6.96... \times 10^{-15}\,\text{m}$$
Therefore nuclear radius $\simeq$ **7.0 × 10⁻¹⁵ m (to 2 s.f.)**.

Page 163 — Fact Recall Questions
Q1 Initial $E_k = E_{elec} = \dfrac{Q_{nucleus} q_{alpha}}{4\pi\varepsilon_0 r}$ or $r = \dfrac{Q_{nucleus} q_{alpha}}{4\pi\varepsilon_0 E_{elec}}$

Q2 $\lambda \simeq \dfrac{hc}{E}$

Q3 $\sin\theta \simeq \dfrac{1.22\lambda}{2R}$

Q4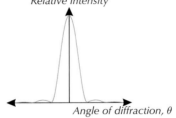

3. Nuclear Radius and Density
Page 166 — Application Questions
Q1 Mass $= m = (21 + 24) \times (1.7 \times 10^{-27})$
$$= \textbf{7.7 × 10⁻²⁶ kg (to 2 s.f.)}$$

Q2 $R = R_0 A^{1/3} = (1.4 \times 10^{-15}) \times (31)^{1/3} = 4.397... \times 10^{-15}$
$$= \textbf{4.4 × 10⁻¹⁵ m (to 2 s.f.)}$$

Q3 Rearrange $R = R_0 A^{1/3}$ for A:
$$A = \left(\dfrac{R}{R_0}\right)^3 = \left(\dfrac{4.2 \times 10^{-15}}{1.4 \times 10^{-15}}\right)^3 = \textbf{27}$$
Don't forget to convert the radius from fm to m.

Q4 $R = R_0 A^{1/3} = (1.4 \times 10^{-15}) \times (23)^{1/3} = 3.981... \times 10^{-15}$ m
$$V = \dfrac{4}{3}\pi R^3 = \dfrac{4}{3}\pi \times (3.981... \times 10^{-15})^3$$
$$= 2.643... \times 10^{-43}$$
$$= \textbf{2.6 × 10⁻⁴³ m³ (to 2 s.f.)}$$

Q5 $R = R_0 A^{1/3} = (1.4 \times 10^{-15}) \times (207)^{1/3} = 8.281... \times 10^{-15}$ m
$$V = \dfrac{4}{3}\pi R^3 = \dfrac{4}{3}\pi \times (8.281... \times 10^{-15})^3 = 2.379... \times 10^{-42}\,\text{m}^3$$
$$m = (1.7 \times 10^{-27}) \times 207 = 3.519 \times 10^{-25}\,\text{kg}$$
$$\rho = m/v = (3.519 \times 10^{-25}) \div (2.379... \times 10^{-42})$$
$$= 1.479... \times 10^{17}$$
$$= \textbf{1.5 × 10¹⁷ kg m⁻³ (to 2 s.f.)}$$

Page 166 — Fact Recall Questions
Q1 a) 10^{-15} m (or 1 fm)
   b) 5×10^{-11} m (or 0.05 nm)

Q2 The particles that make up the nucleus — protons and neutrons.

Q3 a) Nuclear radius is directly proportional to the cube root of the nucleon number. $R = R_0 A^{1/3}$.
   b) To show this relationship from experimental data — plot a graph of the cube root of nucleon number against nuclear radius to show that this gives a straight line.
   c) $R \propto A^{1/3}$ means that $R^3 \propto A$. This means each nucleon takes up roughly the same volume in the nucleus, and since each nucleon has roughly the same mass, nuclear density $\left(= \dfrac{\text{mass}}{\text{volume}}\right)$ is constant, regardless of the number of nucleons.

4. Properties of Nuclear Radiation

Page 170 — Application Question

Q1 Since the radiation is blocked by aluminium, but not by paper, it must be beta-minus radiation.

Q2 E.g. Gamma radiation can be used to treat cancer. The radiation damages cancerous cells in tumours which can sometimes cure patients of cancer. Radiation damages all cells though, whether they're cancerous or not. The damage to healthy cells can be minimised (e.g. by using shielding and rotating the angle of the beam of radiation), but patients can suffer side effects such as tiredness, sore/reddened skin and infertility.

Page 170 — Fact Recall Questions

Q1 When an unstable atomic nucleus breaks down to become more stable, by releasing energy and/or particles.

Q2 Alpha (helium nuclei — two protons and two neutrons), beta-plus (positrons), beta-minus (electrons), gamma (high-frequency electromagnetic wave/photons).

Q3 Alpha — strong. Beta — weak. Gamma — very weak. Alpha radiation can't penetrate the skin but is dangerous if it is ingested and is very strongly ionising — causing a lot of damage to body tissue. Beta radiation causes less damage to body tissue compared to alpha radiation, due to lower ionising power. Gamma radiation causes even less damage to body tissue compared to alpha and beta radiation as gamma radiation is very weakly ionising.

Q4 Look at how the radiation is deflected when it passes through a magnetic field perpendicular to its motion. If the radiation is positively charged it will curve one way, if it's negative it'll curve the other way, and if it's not charged it will continue in a straight line. The radius of curvature of its path can also tell you about its charge and mass. Alternatively, use a Geiger-Müller tube and counter to measure the radiation passing through both paper and a few mm of aluminium. If the radiation is blocked by the paper, it's alpha radiation. If it passes through the paper but is blocked by the aluminium, it's beta radiation. If it passes through the paper and the aluminium it's gamma radiation.

Q5 E.g. Alpha radiation is used in smoke detectors, because it is strongly ionising so it will allow a current to flow through the detector, but it is weakly penetrating, so it's blocked by smoke which sets off the alarm. Beta radiation is used to measure the thickness of sheets of material (e.g. paper), because small variations in the thickness of the paper will affect the amount of radiation that can get through (whereas alpha radiation would never be able to pass through and gamma radiation would always pass through). Gamma radiation is used in medical imaging, because it is more weakly ionising than beta radiation and causes much less damage to body tissue, so is safer to use in the body.

5. Background Radiation and Intensity

Page 173 — Application Questions

Q1 20.0 cm = 0.200 m and count rate varies with inverse square law. So count rate at 20.0 cm = $\frac{k}{0.200^2}$ = 54 s^{-1}

Therefore $k = 54 \times 0.200^2 = 2.16$

So count rate at 45 cm (0.45 m) = $\frac{2.16}{0.45^2}$ = 10.666...

= **11 s^{-1} (to 2 s.f.)**

Q2 If the intensity obeys the inverse square law, then $I = \frac{k}{x^2}$ will be true, with the same constant k for both pairs of measurements. Plug the 1st intensity and distance into the inverse square law:

$I = \frac{k}{x^2} \Rightarrow 4.0 \times 10^{-10} = \frac{k}{0.50^2}$

Therefore $k = 4.0 \times 10^{-10} \times 0.50^2 = 1.0 \times 10^{-10}$ (to 2 s.f.)

Do the same with the other values for intensity and distance:

$I = \frac{k}{x^2} \Rightarrow 1.8 \times 10^{-10} = \frac{k}{0.75^2}$

Therefore $k = 1.8 \times 10^{-10} \times 0.75^2 = 1.0 \times 10^{-10}$ (to 2 s.f.)

Both pairs of values give the same k (to 2 s.f.), so the intensity of gamma radiation obeys the inverse square law.

Page 174 — Fact Recall Questions

Q1 E.g. Take three measurements of the background radiation count rate (without the source present). Average these measurements and subtract this average from all measurements of the source's count rate.

Q2 Any three of:
- The air — rocks release radioactive radon gas, which emits alpha radiation.
- The ground and buildings — rocks contain radioactive materials.
- Cosmic radiation — high-energy particles from space collide with particles in the upper atmosphere and produce nuclear radiation.
- Living things — all living organisms contain radioactive carbon-14.
- Man-made medical or industrial sources.

Q3 The intensity is inversely proportional to the square of the distance from the source.

Q4 Take three measurements of the count rate due to background radiation and average them. Take measurements of the count rate with a Geiger-Müller tube at different distances from a gamma radiation source. Take three readings at each distance and calculate an average value for the count rate at each distance. Subtract the average background radiation measurement from each of these averaged values to eliminate background radiation from your data. Plot a graph of corrected count rate against distance — you should get a curve that shows the corrected count rate is proportional to the inverse of the distance squared. The intensity of the gamma radiation is directly proportional to the count rate (as long as the area of the collecting tube remains constant). So the intensity of the gamma radiation is proportional to the inverse of the distance squared.

Alternatively, you could plot a graph of count rate against the inverse of the distance squared. If the data fits a straight line through the origin, then the inverse square law is verified.

Q5 The intensity of gamma radiation is proportional to the inverse square of the distance from the radiation source. When held at arm's length, the intensity of the radiation that reaches your body is much lower than when held closer.

6. Exponential Law of Decay

Page 178 — Application Questions

Q1 Activity = $A = \lambda N = (1.1 \times 10^{-13}) \times (4.5 \times 10^{18})$
= 4.95×10^5
= **5.0 × 10^5 Bq (to 2 s.f.)**

Q2 Time in seconds = $t = 6.5 \times 3600 = 23\,400$ s
Activity at time $t = A = A_0 e^{-\lambda t}$
= $(3.2 \times 10^3) \times e^{-(1.3 \times 10^{-4}) \times 23400}$ = 152.76...
= **150 Bq (to 2 s.f.)**

Q3 Time in seconds = $t = 35 \times 60 = 2100$ s

Number of atoms in sample at time $t = N = N_0 e^{-\lambda t}$
$= (2.5 \times 10^{15}) \times e^{-(9.87 \times 10^{-3}) \times 2100} = 2.49... \times 10^6$
$= \mathbf{2.5 \times 10^6}$ **(to 2 s.f.)**

Q4 a) Number of moles, $n = \dfrac{\text{mass of sample}}{\text{molar mass}} = \dfrac{32.0}{60.0}$
$= 0.5333...$ mol

$N_0 = nN_A = 0.5333... \times 6.02 \times 10^{23}$
$= 3.210... \times 10^{23} = \mathbf{3.21 \times 10^{23}}$ **(to 3 s.f.)**

b) Number of seconds in 3.50 years:
$t = 3.50 \times 365 \times 24 \times 3600 = 1.10376 \times 10^8$ s
Number of remaining unstable atoms after 3.50 years:
$N = N_0 e^{-\lambda t} = (3.210... \times 10^{23}) \times e^{-(4.17 \times 10^{-9}) \times (1.10376 \times 10^8)}$
$= 2.0262... \times 10^{23}$ atoms
Number of moles remaining of ^{60}Co after 3.50 years:
Number of moles = $\dfrac{\text{number of atoms}}{\text{Avogadro's constant}}$
$= \dfrac{2.0262... \times 10^{23}}{6.02 \times 10^{23}} = 0.3365...$ mol
Convert this into grams:
mass of sample = number of moles × molar mass
$= 0.3365... \times 60.0$
$= 20.195... = \mathbf{20.2\ g}$ **(to 3 s.f.)**

Page 179 — Fact Recall Questions

Q1 a) The probability of an atomic nucleus decaying per unit time.

b) The number of atomic nuclei that decay per second.

c) $A = \lambda N$

Q2 a) $\dfrac{\Delta N}{\Delta t} = -\lambda N$

b) E.g. spreadsheet modelling

Q3 $N = N_0 e^{-\lambda t}$

Q4 a) Rolling a 6 on a dice is a random event with a constant probability, similar to the event of an unstable nucleus decaying.

b) E.g. The sample was not big enough.

Q5 The molar mass tells you the mass of one mole of the substance. By dividing the mass present in a sample by the molar mass, you find the number of moles of the substance that there are. You can then multiply this value by the Avogadro constant to find the number of atoms of that substance present in a sample.

Q6 $A = A_0 e^{-\lambda t}$

7. Half-life and its Applications

Page 183 — Application Questions

Q1 Find the number of times the activity halved in 24 hours — count the number of times, n, you have to divide 2400 by 2 to reach 75.
$n = 5$, therefore $T_{1/2} = 24 \div 5 = \mathbf{4.8\ hours}$

Q2 Decay constant $\lambda = \dfrac{\ln 2}{T_{1/2}} = \dfrac{0.693...}{483} = 1.435... \times 10^{-3}$
$= \mathbf{1.44 \times 10^{-3}\ s^{-1}}$ **(to 3 s.f.)**

Q3 The gradient of the graph = $-\lambda$.
gradient = $\dfrac{0.5 - 9.5}{14 - 0} = -0.642...$
So $\lambda = 0.642...$ s^{-1}
$T_{1/2} = \ln 2 \div \lambda = \ln 2 \div 0.642... = 1.07... = \mathbf{1.1\ s}$ **(to 2 s.f.)**

Q4 Half-life of carbon-14 in seconds:
$T_{1/2} = 5730 \times (365 \times 24 \times 60 \times 60) = 1.8070... \times 10^{11}$ s
Decay constant $\lambda = \dfrac{\ln 2}{T_{1/2}} = \dfrac{\ln 2}{1.8070... \times 10^{11}}$
$= 3.8358... \times 10^{-12}$ s^{-1}
Activity when animal died = $A_0 = 1.2$ Bq
Activity now = $A = 0.45$ Bq
Activity decreases exponentially according to: $A = A_0 e^{-\lambda t}$
Rearranging this for time $t = -\dfrac{1}{\lambda} \times \ln\left(\dfrac{A}{A_0}\right)$
$= 2.5569... \times 10^{11}$ s = **8100 years (to 2 s.f.)**

Page 183 — Fact Recall Questions

Q1 The time it takes for the number of unstable nuclei in a sample of an isotope (or the sample's activity or count rate) to halve.

Q2 Read off the activity at $t = 0$. Halve this value, draw a horizontal line to the curve, then a vertical line down to the x-axis. Read off the time — this the half-life. Repeat for a quarter of the original value and divide by two to check.

Q3 E.g. estimating the age of an object using radioactive dating requires a long half-life isotope so that it is still radioactive after thousands of years of decay; using radioactive isotopes as tracers in medical imaging requires a relatively short half-life so as not to remain radioactive inside the patient for long (but long enough to image the body).

8. Nuclear Decay

Page 187 — Application Questions

Q1 Beta-minus decay involves the conversion of a neutron to a proton when an electron is emitted (along with the release of an anti-neutrino). So nucleon number A remains constant, and proton number increases by 1:
$^{137}_{55}\text{Cs} \longrightarrow\ ^{137}_{56}\text{Ba} +\ ^{0}_{-1}\beta +\ ^{0}_{0}\bar{\nu}_e$
You can check your answer by making sure that the total proton number (or charge) on the left of the equation balances the total proton number (or charge) on the right. Do the same for the nucleon number and the lepton number.

Q2 $^{211}_{85}\text{At} \longrightarrow\ ^{207}_{83}\text{Bi} +\ ^{4}_{2}\alpha$

Q3 $^{83}_{37}\text{Rb} +\ ^{0}_{-1}\beta \longrightarrow\ ^{83}_{36}\text{Kr} +\ ^{0}_{0}\nu_e$

Q4 a) Iodine first decays into xenon by emitting a beta-minus particle. This xenon nucleus is excited and unstable so decays further by emitting gamma radiation.

b) The change in energy of the nucleus for each step, which is the energy released for that step.

Page 187 — Fact Recall Questions

Q1 Too many neutrons, too few neutrons, too many nucleons or too much energy.
Too many neutrons and the nucleus will β^- decay. Too few neutrons and it will β^+ decay. Too many nucleons all together and it will α decay. Too much energy and some energy will be released as γ radiation.

Q2

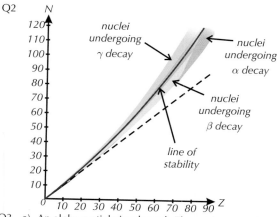

Q3 a) An alpha particle is released. The nucleon number decreases by 4, and the proton number decreases by 2.

b) An electron is released (as well as an antineutrino). The nucleon number remains the same, but the proton number increases by 1.

c) A positron is released (as well as a neutrino). The nucleon number remains the same, but the proton number decreases by 1.

Q4 a) The nucleus has excess energy, it is in an excited state. It loses this energy and becomes stable by emitting gamma radiation.

b) E.g. Medical tracer / medical diagnosis

Q5 Any three of: energy, momentum, charge, nucleon number and lepton number.

9. Mass Defect and Binding Energy

Page 189 — Application Questions

Q1 Mass defect is equivalent to binding energy.
1u is roughly 931.5 MeV. Therefore the binding energy is $0.0989 \times 931.5 = 92.12535 =$ **92.1 MeV (to 3 s.f.)**

Q2 Mass of nucleus = 15.994915 u
Number of protons = 8, number of neutrons = 8
Total mass of nucleons = $(8 \times 1.00728\ u) + (8 \times 1.00867\ u)$
$= 16.1276\ u$
So mass defect = mass of nucleons – mass of nucleus
$= 16.1276\ u - 15.994915\ u =$ **0.132685 u**

Q3 Binding energy = binding energy per nucleon × nucleon number = $8.79\ \text{MeV} \times 56 = 492.24\ \text{MeV}$
Convert from binding energy to mass defect:
Mass defect = binding energy $\div 931.5 =$ **0.528 u (to 3 s.f.)**

Page 189 — Fact Recall Questions

Q1 All energy changes.
Q2 The energy needed to separate all the nucleons in a nucleus. It is equivalent to the mass defect.
Q3 931.5 MeV
Q4

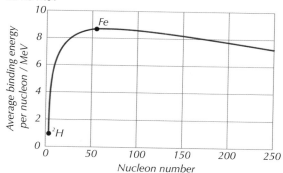

10. Nuclear Fission and Fusion

Page 192 — Application Questions

Q1 $\Delta m = m_p + m_p - m_{H\text{-}2} - m_e - m_v$
$= 1.00728 + 1.00728 - 2.01355 - 0.00055 - 0$
$= 0.00046\ u$
In MeV: $\Delta m = 0.00046 \times 931.5$
$= 0.428... =$ **0.43 MeV (to 2 s.f.)**

Q2 First, calculate the number of neutrons produced by balancing the nucleon number A:
$A_{before} = 1 + 235$, $A_{after} = 94 + 139 + x$
Therefore number of neutrons produced = $x = 3$
So $\Delta m = m_n + m_{U\text{-}235} - m_{Zr\text{-}94} - m_{Te\text{-}139} - 3m_n$
$= 1.00867 + 234.99333 - 93.88431 - 138.90613 - (3 \times 1.00867)$
$= 0.18555\ u$
In MeV: $\Delta m = 0.18555 \times 931.5$
$= 172.83... =$ **172.8 MeV (to 4 s.f.)**

Page 192 — Fact Recall Questions

Q1 Fission is the spontaneous or induced splitting of a larger nucleus into two smaller nuclei.
Q2 Spontaneous fission occurs randomly by itself, while induced fission occurs when a (thermal) neutron hits a nucleus, making the nucleus unstable and causing it to fission.
Q3 The fusing of two smaller nuclei to form one larger nucleus.
Q4 For a reaction to be energetically favourable, energy must be released — this only happens if the binding energy per nucleon increases.
(The products have a greater total mass defect than the initial nuclei, and so more mass is converted to energy that is released.)
Q5 Nuclei are positively charged, so they are repelled by the electrostatic interaction. They must have a lot of energy to overcome this interaction (so that the strong nuclear interaction can take over and attract them together).
Q6

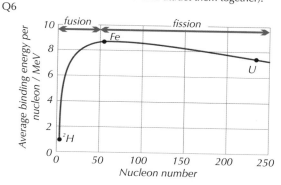

11. Nuclear Fission Reactors

Page 196 — Fact Recall Questions

Q1 When the neutrons released by nuclear fission cause other nuclei to fission and release more neutrons — and so on. In a nuclear reactor, you want the chain reaction to continue on its own at a steady rate, where one fission follows another. The amount of 'fuel' you need to do this is called the critical mass — any less than the critical mass (sub-critical mass) and the reaction will just peter out.

Q2 The material used for the moderator needs to be able to slow down neutrons so they can be absorbed by the uranium nuclei. The material used for the coolant needs to be a liquid or a gas at room temperature so it can be pumped around the reactor. It also needs to be efficient at transferring heat from the reactor to cool it. Water could be used as a moderator and as a coolant.

Q3 Neutrons collide with particles in the moderator — these collisions are elastic. Some energy and momentum is transferred from the neutron to the moderator particle, which slows down the neutron. These neutrons are called thermal neutrons. The neutrons need to be slowed down in order for uranium nuclei to capture them, to induce fission.

Q4 The control rods are made of a material that absorbs neutrons (e.g. boron). By inserting them into the reactor, the number of neutrons in the reactor can be limited, and the rate of reaction can be controlled by controlling the amount that the control rods are inserted into the reactor.

Q5 E.g. reactor shielding — a thick concrete case that prevents radiation escaping and reach the nuclear workers; emergency shut-down by lowering the fuel rods fully into the reactor to slow down the reaction.

Q6 Unused fuel rods only emit weakly-penetrating alpha radiation, which can easily be blocked. Used fuel rods contain different radioactive isotopes which emit beta and gamma radiation too, both of which are more strongly penetrating.

Q7 To limit the radiation that workers are exposed to.

Q8 When material is removed from the reactor, it is initially very hot, so it is placed in cooling ponds until the temperature falls to a safe level. The radioactive waste should then be stored in sealed containers until its activity has fallen sufficiently.

Q9 So society can make informed decisions on nuclear power, e.g. how nuclear waste is dealt with, where it is stored, whether more nuclear power stations should be built etc.
Risk — any one from e.g. A thermal nuclear reactor has the potential to cause a nuclear disaster, endangering people and the environment / The waste is highly radioactive and poses a risk to people and the environment if stored incorrectly.
Benefit — any one from: e.g. There is enough thermal nuclear reactor fuel to keep generating electricity for much longer than burning fossil fuels / No harmful greenhouse gases are released during the process / It's a lot more efficient than burning fossil fuels — more energy is produced per unit mass fuel.

Exam-style Questions – pages 198-200

1   **A** *(1 mark)*
2   **C** *(1 mark)*
1 MeV = 1000 keV, so the change in energy due to the beta decay = 1198 keV. The total energy lost by the nucleus is 1198 + 1294 = 2492 keV.
3   **A** *(1 mark)*
Number of moles present in sample, n,
= mass ÷ relative atomic mass = 0.6 ÷ 60 = 0.01 mol
Number of atoms = nN_A = 0.01 × 6.02 × 10^{23} = 6.02 × 10^{21}
3.01 × 10^{21} is half of 6.02 × 10^{21}, so 1 half life must pass.
4.1   Energy in J = (250 × 10^6) × (1.60 × 10^{-19} eV) = 4.00 × 10^{-11} J
Wavelength $\lambda \simeq \dfrac{hc}{E}$
= (6.63 × 10^{-34}) × (3.00 × 10^8) ÷ (4.00 × 10^{-11})
= 4.9725 × 10^{-15} = **4.97 × 10^{-15} m (to 3 s.f.)**
(3 marks for the correct answer, otherwise 1 mark for correctly calculating the electron energy in joules, and 1 mark for correct working.)

4.2 $\sin\theta \simeq \dfrac{1.22\lambda}{2R}$
Rearrange to get diameter: $R \simeq \dfrac{1.22\lambda}{2\sin\theta}$
= 1.22 × (4.9725 × 10^{-15}) ÷ 2sin(57.4°)
= 3.600... × 10^{-15}
= **3.60 × 10^{-15} (to 3 s.f.)**
(2 marks for the correct answer, otherwise 1 mark for correct working.)

4.3   Radius = $R = R_0 A^{1/3}$ = (1.4 × 10^{-15}) × (23)$^{1/3}$
= 3.98... × 10^{-15}
= **4.0 × 10^{-15} m (to 2 s.f.)**
(2 marks for the correct answer, otherwise 1 mark for attempting to use the correct formula.)

4.4   Density = $m ÷ V$
Assume nucleus is a sphere, so $V = (4/3) × \pi r^3$
$V = (4/3) × \pi × (3.98... × 10^{-15})^3$
= 2.643... × 10^{-43}
Density = (3.8 × 10^{-26}) ÷ (2.643... × 10^{-43}) = 1.437... × 10^{17}
= **1.4 × 10^{17} kg m^{-3} (to 2 s.f.)**
(3 marks for the correct answer, otherwise 1 mark for attempting to use the formula for the volume of a sphere, and 1 mark for attempting to use the formula for density.)

4.5   Positive alpha particles were repelled, so the nucleus must also be positive *(1 mark)*. Only some alpha particles were deflected, while most just passed straight through, so the atom must be mostly empty space *(1 mark)* and very few alpha particles were deflected by an angle greater than 90°, so the nucleus must be tiny *(1 mark)*. Even the fast alpha particles (with high momentum) were deflected by the nucleus, so most of the atom's mass must be in the nucleus *(1 mark)*.

5.1   How to grade your answer (pick the description that best matches your answer):
5-6 marks: All of the components are identified. The functions of at least two components are fully explained. Grammar, spelling and punctuation are used accurately and there is no problem with legibility. An appropriate form and style of writing is used, and information is well structured and logical.
3-4 marks: All of the components are identified. The function of at least one component is fully explained. Only a few errors with grammar, spelling, punctuation and legibility. Answer has some structure, and information is presented logically.
1-2 marks: One or two of the components are identified, but no correct explanation of their functions is given. Several errors with grammar, spelling, punctuation and legibility. Answer has lack of structure and information.
0 marks: There is no relevant information.
Here are some points your answer may include:
A = control rods. They absorb neutrons, to stop them causing any more nuclei to decay. By varying how far they are inserted into the reactor, the rate of reaction can be controlled.
B = shielding/concrete case. It stops radiation from escaping. The radiation is ionising and can cause damage to people and the environment.
C = moderator. This slows down fast-moving neutrons. It does this through elastic collisions, where some of the neutrons' energy is transferred to the moderator. Slowing down the neutrons ensures they can be absorbed by nuclei.

5.2 A critical mass of fuel means there is just enough mass for the fission chain reaction to continue at a steady rate on its own (e.g. without any control rods). A greater than critical (supercritical) mass is used so that the control rods can be used to vary the rate of fission.
(3 marks — 1 mark for stating there is a fission chain reaction, 1 mark for saying a critical mass is the amount needed for the chain reaction to continue at a steady rate on its own, 1 mark for a reason why a greater than critical mass is used.)

5.3 The material used should be a liquid or gas at room temperature *(1 mark)*, and be efficient at transferring heat *(1 mark)*, e.g. water *(1 mark)*.

5.4 In an emergency shut-down, the control rods are lowered fully into the reactor *(1 mark)* to absorb neutrons causing the chain reaction to happen and slow down the reaction as quickly as possible *(1 mark)*.

6.1 The energy needed to pull all the nucleons in a nucleus apart/the energy released when a nucleus forms *(1 mark)*.

6.2
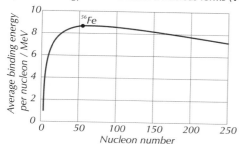
(3 marks — 1 mark for the overall shape, 1 mark for indicating Fe-56 is at the peak of the curve, 1 mark for indicating the peak is in the range 8-9.2 MeV)

6.3 Use the conversion 1 u = 931.5 MeV:
So binding energy per nucleon
$= (0.62065 \times 931.5) \div 66$
$= 8.7596... = \textbf{8.760 MeV (to 4 s.f.)}$ *(1 mark)*

6.4 Use conservation of charge to find proton number of ^{94}Sr:
$a = 92 - 54 = \textbf{38}$
Use conservation of nucleon number to find number of neutrons: $b = (235 + 1) - (140 + 94) = \textbf{2}$
(2 marks — 1 mark for correctly calculating each of a and b.)

6.5 The average binding energy of the final nuclei is greater than the binding energy of the initial nucleus *(1 mark)*.
An increase in binding energy means an increase in the total mass defect, and so more mass is converted to energy.

6.6 $\Delta m = (234.99333 + 1.00867) - (139.89194 + 93.89446 + (2 \times 1.00867)) = 0.19826$ u
so energy $= 0.19826 \times 931.5 = 184.67...$
$= \textbf{184.7 MeV (to 4 s.f.)}$
(3 marks for the correct answer or incorrect answer due to using incorrect answer to 6.4, otherwise 1 mark for correctly calculating the mass defect and 1 mark for attempting to convert it to binding energy.)

7.1 $^{33}_{15}$P $\longrightarrow$ $^{33}_{16}$S $+ ^{0}_{-1}\beta + ^{0}_{0}\overline{\upsilon}_e$
Beta-minus radiation is made up of electrons. By balancing mass (nucleon number), charge (proton number) and lepton number, you can work out the other two products.
(3 marks — 1 for each correct product)

7.2 Half-life in s = $25.4 \times 24 \times 3600 = 2\ 194\ 560$ s
Half-life $T_{1/2} = \ln 2 \div \lambda$
So rearrange: $\lambda = \ln 2 \div T_{1/2} = \ln 2 \div 2194560$
$= 3.158... \times 10^{-7} = \textbf{3.16} \times \textbf{10}^{-7}\ \textbf{s}^{-1}$ **(to 3 s.f.)**
(2 marks for the correct answer, otherwise 1 mark for attempting to use the equation for decay constant.)

7.3 Number of atoms given by: $N = N_0 e^{-\lambda t}$
Rearrange for time: $t = -\ln\left(\dfrac{N}{N_0}\right) \times \dfrac{1}{\lambda}$
So $t = -\ln\left(\dfrac{7.0 \times 10^{13}}{1.6 \times 10^{15}}\right) \times \dfrac{1}{3.158... \times 10^{-7}}$
$= 9.907... \times 10^6$ s $= \textbf{110 days (to 2 s.f.)}$
(2 marks for the correct answer, otherwise 1 mark for correct working.)

7.4 Intensity $I = \dfrac{k}{x^2}$
Rearrange for k: $k = I \times x^2$
At 0.2 m, $k = 3.6 \times 10^{-10} \times 0.2^2 = 1.44 \times 10^{-11}$ J
Intensity at 0.5 m $= (1.44 \times 10^{-11}) \div 0.5^2$
$= 5.76 \times 10^{-11}$
$= \textbf{5.8} \times \textbf{10}^{-11}\ \textbf{Wm}^{-2}$ **(to 2 s.f.)**
(2 marks for the correct answer, otherwise 1 mark for correctly calculating k.)

7.5 Any two of:
Gamma radiation is strongly penetrating, so it can travel through the body and reach the detector.
Gamma radiation is weakly ionising, so it causes little damage to tissues in the body.
It decays to a stable/safe isotope.
It has a half-life long enough for radiation to be picked up by the detector, but short enough to limit the body's exposure to radiation.
(2 marks — 1 mark for each reason.)

Section 7 — Option A: Astrophysics

2. Optical Telescopes

Page 207 — Application Question

Q1 a) $M = \dfrac{f_o}{f_e} = \dfrac{0.52}{0.0010} = \textbf{520}$

b) $M = \dfrac{\text{angle subtended by image at eye}}{\text{angle subtended by object at unaided eye}}$

$\dfrac{\text{angle subtended by image at eye}}{M} = \dfrac{0.331}{520}$
$= 6.365... \times 10^{-4}$

So angle subtended by object at unaided eye is equal to $\textbf{6.4} \times \textbf{10}^{-4}$ **rad (to 2 s.f.)**

Page 207 — Fact Recall Questions

Q1 a) That the light rays are parallel to each other.
b)
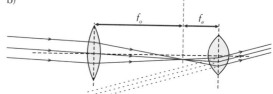

c) It is a virtual image (at infinity).

Q2 $M = \dfrac{f_o}{f_e}$, where f_o is the focal length of the objective lens and f_e is the focal length of the eye lens.

Q3 The length of an astronomical refracting telescope $= f_o + f_e$.
$M = \dfrac{f_o}{f_e}$, so an astronomical refracting telescope needs to be long in order to have a large f_o and therefore a large magnification.

Q4

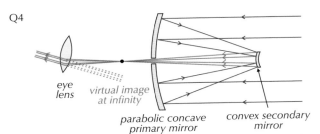

eye
lens   virtual image
         at infinity

parabolic concave   convex secondary
primary mirror   mirror

Q5 a) The quantum efficiency of a sensor is the percentage of
        incident photons that are detected.
    b) The quantum efficiency of a CCD is usually > 80%.
        The quantum efficiency of the eye is about 1%.
Q6 E.g. CCDs have a spatial resolution of around 10 µm,
    whereas the eye only has a spatial resolution of around
    100 µm. CCDs are less convenient than the eye, as looking
    through a telescope requires no extra equipment. However,
    CCDs can be also used to store and share images, which the
    eye cannot do.

3. Comparing Telescopes
Page 210 — Application Questions
Q1 Telescope A
    The minimum angular resolution is inversely proportional to
    the diameter of the dish — so the bigger the diameter, the
    smaller the minimum angular resolution, and so the better the
    resolving power (for a given wavelength).
Q2 a) $\theta \approx \frac{\lambda}{D} = \frac{650 \times 10^{-9}}{3.2} = 2.031... \times 10^{-7}$
                                        $= \mathbf{2.0 \times 10^{-7}}$ **radians (to 2 s.f.)**
    b) $\lambda \approx \theta D = (1.99 \times 10^{-7}) \times 3.2$
        $= 6.368 \times 10^{-7}$ m = **640 nm (to 2 s.f.)**

Page 210 — Fact Recall Questions
Q1 A measure of how much detail a telescope can see.
Q2 The light diffracts and produces a diffraction pattern when
    it passes through a circular aperture. The central circle of
    the pattern is called the Airy disc. A telescope has a poor
    resolution if the Airy discs of the light that makes up the
    image overlap too much.
Q3 $\theta \approx \frac{\lambda}{D}$
Q4 a) Glass refracts different colours of light by different
        amounts and so the image for each colour is in a slightly
        different position. This blurs the image.
    b) E.g. any two from: bubbles and impurities in the lenses
        absorb or scatter some light / building large lenses of
        good quality is expensive and difficult / large lenses are
        heavy and can only be supported at their edges, meaning
        they become distorted / for a large magnification the
        telescope has to be very long (due to a long focal length
        of the objective lens), so they need large buildings to
        house them.
Q5 E.g. any two from: it's easier to make good quality mirrors
    than lenses / mirrors don't suffer from chromatic aberration
    / it's cheaper to make large mirrors than lenses / mirrors can
    be supported from underneath, which means they are less
    likely to distort than lenses which cannot be supported in
    this way.
Q6 The effect is known as spherical aberration. If the shape of
    a mirror isn't quite parabolic (e.g. because the mirror has
    become distorted), parallel rays reflecting off different parts
    of the mirror do not all converge onto the same point. This
    causes the image produced by the mirror to be blurred.

4. Non-optical Telescopes
Page 214 — Application Question
Q1 a) $\dfrac{collecting\ power\ of\ VISTA}{collecting\ power\ of\ optical\ telescope}$
        $= \dfrac{diameter\ of\ VISTA^2}{diameter\ of\ optical\ telescope^2}$
        $= \dfrac{4.1^2}{1.02^2} = 16.15... = \mathbf{16}$ **(to 2 s.f.)**
    b) E.g. The Rayleigh criterion says that $\theta \approx \frac{\lambda}{D}$, so a shorter
        wavelength gives a smaller minimum angular resolution
        and so a better resolving power. Visible light has a
        shorter wavelength than infrared, so the resolving power
        of the telescope will be better in the visible region.
        Remember — the smaller the angle between objects can be,
        the better the resolving power.
    c) The telescope is used to detect infrared radiation. Water
        vapour in the atmosphere absorbs infrared radiation. The
        higher up you are, and the drier the atmosphere, the less
        infrared is absorbed, so infrared telescopes need to be in
        high-up, dry places.
    d) E.g. increase the resolving power of the detector.

Page 214 — Fact Recall Questions
Q1 A radio dish consists of a parabolic fine-wire-mesh dish, an
    antenna and a preamplifier. The dish reflects radio waves
    and focuses them on to a detector, positioned on the end of
    the antenna at the principle focus of the parabolic dish. The
    detector detects the radiation and the preamplifier amplifies
    the signal without adding too much noise.
    Remember, radio waves have long wavelengths that don't
    notice the gaps in the wire mesh, but shorter wavelength
    radiation, e.g. visible light, will pass right through it.
Q2 Radio waves have a much longer wavelength than visible
    light, so for a similar diameter of telescope, the resolving
    power is much lower, because of the Rayleigh criterion.
Q3 Many radio telescopes can be linked together and their data
    can be combined to form a single image. This is equivalent
    to one huge dish the size of the separation of the telescopes.
    Due to the Rayleigh criterion, $\theta \approx \frac{\lambda}{D}$, a huge diameter
    means that very good resolving powers can be achieved
    despite the long wavelength of radio waves.
Q4 E.g. any two from: a mesh dish is easier to construct than
    a lens or a mirror / a mesh dish is cheaper than a lens
    or mirror / radio dishes don't need to be as precisely
    constructed.
Q5 U-V radiation has a shorter wavelength than visible light.
    The shorter the wavelength, the more precise the shape of
    the mirror must be, so U-V telescopes must be more precise
    than optical telescopes.
Q6 I-R telescopes produce their own infrared radiation, which
    will get mixed up with the I-R radiation being observed,
    so they need to be cooled to reduce their temperature and
    minimise the I-R produced by the telescope.
Q7 X-ray radiation is usually either absorbed by a material or
    passes straight through it. X-rays only reflect if they graze
    a mirror's surface. An X-ray telescope must use grazing
    mirrors to gradually alter the direction of the X-rays until
    they are focused.
Q8 a) E.g. strapped to a weather balloon / aeroplane / in space
        in orbit around the Earth.
    b) E.g. on a mountain (at high altitude) somewhere dry /
        strapped to a weather balloon or aeroplane / in space in
        orbit around the Earth.
Q9 Collecting power is proportional to the mirror/dish diameter
    squared.

5. Parallax and Parsecs

Page 218 — Application Questions

Q1 20 pc = 20 × 3.26 = 65.2 = **70 ly (to 1 s.f.)**

Q2 a) The angle that Sirius moves in relation to distant background stars between either end of the Earth's orbit needs to be measured. The Earth orbits the Sun in 12 months, so it will take 6 months to move from one end of it's orbit to the other. The star will appear to move by its maximum amount between these two points.

b) First convert 0.37 arcseconds into degrees and then radians:

$0.37 \times \frac{1}{3600} = 1.0277... \times 10^{-4\,\circ}$

$1.0277... \times 10^{-4\,\circ} \times \frac{\pi}{180} = 1.79... \times 10^{-6}$ rad

Then $d = \frac{r}{\theta} = \frac{1.50 \times 10^{11}}{1.79... \times 10^{-6}}$

$= 8.36... \times 10^{16}$ m = **8.4 × 10¹⁶ m (to 2 s.f.)**

c) The distance to Sirius is 8.36... × 10¹⁶ m. 1 light year = 9.46 × 10¹⁵ m, so Sirius is (8.36... × 10¹⁶) ÷ (9.46 × 10¹⁵) = 8.83... = 8.8 light years away (to 2 s.f.). So light from Sirius will take **8 years and 10 months** to reach Earth.

Q3 a) The radius of Mars.

b) $d = \frac{r}{\theta} = \frac{3389.5}{2.5 \times \frac{1}{3600} \times \frac{\pi}{180}}$ = **2.8 × 10⁸ km (to 2 s.f.)**

You need to use 2.5 arcseconds as you're given the radius of the planet, not the diameter.

Page 218 — Fact Recall Questions

Q1 Half the angle by which a nearby star appears to move in relation to the background stars in 6 months as the Earth moves from one end of its orbit to the other.

Q2 A parsec is a unit of distance equal to 3.08 × 10¹⁶ m. A star is exactly one parsec (pc) away from Earth if the angle of parallax, $\theta = 1$ arcsecond $= \left(\frac{1}{3600}\right)^{\circ}$.

Q3 The distance that light travels in a vacuum in one year.

Q4 Light travels at a constant (finite) velocity. It takes about 8 minutes to travel the distance between the Sun and Earth.

Q5 $\theta \approx \frac{r}{d}$ where r is the radius of the object, d is the distance to the object and θ is half the angle subtended by the object in radians. This assumes that θ is small.

Q6 It makes the distances measured more precise and some methods only work for certain distances.

6. Magnitude

Page 222 — Application Questions

Q1 a) Difference in magnitude is 5.1 − 1.25 = 3.85
So the brightness ratio is:

$\frac{I_A}{I_B} \approx 2.51^{3.85} = 34.573... = 35$ (to 2 s.f.)

So Deneb is approximately **35 times** brighter than Alberio B.

b) Rearranging the formula, $M = m - 5\log\left(\frac{d}{10}\right)$.

So $M = 1.25 - 5\log\left(\frac{975}{10}\right) = -8.69... = $ **−8.70 (to 3 s.f.)**

Q2 a) $m = M + 5\log\left(\frac{d}{10}\right)$

$= -6.39 + 5\log\left(\frac{413}{10}\right) = 1.689... = $ **1.69 (to 3 s.f.)**

b) $M = m - 5\log\left(\frac{d}{10}\right)$

$= 1.64 - 5\log\left(\frac{75}{10}\right) = -2.735... = $ **−2.74 (to 3 s.f.)**

c) Bellatrix

d) Alnilam

Q3 Distance = 3.4 × 10⁶ pc

$m = M + 5\log\left(\frac{d}{10}\right) = -19.3 + 5\log\left(\frac{3.4 \times 10^6}{10}\right)$

$= 8.357... = $ **8.36 (to 3 s.f.)**

Page 222 — Fact Recall Questions

Q1 Luminosity.

Q2 a) The brightness of an object will appear to vary, depending on how far you are from the object.

b) Distance from Earth and power output/luminosity.

Q3 a) The apparent magnitude of a star is how bright a star appears when viewed from Earth. The apparent magnitude scale is logarithmic, so a magnitude 1 star has an intensity 100 times greater than a magnitude 6 star.

b) The dimmest visible stars were given an apparent magnitude of 6.

Q4 2.51 times.

Q5 $\frac{I_2}{I_1} \approx 2.51^{m_1 - m_2}$

Q6 It is very bright (brighter than the star Vega, which has an apparent magnitude of 0).

Q7 What the star's apparent magnitude would be if it were 10 parsecs away from Earth.

Q8 $m - M = 5\log\left(\frac{d}{10}\right)$

Q9 A standard candle is an object for which the absolute magnitude can be calculated directly. Type 1a supernovae all have the same peak in absolute magnitude, so the distance to them can be calculated by comparing how bright they look to how powerful they are known to be.

7. Stars as Black Bodies

Page 226 — Application Questions

Q1 To use the inverse square law, you must assume that the star's power output is the same in all directions, but it will be lower from sunspots because they are cooler ($P \propto T^4$).

Q2 a) $\lambda_{max}T = 2.9 \times 10^{-3}$ so $\lambda_{max} = 2.9 \times 10^{-3} \div T$
$= 2.9 \times 10^{-3} \div 11\,000 = 2.636... \times 10^7$

$= $ **2.6 × 10⁻⁷ m (to 2 s.f.)**

b)

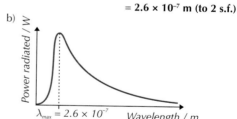

$\lambda_{max} = 2.6 \times 10^{-7}$ Wavelength / m

(y-axis: Power radiated / W)

c) First convert the distance to metres:
773 ly = 773 × (9.46 × 10¹⁵) = 7.31... × 10¹⁸ m

Use the inverse square law $I = \frac{P}{4\pi d^2}$ to find the power output. Rearranging gives

$P = 4\pi I d^2 = 4\pi(3.7 \times 10^{-8})(7.31... \times 10^{18})^2$
$= 2.48... \times 10^{31}$ W

$P = \sigma A T^4$ so $A = \frac{P}{\sigma T^4} = \frac{2.48... \times 10^{31}}{(5.67 \times 10^{-8})(11\,000)^4}$

$= 2.99... \times 10^{22} = $ **3.0 × 10²² m² (to 2 s.f.)**

Page 226 — Fact Recall Questions

Q1 A body that absorbs all electromagnetic radiation of all wavelengths and can emit all wavelengths of electromagnetic radiation.

Q2

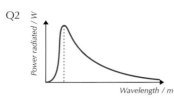

Q3 The temperature of the black body.

Q4 $\lambda_{max} T = 2.9 \times 10^{-3}$ m K

Q5 $P = \sigma A T^4$

Q6 $I = \dfrac{P}{4\pi d^2}$

8. Stellar Spectral Classes
Page 230 — Application Question
Q1 a) Star 1.
   b) Star 2, because it has molecular band absorptions from compounds like TiO. Compounds like this can only form in cooler stars.
   c) Star 1, because it only has absorption lines from hydrogen and helium. Star 2 is not a class B star because it has absorption lines from molecules.
   d) Spectral class M, because it has strong absorption lines for TiO.
   e) 11 000 – 25 000 K.

Page 230 — Fact Recall Questions
Q1 Each atom, ion or molecule in the star absorbs particular wavelengths of radiation that correspond with the differences between its energy levels. So there will only be absorption lines in the spectra at the particular wavelengths corresponding to the ions, atoms or molecules found in the star.

Q2 Because the electron transitions in atomic hydrogen that absorb photons with a wavelength in the visible part of the electromagnetic spectrum are between the n = 2 and higher energy levels.

Q3

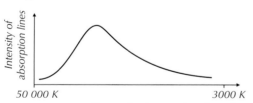

Decreasing temperature (non-linear scale)

Q4 O, B, A, F, G, K, M

Q5 A blue star.

Q6 a) Metal ions and metal atoms.
   b) Neutral atoms, and compounds such as TiO.
   c) He^+, He and H.
   d) Mostly neutral metal atoms.

Q7 F

Q8 A

9. The Hertzsprung-Russell Diagram
Page 232 — Application Question
Q1 a) Star 2, because both stars are at about the same temperature, but Star 2 is much brighter, so it must be bigger due to Stefan's law.
   b) White dwarf
   c) Main sequence star
   d) Star 1

Page 232 — Fact Recall Questions
Q1 Absolute magnitude from –10 to 15.

Q2 Spectral class or temperature (decreasing)

Q3

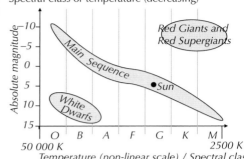

Temperature (non-linear scale) / Spectral class

Q4 Hydrogen is fusing into helium.

10. Evolution of Sun-like Stars
Page 235 — Application Question
Q1 a) **b** is a red giant and **a** is the hottest main-sequence star.
   b) **c**
   c) **c - b - a - d** (or **c - b - d**)

Page 235 — Fact Recall Questions
Q1 Gravity

Q2 The main sequence

Q3 The pressure produced from the hydrogen fusion in the core of the star balances the gravitational force compressing the star, so it is stable.

Q4 When the core of a star runs out of hydrogen, hydrogen fusion, and the pressure caused by it, stops and the core starts to contract and heat up. The heat from the core of the star heats up the layer until it is hot enough for hydrogen to fuse into helium.

Q5 Core hydrogen burning stops in the star and the core of the star begins to contract, causing the outer layers to expand and cool, and the star becomes a red giant.

Q6 Once fusion stops in the core of a low-mass star, the pressure created by fusion is lost, so the core begins to contract under its own weight and heat up. Once the core gets to about the size of the Earth, electron degeneracy pressure stops it contracting any further. The hot dense core that is left over is a white dwarf. A planetary nebula is left behind by the outer layers of the star.

Q7 The remnants of the outer layers of a red giant, that are ejected as the star becomes a white dwarf.

11. Supernovae, Neutron Stars and Black Holes
Page 240 — Application Questions
Q1 No. The photon is within the Schwarzschild radius of the black hole. Inside the Schwarzschild radius, the escape velocity is greater than the speed of light, so the photon is doomed never to escape.

Q2 a) The mass of the Sun is 1.99×10^{30} kg, so the black hole has a mass of $(4.31 \times 10^6) \times (1.99 \times 10^{30})$
   $= 8.57... \times 10^{36}$ kg $= \mathbf{8.58 \times 10^{36}}$ **kg (to 3 s.f.)**.
   b) $R_s = \dfrac{2GM}{c^2} = \dfrac{2(6.67 \times 10^{-11})(8.57... \times 10^{36})}{(3.00 \times 10^8)^2}$
   $= 1.271... \times 10^{10} = \mathbf{1.27 \times 10^{10}}$ **m (to 3 s.f.)**

c) The volume (V) of a sphere of radius r is $\frac{4}{3}\pi r^3$
The Sun has a radius of 6.96×10^8 m.

So $V_{Sun} = \frac{4}{3}\pi(6.96 \times 10^8)^3 = 1.412... \times 10^{27}$ m³

and $V_{black\ hole} = \frac{4}{3}\pi(1.27... \times 10^{10})^3 = 8.6... \times 10^{30}$ m³

$8.6... \times 10^{30} \div 1.412... \times 10^{27} = 6094.0...$

So the volume of the supermassive black hole at the centre of the galaxy is **6090 (to 3 s.f.) times** bigger than that of the Sun.

Page 240 — Fact Recall Questions

Q1 The electron degeneracy pressure cannot withstand the gravitational forces at this mass, so when the fusion reactions have stopped, the core will continue to contract further than a white dwarf.

Q2 The core contracts and the outer layers fall onto the core and rebound, causing a huge shockwave that propels the outer layers into space. The star experiences a brief and rapid increase in absolute magnitude when it explodes.

Q3 A collapsing star leading to a supernova could release a burst of gamma rays of 10^{44} J (or more, depending on the type of supernova) — roughly the entire energy output of the Sun over its whole lifetime.

Q4 A light curve is a graph of absolute magnitude, M, (or sometimes the brightness) plotted against the time since the supernova reached peak magnitude. The light curve for a type 1a supernova is:

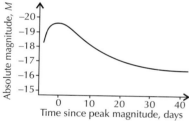

Q5 Neutron stars typically have a density of around 4×10^{17} kg m⁻³. They are made mostly of neutrons.

Q6 The velocity that an object would need to travel at to have enough kinetic energy to overcome the gravitational potential energy of another object (e.g. a star or a planet).

Q7 A black hole is an object whose escape velocity is greater than the speed of light. They are formed when a star of core mass more than 3 solar masses contracts and collapses into an infinitely dense point.

Q8 a) The boundary of a region surrounding an object at which its escape velocity is equal to the speed of light.
b) The Schwarzschild radius of a black hole is the radius of the event horizon, or the radius at which the escape velocity is equal to the speed of light.
c) $R_s = \frac{2GM}{c^2}$

Q9 A supermassive black hole.

12. Doppler Effect and Red Shift

Page 244 — Application Questions

Q1 a) Star A, because the frequency of the absorption line is lower than it should be, so it is red-shifted.
b) $z = \frac{\Delta f}{f} = \frac{(4.57 \times 10^{14}) - (4.37 \times 10^{14})}{4.57 \times 10^{14}}$
$= 0.0437... = \mathbf{0.0438\ (to\ 3\ s.f.)}$

c) $z = \frac{v}{c}$ so $v = zc = 0.0437... \times (3.00 \times 10^8)$
$= 1.312... \times 10^7 = \mathbf{1.31 \times 10^7\ ms^{-1}\ (to\ 3\ s.f.)}$

Q2 a) Positive, because it is receding.
b) $z = \frac{v}{c} = \frac{463 \times 10^3}{3.00 \times 10^8} = 1.543... \times 10^{-3}$
$= \mathbf{1.54 \times 10^{-3}\ (to\ 3\ s.f.)}$
c) $z = -\frac{\Delta\lambda}{\lambda} = -\frac{\lambda - \lambda_{obs}}{\lambda}$, so making λ_{obs} the subject:
$\lambda_{obs} = \lambda(z + 1) = 0.21121(1.54... \times 10^{-3} + 1) = 0.21153...$
$= \mathbf{0.212\ (to\ 3\ s.f.)}$

Remember, always give your answer to the lowest number of significant figures given in the question.

Page 244 — Fact Recall Questions

Q1 The sound waves travelling in the same direction as the police car are 'bunched up' in front of the police car, and those travelling in the opposite direction spread out behind it. The bunching up causes the frequency and therefore the pitch of the sound waves to be higher.

Q2 As a radiation source moves away from an observer, the wavelengths detected by the observer get longer and the frequencies get lower — this effect is called red shift. Blue shift is caused by a source moving towards the observer — the wavelengths of the radiation observed are shorter and the frequencies get higher than when the source is stationary.

Q3 $z = \frac{v}{c}$, assuming $v \ll c$.

Q4 Look for absorption lines with known wavelengths in the spectrum and see how their wavelengths have been shifted. If the absorption lines have a red shift, the source is moving away from us and if the absorption lines have been blue shifted, the source is moving towards us.

Q5 $z = \frac{\Delta f}{f}$ and $z = -\frac{\Delta\lambda}{\lambda}$.

Q6 The red shift of distant objects that is caused by space itself expanding and light waves stretching along with it (rather than the objects moving through space away from us).

13. The Big Bang Theory

Page 248 — Application Questions

Q1 1.4×10^7 pc $= 14 \times 10^6$ pc $= 14$ Mpc.
Use Hubble's law: $v = Hd$
The maximum value of Hubble's constant, 75 km s⁻¹ Mpc⁻¹, gives the maximum recessional velocity:
$v = Hd = 75 \times 14 = \mathbf{1050\ km s^{-1}}$.
The minimum value of Hubble's constant, 65 km s⁻¹ Mpc⁻¹, gives the minimum recessional velocity:
$v = Hd = 65 \times 14 = \mathbf{910\ km s^{-1}}$.

Q2 $z = -\frac{\Delta\lambda}{\lambda} = -\frac{-35 \times 10^{-9}}{725 \times 10^{-9}} = 0.0482...$
$z = \frac{v}{c}$, so $v = zc = 0.0482... \times (3.00 \times 10^8)$
$= 1.448... \times 10^7$ ms⁻¹ $= 1.448... \times 10^4$ kms⁻¹
$v = Hd$, so $d = \frac{v}{H} = \frac{1.448... \times 10^4}{72} = 201.149...$
$= \mathbf{200\ Mpc\ (to\ 2\ s.f.)}$
The star is receding, so $\Delta\lambda$ is negative in the first equation.

Q3 a) H has not been constant since the universe began.
b) If the universe is 13.8 billion years old, the observable universe is limited by the time taken for light to travel to us — some light won't have reached us yet so the actual size of the universe cannot be calculated.

Page 248 — Fact Recall Questions

Q1 On a large scale the universe is homogeneous and isotropic.

Q2 The red shift of distant objects is proportional to their distance.

Q3 $v = Hd$, where v is in $km\,s^{-1}$, H is in $km\,s^{-1}\,Mpc^{-1}$ and d is in Mpc.

Q4 The theory that the universe started off very hot and very dense (perhaps as an infinitely hot, infinitely dense singularity) and has been expanding ever since.

Q5 If you assume the expansion of the universe has been constant, then the age of the universe is $1/H$, where H is Hubble's constant.

Q6 We can only measure the size of the observable universe, because only the light from within that distance has had enough time to reach us.

Q7 The expansion of the universe is thought to be accelerating, not decelerating as astronomers expected — dark energy might explain this.

Q8 Cosmic microwave background radiation is electromagnetic radiation in the microwave region that is found everywhere in the universe, and is largely homogeneous and isotropic. The Big Bang theory predicts that this radiation was produced in the very early universe, and that its wavelengths have been stretched by the expansion of the universe to match those observed today.

Q9 There is a large abundance of helium in the universe. This supports the Big Bang theory as it suggests that at some point the universe was hot enough for hydrogen to fuse into helium.

14. Detection of Binary Stars, Quasars and Exoplanets

Page 253 — Application Questions

Q1 a) No, one of the stars is much brighter than the other. *From the graph you can see that one eclipse leads to a bigger drop in apparent magnitude than the other, so one of the stars must be dimmer.*

b) At point A, both stars can be seen, as they are not in the same line of sight because they are near their maximum separation.
At point B, the stars are in the same line of sight, i.e. one behind the other. The dimmer star is behind the brighter star.
At point C, the stars are in the same line of sight. The brighter star is behind the dimmer star.

c) The time between the two big dips in apparent magnitude — about 3 days.

Q2 $z = 0.12$, and $z = \frac{v}{c}$ so $v = zc = 0.12 \times 3.00 \times 10^8$
$= 3.6 \times 10^7 \ m\,s^{-1} = 3.6 \times 10^4 \ km\,s^{-1}$
Now use Hubble's law: $v = Hd$ and so
$d = \frac{v}{H} = \frac{3.6 \times 10^4}{65} = 553.8... = \textbf{550 Mpc (to 2 s.f.)}$

Q3 a) E.g: They're orbiting stars which are much brighter than them, so are hard to see. / They're too close to their stars to be resolved with most telescopes.

b) An exoplanet orbiting a star has a small effect on the star's orbit. It causes tiny variations in the star's orbit, because the star and the exoplanet are actually orbiting around the centre of mass between them. The star's orbital variations cause tiny red and blue shifts in the star's emissions, which can be detected on Earth and can suggest the presence of an exoplanet.

c) She is not correct — just because there is no red or blue shift detected doesn't necessarily mean there are no exoplanets. It could just be that the star's movement isn't aligned with the observer's line of sight — if movement is perpendicular to the line of sight then there won't be any detectable shift in the light from the star.

Page 253 — Fact Recall Questions

Q1 If you take a spectral line from an element in the spectrum of a spectroscopic binary system, the line will periodically split into two lines (one red-shifted and one blue-shifted) then recombine. The lines which will reach their maximum separation in half the orbital period of the system.

Q2 Measure the time between zero and maximum separation between the red- and blue-shifted absorption lines of an element, and multiply by 2 to get the period.

Q3 As the stars eclipse each other, the apparent magnitude of the system drops because light from one star is blocked out by the other star.

Q4 Stars in our galaxy.

Q5 They have very large red shifts.

Q6 A quasar is a powerful galactic nucleus centred around a supermassive black hole. The strong radiation emitted is due to the mass of whirling gas falling into it.

Q7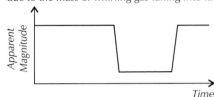

Q8 Advantage: E.g. the transit method allows you to calculate the exoplanet's radius.
Disadvantage: E.g. the transit method only works if the exoplanet passes in front of the star as viewed by the observer.

Exam-style Questions — Pages 255-258

1.1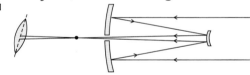

(2 marks — 1 for the primary concave and secondary convex mirrors drawn correctly, 1 mark for the path of both rays drawn correctly to the eyepiece.)

1.2 If a mirror in a Cassegrain arrangement isn't quite the right shape, the light rays refract by the wrong angles and there is no focal point. The image is blurry. *(1 mark)*

1.3 By using a precisely made parabolic mirror *(1 mark)*.

1.4 The Earth's atmosphere absorbs most ultraviolet and some infrared wavelengths so the telescope would be unable to observe those wavelengths from Earth *(1 mark)*.

1.5 How to grade your answer (pick the description that best matches your answer):
0 marks: There is no relevant information.
5-6 marks: Three or more of the advantages and disadvantages of a CCD compared to the eye are described, and there is a good level of detail. Grammar, spelling and punctuation are used accurately and there is no problem with legibility. Uses an appropriate form and style of writing and information.
3-4 marks: Two or three of the advantages and disadvantages of a CCD compared to the eye are described, and there is some detail. Only a few errors with grammar, spelling, punctuation and legibility. Answer has some structure and information.
1-2 marks: One or two of the advantages and disadvantages of a CCD compared to the eye are described, but there is little detail. Several errors with grammar, spelling, punctuation and legibility. Answer lacks structure and information.

Here are some points your answer may include:
- A CCD has a high quantum efficiency (around 80%) whereas the eye has a low quantum efficiency (around 1%). This is the proportion of incident photons that are detected, so more light (around 80% of the incident photons) is detected by a CCD.
- A CCD can detect a wide spectrum of EM radiation whereas the eye can only detect visible light.
- A CCD has a very small spatial resolution (around 10 μm) whereas the eye has a reasonably large spatial resolution (around 100 μm), meaning a CCD can capture finer detail.
- A CCD requires equipment to be able to view the data stored on the CCD. The eye is often more convenient as it doesn't need any equipment to detect EM radiation. However, a CCD can be used in environments (e.g. space) that it would not be possible to detect radiation with the eye in.
- A CCD can store and share images, whereas the eye cannot.

2.1 A black body is a body that absorbs all electromagnetic radiation at all wavelengths and can emit all wavelengths of electromagnetic radiation *(1 mark)*.

2.2 Assuming that Mu Cephei behaves as a black body,
$\lambda_{max}T = 2.9 \times 10^{-3}$ m K so
$T = \frac{2.9 \times 10^{-3}}{\lambda_{max}} = \frac{2.9 \times 10^{-3}}{828.6 \times 10^{-9}} = 3499.8...$ K
$= \mathbf{3500\ K}$ **(to 2 s.f.)**
(2 marks available for correct answer and 1 mark for correct assumption, otherwise 1 mark for using correct values in the equation, 1 mark for correct assumption.)

2.3 Find the surface area of Mu Cephei first:
Assume that Mu Cephei is a sphere and use that the radius of the Sun $= 6.96 \times 10^8$ m
$A = 4\pi r^2 = 4\pi(1500 \times 6.96 \times 10^8)^2$
$= 1.369... \times 10^{25}$
Using Stefan's law:
$P = \sigma AT^4 = (5.67 \times 10^{-8})(1.369... \times 10^{25})(3499.8...)^4$
$= 1.165... \times 10^{32}$ W $= \mathbf{1.2 \times 10^{32}\ W}$ **(to 2 s.f.)**.
(3 marks for correct answer, otherwise 1 mark for substituting the correct values into the formula for a sphere's surface area, 1 mark for substituting the correct values into Stefan's law.)

2.4 Rearranging the inverse square law to make d the subject gives $d = \sqrt{\frac{P}{4\pi I}} = \sqrt{\frac{1.165...... \times 10^{32}}{4\pi \times (5.1 \times 10^{-9})}} = 4.26... \times 10^{19}$ m
In light years,
$d = 4.26... \times 10^{19} \div (9.46 \times 10^{15}) = 4507.3...$
$= \mathbf{4500\ ly}$ **(2 s.f.)**
(2 marks for correct answer, otherwise 1 mark for substituting the correct values into the inverse square law to get distance.)
If you get an answer wrong, and then use the wrong value in the next part, you will get full marks for that part as long as you use the wrong value in the correct way.

2.5 E.g. Mu Cephei is not a perfect black body, so the value for surface temperature may be wrong *(1 mark)*.

3.1 Omicron[1] *(1 mark)* (as it has the highest value of apparent magnitude).

3.2 The absolute magnitude of an object is what the apparent magnitude would be if the object was 10 parsecs away *(1 mark)*.

3.3 Sirius *(1 mark)*, because its apparent magnitude is brighter than its absolute magnitude *(1 mark)*.

3.4 $m - M = 5\log\frac{d}{10}$
so $d = 10 \times 10^{\frac{m-M}{5}} = 10 \times 10^{\frac{1.83 - (-6.87)}{5}}$
$= 549.5... = \mathbf{550\ parsecs}$ **(to 3 s.f.)**
(2 marks for correct answer, otherwise 1 mark for substituting into the correct formula.)

3.5 Wezen is in spectral class F, so it is a white star *(1 mark)* and contains metal ions *(1 mark)*. The main absorption lines that you would expect to see in its line spectrum are lines due to metal ions *(1 mark)*.

4.1 $\theta \approx \frac{\lambda}{D} = \frac{8.5 \times 10^{-3}}{25} = \mathbf{3.4 \times 10^{-4}\ rad}$
(2 marks for correct answer, otherwise 1 mark for substituting into the correct formula.)

4.2 For a 36 m dish, $\theta \approx \frac{\lambda}{D} = \frac{8.5 \times 10^{-3}}{36 \times 10^3}$
$= 2.36... \times 10^{-7}$ rad
$\frac{3.4 \times 10^{-4}}{2.36... \times 10^{-7}} = 1440 = \mathbf{1400\ times\ better}$ **(to 2 s.f.)**
(2 marks for correct answer, otherwise 1 mark for using the correct resolving power of the dish and the VLA.)

4.3 The formula $z = \frac{v}{c}$ only works for $v \ll c$ *(1 mark)*, which means $z \ll 1$. The formula won't work for $z = 5.95$ *(1 mark)*.

4.4 E.g. object A could be a quasar because it is a strong radio source *(1 mark)* with a very large red shift *(1 mark)*.

5.1 Half the angle moved by an object in relation to the background stars over a period of 6 months, when the Earth is at opposite ends of its orbit *(1 mark)*.

5.2 Only nearby stars appear to move, while distant stars appear to be stationary, so parallax only works for nearby stars *(1 mark)*.

5.3 The stars have the same (or very similar) brightness as each other, because the drops in apparent magnitude are by the same amount for each eclipse. *(1 mark)*.

5.4 95 hours *(1 mark)*.

5.5 As the stars orbit, the light from each star will be red-shifted and blue-shifted, relative to the average frequency of the light, as it travels away from and towards us respectively *(1 mark)*.

5.6 When the observed wavelength is at its maximum, the star will be travelling away from us almost in our line of sight, at the orbital velocity, when it is at its minimum, it will be travelling towards us at the orbital velocity. When the star is travelling perpendicular to our line of sight, the observed red shift will be between the maximum and minimum, so it will be 434.051 nm.
Use $\frac{\Delta\lambda}{\lambda} = -\frac{v}{c}$ to get $v = -\frac{c\Delta\lambda}{\lambda}$
$= -\frac{(3.00 \times 10^8)(434.051 - 434.076)}{434.051}$
$= 17\ 279.0...$ m s^{-1}
So the orbital velocity is 17 000 m s^{-1} (to 3 s.f.)
$= \mathbf{17\ km\ s^{-1}}$ **(to 3 s.f.)**
(3 marks for correct answer, otherwise 1 mark for correct rearranging of equation and 1 mark for substituting in the right values.)

5.7 The stars will eventually become red giants *(1 mark)*. The cores will contract further and heat up until helium fusion occurs in the core *(1 mark)*.

6.1

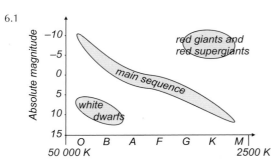

(3 marks available — 1 mark for correct shape of main sequence, 1 mark for drawing the red giants, red supergiants and white dwarfs correctly, 1 mark for labelling temperature and spectral class scale correctly.)

6.2

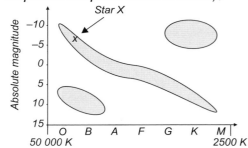

(1 mark for correctly plotting Star X on the H-R diagram at the correct absolute magnitude and between spectral classes O and B.)

6.3 When fusion in the core of the star stops, the core will begin to contract *(1 mark)*. The gravitational forces will be so large that the core will not stop at a white dwarf or a neutron star, but will collapse to an infinitely dense point *(1 mark)*. The gravitational forces of the infinitely dense point are so strong that it becomes a black hole *(1 mark)*.

6.4 The Schwarzschild radius of a black hole is the radius at which the black hole's escape velocity is equal to the speed of light, c *(1 mark)*.

7.1 The universe started off very hot and very dense (perhaps as an infinitely hot, infinitely dense singularity) and has been expanding ever since *(1 mark)*.

7.2 Hubble's law says that the further away a galaxy is, the faster it is moving away from us. It suggests that the universe is expanding so at one point it must have been denser and hotter *(1 mark)*.

7.3 $H = 65 \text{ km s}^{-1} \text{Mpc}^{-1} = \dfrac{65 \times 10^{3}}{3.08 \times 10^{22}} = 2.11... \times 10^{-18} \text{ s}^{-1}$

$t = \dfrac{1}{H} = \dfrac{1}{2.11... \times 10^{-18}} = 4.738... \times 10^{17} \text{ s}$

= 15 billion years (to 2 s.f.)

(3 marks for correct answer, otherwise 1 mark for correctly converting H to s⁻¹ and 1 mark for substituting into correct formula.)

7.4 E.g. Cosmic microwave background radiation: The Big Bang theory predicts that lots of EM radiation was given out in the early universe *(1 mark)*. The cosmic microwave background radiation matched this prediction *(1 mark)*. Relative abundances of H and He: The Big Bang theory predicts that the early universe was so hot that hydrogen fused to make helium *(1 mark)*, which explains the relative abundances of H and He in the universe now *(1 mark)*.

Section 7 — Option B: Medical Physics

1. Lenses
Page 262 — Application Questions
Q1 E.g.

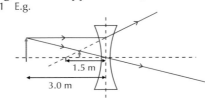

Q2

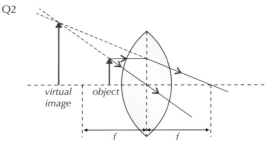

Page 262 — Fact Recall Questions
Q1 A converging lens is convex in shape. It brings rays parallel to its principal axis together at its principal focus. A diverging lens is concave in shape. It causes parallel rays of light to diverge.

Q2 The principal focus of a converging lens is the point on which incident rays parallel to the principal axis converge. The principal focus, of a diverging lens is the point that incident rays parallel to the principal axis appear to have come from.

Q3 Draw two rays coming from the same point on the object: one parallel to the principal axis which refracts through the principal focus, and one that travels straight through the lens' centre and does not refract. Where the rays meet is where the image of that point on the object is formed. If one point on the object is on the principal axis, that point on the image will also be on the principle axis. Otherwise, repeat the above steps for another point on the object. The image of the object can now be drawn

Q4 A real image is formed when light rays from an object are made to pass through another point in space. A real image is formed on the other side of the lens from the object, and can be captured on a screen. A virtual image is formed when light rays from an object diverge and therefore appear to have come from another point in space. Virtual images are formed on the same side of the lens as the object. The rays of light don't actually pass through the point where the image appears to be, and so this type of image cannot be captured on a screen.

2. Calculations with Lenses
Page 265 — Application Questions
Q1 object height = 0.20 m, image height = 0.36 m

$m = \dfrac{\text{image height}}{\text{object height}} = \dfrac{0.36}{0.20} = \textbf{1.8}$

Q2 Rearrange $P = \dfrac{1}{f}$ to get $f = \dfrac{1}{P}$.

$f = \dfrac{1}{P} = \dfrac{1}{2.4} = 0.4166... = \textbf{0.42 m (to 2 s.f.)}$

Q3 $u = 65$ cm, $v = 105$ cm and $\frac{1}{f} = \frac{1}{u} + \frac{1}{v}$.

Substituting in u and v gives:

$\frac{1}{f} = \frac{1}{65} + \frac{1}{105}$ so $f = 40.14... = $ **40 cm (to 2 s.f.)**

Remember you don't need to convert all the distances into metres before plugging them into the lens equation — they just need to all be in the same units.

Page 265 — Fact Recall Questions

Q1 The lens equation. $\frac{1}{f} = \frac{1}{u} + \frac{1}{v}$, where $u = $ object distance from the lens, $v = $ image distance from the lens and $f = $ focal length of the lens.

Q2 $P = \frac{1}{f}$

3. Physics of the Eye

Page 269 — Fact Recall Questions

Q1 The cornea and lens (and other parts of the eye) refract and focus light on to the retina. The eye's lens changes shape depending on the object's distance, which changes the focal length, so that an image is always formed on the retina.

Q2 a) The furthest distance the eye can focus on comfortably.
b) The closest distance the eye can focus on.

Q3 a) At the back of the retina.
b) There is one type of rod and there are three types of cones.
Each type of cone is sensitive to different colours of light — red, green or blue.

Q4

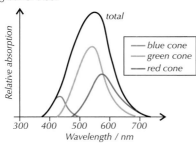

Q5 a) A measure of the ability to form separate images of objects that are close together.
b) There must be at least one rod or cone between the rods or cones detecting the light from each object.

4. Defects of Vision

Page 272 — Application Questions

Q1 a)

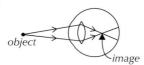

b) Myopia.
c) A person with myopia can't focus on distant objects as their far point is closer than infinity. A diverging lens with its principal focus at the eye's faulty far point can be used to correct this. This means distant objects at infinity, which were blurry, now appear to come from the eye's far point and will be in focus.

Q2 The focal point of the lens used should be –4.2 m, because it should be a diverging lens that makes distant rays appear to come from the uncorrected far point of 4.2 m.

So $f = -4.2$ m and $P = \frac{1}{-4.2} = 0.2380... = $ **–0.24 D (to 2 s.f.)**

Q3 A lens correcting hypermetropia takes an object at 0.25 m and creates a virtual image at the patient's uncorrected near point. So $u = 0.25$ and $f = 0.27$.

$\frac{1}{f} = \frac{1}{u} + \frac{1}{v}$ so $\frac{1}{v} = \frac{1}{f} - \frac{1}{u} = \frac{1}{0.27} - \frac{1}{0.25} = -0.296...$
so $v = -3.375$.

So the uncorrected near point is **3.4 m (to 2 s.f.)**.

Remember, a negative value just means on the same side of the lens as the object, but the near point is always on the same side of the lens as the object, so you can ignore it here.

Page 272 — Fact Recall Questions

Q1 Myopia is short-sightedness, where people can't focus on distant objects. Their far point is closer than infinity.

Q2 a) Hypermetropia is long-sightedness, where people can't focus on near objects.
Their near point is further away than normal (25 cm or more).
b) Converging lens.

Q3 a) Someone who suffers from astigmatism will have different focal lengths for different planes. This may be caused by an irregularly shaped cornea or lens.
b) The prescription will give the power and axis angle of the cylindrical lens.

5. Physics of the Ear

Page 274 — Fact Recall Questions

Q1 The amount of sound energy passing per second per unit area (power per unit area).

Q2 Outer ear — pinna and auditory canal.
Middle ear — ossicles (malleus, incus and stapes) and Eustachian tube.
Inner ear — semicircular canals, cochlea and auditory nerve.

Q3 E.g. the ossicles act as a lever system. They pass on the vibrations of the eardrum to the oval window. They also amplify the force of the vibrations and reduce the energy reflected back from the inner ear.

Q4 The pressure variations at the oval window are about 20 times greater than those at the eardrum.

Q5 Sound waves are funnelled by the pinna into the auditory canal in the outer ear. The variations in air pressure due to the sound waves cause the tympanic membrane (eardrum) to vibrate. The tympanic membrane is connected to the malleus, which passes vibrations via the incus to the stapes (the malleus, incus and stapes are small bones in the middle ear, also known as ossicles). The force of the vibrations is increased by about 50% by the ossicles. The vibration of the stapes causes the oval window to vibrate, which causes pressure waves in the fluid of the cochlea. The pressure waves cause the basilar membrane to vibrate (which part vibrates the most depends on the frequency of the sound), which causes certain hair cells to vibrate enough to trigger an electrical impulse in the auditory nerve which is sent to the brain and interpreted as sound.

6. Intensity and Loudness

Page 277 — Application Questions

Q1 Humans experience hearing loss at high frequencies as a result of aging, so only younger people can hear the very highest frequencies in the range of human hearing. The annoying high frequency noise deters young people but cannot be heard by older people.

Q2   Intensity level $= 10\log\left(\frac{I}{I_0}\right)$ and $I_0 = 1 \times 10^{-12}$ W m^{-2}.

$40 = 10\log\left(\frac{I}{1 \times 10^{-12}}\right)$ so $4 = \log\left(\frac{I}{1 \times 10^{-12}}\right)$.

Raise each side to the power 10, to get rid of the log:

$10^4 = 10^{\log\left(\frac{I}{1 \times 10^{-12}}\right)}$ and so $10^4 = \left(\frac{I}{1 \times 10^{-12}}\right)$

so $I = 10^4 \times (1 \times 10^{-12}) = \mathbf{1 \times 10^{-8}}$ **W m^{-2}**

Page 277 — Fact Recall Questions

Q1   The intensity of the sound and its frequency.

Q2   The ear's response to sound is logarithmic and so it needs a logarithmic scale to reflect this.
The perceived loudness of a sound doesn't increase linearly with intensity, but logarithmically.

Q3   The decibel scale, dB

Q4   It is an adjusted decibel scale which takes into account the ear's response to different frequencies.

Q5   The threshold of hearing is the minimum intensity of sound that can be heard by a normal ear at 1000 Hz. Its value is 1.0×10^{-12} Wm^{-2}.

Q6   Set one signal generator to 1000 Hz at a particular intensity level. This is the control signal generator. Set a second signal generator to a different frequency and alter the intensity level until it appears to have the same loudness as the control signal generator. Record the intensity level of the second signal generator at this point. Repeat this for different frequencies of the second signal generator and plot the results on a graph. To plot another curve on the graph, change the intensity level of the control signal generator and repeat the experiment.

Q7   E.g. Old age and excessive exposure to noise.

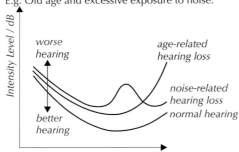

7. Physics of the Heart

Page 279 — Application Question

Q1

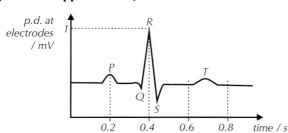

Point P shows the contraction of the atria. The QRS wave shows the contraction of the ventricles (and the relaxation of the atria). Point T shows the relaxation of the ventricles.

Page 279 — Fact Recall Questions

Q1   Electrodes are placed on the body at the points closest to the heart (i.e. the chest), and where the arteries are closest to the surface (i.e. the limbs). The variation in potential difference across the sites where the electrodes are placed is measured. The potential difference is plotted as a function of time.

Q2   Any two from: e.g. amplify the signal using a high impedance amplifier / remove hairs and dead skin cells (e.g. using sandpaper) / use a conductive gel / make sure the patient remains relaxed and still during the procedure / shield leads from interference from possible a.c. sources in the area.

8. Ultrasound

Page 283 — Application Question

Q1   a) $Z = \rho c \Rightarrow c = \dfrac{Z}{\rho} = \dfrac{1.5 \times 10^6}{1000} = \mathbf{1500 \; ms^{-1}}$

      b) $\dfrac{I_r}{I_i} = \left(\dfrac{Z_2 - Z_1}{Z_2 + Z_1}\right)^2$

           $= \left(\dfrac{(1.5 \times 10^6) - (8.0 \times 10^6)}{(1.5 \times 10^6) + (8.0 \times 10^6)}\right)^2$

           $= 0.4681...$

      So the percentage $= 0.4681... \times 100 = \mathbf{47\%}$ **(to 2 s.f.)**

Page 283 — Fact Recall Questions

Q1   Ultrasound is sound with a frequency higher than humans can hear (>20 000 Hz).

Q2   The acoustic impedance, Z, is defined as the density (ρ) multiplied by the speed of sound in the material (c). It affects how much of a wave will be reflected when it passes between materials.

Q3   The piezoelectric effect is when a material produces a potential difference when it's deformed, or vice versa.

Q4   Ultrasound waves are generated and detected by an ultrasound transducer. Inside the transducer, an alternating potential difference is applied to piezoelectric crystals, which vibrate to create a sound wave. When the wave is reflected and returns, it causes the crystals to vibrate, which in turn creates an alternating potential difference that is detected in an adjoining circuit.

Q5   Ultrasound imaging doesn't work if there are any air gaps involved because the difference in acoustic impedance between the air and the soft tissue would cause the majority of an ultrasound signal to be reflected back. Because of this, a coupling medium with similar acoustic impedance to the skin is needed between the transducer and the body.

Q6   a)   Short pulses of ultrasound are sent into the body while an electron beam sweeps across a cathode ray oscilloscope. Reflected pulses are detected and show up as vertical deflections on the oscilloscope screen. The time difference between reflected pulses allows a computer to calculate the depth of an object.

      b)   B-scans are like A-scans, but the electron beam sweeps down the screen rather than across, and the amplitude of reflected pulses is shown as the brightness of spots on the screen. A linear array of transducers allows a 2D image to be formed of inside the body.

      c)   E.g. A-scans can be used to measure the depth of an eyeball. B-scans can be used to form a 2D image of a fetus.

Q7   a)   Any three from: e.g. There are no known hazards or side effects / It's good for imaging soft tissues / It produces real-time images / Ultrasound devices are relatively cheap and portable / An ultrasound scan is quick / The patient can move during the scan.

b) Any three from: e.g. Ultrasound waves can't pass through bone (so can't be used to look at the brain or detect fractures) / Ultrasound waves can't pass through air spaces in the body because of the impedance mismatch (so can't be used to look behind the lungs) / The resolution is poor / It doesn't give any information about any solid masses found.

9. Endoscopy
Page 287 — Application Questions
Q1 a) $\sin \theta_c = \frac{n_2}{n_1} \Rightarrow \theta_c = \sin^{-1}\left(\frac{n_2}{n_1}\right) = \sin^{-1}\left(\frac{1.54}{1.61}\right)$
$= 73.0425...$
$= \textbf{73.0° (to 3 s.f.)}$
b) (i) It would not be totally internally reflected.
(ii) It would be totally internally reflected.
(iii) It would be totally internally reflected.
If the angle of incidence is greater than the critical angle, the light will be totally internally reflected.
Q2 $\sin \theta_c = \frac{n_2}{n_1} \Rightarrow n_2 = n_1 \times \sin \theta_c = 1.70 \times \sin 65°$
$= 1.5407... = \textbf{1.5 (to 2 s.f.)}$
Q3 a) E.g. non-coherent bundles could send down light into the digestive tract, like a flexible torch.
b) E.g. coherent bundles could transmit real-time images back from the digestive tract, like a video camera.

Page 287 — Fact Recall Questions
Q1
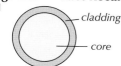
cladding

core
Q2 Total internal reflection
Q3 a) Coherent optical fibre bundles have fibres in the same relative position at both ends, whereas non-coherent optical fibres don't.
b) Coherent
Q4 An endoscope is a long, flexible tube that contains two bundles of optical fibres — a non-coherent bundle to carry light for illumination and a coherent bundle to carry images back.
Q5 E.g. to examine inside the body; for seeing what's happening during keyhole surgery.
Q6 E.g. the hole(s) required are much smaller, so there's less exposure to infection; recovery times are usually much quicker.

10. X-ray Production
Page 291 — Application Question
Q1 $E_{max} = e \times V = 1.60 \times 10^{-19} \times 85.0 \times 10^3 = \textbf{1.36} \times \textbf{10}^{-14}$ **J**
Assume here that the maximum photon energy is the same as the maximum kinetic energy of the electrons.

Page 291 — Fact Recall Questions
Q1 a)
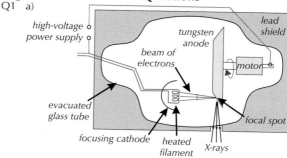
high-voltage power supply

tungsten anode

lead shield

beam of electrons

motor

evacuated glass tube

focal spot

focusing cathode   heated filament   X-rays
b) A lot of the electrons' energy is converted to heat, so rotating the anode prevents it from overheating.
c) E.g. reducing the size of the anode focal point (e.g. by decreasing the slope of the anode) / increasing the distance between the anode and the object / decreasing the distance between the object and screen.
Q2 When electrons smash into the (tungsten) anode, they decelerate and some of their kinetic energy is converted into electromagnetic energy, as X-ray photons. This results in a continuous spectrum of X-ray radiation.
X-rays are also produced when electrons are knocked out of tungsten atoms and outer shell electrons fall inwards to take their place. As they do this, they release energy in the form of X-ray photons with set energies that correspond to energy level transitions, producing the spikes seen in the graph.
Q3 a) Intensity increases with tube voltage, approximately in proportion to the square of the voltage. The maximum photon energy also increases proportionally to the tube voltage.
b) Intensity increases with tube current, approximately in proportion to the current. The maximum photon energy isn't affected by the current.

11. X-ray Imaging
Page 295 — Application Question
Q1 a) $\mu_m = \frac{\mu}{\rho} \Rightarrow \mu = \mu_m \rho = 0.133 \times 1900$
$= 252.7 \text{ m}^{-1} = \textbf{250 m}^{-1} \textbf{(to 2 s.f.)}$
b) $I = I_0 e^{-\mu x} = 30.0 \times e^{-(252.7 \times (0.690 \times 10^{-2}))}$
$= 5.246... = \textbf{5.2 Wm}^{-2} \textbf{(to 2 s.f.)}$

Page 295 — Fact Recall Questions
Q1 E.g. make sure the patient keeps still to reduce blurring of the image / put a lead collimator grid between the patient and the film to stop scattered radiation fogging the film / use an intensifying screen next to the film surface to reduce the exposure to X-rays needed to produce a clear image.
Q2 Intensifying screens contain crystals that fluoresce, so when they absorb X-rays they re-emit visible light.
They're placed close to photographic film, so that the visible light emitted exposes the film in the correct areas and a clear image is produced.
Q3 The half-value thickness is the thickness of material required to reduce the intensity of the X-rays travelling through the material to half its original value.
Q4 X-rays are attenuated (or absorbed) more by materials with a higher atomic number. Bone has a higher average atomic number than soft tissues and so it absorbs more X-rays, and clear images can be formed around bones. Most of the X-rays just pass straight through soft tissue, so it's difficult to produce a clear image.

Q5 Barium has a high atomic number and so it shows up clearly in X-ray images compared to the soft tissue in and around the digestive tract, which have a low atomic number. This helps improve the contrast between tissues on an X-ray image.
A patient can be given a barium meal to swallow so that the barium can be tracked by X-rays as it passes through their body.

Q6 a) A flat panel detector is made up of a scintillator layer, a layer of photodiode pixels and a layer of thin-film transistors. When X-rays hit the scintillator, light is produced. The photons are detected by the photodiode pixels, which generate a voltage. The thin-film transistors then read the electric signal, and these readings are used to produce a digital image.
   b) Any two from: e.g. They are more lightweight and compact / They have a higher resolution / The final image is less distorted / The digital read-out is easy to copy, store and share / They require a lower exposure for a clear image.

Q7 A computed tomography (CT) scan is a scan in which a narrow, monochromatic X-ray beam is rotated around the body and picked up by an array of detectors. These feed a signal to a computer, which works out how much attenuation has been caused by each part of the body and produces a very high quality image.

Q8 a) Any two from: e.g. X-rays used in CT scans can produce a high resolution and so images are clear / Scans are quick / CT scanners are relatively cheap.
   b) Any one from: e.g. CT scans use high doses of X-rays, which are ionising and can cause damage to cells / CT scans use X-rays, which are generally unsuitable for pregnant women.

12. Magnetic Resonance Imaging
Page 298 — Application Question
Q1 a) Yes — kidney is a soft tissue, and so an MR scan would produce a high quality image.
   b) Yes — the brain is a soft tissue and an MR scan can be used to create good contrast and resolution between different tissue types.
   c) No — people with pacemakers usually can't have MR scans because of the strong magnetic fields involved.
   d) No — MR scans give poor images of bones compared to using X-rays, so it would be better to use some form of X-ray scan.
   e) Yes — imaging a blood clot requires contrast between soft tissues (the clot itself and surrounding veins, muscles, etc.).

Page 298 — Fact Recall Questions
Q1 Superconducting magnet — used to create a uniform magnetic field.
Gradient field coils — used to superimpose smaller magnetic fields onto the main one, to create a gradient of magnetic fields throughout the body.
Radio frequency (RF) coils — used to transmit pulses of RF radiation that will excite protons of certain precession frequencies.
RF detector — detects the RF radiation emitted as protons de-excite.
Computer — uses the detected radiation data to build up an image of the body.
Q2 Contrast can be controlled by varying the time between pulses.
Q3 The correct answer is d).

Q4 a) Any four from: e.g. No known side effects / Uses non-ionising radiation so doesn't damage cells / An image can be formed of any slice of the body as well as multi-plane images, all from one scan / High resolution between different tissue types / Contrast between different tissues can be weighted / Some scanners can give real-time images.
   b) Any four from: e.g. Poor imaging of bones / Scans are noisy and can take a long time / Some patients suffer from claustrophobia as scanners are very narrow / Patients with pacemakers or metal implants might not be able to use the scanner due to strong magnetic fields / Scanners cost millions of pounds.

13. Medical Uses of Radiation
Page 301 — Application Question
Q1 The biological half-life is the time it takes for the body to metabolise (use up) half of a substance. The physical half-life is the time it takes for half of a radioactive substance to decay.
$T_B = 22 \times 24 \times 3600 = 1.9008 \times 10^6$ s
$T_P = 9.5 \times 24 \times 3600 = 8.208 \times 10^5$ s
$$\frac{1}{T_E} = \frac{1}{T_B} + \frac{1}{T_P}$$
$$= \frac{1}{1.9008 \times 10^6} + \frac{1}{8.208 \times 10^5}$$
$$= 1.74441... \times 10^{-6}$$
$$T_E = \frac{1}{1.74441... \times 10^{-6}}$$
$$= 5.7325... \times 10^5 \text{ s}$$
$$= \mathbf{5.7 \times 10^5 \text{ s (to 2 s.f.) (= 6.6 days)}}$$
Don't forget — once you've added the two fractions together you need to divide 1 by the answer to find T_E.
You could have done this question without converting the values to seconds and keeping everything as days. Be careful if you use non-SI units though — for this equation all the values must have the same units.

Page 301 — Fact Recall Questions
Q1 The medical tracer labels a compound that is naturally used by the part of the body that is being investigated.
Q2 Any from:
Technetium-99m (half-life = 6 hours, energy = 140 keV) — used as it has a half-life that is long enough to still be emitting radiation when it reaches the organ being investigated, but short enough that the patient isn't exposed to radiation for too long.
Iodine-131 (half-life = 8 days, energy = 360 keV) — useful for investigating the thyroid as iodine is naturally used by the thyroid.
Indium-111 (half-life = 2.8 days, energy = 170 or 250 keV) — used to label antibodies and blood cells in order to detect infections.
Q3 a) Molybdenum bonds strongly with aluminium oxide inside the generator and then decays, producing technetium-99m. The technetium-99m doesn't bond very strongly with aluminium oxide and so a saline solution can be used to wash out any technetium-99m.
   b) Technetium-99m is one of the most useful medical tracers, but its relatively short physical half-life makes it difficult to transport to hospitals. Having the Molybdenum-Technetium generator delivered to the hospital means that the technetium-99m can be made on site. Molybdenum has a longer half-life and so can be transported to the hospital.

Q4 Lead collimator — only lets gamma rays travelling parallel to the holes in the collimator pass through.
Lead shielding — prevents any external radiation interfering with the results.
Sodium iodide crystal (scintillator) — produces a flash of light when hit by a gamma ray.
Photomultiplier tubes — detect the flashes of light and turn them into electrical signals. This is done by releasing electrons at the photocathode by the photoelectric effect. The electrons are then multiplied to produce more electrons which amplifies the signal.
Electronic circuit — collects the electrical signals and sends the information to a computer which forms an image.

Q5 The patient is injected with glucose that is labelled with a radiotracer. Glucose is naturally used in the body and will be used more by cells with an increased metabolism. The radiotracer emits positrons that annihilate with electrons in the areas of metabolic activity, producing gamma photons which are detected outside the body. A computer is used to build up an image.

Q6 High-energy X-rays are fired at the tumour from outside the patient's body. Implants containing beta-emitters are placed next to or inside the tumour inside the patient's body. In both cases, the radiation destroys the tumour cells.

Q7 Healthy cells are damaged by X-ray/gamma radiation, which can lead to mutations and higher risks of future cancers. These risks are reduced by using focused beams and shielding. The radiation source can also be rotated around the patient to minimise the amount of radiation that each healthy cell is receiving. Another way to reduce the risks is to use beta radiation from a source which is placed next to or inside the tumour — the risks are lowered because the range of beta particles is quite short.

14. Comparing Imaging Types
Page 302 — Fact Recall Questions
Q1 Any two from: e.g. there are no known side effects from an MR scan, whereas a CT scan uses dangerous ionising radiation / MR scans give better resolution for soft tissues / radiation used in MR imaging is non-ionising, whereas a CT scan uses dangerous ionising radiation / multi-plane images can be made from the same scan, whereas a CT needs a different scan for each plane.

Q2 Any two from, e.g. the patient might be fitted with a pacemaker / the problem might be bone-related, and MR scans don't image bones very effectively / it might not be affordable or cost-effective / the patient may be claustrophobic.

Q3 a) E.g. ultrasound imaging has no known side effects; ultrasound can produce real-time moving images; ultrasound is non-ionising; the patient is allowed to move during the scan.
b) E.g. ultrasound is much quicker; ultrasound is much cheaper; ultrasound tranducers are portable.

Q4 Any two from: e.g. it gives a high quality image of bones / it is becoming more portable / it is cheaper than a PET scan / the scan is usually quicker / it is not as uncomfortable for the patient.

Exam-style Questions — Pages 304-308
1 **D** *(1 mark)*
2 **B** *(1 mark)*
3 **C** *(1 mark)*
A relatively short range means the beta radiation is less likely to damage the healthy cells around the tumour.

4 **A** *(1 mark)*
5.1 A — The cornea
B — The lens
C — The retina
D — Fovea (or yellow spot)
(2 marks available — 1 mark for every 2 correct answers.)
5.2 Rods and cones contain a chemical pigment that bleaches when light hits them *(1 mark)*. The bleaching stimulates the cell, which sends an electrical signal to the brain *(1 mark)*. Rods are more sensitive to light *(1 mark)*, and cones detect colour *(1 mark)*.
5.3 The eye can distinguish between two objects if there is at least one rod or cone between the images of the objects formed on the retina (as long as the rod or cone between the image doesn't share an optic nerve with any of the rods and cones detecting the images) *(1 mark)*. So a higher density of rods and/or cones will allow for better spatial resolution *(1 mark)*. The yellow spot (or fovea) has a higher density of cones than the rest of the retina has rods or cones, and so spatial resolution is best at the yellow spot *(1 mark)*.
5.4 In an A-scan, a short pulse of ultrasound is sent into the eyeball *(1 mark)*. Part of the pulse is reflected at the boundary at the back of the eye and is picked up on a CRO screen *(1 mark)*. The time it takes for this echo to come back can be used to calculate the depth of the eyeball *(1 mark)*.
6.1 The pinna *(1 mark)*.
6.2 The ossicles amplify the force of the vibrations arriving at the oval window from the eardrum *(1 mark)*. The area of the oval window is also smaller than the eardrum, which results in increased pressure differences at the oval window *(1 mark)*.
6.3 The threshold of hearing I_0 is the minimum intensity of sound that can be heard by normal ears *(1 mark)* at a frequency of 1000 Hz (1 kHz) *(1 mark)*.
6.4 Intensity level $= 10\log\left(\dfrac{I}{I_0}\right)$
$$\Rightarrow I = I_0 \times 10^{\frac{IL}{10}} = (1.0 \times 10^{-12}) \times 10^{\frac{81}{10}}$$
$= 1.2589... \times 10^{-4} = \mathbf{1.3 \times 10^{-4}\ Wm^{-2}}$ **(to 2 s.f.)**
(2 marks for correct answer, otherwise 1 mark for correct rearrangement to make I the subject.)
6.5 The ear responds differently to sound of different frequencies *(1 mark)*, and so the intensity level of a sound in dB might not reflect the perceived loudness. A dBA scale would show how loud (and therefore how much of a nuisance) the machine seems to the workers *(1 mark)*.

7.1
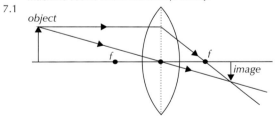
(2 marks in total — 1 mark for ray parallel to the axis passing through the principal focus and 1 mark for a second correct ray and labelled image.)
7.2 $m = \dfrac{h_i}{h_o} \Rightarrow h_o = \dfrac{h_i}{m} = \dfrac{(29.3 \times 10^{-2})}{1.61} = 0.18198...$
$= \mathbf{0.182\ m}$ or **18.2 cm (to 3 s.f.)**
(2 marks for correct answer, otherwise 1 mark for correct rearrangement to make u the subject.)
7.3 A corrective lens will produce a virtual image *(1 mark)* at the person's uncorrected near point, 4.9 m *(1 mark)*.
A converging lens will be used.

7.4 $\text{power} = \frac{1}{f}$, so power $= \frac{1}{u} + \frac{1}{v}$
$$= \frac{1}{0.25} + \frac{1}{-4.9}$$
$$= +3.7959...$$
$$= +3.8\,\text{D (to 2 s.f.)}$$
(2 marks for correct answer, otherwise 1 mark for correct working.)
Remember, v is negative for virtual images.

7.5 Astigmatism **(1 mark)**.

8.1 When the X-rays are produced a lot of heat is also produced at the focal spot — a small part of the anode **(1 mark)**. Rotating the anode means the same part of it is never producing X-rays for more than a few milliseconds, and so it doesn't overheat **(1 mark)**.

8.2 When electrons collide with the anode, they decelerate and some of their kinetic energy is lost. This is converted into X-ray photons with a range of energies that depends on the amount of energy lost by the electrons **(1 mark)**.

8.3 As well as losing kinetic energy, beam electrons can knock other electrons out of inner shells in the anode atoms **(1 mark)**. When this happens, outer shell electrons fill the gaps left by the inner electrons and release X-ray photons with fixed energies that coincide with differences between energy levels of the atom **(1 mark)**.

8.4 The half-value thickness of a material is the thickness of material needed to reduce the intensity (of a beam of X-rays) to half its original value **(1 mark)**.

8.5 $I = I_0 e^{-\mu x} \Rightarrow \frac{I}{I_0} = e^{-\mu x}$
$$\Rightarrow x = -\frac{1}{\mu}\ln\left(\frac{I}{I_0}\right) = -\frac{1}{9.64}\ln\left(\frac{1}{5}\right) = 0.1669...\,\text{m}$$
$$= 16.7\,\text{cm (to 3 s.f.)}$$
(3 marks for correct answer, otherwise 1 mark for correct substitution of $I/I_0 = 1/5$ and 1 mark for correct rearrangement to make x the subject.)

8.6 An FTP detector contains several layers. An X-ray first hits a scintillator material, which produces a flash of light **(1 mark)**. The light from the scintillator then hits the layer of photodiode pixels, which generates a voltage **(1 mark)**. A thin film transistor then reads this digital signal, and these readings are used to create an image of the object being X-rayed **(1 mark)**.

8.7 Any two from: e.g. Use intensifying screens / keep the patient still / put the screen and patient close to each other and the X-ray tube far away / put a lead collimator between the patient and the film **(1 mark for each)**.

9.1 The valves in the heart are found between the atria and the ventricles, as well as in the arteries leaving the ventricles of the heart **(1 mark)**. The valves prevent blood going backwards into the ventricles or atria when the heart contracts **(1 mark)**.

9.2
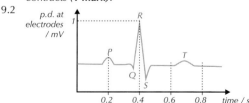
(1 mark for each axis labelled correctly)

9.3 *P wave* — Atria are contracting.
QRS wave — Ventricles are contracting.
T wave — Ventricles are relaxing.
(3 marks available — 1 for each part)

9.4 The sino-atrial node produces electrical signals **(1 mark)**

which cause the heart muscle to contract around 70 times per minute **(1 mark)**.

9.5 Any two from: e.g. placing electrodes on the body near arteries / removing hairs and dead skin cells (e.g. using sandpaper) to get a good contact / using conductive gel to get a good contact / getting the patient to remain still / using a high impedance amplifier.
(2 marks available — 1 for each point.)

10.1 The critical angle of total internal reflection is the smallest angle of incidence above which a wave travelling from a medium with a higher optical density to one with a lower optical density will be totally internally reflected (none of the wave will leave the first medium). **(1 mark.)**

10.2 $\sin\theta_c = \frac{n_2}{n_1} \Rightarrow \theta_c = \sin^{-1}\frac{n_2}{n_1} = \sin^{-1}\frac{1.15}{1.25} = $ **66.9° (to 3 s.f.)**
(2 marks for correct answer, otherwise 1 mark for correct rearrangement to make θ_c the subject.)

10.3 A coherent optical fibre bundle is a bundle of optical fibres where the relative positions of fibres are the same at each end **(1 mark)**.

10.4 Non-coherent bundles are used to send visible light towards the part of the body being investigated/operated on **(1 mark)**. Coherent bundles are used to send images of the part of the body back towards the eyepiece **(1 mark)**.

10.5 E.g. Keyhole surgery requires a much smaller cut than traditional surgery **(1 mark)**, and so exposure to potential infection is reduced **(1 mark)**.
Damage done to the body is reduced and so recovery times are shorter **(1 mark)**.
Because recovery time is shorter, the patient takes up hospital resources for less time, which reduces the overall cost of the procedure **(1 mark)**.

11.1 How to mark your answer (pick the level description that best matches your answer):
5-6 marks:
The answer fully describes how an MR scanner works, including a detailed description of a proton's spin, the use of gradient coils, and the use of radio waves of different frequencies to excite protons with different precession frequencies.
3-4 marks:
The answer describes how an MR scanner works but is lacking important information about how the gradient coils or RF waves are used to excite protons with different precession frequencies.
1-2 marks:
The answer briefly outlines the basic workings of an MR scanner but with very little explanation relating to the excitation of protons and how this is achieved with gradient coils and RF waves.
0 marks:
No relevant information is given.
Here are some points your answer may include:
- Protons have a property called spin.
- The MR scanner creates a uniform magnetic field with a superconducting magnet.
- When protons are placed in a uniform magnetic field the spin axes all align with the magnetic field (normally they are randomly aligned).
- The protons precess about the magnetic field lines with an angular frequency, called the precession frequency.
- Radio frequency coils transmit pulses of radio frequency waves of different frequencies.
- Protons absorb RF waves with the same frequency as their precession frequency.
- Protons are excited by absorbing radio frequency photons, causing their spin state to change.
- Protons de-excite and re-emit electromagnetic energy with

- Protons de-excite and re-emit electromagnetic energy with the same frequency as their precession frequency.
- Re-emitted waves are detected and the computer can tell which part of the body the waves have come from due to the gradient of the magnetic field.
- The computer builds up an image by measuring various quantities of the MR signal.

11.2 Any three from: e.g. no known side effects / no ionising radiation used / images for many different slices of the body can be created from one scan / high resolution of final image / contrast between different tissues can be increased / real time images are possible.
(3 marks — 1 for each advantage)

11.3 A narrow monochromatic beam of X-rays is produced *(1 mark)* and rotated around the body *(1 mark)*. Detectors around the body pick up the X-rays *(1 mark)*. The detectors send the signals to a computer, which builds up an image of a two-dimensional slice through the body *(1 mark)*.

11.4 An MR scanner uses a very strong magnet. Pacemakers contain metal, so this would be very dangerous for the patient *(1 mark)*.

11.5 CT scan or PET scan *(2 marks — 1 mark for each type of scan)*

11.6 E.g. A CT scan has a lower resolution than a PET scan *(1 mark)*, but it is more convenient as the machines used are more portable *(1 mark)*. The CT scan is also much more comfortable for the patient, as they usually only have to lie still for up to 10 minutes *(1 mark)*, whereas for a PET scan, they usually have to lie still for up to 30 minutes *(1 mark)* in a narrow tube, as well as having a tracer injection *(1 mark)*.

12.1 A gamma-emitting radioisotope that can be traced as it moves through the body *(1 mark)*.

12.2 6 hours *(1 mark)*.

12.3 The technetium-99m is combined with a compound that is naturally used by the liver *(1 mark)*.

12.4 The time it takes for the body to metabolise (use up or excrete) half of the substance *(1 mark)*.

12.5 $T_B = 1 \times 24 \times 3600 = 86\,400$ s
$T_P = 6 \times 3600 = 21\,600$ s
$\frac{1}{T_E} = \frac{1}{T_B} + \frac{1}{T_P} = \frac{1}{86\,400} + \frac{1}{21600} = 5.7870... \times 10^{-5}$
$T_E = \frac{1}{5.7870... \times 10^{-5}} = 17\,280$
$= 17\,000$ s (to 2 s.f.) (or 4.8 hours)
(2 marks for correct answer — otherwise one mark for using the correct formula)

12.6 The half-life of technetium-99m is too short for it to be transported to hospitals *(1 mark)*. Molybdenum-Technetium generators are used to produce technetium-99m and can be transported as molybdenum has a much longer half-life (of around 66 hours) *(1 mark)*.

12.7 Molybdenum is combined with aluminium oxide *(1 mark)*. The molybdenum then decays and produces technetium-99m *(1 mark)*. The technetium-99m doesn't bond as strongly with the aluminium oxide as the molybdenum does, so it can be washed out with a saline solution *(1 mark)*.

12.8 A gamma camera is used *(1 mark)*. A lead collimator only allows through gamma rays that are travelling parallel to the holes *(1 mark)*. The gamma rays then hit a sodium iodide crystal which scintillates when hit by a gamma ray *(1 mark)*. Photomultiplier tubes detect the flashes of light produced by the crystal and turn them into pulses of electricity by the photoelectric effect *(1 mark)*. An electronic circuit then collects these electrical signals and sends the results to a computer, which forms an image *(1 mark)*.

Section 7 — Option C: Engineering Physics

1. Inertia and Kinetic energy
Page 311 — Application Question
Q1 a) $I = \frac{1}{2}mr^2$, so $r = \sqrt{\frac{2I}{m}} = \sqrt{\frac{2 \times 790}{470}} = 1.8334...$
$= \textbf{1.8 m (to 2 s.f)}$
b) $I_{child} = mr^2 = 48 \times 0.95^2 = 43.32$ kg m²
$I_{total} = I_{roundabout} + I_{child} = 790 + 43.32 = 833.32$
$= \textbf{830 kgm}^2 \textbf{ (to 2 s.f.)}$
c) $E_k = \frac{1}{2}I\omega^2$, so $\omega = \sqrt{\frac{2E_k}{I}} = \sqrt{\frac{2 \times 1350}{833.32}} = 1.8000...$
$= \textbf{1.8 rad s}^{-1} \textbf{ (to 2 s.f.)}$

Page 311 — Fact Recall Questions
Q1 Moment of inertia is a measure of how difficult it is to rotate an object, or to change its rotational speed.
Q2 $I = \Sigma mr^2$, where Σmr^2 is the sum of the moments of inertia of all the point masses making up the extended object, m is the mass of each point mass, and r is the distance of each point mass from the axis of rotation.
Q3 The magnitude of the object's mass, and its distribution about the axis of rotation.
Q4 Rotational kinetic energy: $E_k = \frac{1}{2}I\omega^2$. Linear kinetic energy: $E_k = \frac{1}{2}mv^2$. The equations are the same, but with m swapped for I and v swapped for ω. This is because moment of inertia, I, is the rotational equivalent of linear mass, m, and rotational speed, ω, is the rotational equivalent of linear speed, v.

2. Rotational Motion
Page 316 — Application Questions
Q1 $\omega = \frac{\Delta\theta}{\Delta t}$. The hour hand makes 1 revolution every 12 hours,
so $\omega = \frac{2\pi}{12 \times 60 \times 60} = 1.454... \times 10^{-4}$
$= \textbf{1.45} \times \textbf{10}^{-4} \textbf{ rad s}^{-1} \textbf{ (to 3 s.f.)}$.

Q2 $\alpha = \frac{\Delta\omega}{\Delta t} = \frac{13.7}{41.2} = 0.33252... = \textbf{0.333 rad s}^{-2} \textbf{ (to 3 s.f.)}$

Q3 $\omega_2^2 = \omega_1^2 + 2\alpha\theta$, so $\omega_1^2 = \omega_2^2 - 2\alpha\theta$
$= 0^2 - 2 \times -630 \times 450 = 567\,000$
$\omega_1 = \sqrt{567\,000} = 752.99...$ rad s⁻¹
$= 752.99... \div (2\pi) = 119.84...$ revolutions s⁻¹
$= 119.84... \times 60 = 7190.5... = \textbf{7200 rpm (to 2 s.f.)}$
Don't forget to convert your answer into revolutions per minute — remember that there are 2π radians in every full revolution.

Q4 a) $\omega_{Earth} = \frac{\Delta\theta}{\Delta t} = \frac{2\pi}{24 \times 60 \times 60} = 7.272... \times 10^{-5}$ rad s⁻¹, so
$t = \frac{\Delta\theta}{\omega_{Earth}} = \frac{7.0 \times 10^{-3}}{7.272... \times 10^{-5}} = 96.25...$ s.
$\alpha = \frac{\Delta\omega}{\Delta t} = \frac{(0.030 - 0.014)}{96.25...} = 1.662... \times 10^{-4}$
$= \textbf{1.7} \times \textbf{10}^{-4} \textbf{ rad s}^{-2} \textbf{ (to 2 s.f.)}$
b)

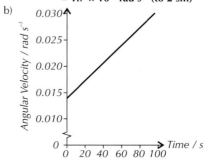

c) $\theta = \dfrac{(\omega_1 + \omega_2)}{2}t = \dfrac{0.014 + 0.030}{2} \times 96.25... = 2.117...$

   $= \textbf{2.1 rad (to 2 s.f.)}$

You could also answer this using $\theta = \omega_1 t + \frac{1}{2}\alpha t^2$ or by rearranging $\omega_2{}^2 = \omega_1{}^2 + 2\alpha\theta$.

Page 316 — Fact Recall Questions

Q1 Angular velocity, ω, is a vector quantity describing the angle a point rotates through per second.

Q2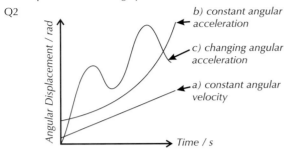

b) constant angular acceleration

c) changing angular acceleration

a) constant angular velocity

Q3 $\omega_2 = \omega_1 + \alpha t$, where ω_1 is initial angular velocity, ω_2 is final angular velocity, α is angular acceleration and t is time. Corresponds to $v = u + at$.

$\omega_2{}^2 = \omega_1{}^2 + 2\alpha\theta$, where symbols are as above and θ is angular displacement. Corresponds to $v^2 = u^2 + 2as$.

$\theta = \omega_1 t + \frac{1}{2}\alpha t^2$, where symbols are as above. Corresponds to $s = ut + \frac{1}{2}at^2$.

$\theta = \left(\dfrac{\omega_1 + \omega_2}{2}\right)t$, where symbols are as above.

Corresponds to $s = \left(\dfrac{u + v}{2}\right)t$.

3. Torque, Work and Power
Page 320 — Application Questions

Q1 $T = I\alpha$, $\alpha = \dfrac{\Delta\omega}{\Delta t} = \dfrac{510}{2.5} = 204$ rad s^{-2}

   $T = (5.4 \times 10^{-4}) \times 204 = 0.11016 = \textbf{0.11 Nm (to 2 s.f.)}$

Q2 a) $W = T\theta = 2.6 \times 32 = 83.2 = \textbf{83 J (to 2 s.f.)}$

   b) $T_{frictional} = T_{applied} - T_{net}$
      $T_{applied} = Fr = 29 \times 0.10 = 2.9$ Nm
      $T_{frictional} = 2.9 - 2.6 = \textbf{0.30 Nm (to 2 s.f.)}$

   c) $P = T\omega$
      $\omega = \dfrac{\Delta\theta}{\Delta t} = \dfrac{32}{0.80} = 40$ rad s^{-1}
      $P = 2.9 \times 40 = 116 = \textbf{120 W (to 2 s.f.)}$

   You need to use the applied torque to work out the power supplied to the cog — you want to know the total power required to both turn the cog and overcome the frictional torque.

Q3 $E_p = E_K + W$
    $E_p = mgh$
    $E_k = \frac{1}{2}mv^2 + E_{K rotational}$
    $W = T_{frictional}\theta$

    So $mgh = \frac{1}{2}mv^2 + E_{K rotational} + T_{frictional}\theta$

    $\Rightarrow \theta = \dfrac{mgh - \frac{1}{2}mv^2 - E_{K rotational}}{T_{frictional}}$

    $= \dfrac{950 \times 9.81 \times 12.1 - \frac{1}{2} \times 950 \times 9.31^2 - 61000}{425}$

    $= 24.928... = \textbf{25 rad (to 2 s.f.)}$

Page 320 — Fact Recall Questions

Q1 a) $T = Fr$, where T is torque in Nm, F is the applied force in N, and r is the perpendicular distance from the axis of rotation in m.

   b) $T = I\alpha$, where T is torque in Nm, I is moment of inertia in kg m^2, and α is angular acceleration in rad s^{-2}.

Q2 Work done = torque × angular displacement, $W = T\theta$.

   Power = $\dfrac{\text{work}}{\text{time}}$, so $P = \dfrac{\Delta(T\theta)}{\Delta t}$. But $\dfrac{\Delta\theta}{\Delta t} = \omega$, so $P = T\omega$.

4. Flywheels
Page 324 — Application Questions

Q1 The engineer should consider e.g. the mass of the replacement flywheel (because more mass means more energy is stored) / whether the wheel has spokes (because whether it is spoked or not affects the moment of inertia and hence the energy storage capacity) / whether it is thinner nearer the centre (because the distribution of mass about the axis of rotation also affects the moment of inertia) / the space available for the wheel (because the flywheel has to be an appropriate size for the machine) / the material used in the flywheel (because the flywheel should be strong enough to spin at the same rate as the original flywheel).

Q2 Flywheels can be used in cars to improve efficiency through regenerative braking. When the brakes are applied, a flywheel is engaged. The flywheel then charges up with some of the energy which would otherwise be lost during braking. When the car is ready to accelerate, the flywheel uses its energy to turn the car's wheels faster, before being disengaged until it's needed again.

Page 324 — Fact Recall Questions

Q1 A flywheel is a heavy wheel that has a high moment of inertia in order to resist changes to its rotational motion. Flywheels can be used to store rotational kinetic energy, which can be used to smooth angular velocity and improve efficiency in vehicles and machinery.

Q2 If the load torque is too high, the flywheel decelerates, delivering an energy top-up to the system. When the load torque is lower than the engine torque, the flywheel accelerates and stores the spare energy until it is needed.

Q3 E.g.
   1. Traditional potter's wheels — A traditional potter's wheel is powered by a foot pedal, making it hard to apply a constant force to it. A flywheel is used to keep the speed of the wheel constant in order to make ceramic pots.
   2. Power grids — When lots of electricity is used in an area, the electricity grid sometimes cannot meet that demand. Flywheels can be used to store surplus power in times of low demand and then provide extra energy while backup power stations are started up in times of high demand.
   3. Wind turbines — Flywheels can be used to store excess power on windy days or during off-peak times, and to give power on days without wind.

5. Angular Momentum
Page 327 — Application Questions

Q1 a) $I_{record} = \frac{1}{2}mr^2 = 0.5 \times 0.180 \times \left(\frac{0.304}{2}\right)^2 = 2.07936 \times 10^{-3}$

$\omega_{turntable} = \frac{33\frac{1}{3}}{60} \times 2\pi = 3.4906...$

$I_{initial}\omega_{initial} = I_{final}\omega_{final}$, so
$I_{turntable}\omega_{turntable} + I_{record}\omega_{record} = I_{combined}\omega_{combined}$

$\Rightarrow \omega_{combined} = \frac{I_{turntable}\omega_{turntable} + I_{record}\omega_{record}}{I_{combined}}$

$= \frac{5.12 \times 10^{-3} \times 3.49... + 2.07... \times 10^{-3} \times 0}{5.12 \times 10^{-3} + 2.07... \times 10^{-3}}$

$= \frac{0.0178...}{7.199... \times 10^{-3}} = 2.4824...$

$= \textbf{2.48 rad s}^{-1}$ **(to 3 s.f)**

b) Angular impulse $= \Delta(I\omega) = I_{final} \times \omega_{final} - I_{initial} \times \omega_{initial}$
$\Delta(I\omega) =$
$7.199... \times 10^{-3} \times 3.49... - 7.199... \times 10^{-3} \times 2.48...$
$= 7.258... \times 10^{-3} = \textbf{7.26} \times \textbf{10}^{-3} \textbf{ Nms (to 3 s.f.)}$

Q2 $\Delta(I\omega) = T\Delta t$, so $T = \frac{\Delta(I\omega)}{\Delta t} = \frac{0.75 \times (1.4 - 1.0)}{25} = \textbf{0.012 Nm}$

Q3 The diver should tuck herself into a ball to minimise her radius. Angular momentum is equal to $I\omega$ and must be conserved. I is proportional to r^2, so tucking in her arms to reduce her radius of rotation will decrease I. Because angular momentum is conserved, if I decreases then ω will increase, and a larger ω means that she can complete more somersaults before she hits the water.

Page 327 — Fact Recall Questions

Q1 Angular momentum = moment of inertia × angular velocity
$= I\omega$
Q2 Angular impulse is the change in angular momentum, $\Delta(I\omega)$.
Q3 Angular impulse = torque × time torque is applied
$= T\Delta t$
Q4 E.g. An ice skater doing a spin. At the start of the spin, the skater's arms are out away from their body. He then pulls his arms closer towards him, and begins to spin faster. This is due to the conservation of angular momentum — as he pulls in his arms, he decreases his moment of inertia, so his angular velocity must increase in order to conserve angular momentum.

6. The First Law of Thermodynamics
Page 330 — Application Questions

Q1 $Q = \Delta U + W = 0.300 + -1.98 = -1.68$ J
So **1.68 J** of heat energy is lost.
Assumption: The system is closed, so no air escapes.
The balloon is being compressed, so the value of W is negative.
Q2 $pV = nRT$

$p = \frac{nRT}{V} = \frac{22 \times 8.31 \times 308}{0.82^3} = 1.02125... \times 10^5$
$= \textbf{1.0} \times \textbf{10}^5 \textbf{ Pa (to 2 s.f.) (= 100 kPa)}$

Page 330 — Fact Recall Questions

Q1 $Q = \Delta U + W$
Q is the energy transferred to the system, ΔU is the increase in internal energy and W is the work done by the system.
Make sure you say that W is the work done <u>by</u> the system and not on the system.
Q2 a) Work is done on the system.
b) Both Q and W are negative.
If internal energy doesn't change while the gas is being compressed, then the work done in compressing the gas must be transferred to the surroundings as heat energy.

7. Non-flow Processes
Page 334 — Application Questions

Q1 $p_1V_1 = p_2V_2$

$p_2 = \frac{p_1V_1}{V_2} = \frac{112 \times 10^3 \times 6.8}{4.3} = 1.77116... \times 10^5$
$= \textbf{1.8} \times \textbf{10}^5 \textbf{ Pa (to 2 s.f.) (= 180 kPa)}$

Q2 $p_1V_1^\gamma = p_2V_2^\gamma$

$p_1 = \frac{p_2V_2^\gamma}{V_1^\gamma} = \frac{43 \times 10^3 \times 0.80^{1.28}}{0.60^{1.28}} = 6.2142... \times 10^4$
$= \textbf{6.2} \times \textbf{10}^4 \textbf{ Pa (to 2 s.f.) (= 62 kPa)}$

Q3 a) $W = p\Delta V$

$\Delta V = \frac{W}{p} = \frac{275}{88 \times 10^3} = 0.003125$ m^3
$\Delta V = V_{final} - V_{initial}$
$V_{initial} = V_{final} - \Delta V = 0.050 - 0.003125$
$= 0.046875 = \textbf{0.047 m}^3 \textbf{ (to 2 s.f.)}$

b) $\frac{V_1}{T_1} = \frac{V_2}{T_2}$

$T_1 = \frac{V_1T_2}{V_2} = \frac{0.046875 \times 281}{0.050} = 263.4375$
$= \textbf{260 K (to 2 s.f.)}$

Q4 $\frac{p_1}{T_1} = \frac{p_2}{T_2}$

$T_2 = \frac{p_2T_1}{p_1} = \frac{124 \times 10^3 \times 115}{24 \times 10^3} = 594.166... = \textbf{590 K (to 2 s.f.)}$

Page 334 — Fact Recall Questions

Q1 A change in which the temperature of the system remains constant.
Q2 A change in which no heat is gained or lost by the system.
Q3 $W = p\Delta V$
Q4 $W = 0$

8. *p-V* Diagrams
Page 339 — Application Questions

Q1 Height of each square represents 0.20×10^5 Pa.
Width of each square represents 0.10 m^3.
For each square, $W = p\Delta V = 0.20 \times 10^5 \times 0.10$
$= 2000$ J
Total number of squares ≈ 8
So, work done $\approx 8 \times 2000 = \textbf{16 000 J (= 16 kJ)}$
Q2 a) Height of each square represents 0.50×10^5 Pa.
Width of each square represents 0.10 m^3.
For each square, $W = p\Delta V = 0.50 \times 10^5 \times 0.10 = 5000$ J
Total number of squares under "top" curve ≈ 20
So, work done by the system $\approx 20 \times 5000$
$= \textbf{100 000 J (= 100 kJ)}$
b) Total number of squares in loop ≈ 12
So, work done by the system $\approx 12 \times 5000$
$= \textbf{60 000 J (= 60 kJ)}$
Q3 E.g.

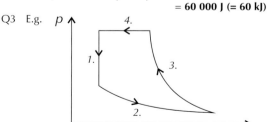

Page 339 — Fact Recall Questions

Q1 A *p-V* curve for an isothermal process.
Q2 The isotherm would move further away from the origin.
Q3 A gas does more work if it expands isothermally.

Q4 The *p-V* line is a vertical straight line, so the area under the line (and therefore the work done) is zero.

9. Four-stroke Engines and Indicator Diagrams
Page 343 — Fact Recall Questions

Q1 <u>Induction</u> — the piston moves downwards, sucking in a fuel-air mixture through the open inlet valve. The volume of the system increases but the pressure remains constant.
<u>Compression</u> — with the inlet valve closed, the piston compresses the air-fuel mixture. The spark plug then creates a spark which ignites the air-fuel mixture. At this point, the temperature and pressure increase and the volume remains almost constant.
<u>Expansion</u> — the air-fuel mixture is now at a higher temperature and expands, doing work on the piston to move it downwards. As the piston reaches the bottom of its stroke, the exhaust valve opens.
<u>Exhaust</u> — the piston moves upwards, pushing the burnt gas through the open exhaust valve. The volume of the system decreases and the pressure remains constant.

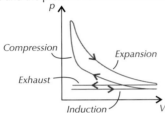

Q2 <u>Induction</u> — the piston moves downwards, sucking in air through the open inlet valve. The volume of the system increases but the pressure remains constant.
<u>Compression</u> — with the inlet valve closed, the piston compresses the air so that it reaches a high temperature. At the end of the stroke, diesel is sprayed into the cylinder through a fuel injector. The temperature of the air is hot enough that it ignites the diesel. At this point, the temperature and pressure increase and the volume remains almost constant.
<u>Expansion</u> — the ignited air-fuel mixture is at a high enough temperature to expand, doing work on the piston to move it downwards. Initially the pressure stays constant as the gas expands. As the piston reaches the bottom of its stroke, the exhaust valve opens.
<u>Exhaust</u> — the piston moves upwards, pushing the burnt gas through the open exhaust valve. The volume of the system decreases and the pressure remains constant.

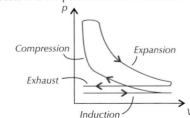

Q3 A — Gas compresses adiabatically.
B — Temperature and pressure increase at a constant volume.
C — Gas expands adiabatically.
D — Temperature and pressure decrease at a constant volume.

Q4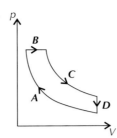

A — Gas compresses adiabatically.
B — Temperature and volume increase at a constant pressure.
C — Gas expands adiabatically.
D — Temperature and pressure decrease at a constant volume.

Q5 Changes within the cylinder are not instantaneous. The valves take time to open and close.

Q6 It is assumed that the same air cycles around the system continuously.

Q7 Any two from: e.g. the theoretical engine assumes the same air is cycled continuously, but a real engine has negative work between the exhaust and induction line / the theoretical engine assumes there is an external heat source, but real engines have an internal heat source (the air-fuel mixture), and fuel is never completely burned in the cylinder, so you can't get the maximum energy out of it / energy is lost through friction of moving parts in the engine.

10. Engine Power and Efficiency
Page 347 — Application Questions

Q1 a) Indicated power = (area of *p-V* loop) × (no. of cycles per second) × (no. of cylinders)
Number of squares in the *p-V* loop ≈ 103
Area of one square = $0.10 \times 10^{-3} \times 0.2 \times 10^{6}$ = 20 J
So area of *p-V* loop ≈ 103 × 20 = 2060 J
Indicated power = 2060 × 24 × 1 = 49 440
= **49 000 W (to 2 s.f.) (= 49 kW)**

b) $\omega = 2\pi f = 2 \times \pi \times 24 = 48\pi$
$P = T\omega = 260 \times 48\pi = 39\,207.0...$
= **39 000 W (to 2 s.f.) (= 39 kW)**

c) (i) Friction power = indicated power – brake power
= 49 440 – 39 207.0... = 10 232.9...
= **10 000 W (to 2 s.f.) (= 10 kW)**

(ii) Mechanical efficiency = $\dfrac{\text{brake power}}{\text{indicated power}}$
$= \dfrac{39\,207.0...}{49\,440} = 0.79302... =$ **0.79 (to 2 s.f.) (= 79%)**

Q2 a) Input power = calorific value × fuel flow rate
= $43.2 \times 10^{6} \times 0.018 = 777\,600$ W
= **780 000 W (to 2 s.f.) (= 780 kW)**

b) Doubling the time taken to complete a cycle halves the amount of fuel that is used per second. So the input power would halve.

Page 347 — Fact Recall Questions

Q1 Indicated power = (area of *p-V* loop) × (no. of cycles per second) × (no. of cylinders)

Q2 A crankshaft converts the up and down motion of a piston into a rotational motion.

Q3 $P = T\omega$ *P* is the output (or brake) power in W
T is the torque in Nm
ω is the angular velocity of the crankshaft in rads⁻¹

Q4 Friction power = indicated power – brake power
Input power = calorific value × fuel flow rate

Q5 Mechanical efficiency = $\dfrac{\text{brake power}}{\text{indicated power}}$

Thermal efficiency = $\dfrac{\text{indicated power}}{\text{input power}}$

Overall efficiency = $\dfrac{\text{brake power}}{\text{input power}}$

11. The Second Law of Thermodynamics

Page 350 — Application Question

Q1 a) maximum theoretical efficiency $= \dfrac{T_H - T_C}{T_H} = \dfrac{460 - 305}{460}$
$= 0.33695...$
$= \mathbf{0.34 \text{ (to 2 s.f.) (= 34\%)}}$

b) Rearrange the equation for efficiency for Q_C:
efficiency $= \dfrac{Q_H - Q_C}{Q_H}$
$Q_H \times$ efficiency $= Q_H - Q_C$
$Q_C = Q_H - (Q_H \times$ efficiency$)$
$= 790 - (790 \times 0.26) = 584.6$
$= \mathbf{580 \text{ J (to 2 s.f.)}}$

c) $W = Q_H - Q_C = 790 - 584.6 = 205.4 = \mathbf{210 \text{ J (to 2 s.f.)}}$
You could also calculate the work done from $W = Q_H \times$ efficiency

Page 350 — Fact Recall Questions

Q1 It's not possible for an engine to convert all of the heat it is supplied into useful energy. Some of the heat supplied will end up heating part of the engine. Once the engine reaches the source temperature, heat energy will stop flowing.

Q2 The second law of thermodynamics says that a heat engine must operate between a heat source and a heat sink. Not all of the energy from the heat sink can be converted into work done by the engine.

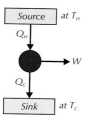

Q3 Efficiency $= \dfrac{W}{Q_H} = \dfrac{Q_H - Q_C}{Q_H}$
Maximum theoretical efficiency $= \dfrac{T_H - T_C}{T_H}$

Q4 Any two from: e.g. friction between components causes energy to be lost / fuel never burns completely, so the value of Q_H is lower / some of the energy is used in moving the components of the engine.

Q5 A combined heat and power plant is a power plant that finds an alternative use for heat energy that would normally be wasted through a heat sink. They are designed to maximise the efficiency of the fuel use.

12. Reversed Heat Engines

Page 353 — Application Questions

Q1 a) $COP_{ref} = \dfrac{T_C}{T_H - T_C} = \dfrac{(4.0 + 273)}{(22.0 + 273) - (4.0 + 273)} = 15.388...$
$= \mathbf{15 \text{ (to 2 s.f.)}}$

b) $COP_{ref} = \dfrac{Q_C}{Q_H - Q_C} = \dfrac{4.9 \times 10^5}{(6.2 \times 10^5) - (4.9 \times 10^5)} = 3.7692...$
$= \mathbf{3.8 \text{ (to 2 s.f.)}}$

Q2 a) Rearrange $COP_{hp} = \dfrac{Q_H}{W}$ to get:
$Q_H = COP_{hp} \times W = 1.40 \times 6250$
$= \mathbf{8750 \text{ J}}$

b) Rearrange $COP_{hp} = \dfrac{T_H}{T_H - T_C}$ to get:
$T_H = \dfrac{COP_{hp} \times T_C}{COP_{hp} - 1} = \dfrac{1.40 \times 280}{1.40 - 1} = \mathbf{980 \text{ K}}$

Page 353 — Fact Recall Questions

Q1 A reversed heat engine transfers heat from a cold space to a hot space. This requires an input of work.

Q2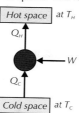

Q3 A refrigerator aims to extract as much heat energy as possible (per joule of work done) from the cold space. A heat pump aims to pump as much heat energy as possible (per joule of work done) into the hot space.

Exam-style Questions — Pages 355-359

1.1 A gas that expands adiabatically does so with no heat being transferred in or out of the system. *(1 mark)*

1.2 Adiabatic expansion, so $p V^\gamma =$ constant.
$p_1 V_1^\gamma = p_2 V_2^\gamma$
$p_2 = \dfrac{p_1 V_1^\gamma}{V_2^\gamma} = \dfrac{1820 \times 10^3 \times (125 \times 10^{-9})^{1.4}}{(986 \times 10^{-9})^{1.4}} = 1.0099... \times 10^5$ Pa
$= \mathbf{1.0 \times 10^5 \text{ Pa (to 2 s.f.)}}$
(2 marks for correct answer, otherwise 1 mark for using the equation $PV^\gamma =$ constant)

1.3 $\dfrac{p_1 V_1}{T_1} = \dfrac{p_2 V_2}{T_2}$
$T_2 = \dfrac{p_2 V_2 T_1}{p_1 V_1} = \dfrac{1.0099... \times 10^5 \times 986 \times 10^{-9} \times 380}{1820 \times 10^3 \times 125 \times 10^{-9}} = 166.34...$
$= \mathbf{170 \text{ K (to 2 s.f.)}}$
(2 marks for correct answer, otherwise 1 mark for forming a correct equation for T_2)

1.4 $Q = \Delta U + W$
Adiabatic expansion, so $Q = 0$ and therefore $\Delta U = -W$.
W is given by the area under the graph between 125 mm³ and 986 mm³.
Value of one small square (area) $= 50 \times 10^{-9} \times 1.0 \times 10^5$
$= 0.005$ J
Number of small squares under curve between 125 mm³ and 986 mm³ $\approx 64 \ (\pm 3)$
Area under curve $= W = 0.0050 \times 64 = 0.32$ J
So $\Delta U = -W = \mathbf{-0.32 \text{ J}}$
(3 marks for the correct answer, otherwise 1 mark for $\Delta U = -W$ and 1 mark for calculating the area under the curve to find W)

1.5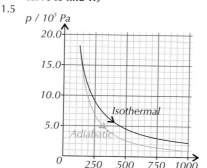

(1 mark for a curved line, 1 mark for curve being higher than adiabatic curve, 1 mark for arrow pointing in the right direction, 1 mark for curve starting at a pressure of 1820 kPa and a volume of 125 mm³)

2.1 $I = \Sigma mr^2$
   $= 1980 + 4 \times (4.20 \times 3.15^2) + 3 \times [(4.20 + 12.7) \times 3.15^2]$
   $= 2649.7...$
   $= \textbf{2650 kg m}^2$ **(to 3 s.f.)**
   (2 marks for correct answer, 1 mark for correct working if answer incorrect)

2.2 $E_k = \frac{1}{2} I \omega^2$
   $\omega = \frac{4.95 \times 2\pi}{60} = 0.518...$ rad s^{-1}
   $E_k = \frac{1}{2} \times 2649.7... \times 0.518...^2 = 355.99...$ J
   $= \textbf{356 J}$ **(to 3 s.f.)**
   (2 marks for correct answer, 1 mark for correct ω if answer incorrect.)

2.3 $T = I\alpha$, so $\alpha = \frac{T}{I} = \frac{3.00}{2649.7...} = 1.13... \times 10^{-3}$ rad s^{-2} **(1 mark)**

   $\theta = \omega_i t + \frac{1}{2}\alpha t^2 = 0.518... \times 95 + \frac{1}{2} \times 1.13... \times 10^{-3} \times 95^2$
   $= 54.353...$ rad **(1 mark)**
   $\frac{54.353...}{2\pi} = 8.6506... = \textbf{8.65 revolutions}$ **(to 3 s.f.) (1 mark)**

2.4 Angular impulse $= T\Delta t = 3.00 \times 95.0 = \textbf{285 Nms}$ **(1 mark)**

3.1 At a constant pressure:
   $W = p\Delta V = 101 \times 10^3 \times ((1930 \times 10^{-6}) - (1620 \times 10^{-6}))$
   $= 31.31$ J
   Substitute the work done into the equation for the first law of thermodynamics:
   $Q = \Delta U + W$
   $\Delta U = Q - W = 35.0 - 31.31 = \textbf{3.69 J}$
   (2 marks for the correct answer, otherwise 1 mark for calculating the work done by the gas)

3.2 Compression is at a constant pressure, so $\frac{V_1}{T_1} = \frac{V_2}{T_2}$.
   Rearranging gives:
   $T_1 = \frac{T_2 V_1}{V_2} = \frac{272 \times 1620 \times 10^{-6}}{1930 \times 10^{-6}} = 228.31... = \textbf{228 K (to 3 s.f.)}$
   (2 marks for the correct answer, otherwise 1 mark for forming a correct equation for T_1)

3.3 $Q = \Delta U + W$
   Isothermal compression, so $\Delta U = 0$ and $Q = W$.
   $Q = -16$ J, so $W = \textbf{--16 J}$ **(1 mark)**
   Work is done on the gas **(1 mark)**.
   Don't forget — Q must be negative as the energy is being transferred <u>away</u> from the system.

3.4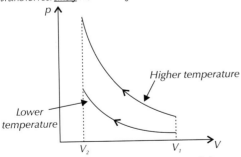
   (1 mark for the right shaped curves, 1 mark for correct labelling, 1 mark for correct direction of arrows)

4.1 Angular momentum $= I\omega = 2.70 \times 10^6 \times \frac{2 \times 2\pi}{60 \times 60}$ **(1 mark)**
   $= 9424.7... = \textbf{9420 Nms}$ **(to 3 s.f.)**
   (1 mark)

4.2 $I_{initial}\omega_{initial} = I_{final}\omega_{final}$, so
   $9424.7... = 3.00 \times 10^6 \times \omega_{final}$
   $\omega_{final} = \frac{9424.7...}{3.00 \times 10^6} = 3.1415... \times 10^{-3}$ rad s^{-1} **(1 mark)**
   $\Delta\omega = \omega_{final} - \omega_{initial} = 3.1415... \times 10^{-3} - \frac{2 \times 2\pi}{60 \times 60}$
   $= -3.490... \times 10^{-4}$
   $= \textbf{--3.49} \times \textbf{10}^{-4} \textbf{ rad s}^{-1}$ **(to 3 s.f.) (1 mark)**

4.3 $\alpha = \frac{\Delta\omega}{t} = \frac{3.49... \times 10^{-4}}{15.0} = 2.327... \times 10^{-5}$ rad s^{-2} **(1 mark)**
   $T = I\alpha = 3.00 \times 10^6 \times 2.327... \times 10^{-5} = 69.813...$
   $= \textbf{69.8 Nm}$ **(to 3 s.f.) (1 mark)**

4.4 $T_{applied} = T_{net} + T_{frictional} = 69.8... + 22.9 = 92.71...$
   $= \textbf{92.7 Nm}$ **(to 3 s.f.) (1 mark)**

4.5 $\theta = \frac{\omega_{final}^2 - \omega_{initial}^2}{2\alpha} = \frac{\left(\frac{2 \times 2\pi}{60 \times 60}\right)^2 - (3.1415... \times 10^{-3})^2}{2 \times 2.327... \times 10^{-5}}$
   $= 0.0497...$ rad **(1 mark)**
   $W = T\theta = 92.71... \times 0.0497... = 4.611...$
   $= \textbf{4.61 J}$ **(to 3 s.f.) (1 mark)**
   $P = \frac{W}{t} = \frac{4.611...}{15.0} = 0.3074... = \textbf{0.307 W}$ **(to 3 s.f.) (1 mark)**

5.1 Work done $=$ area enclosed by loop
   $W = ((50.0 \times 10^5) - (10.0 \times 10^5))$
   $\quad\quad \times ((12.0 \times 10^{-5}) - (2.0 \times 10^{-5}))$
   $= \textbf{400 J}$ **(1 mark)**

5.2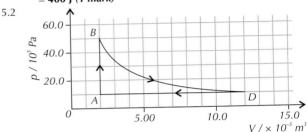
   (1 mark for the correct shape, 1 mark for the correct arrow directions)

5.3 If there are no frictional forces, then the frictional power = 0, and so the output power will be the same as the indicated power.
   Indicated power = (area of p-V loop) × (no. of cycles per second)
   As the time taken to complete one cycle doesn't change, the number of cycles per second stays the same. The decrease in the area of the p-V loop means a decrease in the indicated power, which also means a decrease in the output power.
   (1 mark for stating a decrease in the output power, 1 mark for an appropriate explanation)

6.1 How to mark your answer — pick the description that best matches your answer:
   5-6 marks: A full description of each stroke is explained, with particular reference to what processes the gas in the cylinder is undergoing.
   3-4 marks: A brief outline of each stroke is given. The process that the gas undergoes is only described for two strokes.
   1-2 marks: Only a couple of strokes are explained, with no reference to what process the gas undergoes.
   0 marks: There is no relevant information.
   Here are some points your answer may include:
   • (Induction) Piston moves from the top of the cylinder down, sucking in a mixture of air and fuel through the open inlet valve.
   • (Induction) Volume of the gas increases at a constant pressure.
   • (Compression) Inlet valve closes and the piston moves upwards. The volume of the gas decreases.
   • (Compression) Spark plug creates a spark and ignites the air-fuel mixture.
   • (Compression) When the spark ignites the fuel, the temperature and pressure increase suddenly, almost at a constant volume.
   • (Expansion) Air-fuel mixture expands and moves the piston downwards. The temperature of the air-fuel

mixture is now higher than during the compression stroke, so there is a net output of work.
- (Expansion) The volume increases and the pressure decreases.
- (Expansion) At the end of the expansion stroke, the exhaust valve opens.
- (Exhaust) The piston moves upwards, pushing the burnt air-fuel mixture out through the open outlet valve.
- (Exhaust) Volume of the gas decreases at a constant pressure.

6.2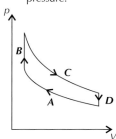

(1 mark for correct graph shape)
A — Gas is compressed adiabatically
B — Gas is heated at a constant volume
C — Gas cools adiabatically
D — Gas is cooled at a constant volume.
(2 marks for all four steps correctly labelled, otherwise 1 mark for two steps correctly labelled)

6.3 Fuel flow rate is the number of kg used in 1 second.
First, calculate the time taken to travel 70.0 km:
$\text{time} = \dfrac{\text{distance}}{\text{speed}}$, where:
distance = 70.0 km = 70 000 m
speed = 110 kmh^{-1}
$= 110\,000$ mh^{-1}
$= \dfrac{110\,000}{3600}$ ms^{-1}
$= 30.55...$ ms^{-1}
So, time $= \dfrac{70\,000}{30.55...} = 2290.90...$ s
3.2 kg of fuel is used in 2290.90... s, so:
Fuel flow rate $= \dfrac{3.2}{2290.90...}$
$= 0.001396... = \textbf{0.0014 kgs}^{-1}$ **(to 2 s.f.)**
(2 marks for the correct answer, otherwise 1 mark for calculating the time taken to travel 70.0 km)

6.4 Input power = calorific value × fuel flow rate
$= 45.4 \times 10^6 \times 0.001396... = 63\,415.8...$
$= \textbf{63 000 W (to 2 s.f.) (= 63 kW)}$ *(1 mark)*

Thermal efficiency $= \dfrac{\text{indicated power}}{\text{input power}}$

Rearranging for indicated power gives:
Indicated power = thermal efficiency × input power
$= 0.428 \times 63\,415.8...$
$= 27\,141.9...$
$= \textbf{27 000 W (to 2 s.f.) (= 27 kW)}$ *(1 mark)*

Mechanical efficiency $= \dfrac{\text{brake power}}{\text{indicated power}}$
$= \dfrac{24 \times 10^3}{27\,141.9...}$
$= 0.8842...$
$= \textbf{0.88 (to 2 s.f.) (= 88\%)}$ *(1 mark)*

Don't forget — brake power is just another name for output power.

Overall efficiency $= \dfrac{\text{brake power}}{\text{input power}} = \dfrac{24 \times 10^3}{63\,415.8...} = 0.3784...$
$= \textbf{0.38 (to 2 s.f.) (= 38\%)}$ *(1 mark)*

6.5 Indicated power = (area of p-V loop) × (no. of cycles per second) × (no. of cylinders)
Rearranging for no. of cycles per second gives:
no. of cycles per second =
$$\dfrac{\text{indicated power}}{(\text{area of }p\text{-}V\text{ loop}) \times (\text{no. of cylinders})}$$
To find the area of the p-V loop, first calculate the value of the area of one small square:
$0.04 \times 10^{-3} \times 0.05 \times 10^6 = 2$ J
No. of small squares enclosed by loop ≈ 117
So area enclosed by loop ≈ 234 J
So number of cycles per second is:
$$= \dfrac{\text{indicated power}}{(\text{area of }p\text{-}V\text{ loop}) \times (\text{no. of cylinders})} = \dfrac{27\,141.9...}{234 \times 4}$$
$= 28.997... = \textbf{29 cycles per second (to 2 s.f.)}$
(3 marks for the correct answer, otherwise 1 mark for rearranging the equation for indicated power for no. of cycles per second and 1 mark for using a sensible method to estimate the area enclosed by the loop)
As you are estimating the area enclosed by the loop, your answer may be in the range of 29 ± 3 cycles per second.

6.6 A flywheel is a heavy wheel with a high moment of inertia *(1 mark)*. A flywheel is able to store energy *(1 mark)*, so if there are any variations in the rotation of the crankshaft from the translational motion of the pistons, the flywheel uses its stored energy to smooth out the angular motion *(1 mark)*.

6.7 $P = T\omega$, where $\omega = 2\pi f = 2 \times \pi \times 28.997...$
$= 182.19...$ rads^{-1}
Rearrange for **T**:
$T = \dfrac{P}{\omega} = \dfrac{24\,000}{182.19...} = 131.72... = \textbf{130 Nm (to 2 s.f.)}$
(2 marks for the correct answer, otherwise 1 mark for calculating ω)
Your answer may be different to the one calculated here if you found a different value for f in 6.5. As long as your value of f is within the suitable range (29 ± 3 cycles per second) and you substituted into the equation for ω correctly, you still get full marks. This also applies to question 6.8.

6.8 $W = T\theta$
Number of radians the flywheel travels through in two minutes:
$\theta = \omega t = 182.19... \times 120$
$= 21\,863.8...$
$W = T\theta = 131.72... \times 21\,863.8... = 2.88 \times 10^6$
$= \textbf{2.9} \times \textbf{10}^6$ **J (to 2 s.f.) (= 2.9 MJ)**
(2 marks for the correct answer, otherwise 1 mark for calculating the number of radians the flywheel travels through in 2 minutes)

7.1 Maximum theoretical efficiency $= \dfrac{T_H - T_C}{T_H} = \dfrac{422 - 350}{422}$
$= 0.1706...$
$= \textbf{0.17 (to 2 s.f.)}$ *(1 mark)*
The efficiency given by the company is higher than the maximum theoretical efficiency, so the company's claim can't be true. *(1 mark)*

7.2 $W = Q_H - Q_C = 29 - 25 = \textbf{4 J}$ *(1 mark)*

7.3 efficiency $= \dfrac{W}{Q_H} = \dfrac{4}{29} = 0.1379...$
$= \textbf{0.14 (to 2 s.f.) (= 14\%)}$ *(1 mark)*

7.4 The second law of thermodynamics says that an engine must operate between a heat source and a heat sink, otherwise the engine will reach the temperature of the heat source, and no more heat energy will flow *(1 mark)*. Some energy is always lost to the heat sink, so not all of the heat extracted from the heat source is converted into work done *(1 mark)*.

7.5

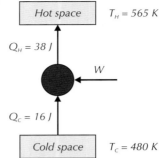

(1 mark for correct arrow directions, 1 mark for correct labelling of temperatures and 1 mark for correct labelling of Q_H, Q_C and W)

7.6 $COP_{ref} = \dfrac{Q_C}{Q_H - Q_C} = \dfrac{16}{38 - 16} = 0.7272...$
= **0.73 (to 2 s.f.) (1 mark)**

$COP_{hp} = \dfrac{Q_H}{Q_H - Q_C} = \dfrac{38}{38 - 16} = 1.727...$
= **1.7 (to 2 s.f.) (1 mark)**

7.7 $COP_{hp} = \dfrac{T_H}{T_H - T_C} = \dfrac{565}{565 - 480} = 6.647...$
= **6.6 (to 2 s.f.) (1 mark)**

7.8 Use the maximum theoretical efficiency already calculated to find A (the work done by the heat engine):
efficiency $= \dfrac{W}{Q_H}$
A $= W =$ efficiency $\times Q_H$
= $0.1706... \times 32$ **(1 mark)**
= $5.4597...$
= **5.5 J (to 2 s.f.) (1 mark)**
Use this answer to find B (= Q_C of the heat engine)
$W = Q_H - Q_C$
B $= Q_C = Q_H - W$
= $32 - 5.4597...$
= $26.540...$
= **27 J (to 2 s.f.) (1 mark)**
The work output by the heat engine is the input of work of the reversed heat engine, so the value found for A can be used to find C (= Q_H of the reversed heat engine):
$W = Q_H - Q_C$
C $= Q_H = W + Q_C = 5.4597... + 16 = 21.45...$
= **21 J (to 2 s.f.) (1 mark)**

Option D — Turning Points in Physics

1. Discovering Electrons
Page 362 — Application Question
Q1 a) The equation $\frac{1}{2}mv^2 = eV$ shows that if you increase the potential difference, V, the velocity of the electrons will also increase.
  b) kinetic energy $= \frac{1}{2}mv^2 = eV$
    = $(1.60 \times 10^{-19}) \times 12.0 \times 10^3 =$ **1.92×10^{-15} J**
  c) $\frac{1}{2}mv^2 = \Delta W = Fs$
    so $F = (1.92 \times 10^{-15}) \div 0.250 =$ **7.68×10^{-15} N**

Page 362 — Fact Recall Questions
Q1 The invisible rays that were thought to cause the glow that appears on the wall of a discharge tube when a potential difference is applied across the terminals.
Q2 When heat gives the free electrons in a metal enough energy to break free from the metal's surface.

Q3 E.g.

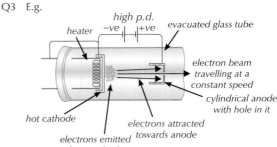

Q4 Electrons are tiny compared to the smallest atom, so they are easily stopped or deflected by atoms. So the glass tube in an electron gun has to be evacuated of air so that the electrons can travel freely in the electric field.
Q5 The formula for the work done in moving a charge through a potential difference is $\Delta W = Q\Delta V$. When the charge, Q, is the charge on an electron, e, and the ΔV is the potential difference between an anode and cathode, V, then $\Delta W = eV$. The kinetic energy, $\frac{1}{2}mv^2$, that an electron will have when it has been accelerated (from rest) is equal to the work done in accelerating it through the potential difference. So $\frac{1}{2}mv^2 = \Delta W = eV$.

2. Specific Charge of an Electron
Page 365 — Application Question
Q1 a) The beam will deflect towards the positive plate.
  b) The work done by an electric field in moving a charge is given by $Fd = Q\Delta V$, so $F = \dfrac{Q\Delta V}{d}$. For an electron, charge $Q = e$, so the electric force on the electron is $F = \dfrac{eV}{d}$ and the magnetic force on the electron is $F = Bev$. When the beam is not deflecting, the forces are equal, so $\dfrac{eV}{d} = Bev$. So $B = \dfrac{V}{vd}$.
  c) The kinetic energy of an electron entering the field is 4.2 keV, which is 4200e in J. So $4200e = \frac{1}{2}mv^2$ and $\dfrac{e}{m_e} = \dfrac{\frac{1}{2}v^2}{4200}$.
    Rearranging the equation given in part b) to find v,
    $v = \dfrac{V}{Bd} = \dfrac{5.00 \times 10^3}{2.60 \times 10^{-3} \times 50.0 \times 10^{-3}}$
    = $3.846... \times 10^7$ ms^{-1}
    $\dfrac{e}{m_e} = \dfrac{\frac{1}{2}v^2}{4200} = \dfrac{\frac{1}{2}(3.846... \times 10^7)^2}{4200} = 1.7610... \times 10^{11}$
    = **1.76×10^{11} C kg^{-1} (to 3 s.f.)**

Page 365 — Fact Recall Questions
Q1 E.g. Use a fine-beam tube (a glass bulb with magnetic field coils either side of it, filled with low-pressure hydrogen and containing an electron gun). Direct the electron beam at right angles to the magnetic field so that the beam travels in a circle. The electron beam will pass through the low-pressure gas, exciting hydrogen atoms. The excited hydrogen atoms will de-excite and emit light, so that the path of the electron beam can be seen. Measure the radius of the circle formed by the beam, the voltage of the electron gun and the magnetic field strength of the magnetic field.
Q2 The magnetic force on the electron is acting as a centripetal force so $\dfrac{mv^2}{r} = Bev$ and so $r = \dfrac{mv}{Be}$.

Q3 The specific charge of the electron is roughly 1800 times greater than the specific charge of a hydrogen ion, so it either has a much bigger charge, or is very light. This led physicists to believe there may be particles smaller than the atom.

3. Millikan's Oil-Drop Experiment
Page 369 — Application Question
Q1 a) The drop is moving at a steady speed, so it is at its terminal velocity and the two forces acting on it, weight and viscous drag, are equal.

So $mg = 6\pi\eta rv \Rightarrow \frac{4}{3}\pi r^3 \rho g = 6\pi\eta rv \Rightarrow r = \sqrt{\frac{9\eta v}{2\rho g}}$

$r = \sqrt{\frac{9 \times (1.8 \times 10^{-5}) \times (2.5 \times 10^{-5})}{2 \times 980 \times 9.81}}$

$= 4.589... \times 10^{-7}$ m $= \mathbf{4.6 \times 10^{-7}}$ **m (2 s.f.)**

b) The electric force is equal to the weight of the drop, so

$\frac{QV}{d} = \frac{4}{3}\pi r^3 \rho g \Rightarrow Q = \frac{\frac{4}{3}\pi r^3 \rho g d}{V}$

$= \frac{\frac{4}{3} \times \pi \times (4.589... \times 10^{-7})^3 \times 980 \times 9.81 \times 4.0 \times 10^{-3}}{98}$

$= 1.5889... \times 10^{-19} = \mathbf{1.6 \times 10^{-19}}$ **C (to 2 s.f.)**

c) In part b), $\frac{\frac{4}{3}\pi r^3 \rho g d}{V} = 1.5889... \times 10^{-19}$ C
The distance has been doubled and the p.d. halved, so in this experiment:

$Q = \frac{\frac{4}{3}\pi r^3 \rho g(2d)}{\frac{V}{2}} = 4 \times \frac{\frac{4}{3}\pi r^3 \rho g}{V}$

$= 4 \times (1.5889... \times 10^{-19}$ C$) = 6.3558... \times 10^{-19}$
$= \mathbf{6.4 \times 10^{-19}}$ **C (to 2 s.f.)**

Page 369 — Fact Recall Questions
Q1 $F = 6\pi\eta rv$, where F is the viscous drag force on a sphere moving through a fluid, η is the viscosity of the fluid, r is the radius of the sphere and v is the speed of the sphere.

Q2

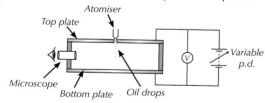

Q3 Electric force, $F = \frac{QV}{d}$.

Q4 a) The weight of the drop and the viscous drag force.
b) The weight of the drop, the viscous drag force and the electric force.
c) The weight of the drop and the electric force.

Q5 The weight of the drop is equal to the electric force on the drop, so $\frac{QV}{d} = mg = \frac{4}{3}\pi r^3 \rho g$. So $\frac{QV}{d} = \frac{4}{3}\pi r^3 \rho g$.

Q6 He found that the charge on oil droplets was always an integer multiple of 1.60×10^{-19} C. He concluded that charge cannot exist in smaller 'packets' and that this must be the size of the charge on an electron.

Q7 Millikan's discovery meant that the mass of an electron could be calculated and its charge deduced.

4. Light — Newton vs Huygens
Page 373 — Application Question
Q1 a) According to the corpuscular theory, the corpuscles of light travel in straight lines. So if the pinhole is made narrower, the beam of light corpuscles will be narrowed too.
b) Light is a wave and can diffract. Making the pinhole smaller will make the size of the gap closer to the wavelength of the light and cause it to diffract more, so the beam will get wider.

Page 373 — Fact Recall Questions
Q1 Newton believed that light was made up of a stream of tiny particles called corpuscles. Reflection was due to a force that pushed the particles away from a surface. Refraction worked if the corpuscles travelled faster in a denser medium.

Q2 Huygens' principle says that every point along a wavefront can be considered as a point source of secondary wavelets (hemispherical wavelets that come from each point) that spread out in the forward direction at the speed of the wave. The new wavefront is the surface that is tangential to all of these secondary wavelets.

Q3 E.g. Newton was very successful and respected / Huygens' theory couldn't explain light polarisation / double refraction / Huygens' theory couldn't explain why sharp shadows were formed by light.

Q4 They struggled to get two coherent light sources, because light is usually emitted in random bursts.

Q5 Young used one light source shining through a slit to create a point light source, and put a slide with two thin slits in front of this light source to create two coherent light sources, one from each slit, avoiding the problem described in Q4. He projected the two coherent light sources onto a screen and observed bright and dark fringes being formed.

Q6 It gave evidence of light diffracting and showing interference, both properties of waves and not particles.

Q7 The discovery that light behaved as a transverse wave.

5. Electromagnetic Waves
Page 377 — Application Question
Q1 a) The wavelength is double the distance between two adjacent nodes, so $\lambda = 0.01 \times 2 = \mathbf{0.02}$ **m**.
b) $c = f\lambda = 1.5 \times 10^{10} \times 0.02 = \mathbf{3 \times 10^8}$ **ms^{-1}**

Page 377 — Fact Recall Questions
Q1 Electromagnetic waves are transverse waves, made up of oscillating electric and magnetic fields that are perpendicular to each other and the direction of travel.

Q2 $c = \frac{1}{\sqrt{\mu_0\varepsilon_0}}$

Q3 $\mu_0 =$ the permeability of free space $= 4\pi \times 10^{-7}$ Hm^{-1}. It relates to the magnetic flux density due to a current-carrying wire in free space.
$\varepsilon_0 =$ the permittivity of free space $= 8.85 \times 10^{-12}$ Fm^{-1}. It relates to the electric field strength due to a charged object in free space.

Q4 Fizeau measured the speed of light by passing a beam of light through a gap between cog teeth to a reflector about 9 km away. The cog was rotated at exactly the right speed so that the reflected beam passed through the next gap in the cog teeth. Fizeau's experiment provided an estimate of the speed of light. This was very close to the value later calculated by Maxwell, whose theory assumed light was a wave. This suggested that light is an electromagnetic wave.

Q5 Hertz used an induction coil and a capacitor to produce a high voltage, and showed that radio waves were produced when high voltage sparks jumped across a gap of air.

Q6 Hertz showed that radio waves had all the properties of electromagnetic waves, including reflection, refraction, diffraction, interference, polarisation and the same velocity.

6. The Photoelectric Effect
Page 382 — Application Question

Q1 a) $E = \frac{hc}{\lambda} = \frac{(6.63 \times 10^{-34})(3.00 \times 10^8)}{(550 \times 10^{-9})} = 3.616... \times 10^{-19}$ J
$= \textbf{3.6} \times \textbf{10}^{-19}$ **J (to 2 s.f.)**

b) Electrons emitted from deeper down in the metal lose more energy as they escape than electrons on the surface, so the photoelectrons have a range of energies.

Page 382 — Fact Recall Questions

Q1 For a given metal, no photoelectrons are emitted if the radiation has a frequency below the threshold frequency. The photoelectrons are emitted with a variety of kinetic energies ranging from zero to some maximum value. (This value of maximum kinetic energy increases with the frequency of the radiation, but is unaffected by the intensity of the radiation.) The number of photoelectrons emitted per second is directly proportional to the intensity of the radiation.

Q2 Einstein believed that EM waves (and the energy that they carry) can only exist in discrete packets called photons, and that photons have a one-on-one, particle-like interaction with an electron in a metal surface. A photon will transfer all its energy to that one electron.

Q3 Max Planck suggested that EM waves can only be released in discrete packets, or quanta.

Q4 Wave theory predicts that the electrons should gradually gain energy from the incident waves until they have enough energy to escape — there is no explanation for why this never happens. The photon theory of light says that photons have a one-on-one, particle-like interaction with an electron — each photon transfers all its energy to one specific electron. Electrons cannot build up energy — they either get enough from one photon of the correct frequency to escape, or don't get any. This is why there is a threshold frequency.

Q5 The work function of a metal is the minimum amount of energy that an electron needs to break the bonds holding it in the metal.

Q6 Wave theory predicts that the kinetic energy of a photoelectron should increase with light intensity. In reality, the kinetic energy of a photoelectron is only affected by the frequency of the light, and the intensity has no effect at all. Photoelectrons have a kinetic energy equal to the energy of the photon that collided with them, hf, minus the energy used to break the bonds of the metal, the work function, ϕ, and any energy they lose whilst leaving the metal. The maximum kinetic energy a photoelectron can have is just the energy supplied by the photon, minus the work function.

Q7 Classical wave theory could explain the slope of black body radiation curves at long wavelengths (low frequencies) — it suggested that the power radiated was proportional to λ^{-4}, but this meant that the power output was predicted to head towards infinity in the ultraviolet region. This was the ultraviolet catastrophe — wave theory, then widely accepted, had predicted something that was impossible.

7. Wave-Particle Duality
Page 385 — Application Questions

Q1 a) $\lambda = \frac{h}{p} = \frac{h}{mv}$, so the de Broglie wavelength is inversely proportional to the velocity of the electron. Therefore increasing the velocity will decrease the de Broglie wavelength of the electrons.

b) Increasing the electron velocity will cause the diffraction pattern circles to squash together towards the middle, i.e. the amount of diffraction decreases.

Q2 $\lambda = \frac{h}{\sqrt{2meV}} = \frac{6.63 \times 10^{-34}}{\sqrt{2(9.11 \times 10^{-31})(1.6 \times 10^{-19})(4200)}}$
$= 1.894... \times 10^{-11}$ m $= \textbf{1.9} \times \textbf{10}^{-11}$ **m (to 2 s.f.)**

Q3 a) $p = \frac{h}{\lambda} = (6.63 \times 10^{-34}) \div (1.71 \times 10^{-10})$
$= 3.87719... \times 10^{-24}$ kg ms^{-1}
$= \textbf{3.88} \times \textbf{10}^{-24}$ **kg ms^{-1} (to 3 s.f.)**

Make sure you always include units with your answer — it could get you some precious extra marks in the exam.

b) $m_e = 9.11 \times 10^{-31}$ kg,
$v = p \div m_e = 3.87719... \times 10^{-24} \div 9.11 \times 10^{-31}$
$= 4.2559... \times 10^6$ ms^{-1}
$E_k = \frac{1}{2}mv^2 = \frac{1}{2} \times 9.11 \times 10^{-31} \times (4.2559... \times 10^6)^2$
$= 8.250... \times 10^{-18}$ J
$= \textbf{8.25} \times \textbf{10}^{-18}$ **J (to 3 s.f.)**

Q4 For the alpha particle, $\lambda = \frac{h}{p} = \frac{h}{mv} = \frac{6.63 \times 10^{-34}}{6.64 \times 10^{-27} \times 75.0}$
$= 1.331... \times 10^{-9}$ m

For the electron, $\lambda = 1.331... \times 10^{-9}$ m
$\lambda = \frac{h}{p} = \frac{h}{mv}$ so $1.331... \times 10^{-9} = \frac{6.63 \times 10^{-34}}{9.11 \times 10^{-31} \times v}$
$v = \frac{6.63 \times 10^{-34}}{9.11 \times 10^{-31} \times 1.331... \times 10^{-9}} = 5.4665... \times 10^5$
$= \textbf{5.47} \times \textbf{10}^5$ **ms^{-1} (to 3 s.f.)**

Because the de Broglie wavelengths are the same, you could just equate mv for each particle to calculate v.

Page 385 — Fact Recall Questions

Q1 All particles have both particle and wave properties. Waves can also show particle properties.

Q2 E.g. Diffraction shows light has wave properties, and the photoelectric effect shows light has particle properties.

Q3 Electron diffraction

8. Electron Microscopes
Page 387 — Fact Recall Questions

Q1 Around 150 V.

Q2 An electron gun produces a beam of electrons with a certain de Broglie wavelength (dependent on the anode potential). A set of electromagnetic coils known as a condenser lens focuses the electrons into a thin, straight beam onto the sample, the structure of which may cause some of the electrons to diffract. Two more sets of electromagnetic coils use magnetic fields to deflect the electrons so that they form a magnified image, which is projected onto a fluorescent screen.

Q3 A very fine probe is positioned very close (around 1 nm) to a sample's surface. A potential is applied so that the probe is negatively charged in relation to the sample. Electrons 'tunnel' from the probe to the surface, which produces a small electric current. A bigger distance results in a smaller current. The probe scans the surface of the sample and produces a 3D image of it, by measuring either the current at a set probe height, or the height of the probe from the sample at a set current.

Q4 Piezoelectric transducers are used. When a p.d. is applied to them, they experience a tiny change in length, which moves the probe.

9. Michelson-Morley Experiment
Page 389 — Fact Recall Questions
Q1 Everything, including light, moves relative to a fixed background known as the ether.
Q2 Michelson and Morley hoped to compare the speed of light parallel and perpendicular to the motion of the Earth, and find a difference between them. With this difference, they would be able to measure the absolute speed of the Earth.
Q3 a) To split the light source into two beams, by reflecting some light and transmitting the rest.
b) To reflect the light back to the semi-silvered glass block, where the two beams will once again converge, forming an interference pattern.
c) To make sure that both beams of light travel through the same amount of glass and air.
Q4 It was expected that rotating the interferometer would result in a shift in the interference pattern seen, due to the change in the angle at which each beam was moving relative to the absolute motion of the Earth.
Q5 The failure of the Michelson-Morley experiment to detect absolute motion is significant because it showed that the speed of light has the same value for all observers (it is invariant), and that it's impossible to detect absolute motion — the ether doesn't exist.

10. Special Relativity
Page 395 — Application Questions
Q1 $l = l_0\sqrt{1 - \frac{v^2}{c^2}} = (6.7 \times 10^{-2})\sqrt{1 - \frac{(0.98c)^2}{c^2}}$

$= (6.7 \times 10^{-2})\sqrt{1 - 0.98^2}$

$= 0.0133...\,\text{m} = \mathbf{1.3 \times 10^{-2}\,m\,(to\,2\,s.f.)}$

Q2 a) (i) $d = v \times t = (2.8 \times 10^8) \times (77 \times 10^{-9})$

$= 21.56\,\text{m} = \mathbf{22\,m\,(to\,2\,s.f.)}$

(ii) $l = l_0\sqrt{1 - \frac{v^2}{c^2}}$

so $d = d_0 \times \sqrt{1 - \frac{v^2}{c^2}} = 21.56 \times \sqrt{1 - \frac{(2.8 \times 10^8)^2}{(3.00 \times 10^8)^2}}$

$= 7.74...\,\text{m} = \mathbf{7.7\,m\,(to\,2\,s.f.)}$

b) $m = \frac{m_0}{\sqrt{1 - \frac{v^2}{c^2}}} = \frac{9.11 \times 10^{-31}}{\sqrt{1 - \frac{(2.8 \times 10^8)^2}{3.00 \times 10^8)^2}}} = 2.537... \times 10^{-30}$

$= \mathbf{2.5 \times 10^{-30}\,kg\,(to\,2\,s.f.)}$

c) $E = mc^2 = (2.537... \times 10^{-30})(3.00 \times 10^8)^2$

$= 2.283.. \times 10^{-13} = \mathbf{2.3 \times 10^{-13}\,J\,(to\,2\,s.f.)}$

Q3 You know the half-life, so you need to calculate the time taken in the reference frame of the particles.
First, the distance in the reference frame of the particles is

$d = d_0\sqrt{1 - \frac{v^2}{c^2}}$ (from length contraction).

In the reference frame of the particles,

$t = \frac{d}{v} = d_0\sqrt{1 - \frac{v^2}{c^2}} \div v = 20 \times \sqrt{1 - \frac{(0.95c)^2}{c^2}} \div 0.95c$

$= 20 \times \sqrt{1 - 0.95^2} \div (0.95 \times 3.00 \times 10^8) = 2.19... \times 10^{-8}\,\text{s}$

Half-life = $11.0 \times 10^{-9}\,\text{s}$, so

$2.19... \times 10^{-8}$ is equal to $\frac{2.19... \times 10^{-8}}{11.0 \times 10^{-9}} = 1.992...$ half-lives

So the time taken in the reference frame of the particles to get between the two detectors is around 2 half-lives. So the ratio is around $1 \div (2^2) = \mathbf{0.25}$.

Page 395 — Fact Recall Questions
Q1 a) A space or system of coordinates that we use to describe the position of an object.
b) A frame of reference in which Newton's first law is obeyed.
Q2 Physical laws have the same form in all inertial frames. The speed of light in free space is invariant.
Q3 An observer of two events that is moving at a constant velocity, v, in relation to the events will measure a longer time interval, t, between the two events than the time measured by an observer who is stationary relative to the events, t_0. The equation for this time is $t = \frac{t_0}{\sqrt{1 - \frac{v^2}{c^2}}}$.
Q4 Muons travel from the upper atmosphere towards the ground at speeds close to c. They have a short half-life, and we expect the intensity of muons to decrease a certain amount as they travel between two points in the atmosphere. But the time taken in their reference frame is shorter than the time taken in our reference frame, so the intensity of the muons drops less than we would expect (if we ignore relativity). This is evidence for time dilation.
Q5 $l = l_0\sqrt{1 - \frac{v^2}{c^2}}$
Q6 The faster an object moves, the more massive it gets. The relativistic mass of an object is the mass of the object when it is moving at a certain velocity.
Q7 $E = mc^2$ and $E = \frac{m_0c^2}{\sqrt{1 - \frac{v^2}{c^2}}}$.
Q8 E.g.

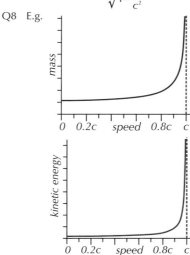

Q9 Bertozzi used particle accelerators to accelerate pulses of electrons to a range of energies from 0.5 MeV to 15 MeV. The particles were smashed into an aluminium disc a set distance away. The time taken by electrons of each energy to reach the aluminium disc was measured so that their speeds could be calculated. As the energy of the electrons was increased, the speed of the electrons didn't increase as you would expect from $E = \frac{1}{2}mv^2$, but instead tailed off towards a maximum value approaching $3 \times 10^8\,\text{ms}^{-1}$ (c). This showed that as the energy increased, the mass increased. To check that the electrons had the energy he thought they had, Bertozzi used the heat generated by the collisions at each energy to calculate the kinetic energy of the particles immediately before impact. Bertozzi found that plotting a graph of kinetic energy against speed gave a curve that closely matched that predicted by Einstein's formula. This was the first direct evidence for special relativity.

1.1 Thermionic emission is the release of electrons from a metal's surface *(1 mark)* when the metal is heated *(1 mark)*.

1.2 E.g. So the electrons only interact with the sample / only get diffracted by the sample and can form a clear image of it. *(1 mark)*

1.3 The electrons diffract around the atoms and molecules of a sample's structure to produce an image of it — this will only happen with waves *(1 mark)*. The electrons would not be able to do this as particles, as they would either pass straight through the sample or be absorbed/blocked by it *(1 mark)*.

1.4 Use the equation for the de Broglie wavelength of an electron: $\lambda = \dfrac{h}{\sqrt{2meV}}$ and rearrange for V:

$$\sqrt{V} = \frac{h}{\lambda\sqrt{2me}}$$

so $V = \dfrac{h^2}{2me\lambda^2}$

$$= \frac{(6.63 \times 10^{-34})^2}{2(9.11 \times 10^{-31})(1.60 \times 10^{-19})\lambda^2}$$

$$= \frac{1.51 \times 10^{-18}}{\lambda^2} \text{ (to 3 s.f.)}$$

(2 marks available — 1 mark for rearranging for V, 1 mark for the final answer.)

1.5 The de Broglie wavelength should be similar in size to the diameter of the atom, 2.20×10^{-10} m.

So $V = \dfrac{1.51 \times 10^{-18}}{(2.20 \times 10^{-10})^2} = 31.1983... = \mathbf{31.2\ V\ (to\ 3\ s.f.)}$

(1 mark)

1.6 $p = mv = h \div \lambda$, so increasing the speed decreases the de Broglie wavelength of the electrons *(1 mark)*. This means they will diffract around smaller objects, and so will be able to form a more detailed image *(1 mark)*.

2.1 Physical laws have the same form in all inertial frames *(1 mark)*.
The speed of light in free space is invariant *(1 mark)*.

2.2 An inertial frame of reference is a space or system of coordinates in which Newton's first law is obeyed *(1 mark)*.

2.3 $m = \dfrac{m_0}{\sqrt{1 - \dfrac{v^2}{c^2}}} = \dfrac{m_0}{\sqrt{1 - \dfrac{(2.9 \times 10^8)^2}{(3.00 \times 10^8)^2}}}$

$= 3.905... m_0$

$\approx \mathbf{4m_0}$
(1 mark)

2.4 The time measured by the observer would be longer *(1 mark)* — this is known as time dilation *(1 mark)*.

3.1 Two bright fringes formed on the screen behind the two slits *(1 mark)*. Newton's corpuscular theory said that light behaved as a particle and naturally travelled in straight lines, so only two fringes would be formed *(1 mark)*.

3.2 A pattern of many fringes *(1 mark)* will be seen, with a bright central fringe and decreasingly bright fringes further out. In this experiment light shows the wave properties interference and diffraction *(1 mark)*, which supports Huygens' wave theory of light *(1 mark)*.

3.3 E.g. any of: Newton had many correct theories in the past / Newton was very well respected so his theory was accepted / Huygen's theory was based on longitudinal waves which did not explain the polarisation of light.
(1 mark for any correct answer)

3.4 E.g. any of: the photoelectric effect / black body radiation curves *(1 mark for any correct answer)*.

4.1 How to grade your answer (pick the description that best matches your answer):
0 marks: There is no relevant information.

1-2 marks: Some of the details of the cathode ray discharge tube and the properties discovered are covered, but there is no overall description. Several errors with grammar, spelling, punctuation and legibility. Answer lacks structure and information.
3-4 marks: The cathode ray discharge tube is covered, and at least one relevant discovery is described. Only a few errors with grammar, spelling, punctuation and legibility. Answer has some structure and information.
5-6 marks: The cathode ray discharge tube is fully described, and a full explanation of the key properties that led scientists to predict subatomic particles is given. Grammar, spelling and punctuation are used accurately and there is no problem with legibility. Uses an appropriate form and style of writing and gives relevant information.
Here are some points your answer may include:
- A discharge tube is an evacuated glass tube with a cathode and an anode and a potential difference applied between them.
- When the potential difference is applied, a glow appears on the wall of the discharge tube.
- Cathode rays were shown to have energy, momentum and mass, showing that they were made up of particles.
- Cathode rays were shown to have a negative charge, also showing that they were made up of particles.
- They were shown to have the same properties no matter what gas was in the tube and what the cathode was made of.
- They were shown to have a huge charge-to-mass ratio, showing they were either highly charged or subatomic.

5.1 The circular motion of the electron beam means that the magnetic force on the electron is acting as a centripetal force. Magnetic force is given by $F = Bev$ and centripetal force is given by $F = \dfrac{mv^2}{r}$.
So $Bev = \dfrac{mv^2}{r}$ and rearranging, $\dfrac{e}{m} = \dfrac{v}{Br}$.

(2 marks for correct answer, otherwise 1 mark for correctly identifying $F = Bev$ and $F = \dfrac{mv^2}{r}$.)

5.2 $\dfrac{e}{m_e} = \dfrac{v}{Br} = \dfrac{(8.44 \times 10^6)}{(1.20 \times 10^{-3})(4.00 \times 10^{-2})} = 1.7583... \times 10^{11}$

$= \mathbf{1.76 \times 10^{11}\ C\,kg^{-1}}$ **(to 3 s.f.)** *(1 mark for correct answer, 1 mark for giving answer to three significant figures)*

5.3 The specific charge of an electron is much greater than the highest specific charge known at the time, that of the H^+ ion, meaning that it either has much more charge, or is much lighter *(1 mark)*. He assumed that electrons had the same charge and that they were much lighter, suggesting that they were subatomic particles *(1 mark)*.

6.1 The distance between two antinodes is half the wavelength of the wave *(1 mark)*. Use this wavelength and the frequency in the equation $c = f\lambda$ to calculate the wave speed *(1 mark)*.

6.2 $c = f\lambda = (4.00 \times 10^7) \times (3.75 \times 2) = \mathbf{3.00 \times 10^8\ m\,s^{-1}}$
(1 mark for correct answer)

6.3 $c = \dfrac{1}{\sqrt{\mu_0 \varepsilon_0}} = \dfrac{1}{\sqrt{(4\pi \times 10^{-7}) \times (8.85 \times 10^{-12})}}$

$= 2.998... \times 10^8\ m\,s^{-1} = \mathbf{3.00 \times 10^8\ m\,s^{-1}}$ **(to 3 s.f.)** *(1 mark)*

6.4 The results of Hertz's experiment showed that radio waves travel at the same speed as electromagnetic waves. This provided evidence that radio waves were electromagnetic waves and was evidence for Maxwell's electromagnetic theory of light *(1 mark)*.

Glossary

A

Absolute magnitude (*M*)
What an object's apparent magnitude would be if it were 10 parsecs away from Earth.

Absolute uncertainty
The uncertainty of a measurement given as a fixed quantity.

Absolute zero
The lowest possible temperature a substance can have, equal to 0 K or −273 °C.

Absorption line
A dark line in the continuous spectrum of a source that corresponds to a wavelength of light that has been absorbed by electrons as they are excited into higher energy levels.

Accurate result
An accurate result is really close to the true answer.

Activity
The number of unstable nuclei in a radioactive sample that decay each second.

Adiabatic
An adiabatic process is one in which no heat is gained or lost by the system.

Alpha particle
A particle formed of two protons and two neutrons (the same as a helium nucleus).

Alpha radiation
Nuclear radiation made up of alpha particles.

Alternating current
A current that changes with time in a regular cycle.

Angular acceleration
The rate of change of angular velocity.

Angular frequency
The equivalent of angular speed for an object moving with simple harmonic motion.

Angular impulse
Angular impulse is the change in angular momentum of a rotating object.

Angular magnification
The measure of a telescope's power, equal to the angle subtended by the image at the eye divided by the angle subtended by the object at the unaided eye.

Angular momentum
The angular momentum of a rotating object is the product of its moment of inertia and angular velocity.

Angular speed or velocity
The angle through which a point on a rotating object rotates each second.

Anode
An electrode which becomes positively charged.

Anomalous result
A result that doesn't fit in with the pattern of the other results in a set of data.

Apparent magnitude (*m*)
A measure of brightness of an object as observed from Earth.

Astronomical refracting telescope
A telescope made up of two converging lenses.

Avogadro constant, N_A
The number of particles in a mole, defined as the number of atoms in exactly 12 g of the carbon isotope $^{12}_{6}C$. It is equal to 6.02×10^{23} mol⁻¹.

Axial ray
A ray that is parallel to the principal axis of a lens or mirror.

B

Background radiation
The weak level of nuclear radiation found everywhere.

Balmer series
The absorption lines in an absorption line spectrum caused by the electrons in atomic hydrogen moving between the first excitation level (n = 2) and higher energy levels.

Beta-minus radiation
Nuclear radiation made up of electrons.

Beta-plus radiation
Nuclear radiation made up of positrons.

Binding energy
The energy released when a nucleus forms, or the energy required to separate all the nucleons in that nucleus. Equivalent to the mass defect of the nucleus.

Black body
A body that absorbs all electromagnetic radiation of all wavelengths and can emit all wavelengths of electromagnetic radiation.

Black hole
An object whose escape velocity is greater than the speed of light.

Blue shift
The shift in wavelength and frequency of a source moving towards us towards (or beyond) the blue end of the electromagnetic spectrum.

Boltzmann constant, *k*
A constant used in the ideal gas equation for N molecules, equal to 1.38×10^{-23} JK⁻¹ or $\frac{R}{N_A}$.

Boyle's law
For an ideal gas at a constant temperature, the pressure *p* and volume *V* are inversely proportional.

Brake power
The power that is output at the crankshaft of an engine. (Also called output power.)

Brightness
The power received from an object per unit area at Earth. Also known as intensity.

Brownian motion
The zigzag, random motion of particles suspended in a fluid.

C

Calibration
Marking a scale on a measuring instrument or checking a scale by measuring a known value.

Capacitance
The amount of charge an object is able to store per unit potential difference (p.d.) across it.

Capacitor
An electrical component that can store charge, made up of two conducting plates separated by a dielectric.

Categoric data
Data that can be sorted into categories.

Cathode
An electrode which becomes negatively charged.

Centripetal force
The force on an object moving with circular motion. It's directed towards the centre of the circle, and is responsible for the object's curved path.

Chain reaction (nuclear)
When the neutrons released by a nuclear fission cause other nuclei to fission and release more neutrons — and so on.

Charles' law
For an ideal gas at constant pressure, the volume V is directly proportional to its absolute temperature T.

Chromatic aberration
A problem with using lenses caused by the lens refracting different colours of light by different amounts, resulting in the image for each colour being in a slightly different position.

Closed system
A system that doesn't allow any transfer of matter in or out.

Collecting power
A measure of how much radiation a telescope can collect, proportional to its collecting area.

Constructive interference
When two waves interfere to make a wave with a larger amplitude.

Continuous data
Data that can have any value on a scale.

Control rod
A rod inserted into a nuclear reactor to control the rate of fission by absorbing neutrons.

Control variable
A variable that is kept constant in an experiment.

Converging lens
A lens which brings parallel light rays together. They are convex in shape.

Cosmological principle
On a large scale, the universe is homogeneous and isotropic.

Couple
A pair of forces of equal size which act parallel to each other but in opposite directions. A couple causes a rotation, but no resultant linear motion.

Critical damping
Damping such that the amplitude of an oscillation is reduced in the shortest possible time.

Critical mass
The amount of fuel needed for a fission chain reaction to continue at a steady rate on its own.

Damping
A force which causes an oscillating object to lose energy and so causes the amplitude of the object's oscillation to decrease.

Dark energy
A type of energy that fills the whole of space, and might explain the accelerating expansion of the universe.

de Broglie wavelength
The wavelength associated with a particle, as part of the theory of wave-particle duality.

Decay constant
The probability of an unstable nuclei decaying in unit time. A measure of how quickly an isotope will decay.

Dependent variable
The variable that you measure in an experiment.

Destructive interference
When two waves interfere to make a wave with a reduced amplitude.

Dielectric
A material that acts as an electric insulator. It is used to separate the two conducting plates in a capacitor.

Diffraction
When waves spread out as they pass through a narrow gap or go round obstacles.

Discrete data
Data that can only take certain values.

Doppler effect
The change in the frequency and wavelength of a wave for a source moving towards or away from an observer.

E

e.m.f.
Short for electromotive force. The voltage across a conductor that produces a current if the conductor is connected in a circuit.

Eddy current
A looping current induced by the changing magnetic flux in the core of a transformer.

Elastic collision
A collision that conserves both linear momentum and kinetic energy.

Electric field strength
The force per unit positive charge experienced by a body in an electric field.

Electric potential
The electric potential energy that a unit positive charge (+1 C) would have at a specific point.

Electric potential energy
The energy stored by a charge due to its position in an electric field. It equals the work done moving a charge from infinity to that position.

Electromagnetic induction
The process of inducing an e.m.f. in a conductor with relative motion to a magnetic field (cutting field lines).

Electron capture
When an atomic nucleus captures an electron, causing a proton to change into a neutron, and a neutrino to be released.

Electron degeneracy pressure
The pressure that stops electrons being forced into the atomic nucleus, e.g. in the dense core of a star.

Electron gun
A device that uses thermionic emission and a high potential difference to produce a beam of accelerated electrons.

Electronvolt
The kinetic energy carried by an electron after it has been accelerated

from rest through a potential difference of 1 volt.

Empirical law
A rule based on observations and evidence that predicts what will happen but which doesn't explain why.

Equipotential
A line (in 2D) or surface (in 3D) that joins together all of the points with the same gravitational potential or electric potential.

Error bar
Used when plotting a graph to show the range of values a data point may lie in.

Escape velocity
The minimum speed required for an unpowered object to leave the gravitational field of another object.

Event horizon
The boundary of the region around an object in which its escape velocity is greater than c.

Evidence
Valid data arising from an experiment, which can be used to support a conclusion.

Exoplanet
Any planet that is not inside our solar system.

Exponential relationship
A relationship in which the rate of change in a quantity is proportional to the amount of that quantity left.

Fair test
An experiment in which all variables are kept constant apart from the independent and dependent variables.

Farad
The standard unit of capacitance. 1 farad (F) = 1 coulomb per volt (CV^{-1}).

Faraday's law
The induced e.m.f. is directly proportional to the rate of change of flux linkage.

Field lines
A way of representing a force field. Also known as flux lines in a magnetic field.

Flux linkage
The magnetic flux in a coil multiplied by the number of turns on the coil.

Flywheel
A heavy wheel that has a high moment of inertia in order to resist changes to its rotational motion. Often used to store energy or smooth angular velocity.

Focal length
The focal length of a lens is the perpendicular distance between the lens axis and its focal plane.

Focal plane
The plane perpendicular to the principal axis of a lens, on which the principal focus lies.

Force field
A region in which a body experiences a non-contact force.

Forced vibration
The oscillation of an object due to an external driving force.

Four-stroke engine
An engine that burns fuel once every four strokes of a piston.

Fractional uncertainty
The uncertainty given as a fraction of the measurement taken.

Frame of reference
A space or coordinate system that we decide to use to describe the position of an object.

Free vibration
The oscillation of an object with no transfer of energy to or from the surroundings.

Frequency
The number of complete revolutions or cycles that a rotating or oscillating object makes per second.

Friction power
The amount of power needed to overcome friction within an engine.

Frictional torque
The torque due to friction that opposes rotational motion in a system.

Gamma radiation
Nuclear radiation made up of high-frequency electromagnetic waves/photons (known as gamma rays).

Geiger-Müller tube
An instrument used to detect nuclear radiation. It is attached to a counter that measures the amount of radiation.

Geostationary satellite
An Earth satellite that orbits directly over the equator and is always above the same point on Earth. Its orbital period is approximately 24 hours.

Gravitational field strength
The force per unit mass, g, experienced by a body in a gravitational field.

Gravitational potential
The gravitational potential energy that a unit mass would have at a specific point.

Gravitational potential energy
The energy stored by an object due to its position in a gravitational field. It equals the work done moving an object from infinity to that position.

Half-life
The average time it takes for the number of unstable nuclei (or the activity or count rate) in a sample of a radioactive isotope to halve.

Heat engine
An engine that operates between a heat source and a heat sink to convert heat energy into useful work.

Heat pump
A reversed heat engine that aims to pump as much heat energy per joule of work done into the hot space.

Heat sink
A region that absorbs heat energy from a heat engine. It must be at a lower temperature than the heat source for the engine to do work.

Heat source
The region from which heat energy is extracted by a heat engine.

Homogenous
Meaning every part of something is the same as every other part.

Hubble's law
The recessional velocity of a distant object in space is proportional to its distance, $v = Hd$.

Hypothesis
A suggested explanation for a fact or observation.

I

Ideal gas
A theoretical gas that obeys the three gas laws.

Ideal gas equation
A combination of the three gas laws, given by $pV = nRT$ or $pV = NkT$.

Independent variable
The variable that you change in an experiment.

Indicated power
The net work done per second by the cylinders in an engine.

Inelastic collision
A collision that conserves linear momentum, but not kinetic energy.

Inertia
A measure of how much an object resists a change in velocity. The larger the inertia of an object, the larger the applied force needed to change its velocity by a given amount.

Inertial frame
A frame of reference in which Newton's first law is obeyed. (Newton's first law says that objects won't accelerate unless they're acted on by an external force.)

Intensity
The power received from an object per unit area. Also known as brightness.

Internal energy
The sum of the potential and kinetic energies of all of the particles in a system.

Inverse square law
A law that relates two variables by a factor of $1/r^2$.

Ionising
The process of an atom losing or gaining an electron (often when hit by ionising radiation).

Isothermal
An isothermal process is one in which the temperature of the system remains constant.

Isotope
Isotopes of an element have the same number of protons, but different numbers of neutrons in their nuclei.

Isotropic
Meaning everything looks the same in every direction.

K

Kinetic energy
The energy of an object due to its motion.

Kinetic theory
The term given to explaining an object's properties by considering the motion of its particles.

L

Length contraction
A consequence of Einstein's theory of special relativity which says that an object moving in the same direction as its length looks shorter to an external observer.

Lens
A shaped piece of glass (or other material) that changes the direction of rays of light by refraction.

Lens axis
The axis passing through the centre of a lens and perpendicular to the principal axis.

Lenz's law
The induced e.m.f. is always in such a direction as to oppose the change that caused it.

Light curve
A graph which shows the absolute magnitude (or apparent magnitude or brightness) of an object against time.

Light year
The distance that electromagnetic waves travel through a vacuum in one year.

Line of best fit
A line that is drawn on a scatter graph to show the correlation between two variables. There should be about as many data points above the line as there are below.

Low orbiting satellite
A satellite that orbits between 180 and 2000 km above the Earth's surface and at a higher angular speed than Earth.

Luminosity
The total amount of energy emitted by an object in the form of electromagnetic radiation each second. Also known as power output.

M

Magnetic field
A region in which a force acts on magnetic materials or magnetically susceptible materials.

Magnetic flux
The magnetic flux (in Wb) passing through an area is given by the magnetic flux density multiplied by the area. It can also be thought of as the number of magnetic field lines passing through the area.

Magnetic flux density
The value of magnetic flux density, in T, is given by the force on one metre of wire carrying a current of one amp at right angles to the magnetic field. Also called magnetic field strength.

Main sequence
A phase of a star's evolution in which the star is fusing hydrogen in its core.

Mass defect
The difference between the mass of a nucleus, and the sum of the individual masses of the nucleons. Equivalent to the binding energy of the nucleus.

Mean square speed, $(c_{rms})^2$
The average of the squared speeds of the particles in a gas.

Mechanical energy
The sum of the potential and kinetic energy of an object.

Model
A simplified picture or representation of a real physical situation.

Moderator
A material (often water) in a nuclear reactor that slows down neutrons so they can be captured by uranium nuclei (or other fissionable nuclei, e.g. 239Pu).

Molar gas constant, R
A constant used in the ideal gas equation, equal to $8.31 \, \mathrm{J \, mol^{-1} \, K^{-1}}$.

Molar mass
The mass that 1 mole of a substance would have.

Mole
An amount of substance containing N_A particles, all of which are identical. N_A is the Avogadro constant.

Molecular mass
The average mass of a molecule relative to an atom of carbon-12,

which has a relative atomic mass of exactly 12.

Moment of inertia
A measure of how much an object resists a change in rotational velocity. The larger the moment of inertia of an object, the larger the torque needed to change its rotational velocity by a given amount.

Momentum (linear)
The momentum of an object is the product of its mass and velocity.

Natural frequency
The frequency of an object oscillating freely.

Neutron star
A star made up of neutrons, formed by the collapse of a red supergiant with a high core mass.

Non-axial ray
A ray that is not parallel to the principal axis of a lens or mirror.

Non-flow process
A process that occurs in a closed system — no gas flows in or out of the system.

Nuclear fission
The spontaneous or induced splitting of a larger nucleus into two smaller nuclei.

Nuclear fusion
The fusing of two smaller nuclei to form one larger nucleus.

Nuclear radiation
Particles or energy released by an unstable atom as it decays. Made up of alpha, beta-minus or beta-plus particles, or gamma rays.

Nucleon
A particle that makes up an atomic nucleus (a proton or neutron).

Nucleus
The small, positive core of an atom where most of the atom's mass and all of its positive charge is concentrated.

Objective lens/mirror
The first lens/mirror in an optical telescope that the light from an object hits.

Orbital period
The time taken for a satellite to complete a full orbit.

Orbital speed
The speed at which a satellite travels.

Ordered / ordinal data
Categoric data where the categories can be put in order.

Overdamping
Heavy damping such that the system takes longer to return to equilibrium than a critically damped system.

Parallax
A measure of how much a nearby object (e.g. a star) appears to move in relation to a distant background due to the observers motion (e.g. as the Earth orbits the Sun). Measured as an angle of parallax.

Parsec
The distance of an object from Earth if its angle of parallax is equal to 1 arcsecond = $\left(\frac{1}{3600}\right)°$.

Percentage uncertainty
The uncertainty given as a percentage of the measurement taken.

Period
The time taken for a rotating or oscillating object to complete one revolution or cycle.

Permittivity
The permittivity of a material is a measure of how difficult it is to create an electric field in that material.

Phase difference
A measure of how much one wave lags behind another wave. It can be measured in degrees, radians or fractions of a cycle.

Photoelectric effect
The emission of electrons (photoelectrons) from a metal when light of a high enough frequency is shone on it.

Photon
A discrete wave packet of electromagnetic radiation.

Planetary nebula
The remnants of the outer layers of a red giant, that are ejected as the star becomes a white dwarf.

Point charge
A charge with negligible volume, or a uniform sphere whose charge acts as if it is concentrated at the centre.

Point mass
A mass with negligible volume, or a uniform sphere whose mass acts as if it is concentrated at the centre.

Potential energy
Energy that is stored (e.g. elastic strain energy is energy stored in something that has been stretched or compressed, like a spring).

Power
The rate of transfer of energy or the rate of doing work. It's measured in watts (W), where 1 watt is equivalent to 1 joule per second.

Precise result
The smaller the amount of spread of your data from the mean, the more precise it is.

Prediction
A specific testable statement about what will happen in an experiment, based on observation, experience or a hypothesis.

Pressure law
For an ideal gas at constant volume, the pressure p is directly proportional to its absolute temperature T.

Principal axis
An axis of a lens that passes through its centre, perpendicular to its surface on both sides.

Principal focus
The principal focus of a converging lens (or mirror) is the point where incident rays parallel to the principal axis of the lens (or mirror) converge.

Proper time
The time interval between two events as measured by an observer that is stationary in relation to the events.

Quantum efficiency
The proportion of incident photons falling on a light detector (e.g. a CCD) that are detected.

Quasar
A quasi-stellar object that is thought to be a galactic nucleus centred around a black hole.

R

Radial field
A field where the field lines all point towards or away from the central point.

Radian
A unit of measurement for angles. There are 2π radians in a complete circle.

Radioactive decay
When an unstable atom breaks down to become more stable, by releasing energy and/or particles.

Radioactive source
A material that emits alpha, beta or gamma radiation as it decays.

Radioactive tracer
A radioactive substance that is used in medicine for imaging inside the body.

Random error
An error introduced by variables which you cannot control.

Ray diagram
A diagram showing rays of light passing through an optical system, sometimes showing image formation.

Rayleigh criterion
The minimum angular resolution of a telescope is roughly equal to the wavelength of the radiation it is detecting divided by the aperture diameter: $\theta \approx \frac{\lambda}{D}$.

Real image
An image formed when light rays from a point on an object are made to pass through another point in space. The light rays are actually there, and the image can be captured on a screen.

Recessional velocity
The speed at which an object is receding from Earth.

Red giant
Stars with a high luminosity and a low temperature. A red giant is a phase of a star's evolution in which the star is fusing larger elements than hydrogen in its core or there is fusion in its shells. The fusion energy causes its outer layers to expand and cool, making it appear red.

Red shift
The shift in wavelength and frequency of a source moving away from us towards (or beyond) the red end of the electromagnetic spectrum.

Reflecting telescope
A telescope that uses mirrors to focus the rays of light.

Refraction
When a wave changes direction and speed as it enters a medium with a different optical density.

Relative atomic mass
The mass of an atom relative to an atom of carbon-12, which has a relative atomic mass of exactly 12.

Relative permittivity
The ratio of the permittivity of a material to the permittivity of free space. It is also known as the dielectric constant.

Relativistic mass
The mass of a body in motion, which is predicted by Einstein's theory of special relativity to increase with velocity and approach infinity as the velocity approaches the speed of light.

Repeatable result
A result is repeatable if you can repeat an experiment multiple times and get the same result.

Reproducible result
A result is reproducible if someone else can recreate your experiment using different equipment or methods, and get the same result you do.

Resolution
The smallest change in what's being measured that can be detected by the equipment.

Resolving power
A measure of how much detail a telescope can see.

Resonance
When an object, driven by a periodic external force at a frequency close to its natural frequency, begins to oscillate with a rapidly increasing amplitude.

Root mean square current (I_{rms})
A measure of the 'average' current of an alternating current supply, found by dividing the peak current by $\sqrt{2}$.

Root mean square speed (c_{rms})
The square root of the average of the squared speeds of the particles in a gas.

Root mean square voltage (V_{rms})
A measure of the 'average' voltage of an alternating current supply, found by dividing the peak voltage by $\sqrt{2}$.

Rotational kinetic energy
The energy of a rotating object due to its rotational motion.

S

Satellite
A smaller mass that orbits a larger mass.

Schwarzschild radius (R_s)
The radius of the event horizon of a black hole.

Simple harmonic motion (SHM)
The oscillation of an object where the object's acceleration is directly proportional to its displacement from its equilibrium position, and is always directed towards the equilibrium.

Solenoid
An electromagnet consisting of multiple coils of wire with length.

Spatial resolution
A measure of the ability to form separate images of objects that are close together.

Specific charge
The ratio of charge to mass for a particle.

Specific heat capacity
The amount of energy needed to raise the temperature of 1 kg of a substance by 1 K (or 1 °C).

Specific latent heat
The quantity of thermal energy required to change the state of 1 kg of a substance (e.g. to melt or vaporise it without changing its temperature).

Spectral class
Groups which stars are classified into according to their temperature and the relative strength of certain absorption lines.

Spherical aberration
A problem with using mirrors to focus light, caused by the mirror being not quite the right shape so that there is no clear focal point and the image is blurry.

Stationary wave
A wave with fixed positions of minimum and maximum oscillation (nodes and antinodes) created by the superposition of two progressive waves with the same frequency (or wavelength) and amplitude, moving in opposite directions.

Stefan's law
The power output of a black body is proportional to the fourth power of the body's temperature and is directly proportional to its surface area:
$P = \sigma A T^4$

Step-down transformer
A transformer that decreases an alternating voltage.

Step-up transformer
A transformer that increases an alternating voltage.

Supermassive black hole
A black hole with a mass larger than 10^6 solar masses.

Supernova
The explosion of a high-mass star after its red supergiant phase, caused by the core collapsing and the outer layers of the star falling in and rebounding, creating huge shockwaves.

Synchronous orbit
An orbit in which the orbiting object has the same orbital period as the rotational period of the object it is orbiting.

Systematic error
An error that can be introduced by the experimental apparatus or method.

Terminal velocity
The maximum velocity of an object through a fluid, reached when the driving force is matched by the frictional force.

Theory
A possible explanation for something. (Usually something that has been observed.)

Thermal efficiency
The efficiency of an engine based on the amount of heat energy that is transferred into useful work.

Thermal neutron
Neutron in a nuclear reactor that has been slowed down enough by a moderator that it can be captured by uranium nuclei (or other fissionable nuclei, e.g. 239Pu).

Thermionic emission
The release of free electrons from a metal's surface when the metal is heated.

Threshold frequency
The lowest frequency of light that when shone on a metal will cause the emission of photoelectrons.

Time constant
The time taken for the charge on a discharging capacitor, the potential difference across the capacitor, or the current in the circuit to fall to $\frac{1}{e}$ (about 37%) of their initial value, or for the charge of a charging capacitor to rise to about 63% of the full charge.

Time dilation
A consequence of Einstein's theory of special relativity which says that an observer of two events that is moving at a constant velocity in relation to the events will measure a longer time interval between the two events than the proper time.

Torque
When a force (or couple) causes an object to turn, the turning effect is known as torque.

Transformer
A device that makes use of electromagnetic induction to change the size of the voltage of an alternating current.

Transit method
A method used to detect exoplanets — the apparent magnitude of the system (the planet and its star) is measured as the planet travels in front of its star.

Type 1a supernova
A type of supernova that always has the same peak in absolute magnitude. Often used as standard candles.

Valid result
A valid result arises from a suitable procedure to answer the original question.

Validation
The process of repeating an experiment done by someone else, using the theory to make new predictions, and then testing them with new experiments, in order to support or refute the theory.

Variable
A quantity in an experiment or investigation that can change or be changed.

Virtual image
An image formed when light rays from a point on an object appear to have come from another point in space. The light rays aren't really where the image appears to be, so the image can't be captured on a screen.

Viscosity
A measure of how thick a fluid is.

Wave-particle duality
The idea that particles and waves can each display both particle and wave-like behaviour.

Wavelets
Wave-like oscillations that were thought to originate from every point along a wavefront and spread out in the forward direction, according to Huygens' principle.

White dwarf
Stars with a low luminosity and a high temperature, left behind when a low-mass star stops fusing elements and contracts.

Work
Work is the amount of energy transferred from one form to another when a force moves an object a distance or when a torque turns an object through an angle.

Work function
The minimum amount of energy required for an electron to escape a metal's surface.

Worst line
Line of best fit which has either the maximum or minimum possible slope for the data and which should go through all of the error bars.

Zero error
When a measuring instrument falsely reads a non-zero value when the true value being measured is zero.

Acknowledgements

AQA Specification material is reproduced by permission of AQA.

Photograph acknowledgements

Cover Photo **Richard Kail**/Science Photo Library, p 3 Science Photo Library, p 5 **Bell Labs/Science Source**/Science Photo Library, p 9 **TimAwe**/iStockphoto.com, p 10 **Simon Whiteley**, p 12 **GIPhotoStock**/Science Photo Library, p 15 Science Photo Library, p 20 **Lawrence Berkeley Laboratory**/Science Photo Library, p 25 **Steve Allen**/Science Photo Library, p 28 **GIPhotoStock**/Science Photo Library, p 43 **GoranQ**/iStockphoto.com, p 48 **Martyn F. Chillmaid**/Science Photo Library, p 50 **Martin Dohrn**/Science Photo Library, p 54 (top) Science Photo Library, p 54 (bottom) **Sheila Terry**/Science Photo Library, p 57 Science Photo Library, p 59 Science Photo Library, p 67 **National Library of Medicine**/Science Photo Library, p 68 **Natural History Museum, London**/Science Photo Library, p 73 **Victor de Schwanberg**/Science Photo Library, p 76 **Henn Photography**/Science Photo Library, p 80 **Adam Hart-Davis**/Science Photo Library, p 85 **Hughes Aircraft Company**/Science Photo Library, p 88 **Ted Kinsman**/Science Photo Library, p 89 **Trevor Clifford Photography**/Science Photo Library, p 93 **Charles D. Winters**/Science Photo Library, p 102 **GIPhotoStock**/Science Photo Library, p 105 **Charles Doswell**, **Visuals Unlimited**/Science Photo Library, p 106 **Sheila Terry**/Science Photo Library, p 108 **GIPhotoStock**/Science Photo Library, p 117 **Chris Knapton**/Science Photo Library, p 122 **GIPhotoStock**/Science Photo Library, p 123 **GIPhotoStock**/Science Photo Library, p 124 **Emilio Segre Visual Archives/American Institute of Physics**/Science Photo Library, p 129 **Lawrence Berkeley Laboratory**/Science Photo Library, p 131 **CC Studio**/Science Photo Library, p 139 **Chemical Heritage Foundation**/Science Photo Library, p 143 **Andrew Lambert Photography**/Science Photo Library, p 144 **Martyn F. Chillmaid**/Science Photo Library, p 148 (top) **National Physical Laboratory © Crown Copyright**/Science Photo Library, p 148 (bottom) **GIPhotoStock**/Science Photo Library, p 150 **Chris B Stock**/Science Photo Library, p 158 **Royal Astronomical Society**/Science Photo Library, p 162 Science Photo Library, p 168 **Andrew Lambert Photography**/Science Photo Library, p 169 **Tim Beddow**/Science Photo Library, p 171 **James King-Holmes**/Science Photo Library, p 177 **Trevor Clifford Photography**/Science Photo Library, p 183 **Patrick Landmann**/Science Photo Library, p 186 **ISM**/Science Photo Library, p 193 **Patrick Landmann**/Science Photo Library, p 201 **GIPhotoStock**/Science Photo Library, p 202 (top) **B. G. Thomson**/Science Photo Library, p 202 (bottom) **Zennie**/iStockphoto.com, p 205 **Emilio Segre Visual Archives/American Institute of Physics**/Science Photo Library, p 206 (top) **Babak Tafreshi**/Science Photo Library, p 206 (bottom) **GIPhotoStock**/Science Photo Library, p 209 **Andrew Lambert Photography**/Science Photo Library, p 211 (top) **Dr P P Kronberg**/Science Photo Library, p 211 (bottom) **Joseph Sohm**/Science Photo Library, p 212 **David Ducros**/Science Photo Library, p 213 (top) **Simon Fraser**/Science Photo Library, p 213 (bottom) **NASA/Carla Thomas**/Science Photo Library, p 221 **NASA/ESA/STSCI/High-Z Supernova Search Team**/Science Photo Library, p 224 **National Library of Congress**/Science Photo Library, p 228 **National Optical Astronomy Observatories**/Science Photo Library, p 233 **NASA/ESA/STSCI/J. Hester & P. Scowen, ASU**/Science Photo Library, p 234 (top) **Royal Observatory, Edinburgh**/Science Photo Library, p 234 (bottom) **Kim Gordon**/Science Photo Library, p 237 **CXC/SAO/F. Seward Et Al/NASA**/Science Photo Library, p 238 **NASA/ESA/STSCI/J. Bahcall, Princeton IAS**/Science Photo Library, p 239 **Emilio Segre Visual Archives/American Institute of Physics**/Science Photo Library, p 245 **NASA/2MASS/J. Carpenter, T. H. Jarrett, R. Hurt**/Science Photo Library, p 247 **NASA/WMAP Science Team**/Science Photo Library, p 250 Science Photo Library, p 251 **NASA/ESA/STSCI/Hubble Heritage Team**/Science Photo Library, p 259 **David Parker**/Science Photo Library, p 261 **Berenice Abbott**/Science Photo Library, p 267 **Eye Of Science**/Science Photo Library, p 271 (top) **Patrick Dumas/Look At Sciences**/Science Photo Library, p 271 (bottom) **JOTI**/Science Photo Library, p 273 **Steve Gschmeissner**/Science Photo Library, p 277 **Kate Jacobs**/Science Photo Library, p 279 **AJ Photo**/Science Photo Library, p 281 **Doncaster And Bassetlaw Hospitals**/Science Photo Library, p 282 **Zephyr**/Science Photo Library, p 284 **Sam Ogden**/Science Photo Library, p 285 **Kevin Curtis**/Science Photo Library, p 286 **Dr P. Marazzi**/Science Photo Library, p 287 **Mark Thomas**/Science Photo Library, p 290 **General Electric Research and Development/Emilio Segre Visual Archives/American Institute of Physics**/Science Photo Library, p 293 **Miriam Maslo**/Science Photo Library, p 294 **Alfred Pasieka**/Science Photo Library, p 295 **Maximilian Stock Ltd**/Science Photo Library, p 296 **Spencer Grant**/Science Photo Library, p 297 **Alfred Pasieka**/Science Photo Library, p 301 (top) **National Institute on Aging**/Science Photo Library, p 301 (bottom) **Stevie Grand**/Science Photo Library, p 302 (top) **Zephyr**/Science Photo Library, p 302 (bottom) **Mehau Kulyk**/Science Photo Library, p 309 (top) **Serghei Starus**/iStockphoto.com, p 309 (bottom) **Sheila Terry**/Science Photo Library, p 318 **AngiePhotos**/iStockphoto.com, p 326 **YinYang**/iStockphoto.com, p 332 **Mark Williamson**/Science Photo Library, p 335 Science Photo Library, p 340 **Miriam and Ira D. Wallach Division Of Art, Prints And Photographs/New York Public Library**/Science Photo Library, p 341 **Luxizeng**/iStockphoto.com, p 342 **Mark Sykes**/Science Photo Library, p 344 **David Leah**/Science Photo Library, p 348 **Johnny Greig**/iStockphoto.com, p 349 **Robert Brook**/Science Photo Library, p 364 **Library of Congress**/Science Photo Library, p 369 Science Photo Library, p 370 **Sheila Terry**/Science Photo Library, p 371 Science Photo Library, p 372 **GIPhotoStock**/Science Photo Library, p 374 **Emilio Segre Visual Archives/American Institute of Physics**/Science Photo Library, p 375 **Emilio Segre Visual Archives/American Institute of Physics**/Science Photo Library, p 376 Science Photo Library, p 379 **US Library of Congress**/Science Photo Library, p 380 **Pierre Marchal/Look At Sciences**/Science Photo Library, p 386 **Northwestern University**/Science Photo Library, p 387 **Philippe Plailly**/Science Photo Library, p 394 **Tom McHugh**/Science Photo Library, p 401 **Andrew Lambert Photography**/Science Photo Library, p 403 **Wavebreak**/iStockphoto.com, p 405 **Sorendls**/iStockphoto.com, p 409 **Wladimir Bulgar**/Science Photo Library.

Index

Data Tables

This page summarises some of the constants, values and properties that you might need to refer to when answering questions in this book. Everything here will be provided in your exam data and formulae booklet somewhere... so you need to get used to looking them up and using them correctly. If a number isn't given on this sheet — unlucky... you'll need to remember it as it won't be given to you in the exam.

Fundamental constants and values

Quantity	Value
acceleration due gravity, g	$9.81 \ ms^{-2}$
atomic mass unit, u	$1.661 \times 10^{-27} \ kg \ (1u \approx 931.3 \ MeV)$
Avogadro constant, N_A	$6.02 \times 10^{23} \ mol^{-1}$
Boltzmann constant, k	$1.38 \times 10^{-23} \ JK^{-1}$
electron charge/mass ratio, e/m_e	$1.76 \times 10^{11} \ Ckg^{-1}$
electron rest mass, m_e	$9.11 \times 10^{-31} \ kg$
gravitational constant, G	$6.67 \times 10^{-11} \ Nm^2 kg^{-2}$
gravitational field strength, g	$9.81 \ Nkg^{-1}$
Hubble constant, H	$65 \ kms^{-1} Mpc^{-1}$
magnitude of the charge of electron, e	$1.60 \times 10^{-19} \ C$
molar gas constant, R	$8.31 \ JK^{-1} mol^{-1}$
neutron rest mass, m_n	$1.67(5) \times 10^{-27} \ kg$
Permeability of free space, μ_0	$4\pi \times 10^{-7} \ Hm^{-1}$
Permittivity of free space, ε_0	$8.85 \times 10^{-12} \ Fm^{-1}$
Planck constant, h	$6.63 \times 10^{-34} \ Js$
proton rest mass, m_p	$1.67(3) \times 10^{-27} \ kg$
speed of light in vacuo, c	$3.00 \times 10^8 \ ms^{-1}$
Stefan constant, σ	$5.67 \times 10^{-8} \ Wm^{-2} K^{-4}$
threshold of hearing, I_0	$1.0 \times 10^{-12} \ Wm^{-2}$
Wien constant, α	$2.90 \times 10^{-3} \ mK$

Solar system data

Body	Mean radius	Mass
Earth	$6.37 \times 10^6 \ m$	$5.98 \times 10^{24} \ kg$
Sun	$6.96 \times 10^8 \ m$	$1.99 \times 10^{30} \ kg$

Astronomical distances

Quantity	Distance
1 astronomical unit (AU)	$1.50 \times 10^{11} \ m$
1 light year	$9.46 \times 10^{15} \ m$
1 parsec	$3.08 \times 10^{16} \ m = 2.06 \times 10^5 \ AU = 3.26 \ light \ years$

PATB62